Sapphire

Elena R. Dobrovinskaya • Leonid A. Lytvynov
Valerian Pishchik

Sapphire

Material, Manufacturing, Applications

Springer

Elena R. Dobrovinskaya
Rubicon Technology
Franklin Park, IL
USA

Leonid A. Lytvynov
Scientific Technological Complex
Institute for Single Crystals
Kharkov
Ukraine

Valerian Pishchik
Gavish, Ltd.
Sapphire Products
Omer
Israel

ISBN: 978-0-387-85694-0 e-ISBN: 978-0-387-85695-7
DOI: 10.1007/978-0-387-85695-7

Library of Congress Control Number: 2008934333

Printed on acid-free paper

springer.com

Preface

By the second half of the twentieth century, a new branch of materials science had come into being — crystalline materials research. Its appearance is linked to the emergence of advanced technologies primarily based on single crystals (bulk crystals and films).

At the turn of the last century, the impending onset of the "ceramic era" was forecasted. It was believed that ceramics would play a role comparable to that of the Stone or Bronze Ages in the history of civilization. Naturally, such an assumption was hypothetical, but it showed that ceramic materials had evoked keen interest among researchers.

Although sapphire traditionally has been considered a gem, it has developed into a material typical of the "ceramic era." Widening the field of sapphire application necessitated essential improvement of its homogeneity and working characteristics and extension of the range of sapphire products, especially those with stipulated properties including a preset structural defect distribution.

In the early 1980s, successful attainment of crystals with predetermined characteristics was attributed to proper choice of the growth method. At present, in view of the fact that the requirements for crystalline products have become more stringent, such an approach tends to be insufficient. It is clear that one must take into account the physical–chemical processes that take place during the formation of the real crystal structure, i.e., the growth mechanisms and the nature and causes of crystal imperfections.

In recent years, certain successes have been achieved in the understanding of crystal formation mechanisms, the morphological stability of the crystallization front, the role of impurities, thermal and concentration flows in the melt, and other factors that influence the formation of structural defects. However, it is the establishment of the relation between the parameters of the real crystal article and the conditions of its attainment that remains the task of paramount importance in obtaining crystals possessing predetermined characteristics. This task necessitates a detailed analysis of the raw material, crystal growth medium, heat and mass transfer at both the liquid–solid interface and in the bulk of these phases, as well as the processes of cooling of the crystals and their subsequent thermal and mechanical treatment.

Theoretical investigations into the process of crystallization, performed on different substances including sapphire, have not yet resulted in creation of a general theory of real crystal formation, and there is a large gap between the theory and practice of crystal growth. Now, it is undoubted that the real crystal structure stores the "genetic" information on the process of formation of crystals and the forecast of their future behavior during treatment and service. The authors' ideas of the embodiment of this information in the "granular" substructure of the crystals are considered in the present book.

This book considers all known methods for the growth of sapphire and modification of its properties, the fields of sapphire application, as well as the most exhaustive data on the crystal structure and physical–chemical properties. The authors believe that this book, which helps to estimate the unique potential of sapphire, will be useful for specialists in crystal growth technologies, designers of new apparatuses, sales managers, scientists, and engineers who use sapphire articles.

Franklin Park, IL, USA — Elena R. Dobrovinskaya
Kharkov, Ukraine — Leonid A. Lytvynov
Omer, Israel — Valerian Pishchik

Contents

Symbols

a,c	Lattice parameters (Á)
a_e	Size of grain structure element (mm)
a_c	Size of an average crystallite (Á)
b	Burgers vector
c_p	Specific heat (kJ (kmol K)$^{-1}$)
C_L	Concentration of impurity in the melt (by mass)
d	Size of abrasive grain, diameter of crystal
D	Diffusion coefficient (cm^2 s^{-1})
D_S	Surface diffusion coefficient
E	Young's modulus (Pa)
E^*	Activation energy (kJ mol^{-1})
E_p	Pulse energy (J)
E_{th}	Threshold generation energy (J)
K_d	Distribution coefficient
k_c	Fracture toughness coefficient (MN m$^{-3/2}$)
k_f	Friction coefficient
g	Gravitation constant
H_h	Hardness (Pa)
M	Molecular mass
N	Number of abrasive particles, number of samples
P	Load, applied normal pressure (Pa)
Q	Grinding productivity
T_a	Annealing temperature (K)
T'_r	Radial temperature gradient (K cm^{-1})
T'_n	Axial temperature gradient (K cm^{-1})
T'_m	Temperature gradient in the melt (K cm^{-1})
T_m	Melting temperature (K)
T_{melt}	Temperature of the melt (K)
T_c	Crystallization temperature (K)
t_p	Pulse time (s(h))
V_{Al}^{3-}, V_{Al}	Aluminum vacancy
V_O^{2+},V_O	Oxygen vacancy
V	Volume (cm^3)

ν	Velocity (cm s^{-1})
ν_t	Tool rotation velocity
Δh	Impurity microband width
ε_p	Plastic deformation
ε^*	Oxidation–reduction potential of atmosphere
Γ	Anomalous birefringence index
η	Kinematics' viscosity (P)
ϕ_l	Slope of liquidus line (deg)
λ_{μ_s}	Period of impurity banding (μm)
λ_T	Head conductivity (W (m K)$^{-1}$)
μ	Grindability of material
ρ	Dislocation density (cm^{-2})
ρ_0	Initial dislocation density (cm^{-2})
ρ_{cr}	Critical dislocation density (cm^{-2})
$\rho_{(0001)}$	Basal dislocation density (cm^{-2})
σ_s	Surface tension
σ	Poisson's ratio, Stefan-Boltzman constant
τ	Stress (Pa)
Σp	Extension of block boundaries (mm^{-1})

Abbreviations

CAST	Capillary action shaping technique
CC	Color center
CF	Crystallization front
EFG	Edge-defined film-fed growth
HDS, HDSM	Horizontal directed solidification method
HEM	Heat exchange method
IR	Infrared region of the spectrum
IA	Induced absorption
MPS	Melt–polycrystal–single crystal
NCS	Noncapillary shaping method
RD	Radiation defect
RE	Ruby element
RLX	Ray luminescence
SI	Sapphire implant
TL	Thermoluminescence
TLD	Thermoluminescent dosimeter
TSL	Thermally stimulated luminescence
UV	Ultraviolet region of the spectrum
VLC	Vapor–liquid–crystal
μ-PD	Micropulling-down method

Chapter 1
Application of Sapphire

As far back as the tenth century BC, sapphires and rubies were valued as gems on the level of diamonds. Artificial sapphires were first used in jewelry art as well, but from the beginning of the twentieth century sapphire has played an increasingly significant role in engineering. At present one can hardly find a branch of science or technology where this crystal is not used Demand for sapphire grows year after year, almost exponentially.

Devices and their components applied in aviation and space industries, in chemical processing, and in many other fields are simultaneously subjected to the action of aggressive media, radiation, high temperatures, pressures, and mechanical loads. Under such extreme conditions any material is prone to intense corrosion and erosion. High-strength alloys have reached the practical limits of their capabilities. The structure of polycrystalline materials and consequently their mechanical properties essentially change under extreme conditions due to recrystallization, corrosion of the grain boundaries, and so forth The rate of diffusion via the grain boundaries grow with increasing temperature, radiation dose, and operation time. As a result, the material breaks down. Such drawbacks are inherent in sapphire components and assemblies to a considerably lesser extent.

There are two classes of modern articles made of sapphire: constructional and functional. Constructional sapphire is used for the creation of products with high mechanical stability. Functional sapphire has a specific structure and electrical, optical, and thermal properties. This classification is arbitrary, as this material is multifunctional, and sapphire articles (such as rocket nose cones) often combine the two functions. Constructional and functional sapphires are comparatively new materials, so the scale of their production compares unfavorably with that of traditional materials. At the same time, the rate of growth in their production far exceeds that of steel, aluminum, and some other materials. It also should be emphasized that sapphire is used in complex technical systems with costs several times as high as that of the sapphire elements.

At present, the greater part of artificial sapphire is used as a functional material. However, extremely high production rates are predicted for constructional sapphire as well. This follows from estimations made by specialists in the world's leading firms. According to the performed analysis, in 70% of cases future prospects for the use of sapphire are tied to its mechanical, thermal, and chemical properties. For a

E.R. Dobrovinskaya et al., *Sapphire: Material, Manufacturing, Applications*,
DOI: 10.1007/978-0-387-85695-7_1, © Springer Science+Business Media, LLC 2009

long time sapphire has not been considered by material scientists to be a possible constructional material due to the fact that its treatment is both complicated and expensive. After the appearance of a wide variety of profiled sapphire, such a problem is not as relevant. However, purely physical obstacles in the use of sapphire as a constructional material still exist. First, this is the main disadvantage of sapphire – its brittleness. Concerning other significant working parameters such as thermal stability, hardness, corrosion resistance, and density, as well as availability and low price of raw material, sapphire compares favorably with metals, alloys, and some ceramics. The tendency of sapphire to brittle failure is associated with low defect mobility, which is a primary result of the specific ionic-covalent bonding of this material. Therefore, a good deal of effort now is being undertaken by researchers to eliminate the microscopic defects that function as crack nucleation centers in sapphire upon loading.

As shown by the last decade of the development of the sapphire industry and the corresponding application fields, interest in both constructional and functional sapphire has grown to a great extent. One may even say that such a "renaissance" of sapphire is one of the most significant tendencies in modern materials science. This is caused by many factors. Here are several:

- Sapphire is a multifunctional material
- The raw material for sapphire production is readily available and cheap
- As a rule, the technology of sapphire production is less energy-intensive than alternative materials and allows the growth of large-size crystals
- Sapphire production does not pollute the environment. The growth of sapphire is less harmful than that of alternative materials due to the absence of processes such as electrolysis, pyrometallurgy, the action of aggressive media, and other factors
- Compared to other materials, sapphire possesses higher corrosion and radiation resistance, which results in operational longevity of sapphire articles in aggressive media
- Sapphire is characterized by higher biological compatibility than metals and polymers; therefore it is used in medical implants, as a constructional material in biotechnology, in medical instrument-making, and in genetic engineering

Today, sapphire primarily is used for:

- Fabrication of substrates for light-emitting diodes, new generations of TV receivers, projectors, and microwave devices
- Fabrication of windows for civilian and military equipment
- Production of bearings and windows for watches and devices
- Making of precious jewels for the jewelry industry

The light sources in which light-emitting diodes are used instead of filament lamps allow reduction of electric power consumption by approximately ten times. It has been calculated that the funds that could be saved due to the substitution of conventional light sources by light diodes in all the world's lighting systems would be equivalent to the cost of several hundred nuclear power stations! The immense output of present-day TV receivers, mobile phones, and other household appliances that contain or will contain sapphire components speaks for itself.

The statement to the effect that the technologies using sapphire articles have penetrated into practically all the spheres of science and engineering is confirmed by an analysis of the recently published book "History of Science and Engineering" (2004, Houghton Mifflin) and the patents granted in the world from 1990 to present. Approximately 5–7% of these publications are in some way related to sapphire.

Over the past few years, the range of manufactured sapphire items has sharply increased and the requirements for the quality of their working surfaces have become more stringent. Therefore, special attention is being concentrated on the possibility of controlling the structure perfection not only in the process of sapphire growth, but also during subsequent thermo- and mechanochemical treatment. Much effort also is being undertaken to investigate thermal power effects on the structure evolution of sapphire articles to develop basically new sapphire treatment methods.

Now let us outline the near future for sapphire. The present-day market for synthetic stones is estimated to exceed $6 billion. In the world ~300 tons of rubies and colored sapphires and 100 tons of sapphire are produced. In the opinion of specialists, by 2008 the world market of synthetic crystals will reach $11.3 billion; the share of sapphire being about one quarter of this value. The growth rate of sapphire production is extremely high. One should expect that during the next 20 years the world production of sapphire will increase approximately tenfold. Currently, the leading manufacturers of sapphire are the United States and Russia. In the United States, sapphire crystals are grown by the Czochralski, Kyropoulos, HEM, and EFG methods. In Russia the Kyropoulos and Stepanov methods dominate. The rate of sapphire production is being stepped up in the Ukraine and Japan. China is endeavoring to hit the world market, too. At present, the annual increase in the production rate of sapphire articles in China is ~20%. However, the quality of these products is still insufficient.

Since the beginning of 2004, contradictory tendencies have been observed for the world sapphire market. During the past 5 years the world production of sapphire has increased by ~10% annually, whereas during the past 2 years the average growth in demand for this material and articles based on it have exceeded 15–20%. At the same time, technical requirements for the quality of the material itself and the working surfaces of sapphire articles become ever more stringent.

The performed analysis indicates that one of most important problems for the development of sapphire production is the increase in the size of the crystals. This is explained not only by economic considerations, but also by technical requirements. Large-size sapphire crystals are needed for optics and airborne windows with apertures of 600 mm and larger (for the medium-wave IR region of the spectrum). At present, the only material used for the windows with a diameter greater than 300 mm is ZnSe. Its ultimate strength reaches 69 MPa (10,000 psi). Large-size windows have a thickness of 20 mm or more, and such a thickness makes the windows heavy and expensive. Moreover, it gives rise to an essential optical scattering.

In contrast to ZnSe windows, those made of sapphire may have a thickness around 5 mm with a complete absence of optical scattering. As shown by recent investigations, large area (500 × 500 mm and 1,000 × 1,000 mm) is necessary for making transparent armor used for the windows of helicopters, special-purpose vehicles, and so forth Work in the creation of this kind of armor has been started in the United States, Russia, Ukraine, Czech Rep., and other countries. The strength of a "glass sandwich" with a thickness of 25–35 mm will correspond to that of ~100-mm-thick armored glass. The upper layer of such a "glass sandwich" would contain a sapphire sheet meant to absorb the greater part of the kinetic energy of bullets. The rest of this energy will be absorbed in the layers of glass and plastic.

The growth of large-size sapphire crystals is an urgent problem. Required for this purpose are large crystallization units that provide not only high perfection of the crystals, but also quantity production. The use of such units essentially decreases consumption of both energy and materials, so the cost of the crystal diminishes. The growth of one crystal with a weight of 30, 100, 200, and even 500 kg is more profitable than the growth of the corresponding quantity of smaller crystals.

Thus, it is to be concluded that in the near future sapphire, as with silicon and germanium, will become one of the strategic materials for materials science. The main trends in the development of the sapphire industry will be the increase in the size of the grown crystals and the creation of technologies for obtaining permanent sapphire joints.

Now let us consider some significant applications of sapphire in detail.

1.1 Use in Jewelry Industry

Not to dwell on the traditional use of sapphire in jewelry industry [1, 2], let us only consider methods of changing the color of the crystals. The brightest colors with nice hues are obtained by introducing a mixture of components into the starting material for the growth of color corundum crystals (Table 1.8 in Appendix). However, the color or hue required cannot always be achieved.

Ennoblement (intensification or change of the color) is a widespread procedure, as most natural crystals have faded coloration that decreases their value. It was many centuries ago that Aristotle noted the ability of ruby to improve its color in oxidizing medium: "The red color of this gem becomes nicer if it is placed in a fire and ash-pit..." [3]. In the Middle Ages Birunee described a method for improvement of the color of gems by slow heating.

At present, most natural gems are subjected to procedures of color ennoblement after which their value increases by 1–2 orders. As a rule, improvement of aesthetic and decorative quality in gems is a result of either the formation of color centers, which have not been realized under natural conditions due to certain causes or supplements to such processes. The color of sapphire can be changed either within the whole of the crystal or in its surface layer. The crystals from different deposits are distinguished by variable combinations of impurities and specific effects. Therefore, different methods of affecting the crystals are used. The most widespread ways for changing the color of the crystals are the following:

- Irradiation
- Implantation of ions
- Thermal treatment in different gaseous media
- Thermochemical treatment

In a number of cases the best results are obtained using a combination of the above methods.

Irradiation with X-rays, cyclotron, reactor, and other kinds of radiation changes the valences of some impurities or favors intensification of the existing color centers or formation of new ones. After irradiation, colorless and pink sapphires acquire orange color. There is no point in irradiating ruby, since this worsens the purity of its color.

Implantation of cobalt ions with $5 \cdot 10^{16}$ to $5 \cdot 10^{17}$ ion/cm^2 doses and an energy of 20 keV into leucosapphire leads to the appearance of gray color. Subsequent heating in air transforms the color from green (1,070 K, 3 h) to light blue (1,270 K, 3 h). The X-ray and photoelectron spectra of Co also change: the spectrum of the nonannealed sample nearly coincides with that of the pure metal. The spectrum of the annealed sample has two maxima corresponding to bands at 779 eV and 781.4 eV. This fact indicates the presence of two types of electronic configuration of Co in colored sapphire.

Thermal treatment in different gaseous media changes the valent states of impurities and the color of the crystal. For instance, Al_2O_3:Ni crystal (yellow sapphire or "oriental topaz") can reversibly acquire a smoky hue depending on the annealing medium. The color of ruby becomes more intense with the transition of chromium of different valences into trivalent chromium. The sapphires with too dark a blue color can be brightened by heating until some part of the Ti–Fe complexes decompose. The color of spotted stones may become more homogeneous due to diffusion processes.

Thermochemical treatment allows changes in the color of crystal surface layer. The results of thermochemical treatment of pink sapphires from Vietnamese deposits and of artificial leucosapphires are reported [4]. Pink sapphires (~0.01% of Cr) were heated in an atmosphere of saturated Co vapors at 1,270–1,470 K. The spectra of the obtained samples were compared with the absorption spectrum of natural blue sapphire in the visible region.

The interaction between aluminum and cobalt oxides at 1,670 K was investigated [5]. The compound of $CoAl_2O_4$ spinel was revealed. The width of the solid solution region is almost temperature-independent. The comparison of the absorption spectra of blue ennobled and natural sapphires (Fig. 1.1) testifies to the coincidence of the absorption band maxima for these two objects.

Ennobled blue sapphires have a faint pinkish hue typical of natural sapphires from Sri Lanka due to the presence of a narrow resonance *R*-line in the spectrum. Such a peculiarity is inherent in crystals with an optical density lower than 1–1.5 cm^{-1} in the absorption maximum. The absorption band of spinel $CoAl_2O_4$ grown by

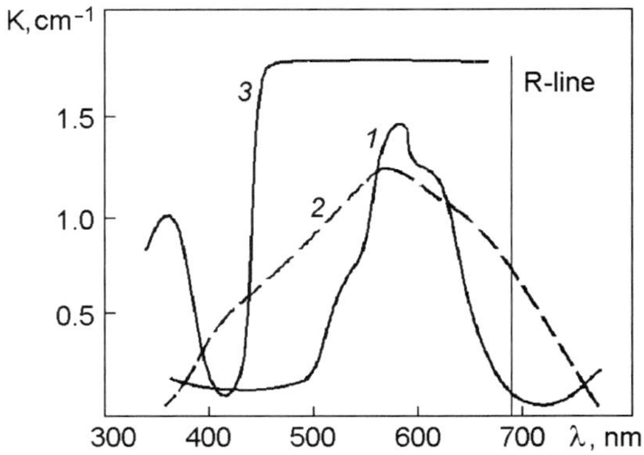

Fig. 1.1 Absorption spectra of ennobled Vietnamese sapphire (*1*), natural blue sapphire (*2*), and cobalt-aluminum spinel (*3*)

the Verneuil method lies in the shorter wavelength region of the spectrum and the crystals are violet. Experiments show that the color crystals acquire as a result of ennoblement is resistant to any action of light, temperature, and radiation that occurs in nature.

Combination of different methods implies the change of the valence of chromophores or their positioning in the structure of corundum, as well as dissolution of foreign phases followed by diffusion distribution of their components. For instance, the conditions of a positive action of different factors on sapphire (0.01–0.15% Ti and 0.03–0.6% Fe) have been reported: heating in vacuum furnace up to 1,770–1,870 K, isothermal annealing during 24–30 h, and fast cooling [6]. Off-grade samples acquired transparency and blue color. Gamma-irradiation with a dose of $2 \cdot 10^6$ rad during 2 h stimulated either the gain of dark-amber color for some samples or weakening of the intensity of dark-blue color in other cases.

1.2 Use in Engineering

The watch industry was historically the first application field for artificial corundum crystals. As far back as 1704, Debraute and Fatio, mathematicians from Geneva, proposed to use gems in watch mechanisms, but this idea was economically unfeasible at that time. As soon as ruby had been put into industrial production, the well-known Swiss watch companies "Breget," "Denis Blondel," and "Luis Odemar" took an interest in it, since this material – with its reasonable cost, stable quality, hardness, and wear resistance – turned out to be suitable for friction pairs. The coefficient of dry friction of a "ruby–steel" pair is 0.25 at a pressure on the order of 3.7 GPa. The coefficient of friction of artificial ruby on steel is lower and more stable in comparison with that of the natural stone. In collaboration with watchmakers, the authors tested a watch mechanism with ruby elements in the friction pairs that continuously operated for 50 years and did not reveal any wear of the ruby components [7].

Plants producing precise technical stones make billions of ruby and leucosapphire elements annually for watches and other precise devices. Mechanical watches contain flat stones with a cylindrical hole, stones with lubricator (spherical hollow), flat or spherical balancing stones with a noncylindrical hole, and lubricator. Laid-on spherical stones are used as supports for the shafts. For reliable functioning of anchor escapement input and output palettes are applied. Pulses from the anchor fork to the balance are transferred via ellipsoid stones.

For watch stones dark-red ruby conventionally is used. Such a tradition originated in those times when watches were assembled by hand. Red stones are beautifully seen and less tiresome to the eye.

Watch "glasses" manufactured on the base of leucosapphire and light-colored sapphires are used by all well-known manufacturers of high-quality watches. This is not a fashion. You will see no scratches on such "glasses."

Stones for devices made of sapphire and ruby are most reliable. Sapphire pivots produced commercially in large quantities serve as supports for the shafts of those mechanisms that must work continuously for tens of years (e.g., water and gas meters). Such pivots are either flat or supplied with conical craters of different shape. Plain bearings are made in the form of bushes with cylindrical or noncylindrical holes. Some examples of serial stones for watches and devices are given in Fig. 1.29 in the Appendix. At present, bearings of larger size are being developed for engines, pumps for transferring aggressive liquids or gases, and other devices.

The main requirements imposed on bearings – homogeneity and wear resistance – can be satisfied only with artificial sapphires, since natural crystals with homogeneity similar to that of artificial sapphire are hardly ever encountered. Inhomogeneity of the stone leads to elevated wear of friction pairs; therefore, the service life of bearings made with artificial sapphire is longer.

The accuracy of scales depends on the sharpness of the prisms on which the levels balance. Nowadays, sapphire prisms with a vertex angle of 35° are produced for scales.

Wear-resistant sapphire elements are unrivaled. The automated lines meant for the production of machine components automatically control dimensions by means of probes. The contacting sections of such probes are made of sapphire or ruby shaped as spheres, hemispheres, cones, or cylinders.

High-wear resistance must be a characteristic of wire die. Therefore, for changing the diameter of gold, aluminum, or silver wire sapphire and ruby wire die are used.

Steel nozzles of powerful sand blasters wear down during one working day, whereas sapphire nozzles serve up to 30 working days. Water-jet cutting is one of the modern techniques for cutting ceramics, solid alloys, rocks, and dangerous articles such as nuclear warheads, military rockets, and gas containers. The quality of cutting and the productivity of hydraulic monitors depend upon nozzle size stability. In a number of cases sapphire jet-forming nozzles successfully substitute their hard-alloy analogs.

Under the action of synthetic threads, sapphire thread carriers used in high-speed looms corrode and wear out essentially slower than the ones made of steel or glass. The thread carriers may have different configurations: straight or bent at different angles by plastic deformation. The passing rings used in spinnings are another type of thread carrier. They must possess high thermal conduction (at insufficient heat removal fishing-lines wear out more rapidly) and wear resistance. Therefore, the leading firms producing spinnings utilize polished sapphire passing rings.

The machines meant for electric erosion treatment of solid alloys are supplied with sapphire guides. Several sapphire rods with grooves for thread are mounted in such a manner that they form a well. The grooves set the direction of sapphire thread, which moves at a high speed.

Sapphire cutters are used for high-quality and fine treatment of copper, aluminum, and their alloys, as well as for nonmetallic materials. In comparison with their hard-alloy analogs, sapphire cutters provide better treatment accuracy and a higher class of surface finish at smooth feed. The sharpening angles of the cutting plate and the working conditions for one of the types of cutters are presented in Fig. 1.30. Sapphire microcutters are used in microinstrument making.

Constructional sapphire elements are applied in the growth of other crystals. The methods of seed fixation in the process of the growth of large-size BGO crystals are shown in Fig. 1.2 [8]. The use of such facilities diminishes the consumption of platinum and increases the service life. While growing high-temperature crystals by the method of moving heater, the use of massive sapphire supports for the crucible fixed on water-cooled rods allow effective removal of heat from the crucible bottom due to the high thermal conductivity of sapphire [9].

Sapphire substrates belong to one of the most significant aspects of the constructional application of this material. They are used for epitaxy of semiconductor films such as Si, GaN, AlGaN, and for making integrated circuits. Sapphire substrates are inert, work at high temperatures and mechanical loads, and can be obtained in large size. Therefore, they are used even in those cases when the lattice parameters do not completely coincide with the parameters of heteroepitaxial structures.

For convenience of positioning, the substrates are supplied with one or two additional (marking) profile planes (Fig. 1.3). The *C* plane is used for the coating of sapphire with CdS, CdTe, CdSe, GaN, SiC, InAlGaN, and $LiNbO_3$ as well as for the epitaxial growth of some oxide (e.g., ZnO) and metal ($Fe_{(110)}$) films. The *A* plane is used for making hybrid microcircuits, devices that possess high-temperature superconductivity, and for coating the crystals with Co, $Fe_{(110)}$, $W_{(110)}$, $Au_{(111)}$, and $V_{(01\bar{1})}$. The *R* plane is suitable for coating sapphire with MgO, α-ZnO, and Si by the method of heteroepitaxy (Fig. 1.4). However, in the case when MgO is applied through a sublayer, other planes are also used:

$$(0001)_{\text{sapphire}}/(111)_{\text{spinel}}/(111)_{\text{MgO}}$$
$$(2\bar{1}\bar{1}0)_{\text{sapphire}}/(111)_{\text{spinel}}/(111)_{\text{MgO}}$$

While making "silicon-on-sapphire" structures, *n*-type silicon is applied on the *R* plane $(10\bar{1}2)$ of sapphire directly or through a buffer layer. The barrier for elec-

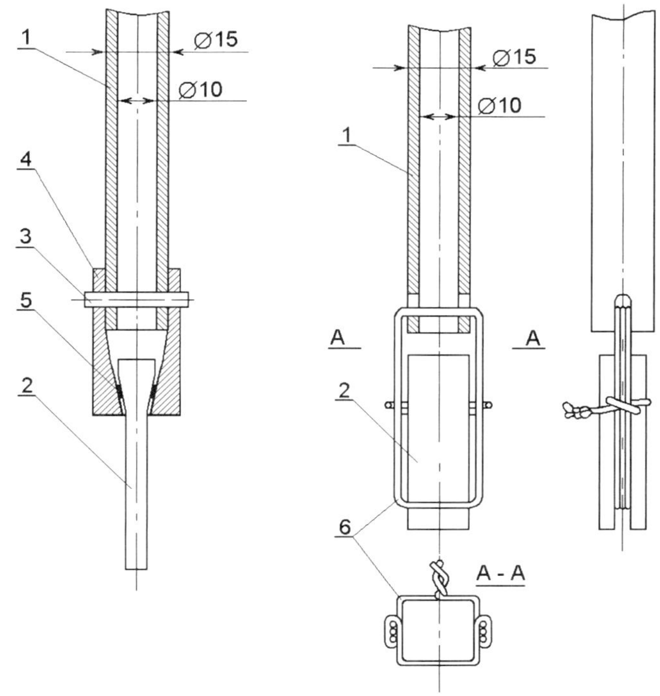

Fig. 1.2 Seed fixation facility: (*1*) sapphire tube, (*2*) seed, (*3*) sapphire rod, (*4*) corundum cone, (*5*) platinum gasket, (*6*) platinum wire

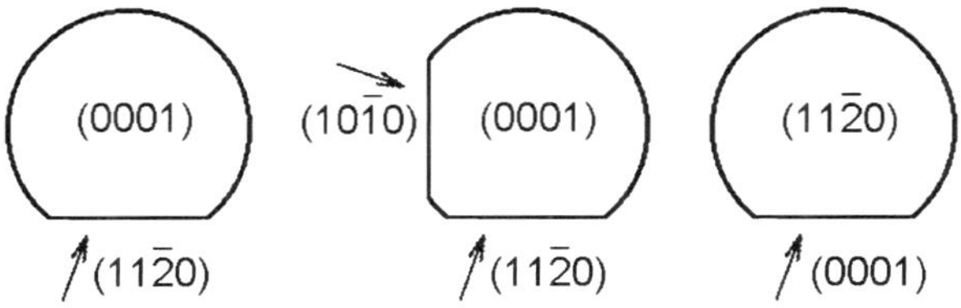

Fig. 1.3 Relative position of the basal and additional planes

trons (potential barrier) between Si and sapphire is 3.25 eV. The layers are deposited from a sublimating silicon source heated up to ~1,650 K, the substrate temperature being 720–1,020 K. At low temperatures the orientation of the growing layers is [110]; at $T > 820$ K they are oriented in the direction [100]. At 820 K there appears a binary orientation: $Si_{(100),(110)}$ [10].

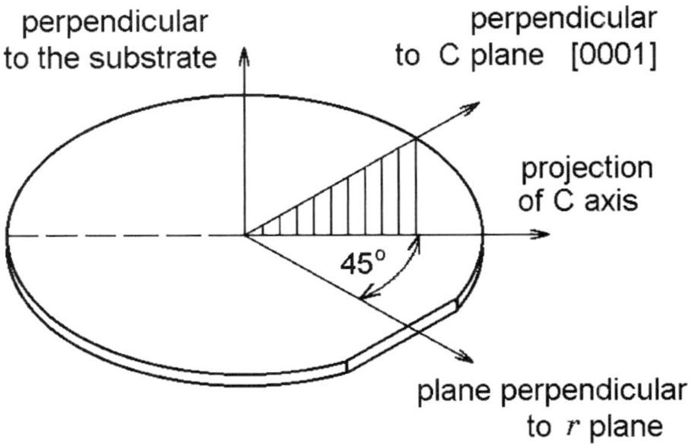

Fig. 1.4 Scheme of (100) Si film on *r* plane of sapphire

For ZnO with the structure of wurtzite the combinations $(11\bar{2}0)_{ZnO}/(01\bar{1}2)_{sapphire}$ and $(0001)_{ZnO}/(0\bar{1}11)_{sapphire}$ are used. Sapphire substrates are employed in sensors measuring pressure, mass, and humidity, as well as in IR-radiation detectors (HgCdTe films on sapphire) and other devices.

Bicrystalline substrates are used for the creation of the Josephson transitions on high-temperature superconductor films. The tapes for the bicrystals are grown onto a bicrystalline seed. The orientation of the wide plane is $[1\bar{1}02]$ the growth direction being $[1\bar{1}02]$. The bicrystalline angle is equal to $2 \times 14°$ [10]. The production of laser diodes and super-bright blue and green light-emitting diodes (LEDs)is in rapid development. Similar rapid growth is observed for their application fields – from traffic lights to the treatment of patients suffering from Alzheimer's disease (irradiation with a light intensity of 30 lx by means of a pilot unit incorporating blue LEDs diminishes the degree of disturbance of sleeping and waking periods, evidently due to a decrease in the patients' body temperature [11]). For blue LED the heterostructures GaN/sapphire are used (despite the fact that the lattice mismatch runs about 13%) and $AlGaN/InGaN/GaN/(0001)_{sapphire}$.

In the nitride-sapphire system several orientation relations such as the following are realized:

$$(0001)_{AlN}/(0001)_{sapphire}$$
$$(0001)_{GaN}/(0001)_{sapphire}$$
$$(11\bar{2}0)_{AlN}/(\bar{1}012)_{sapphire}$$
$$(11\bar{2}0)_{GaN}/(1012)_{sapphire}$$
$$(11\bar{2}6)_{AlN}/(\bar{1}012)_{sapphire}$$

In addition to other conditions, the concentration of the phases $(11\bar{2}6)$ and $(11\bar{2}0)$ is influenced by the substrate disorientation.

The concentration of carriers (holes) in the GaN layer reaches 10^{16}–10^{18} cm^{-3}. The currently used substrates have a diameter of 2, 3, and 4 in. The permitted deviation of the substrate from the plane is $(0001) \pm 0.1°$.

The blue light diodes are used as a base for making white light sources applied in place of filament lamps and luminescent lamps. These new light sources are much more economical; their service life is several tenfold longer. A high quantum radiation yield close to 100% in the blue and green regions of the spectrum makes LEDs applicable in information screens, displace, signaling, and other devices.

Among the promising trends in the use of sapphire substrates, one should mention the technology of carbon nanotube growth on sapphire. Such a method is being developed at North California University. This new material seems to be promising for nanotransistors and sensors.[1] Researchers have found that a-plane sapphire surfaces spontaneously arrange single-walled carbon nanotubes into useful patterns. No template has to be provided to guide this structuring; it is formed automatically.

Chemical ware such as crucibles, boats, measuring glasses, reactor housings, and other articles (Fig. 1.5) came into use after the development of shaped sapphire growth methods. Owing to their high corrosion and erosion resistance (see Chap. 2.2), these articles can be employed in obtaining high-purity substances where even the slightest traces of the material from the chemical ware used are not permitted to be obtained. They also can be used for the dispersion of hard materials. For grinding sapphire components of disintegrators, milling bodies (spheres, cylinders), mortars, pestles, and so forth are produced.

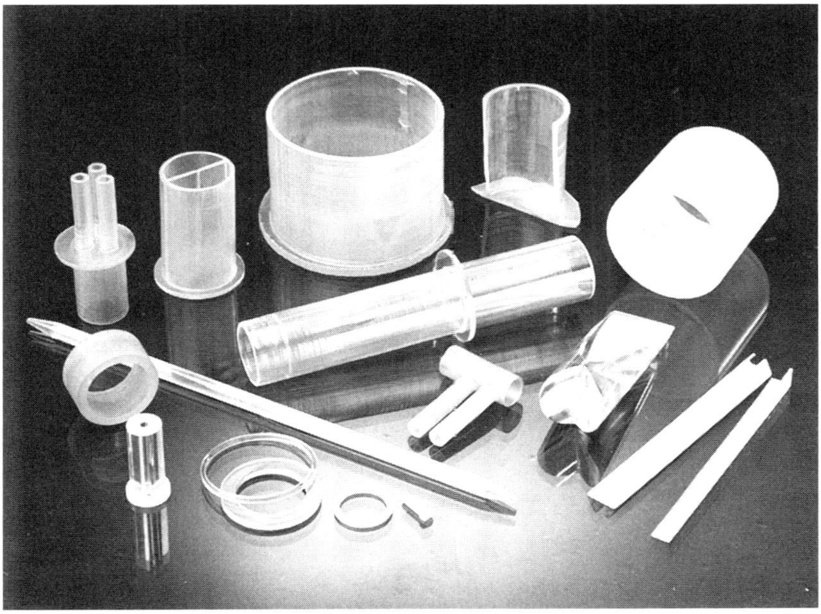

Fig. 1.5 Chemical ware made of sapphire

[1]www.physorg.com/news3863.html.

Rectangular sapphire cuvettes have no alternative for spectroscopy of hot fluoride melts. In a number of chemical processes sapphire articles successfully substitute for platinum ones.

Sapphire capillaries are used for investigating microexplosions, the growth of biocrystals, micro-dose measuring facilities, and other applications. During several years of technological investigations carried out in this direction, the diameter of as-grown capillary channels has diminished from 1.0 to 0.4 mm.

Sapphire fibers are used for the creation of compositional materials; they reinforce metallic and ceramic composites. The main requirements imposed on the fibers are high strength and reasonable cost. The fibers grown for composites have a diameter of 0.4–0.7 mm. Their length varies within the limits of approximately ten diameters. A $SiC_{0.99}O_{0.01}$ coating with a thickness of 4–6 μm raises the strength of the fibers (Fig. 1.6) [12]. Naturally, the strength characteristics of the fibers used in composites may change due to the interaction with the matrix. Such changes are characteristic of thermally stable composites based on Ni-matrixes.

In addition to the fibers for composites, sapphire whiskers with a diameter on the order of 1–10 μm are grown. At temperatures up to 1,270 K the strength of these whiskers is higher than that of bulk crystals. The average whisker strength decreases from 950 kg/mm^2 at room temperature to 150 kg/mm^2 at 1,870 K. Reinforcement with sapphire whiskers raises the strength of Ag matrix by a factor of 5; the composite preserves a considerable strength at $0.9T_m$ of the matrix. The ultimate strength of the composite, which consists of aluminum powder (83.7%), silicon powder (6.3%) and sapphire whiskers (10%) pressed at 810 K, reaches 67.5 kg mm^2 (without whiskers the ultimate strength is 33.5 kg/mm^2). Sapphire whiskers also are used to produce heat-insulating mats with a density of 0.02 g/cm^3 meant for high temperatures (up to 2,270 K).

Sapphire abrasive (α-Al_2O_3) crystal particles are an extremely sharp grinding, long-lasting, blasting abrasive that can be recycled many times. These particles are widely used in blast finishing and surface preparation because of their cost, longevity,

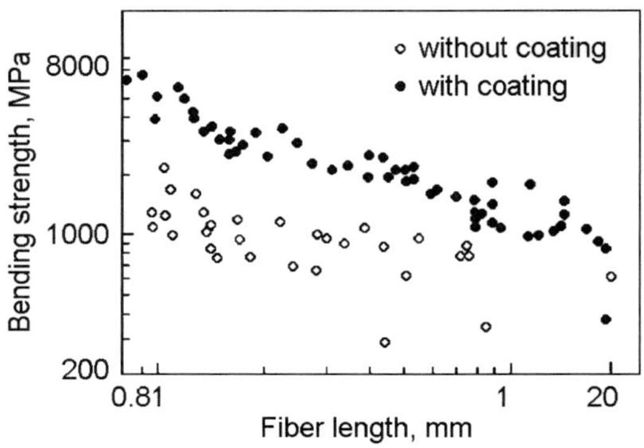

Fig. 1.6 Bending strength of sapphire fibers

and hardness. Harder than other commonly used blasting materials, aluminum oxide grains penetrate and cut even the hardest metals and sintered carbide.

Approximately 50% lighter than metallic media, aluminum oxide has twice as many particles per pound. Fast cutting minimizes the damage to thin materials by eliminating the surface stresses caused by heavier, slower-cutting media.

Aluminum oxide has a wide variety of applications, from cleaning engine heads, valves, pistons, and turbine blades in the aircraft industry to lettering in monument and marker inscriptions. It also is commonly used for matte finishing, as well as cleaning and preparing parts for metallizing, plating, and welding. The larger the grit size, the faster it will cut; but consider the size of perforations or holes through which the grit must pass when screen-separating parts from media after use: Standard mesh sizes for brown aluminum oxide: 12; 36; 60; 80; 120; 220. Standard mesh sizes for white aluminum oxide: 36; 60; 80; 100; 120; 220; 240.

Dispersionally hardened composites consist of a matrix (e.g., aluminum or nickel) in which particles of elements with a higher melting point and a higher strength (e.g., corundum particles) are introduced. The quantity of fine-dispersed corundum particles introduced in Al or Ni matrixes is typically 2.5–10 mass%.

CD disks are fabricated of polycarbonate. In the opinion of the authors, in the near future sapphire disks will be developed for long-term storage of significant data.

Ball lenses are great tools for improving signal coupling between fibers, emitters, and detectors. Polished sapphire and ruby balls are used with $D = 0.3$–6.35 mm ± 2.54 μm and a sphericity to within 0.64 μm.

1.3 Use in Optics

Sapphire of different quality is employed in optical articles. The most recognized classification of sapphire optical quality is presented in the Appendix.

Sapphire windows with diameters from several millimeters to several hundreds of millimeters are applied in various devices that work on the ground, under water, and in outer space in a wide range of pressures and temperatures – from cryogenic to high temperatures. An example design of a sealed-in window is presented in Fig. 1.31. It is meant for use at temperatures starting from cryogenic and up to 720 K, with high pressures both internal and external.

Sapphire windows are applied in gas cryostats and helium microstats with $D(12\ldots25) \times 2$ mm, sealed in metallic flanges or gasketed in the sockets of devices. Copper gaskets or indium wires provide a vacuum-tight seal to the cryostat.

According to *Oxford Instruments Superconductivity,* the place occupied by sapphire windows in comparison with those made of other materials can be represented by the diagram shown in Fig. 1.32. The diagram allows selection of a window material that satisfies the required working conditions. Now sapphire is also used for the scanner windows of cash registers that read bar codes.

Windows with a wedge of 1° and 3° are used in laser engineering (Fig. 1.33) as partially reflecting mirrors or prisms for laser beam control.

Sapphire energy withdrawal windows can be used. One of the factors that limit the withdrawal of energy from devices is secondary electron resonance discharge in vacuum proportional to the secondary electron emission coefficient (Table 1.1). As seen from this table, sapphire compares favorably with other dielectrics in this characteristic.

Sapphire lenses and prisms are used under extreme conditions that glass analogs cannot withstand, or due to high refractive index characteristic of sapphire.

Sapphire light guides incorporated in continuous temperature control devices yield the most unbiased data concerning the state of a melt in melting, dispensing, and mixing furnaces. They can work in molten steels and alloys, or in strongly aggressive melts. Sapphire is the only existing optical material that preserves its optical properties and serviceability in cast iron and steel melts. The light guide is fed through the furnace lining in such a way that one of its ends is put into the melt and the other end is located outside the lining. The intensity of the light guide glow is proportional to the temperature of the melt. By means of a pyrometric converter the thermal radiation is transformed into a signal for the automated temperature control system[2]. To raise the strength of the sapphire rod, it is fixed into a ceramic tube using a high-temperature ceramic compound, and the resulting bar is then installed into the lining [13, 14]. The light guides incorporated into the bar work during the entire technological cycle until the lining is replaced. In melted gold, sapphire light guides with a diameter of 4 mm withstand heating up to 2,070 K and subsequent cooling at a rate of 2 deg/s.

Naturally, the lowest losses by light absorption and scattering are characteristic of leucosapphire. However, the flow of light from metallurgical furnaces is so intense that there is no need to "economize" it. The use of leucosapphire light guides with ruby tips introduced into the melt allows increased measurement accuracy: the transmitted light intensity remains proportional to the temperature, but is clarified by the additional measurement of chromium luminescence intensity.

Sapphire optical fibers are also classified as light guides. In fiber grown along the *C* axis the propagation of light does not depend on the polarization. However,

Table 1.1 Secondary electron emission coefficient at a frequency of 10^9–$1{,}0^{10}$ KHz (SEEC)

	SEEC		
Energy (keV)	Sapphire	Ceramics	BeO
0.5	2.7		6.1
1	3.3	3.7–4	5.7
2	2.8		4.5

[2] This system was developed by the Institute for Single Crystals in collaboration with the Institute for Casting Problems (Kiev). The measurement range is 720–2,070 K, the indication error being 1%.

in real fibers a polarization anisotropy caused by circular symmetry breaking or local stresses connected with bending, compression, and twisting arises. Induced birefringence gives rise to the appearance of a difference between the phase and group rates of light-wave propagation in fiber polarized along the two axes. Such a difference leads to lagging of the pulse polarized along the "slow" axis of birefringence, as well as to the appearance of a difference between the phases of two polarized components (Fig. 1.7).

Focusing cones (focones) also belong to light guide facilities. By means of sapphire focones the emission of high-power xenon lamps is transformed into a point or linear source ("light knife"). For the latter transformation the shape of the input face is calculated (Fig. 1.8), allowing redistribution of incident light flux in the required way [15]. Light guides and focones of different designs are shown in Fig. 1.9.

To raise the photoresistance of focones to high-power UV radiation, the crystals are saturated with anionic vacancies [16]. In the course of long-term operation of the crystal, induced absorption may arise and the temperature of the output face will decrease. Annealing under reducing conditions completely restores the optical quality of the article.

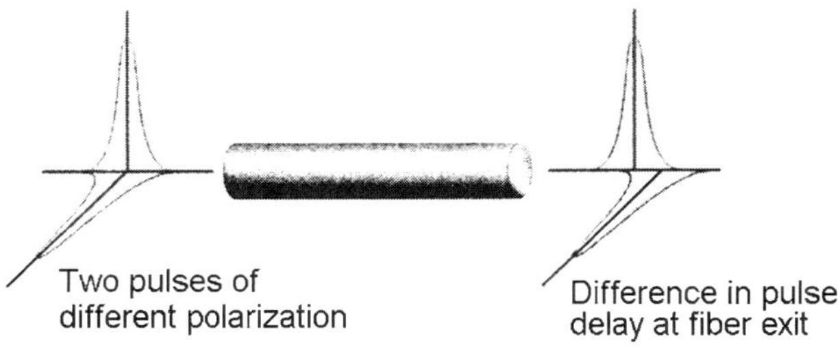

Fig. 1.7 Light pulse propagation in a fiber with determined birefringence

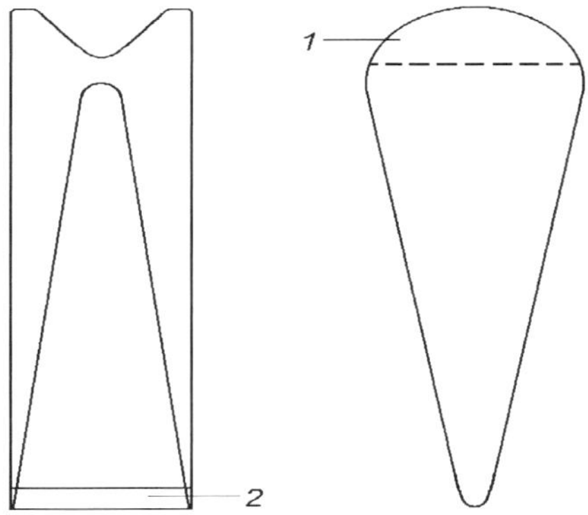

Fig. 1.8 Two projections of a facility for the transformation of spherical light source into linear source: (*1*) entrance face, (*2*) exit face

Fig. 1.9 Sapphire light guides and focones

In some cases it is easier to carry out temperature measurements by means of traditional thermocouples, but in aggressive medium these quickly break down.

Thermocouple casings made of sapphire in the form of tubes with hermetically sealed bottoms provide reliable protection of metallic thermocouples from aggressive media.

Sapphire meniscuses (domes) shaped as hemispheres are used for making the heads of IR-radiation guided missiles. Here, sapphire is employed due to the combination of its sufficient transmission in the region of 3–5 μm and its mechanical strength. The temperature of the working article, and consequently the value of transmission and guiding accuracy, are defined by the outer surface roughness.

Sapphire shells for sodium high-pressure lamps are competitive with polycrystalline shells. The grain boundaries in polycrystalline shells promote the diffusion of sodium (at working temperatures the coefficient of diffusion along the grain boundaries exceeds the bulk coefficient of diffusion by approximately three orders of magnitude), thus shortening the service life of the polycrystalline articles. Although polycrystalline shells are cheaper, sapphire shells possess higher transparency, consume less energy, have a longer service life, and maintain a higher stability of light flux (Fig. 1.10) [17].

The lamps with quartz shells used for pumping solid-state laser elements also have a number of drawbacks. Sapphire shells possess higher thermal conductivity and strength; sapphire has an immeasurably higher resistance to alkali metals. The lamps with sapphire shells withstand higher temperatures.

With other conditions being equal, the value of mechanical strength and ultimate electrical loads on the shell depend on the quantity of blocks in the tube. Shells without blocks withstand loads exceeding 390 W/cm^2, while the ultimate power

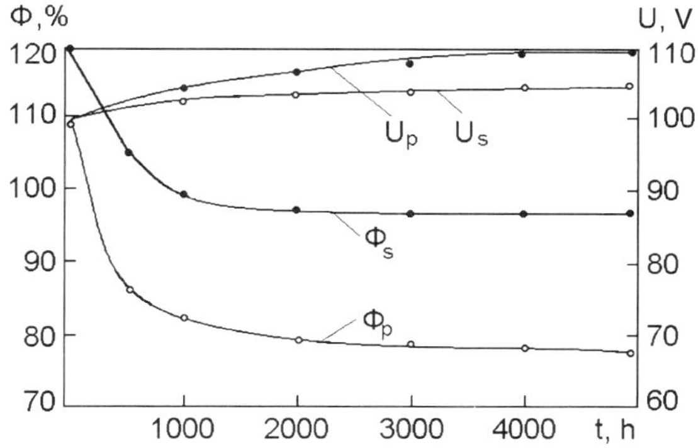

Fig. 1.10 Changes in the parameters of 400-W sodium lamp with time. Φ_s, U_s denote the light flux and tension on the lamp with sapphire shell. Φ_p, U_p are the same parameters for the lamp with polycor shell

decreases to 150 W/cm^2 for those containing 5 blocks and 60 W/cm^2 for those with 7–15 blocks [18].

Sapphire shells for lamps, absorption cells, and filters of atomic (rubidium) frequency standards raise the characteristics of the devices: they diminish the relative frequency change corresponding with the time of work and atmospheric pressure, decrease power consumption, and increase the mean time between failure of devices. Due to their high thermal conductivity, sapphire shells allow withdrawal of heat from laser elements made of materials with low thermal conductivity. Transparent sapphire shells are able to protect Mo-heaters from oxidation or aggressive media [19].

X-ray interferometers of the Fabry-Perot type at first contained silicon crystals as X-ray grids [20]. Then a need arose for crystals possessing higher stability in a wide range of temperatures and pressures, so for this purpose sapphire was chosen. It should be expected that the creation of such interferometers will lead to the appearance of a new generation of optical X-ray devices such as filters, X-ray clocks, and most significantly a new method of the measurement of atomic scales (i.e., high-precision length standards) [21]. The next step in this direction may be bound to the development of devices on the basis of synchrotron radiation. For this class of high-precision apparatus, grids (mirrors) made of dislocation-free sapphire are required.

X-ray monochromators using sapphire [22]. For X-ray diffractometers (Cu–K_α radiation) the planes $(10\bar{1}4)$ and $(03\bar{3}0)$ are used. High resolution is achieved with the plane $(14\bar{5}6)$, for which $2d = 1.660$ Å.

Sapphire-based scintillators are promising for the registration of α- and γ-particles, as well as low-energy X-rays, owing to their high mechanical and optical characteristics [23].

In Al_2O_3:Ti^{3+},Me^{2+} creates shallow levels of charge capture (where Me is Mg or Ca, the bivalent coactivator). Holes formed at the capture of electrons by Ti^{4+} ions settle on such levels, and this is followed by the recombination of electron-hole

$$Mg^{2+} + h^{+} \rightarrow Mg^{3+} \frac{220\,K}{10\,\mu s} \rightarrow Mg^{2+} + h^{+}$$

The scintillator's light output decreases as the matrix stoichiometry is violated due to recombination of the charges on point defects. Therefore, the annealing medium potential needs to be maintained within the limits of –20 to –150 kJ/mol, depending on the content of Ti^{3+}.

Sapphire scintillators have the following advantages: emission in the red region of the spectrum (690–900 nm) rather high light output, low levels of afterglow (Table 1.2), and insignificant absorption of their own radiation (~0.15 cm^{-1}). Due to the radiation resistance of the matrix, sapphire stability is greater in comparison to other scintillators (Table 1.3). After the cessation of radiation interactions, the scintillation characteristics of sapphire scintillators are restored in several seconds.

The methods used for detection and identification of radionuclides and fission products of radioactive materials, including those with complex compositions (e.g., radioactive wastes), by analysis are composed of the parameters of β-particles, internal conversion electrons (ICEs), α-particles, and so forth. Due to its low effective atomic number, Z_{ef}, and small cross-section of radiative capture of thermal neutrons, Al_2O_3:Ti is a promising material for the measurement of β-particle fluxes in fields of thermal neutrons. Table 1.4 presents the characteristics of Al_2O_3:Ti compared to those of other scintillators that are functionally close to it.

Table 1.2 Comparison of scintillator characteristics

Scintillator	Light output (relative units)	Decay time (μs)	Afterglow in 10 μs (%)	Maximum of emission (nm)
CsI:Tl	100	1	0.5–2	550
Al_2O_3:Ti, Ca	30–40	3–4	0.05–0.08	750

Table 1.3 Radiation stability of scintillators under irradiation (the dose rate 600 rad/(min m^2), the energy of Bremsstrahlung γ-radiation quanta = 4 MeV)

	Output signal (relative units) at irradiation duration (min)								
Scintillator	1	2	5	10	pause	21	22	25	30
CsI:Tl	100	97	87	94		90	87	84	77
Al_2O_3:Ti, Ca	100	99	98	97		99	99	98	97

Table 1.4 Spectrometric characteristics of scintillators [24, 27]

Crystal	λ_{max}, (nm)	τ (μs)	α (cm^{-1})	Z_{ef}	S (a.u)	T_{max} (K)
ZnSe:Te	600–620	2–20	0.05–0.15	33	100	400–450
	630–640	>20			170	
Al_2O_3:Ti	750	3–4	0.002–0.05	12	16–20	560
CsI:Tl	550	0.63–1	>0.05	54	100	350–400

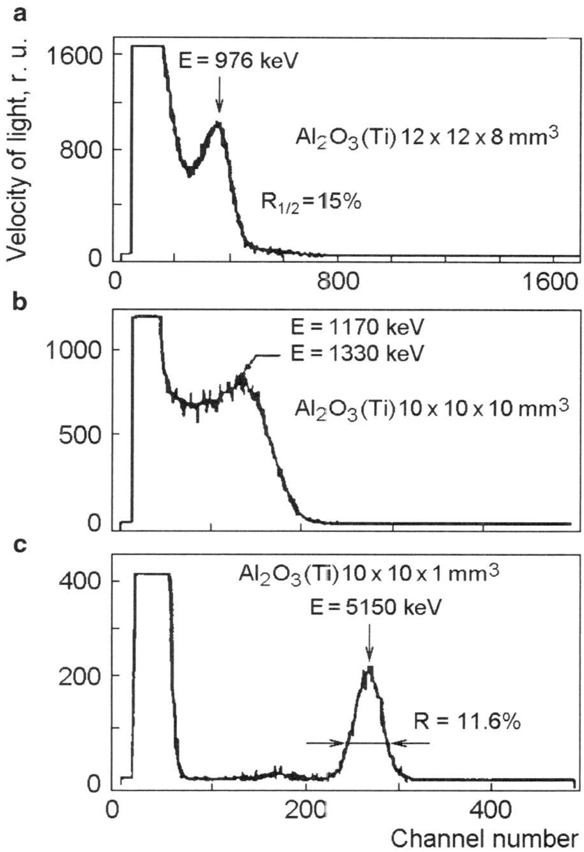

Fig. 1.11 Spectra of (**a**) internal conversion electrons from ^{207}Bi, (**b**) γ-quanta from ^{60}Co, and (**c**) α-particles from ^{239}Pu obtained using Al_2O_3:Ti-based detector and Si *p*-i-*n* photodiode of S 3,590 type

The peak of γ-quanta from ^{241} Am ($E\gamma$ = 59.6 keV), which has appeared in the source as an impurity after β-decay, has been observed in the spectra of ICE (Fig. 1.11) at the left of the peak of ^{239}Pu α-particles. The energy resolution for α-particles with E_α = 5,150 keV is 11.6%.

It has been found that for Al_2O_3:Ti the ratio α/β equals 0.28 [24]. The energy spectrum of scintillation signals excited by different sources is shown in Fig. 1.12 [25].

The range of scintillator dimensions is estimated by taking into account Z_{ef}, the quantity of atoms per unit volume, and the path length of X- or γ-radiation in the crystal. The upper boundary of crystal dimensions corresponds to the case when the photoreceiver is connected to the crystal face perpendicular to incident radiation; it is calculated for the condition when the light from the most distant point exceeds the discrimination threshold (10 keV). In Fig. 1.13, the upper boundary for different values of the absorption coefficient *K* is shown by thin lines [25]. The bold line denotes the lower boundary

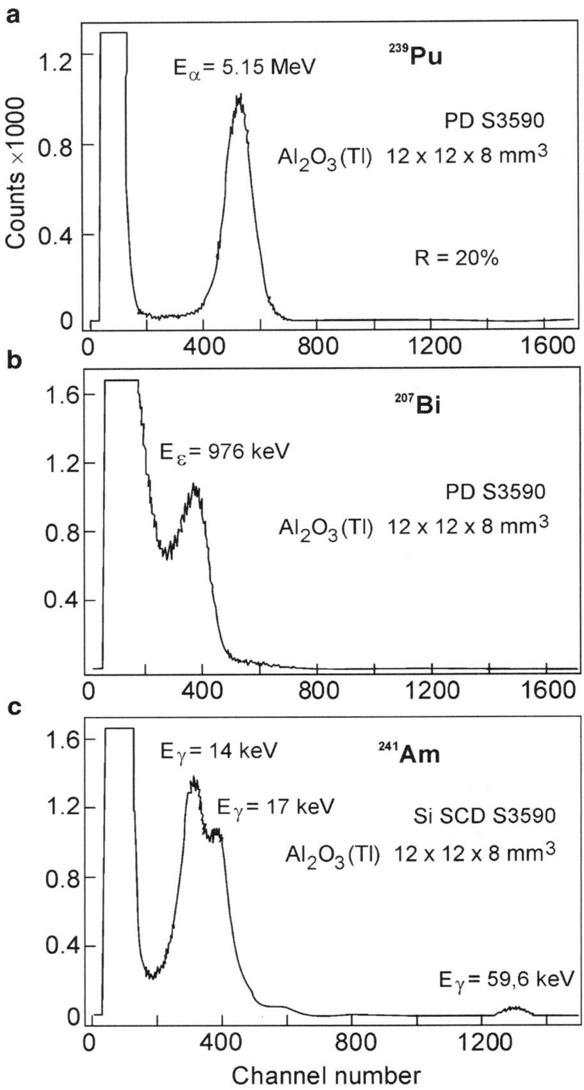

Fig. 1.12 Pulse amplitude spectra of corundum scintillation crystals under exitation by different sources: (**a**) α-particles (5.15 MeV, ^{239}Pu source); (**b**) β-particles (976 keV, ^{207}Bi source); (**c**) γ-photons (14 and 17 keV, ^{241}Am source) [25]

of crystal dimensions calculated for the interval $l(E) < L < L_{max}$, where $l(E)$ is the path length of hard γ-radiation.

A combination detector for simultaneous registration of charged quanta and γ-quanta has been described [26]. The detector contains a ZnSe:Te crystal placed

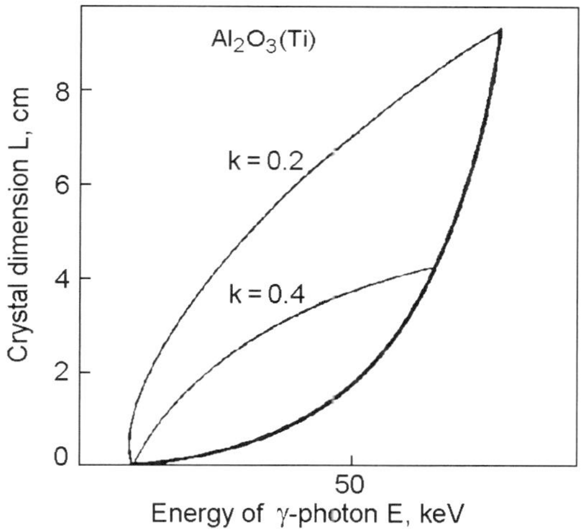

Fig. 1.13 Feasible dimensions of Al_2O_3:Ti as a function of γ-radiation energy

on the input window of the scintillation light guide made of CsI:Tl and Al_2O_3:Ti and shaped as a truncated pyramid. Such a detector was used for the separate registration of α- and γ-radiation at simultaneous excitation by ^{239}Pu and ^{241}Am.

Another combination of detecting materials for the registration of mixed radiation of thermal neutrons and γ-rays can be used. A new compositional material based on fine-crystalline Al_2O_3: Ti and LiF has been proposed for a long-wavelength scintillator [27].

Luminophor screens based on doped sapphire have been developed for accelerators. They are meant for the registration of intense beam parameters by means of digital TV facilities [28]. In comparison with other materials, such screens provide higher light output (Table 1.5). Their light output is linear up to $2 \cdot 10^{14}$ proton/cm^2 and decreases at Cr content < 0.066 mass%. The highest light output was recorded for screens with a thickness of 1.5 mm made of Al_2O_3:Cr^{3+} crystal in which the content of chromium was 0.066 mass%.

Thermoluminescent detectors (TLDs) are made of sapphire with anionic nonstoichiometry. The Chernobyl catastrophe has shown that for radiation environmental control, ionizing radiation TLDs are the most promising. Sapphire-based structures have a number of advantages. Their sensitivity to γ-radiation is an order of magnitude higher than that of LiF(Mg, Ti) and allows registration of radiation doses on the level of 8 μGy. Their sensitivity is independent of the dose in 10^{-11}–10^{-2} Gy·s^{-1} range.

For dosimetry the TL peak at 440 K, which is associated with the liberation of holes captured by impurity ions and their recombination upon reaching the valence zone, is used. The value of the peak depends on the degree of anionic nonstoichiometry, which can increase up to 10^{17} vacancy/cm^3 under the influence of reducing and annealing medium as well as doping by some elements.

Table 1.5 Characteristics of luminophor screens (measured at the Institute of High-Energy Physics, Serpukhov, Russia, for proton beam with 70 GeV energy, 10^{12}–10^{14} proton/cm^2 density, duration 10^{-6}–0.4 s, 8-s period)

Screen material	Activator content (mass%)	Threshold sensitivity (proton/cm^2)	Light output (photon/prot. s)	Afterglow in 60 μs (%)	Ultimate radiation resistance (prot./cm^2)
Al_2O_3:Cr^{3+} single crystal	0.066	$1.7 \cdot 10^6$	440	8–10	10^{18}–10^{19}
	0.08	$2.0 \cdot 10^6$	377	2–4	10^{18}–10^{19}
	1.16	$2.8 \cdot 10^6$	263	50	10^{18}–10^{19}
Al_2O_3:Cr ceramics AF-995	0.1	$4.5 \cdot 10^6$	159		10^{18}–10^{19}
Al_2O_3:Ti^{3+} single crystal	0.011	$1.9 \cdot 10^7$	39	1	10^{18}–10^{19}
	0.047	$2.7 \cdot 10^7$	27	1	10^{18}–10^{19}
ZnS	–	$3.0 \cdot 10^7$	8–16		$<10^{15}$
BeO ceramics	–	$4.7 \cdot 10^8$	1.6		$>10^{18}$

Sapphire TLDs are characterized by extremely low glow of nonradiation origin. Thermal emission at the temperature of the main maximum is equivalent to 0.6 μGy, chemiluminescence is absent, and the level of triboluminescence does not exceed 2.5 μGy [29] (the intrinsic glow of LiF:Mg, Ti-based TLD corresponds to 10 μGy, its triboluminescence is on the order of ~100 μGy).

The crystals intended for use as TLD are grown in the form of calibrated rods by the Stepanov method in a strongly reducing medium. The rods are then cut into 1-mm-thick tablets with a diameter of 5 mm, after which their own background is measured. To make a set of detectors with a preset spread of sensitivity values, this testing should be preceded by irradiation with a known dose. The corresponding data are entered into a computer and, if necessary, displayed; the accumulated doses are registered by a thermoluminescent analyzer.

The crystals grown in the strongly reducing medium mentioned above have a high density of F and F^+ centers, the ratio of which may vary from one sample to another. The neutral F centers are characterized by 6.05 eV absorption bands; those positively charged possess 4.8 and 5.4 eV bands.

The crystals excited using 10^{-2} Gy/s dose power of ^{90}Sr radiation show radioluminescence (RL) with a maximum at 3 eV, which is caused by the recombination processes – in particular by the relaxation of excited F centers in the ground state. The intensity of RL depends weakly on the concentration of anionic vacancies: when the concentration of F centers changes by an order of magnitude, the intensity of RL changes within ±10% limits. The F^+ centers and some impurities compete for the recombination of electron-hole pairs, and they tend to quench F center RL. The energy output of RL is ~3.5% and serves as a measure of irradiation intensity.

A single TL peak is narrow and equivalent to the absorbed dose. The value of TL depends on the conditions of thermal treatment (Fig. 1.14). The detectors have a linear dependence of the accumulated light sums in the main TL peak on the irradiation dose. The sensitivity of the detectors increases as the irradiation doses grow (Fig. 1.15).

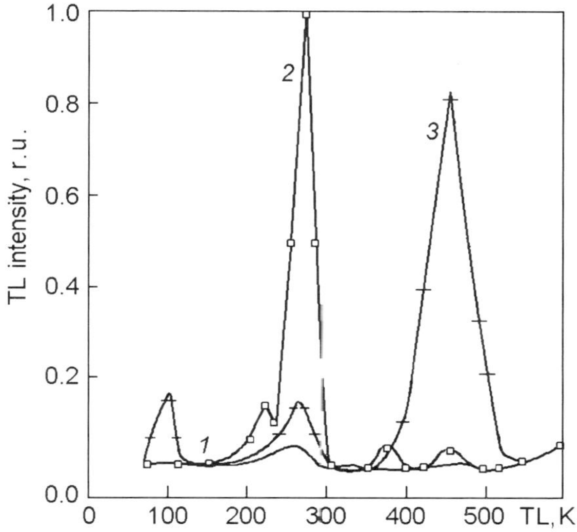

Fig. 1.14 Dependence of sapphire TL on the conditions of thermochemical treatment: (*1*) initial crystal, (*2*) 10^{-3} torr vacuum, 1,800 K, 10 h, (*3*)10^{-3} torr vacuum, 1,800 K, 10 h, graphite [26]

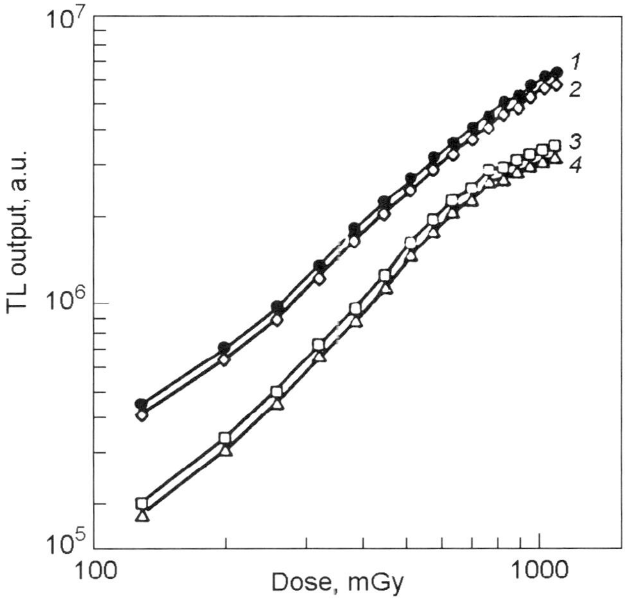

Fig. 1.15 Actual (*1, 3*) an calculated dose dependences of TL for anion-defective saphire at heating rates (K/s): 2 (*1, 2*) and 6 (*3, 4*)

A zone scheme of an interactive system of traps has been proposed in the crystal, which includes a TL-active trap, a luminescence center, and a deep trap [30]. One of the mechanisms responsible for the dose-characteristic linearity is the interactive coupling of the dosimetric and deep traps. Practical recommendations are proposed for raising the dosimeter efficiency. They are based on the decrease in the trap interaction contribution to the formation of the TL dose dependence. Such a decrease can be achieved if the deep traps are either maximally filled or completely emptied in the process of high-dose measurements. Preliminary filling of the deep traps is provided by irradiation at a temperature higher than the temperature range of the dosimetric peak. The deep traps are emptied during the process of annealing at 1,000 K after each event of high-dose irradiation. As seen from Fig. 1.15, it is expedient to use rather low rates of irradiated crystal heating. Impurities may widen the peak of TL and change the dependence of radiation sensitivity on the rate of irradiated sample heating [31]. The characteristics of the sapphire detector TLD-500 K [32] are presented in Table 1.9.

Al_2O_3:Cr^{3+}-based luminescent pressure transducers. The intensity of Al_2O_3:Cr^{3+} luminescence is pressure-dependent. At concentrations of chromium lower than 0.1 mass%, Cr^{3+} luminescence is dominated by two strong lines (R_1 and R_2) of the isolated single Cr^{3+} ions. The pressure dependence of the ratio of these lines is presented in Fig. 1.16 [33]. If the concentrations are higher than 0.1 mass%, additional sharp lines appear (*N* lines) from the exchange-coupled pairs of Cr^{3+}–Cr^{3+}. With increasing Cr^{3+} concentration, the *N*-line intensity increases with respect to that of *R* lines, thus reflecting the increasing probability of pair formation and the energy transfer from single ions to the pairs (Fig. 1.17) [33]. At concentrations exceeding 1 mass%, the luminescence is dominated by broadband emission with a peak at

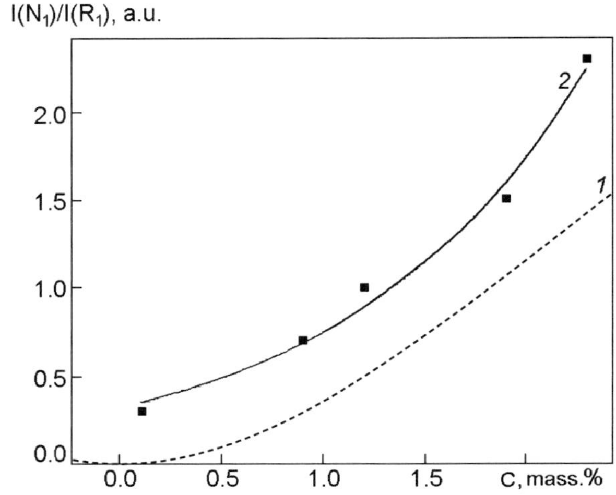

Fig. 1.16 Ratio N_1 (701.6 nm) intensity to R_1 line 1. Dashed line corresponds to calculated results, solid line is the best-fitting line, points represent experimental results

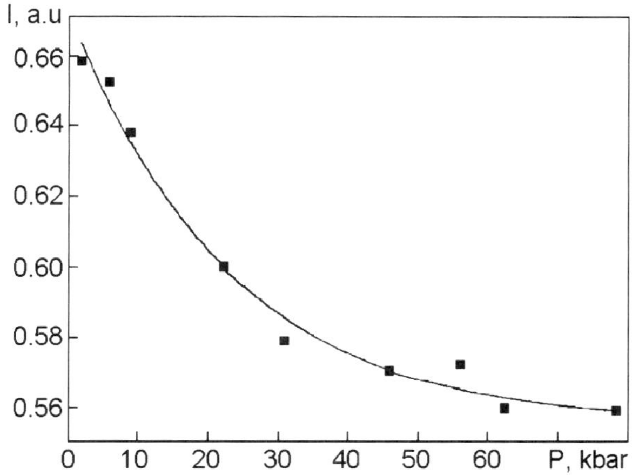

Fig. 1.17 Pressure dependence of R_z/R_1 line intensity ratio

around 750 nm, which has been attributed to clusters of chromium ions [34]. The dependence of the position of the lines on chromium concentration and pressure are presented in Table 1.10.

Luminescent pressure transducers are used in the investigation of equilibrium thermophysical properties of high-temperature superconducting substances, including T_m, as well as for the construction of phase-state diagrams. For this purpose, the studied material and a thin ruby plate are clamped between sapphire plates. Under the influence of electric current or laser heating on the investigated plate, a pressure of ~100 kbar arises during 1–2 ms. The pressure is measured from the shift in the ruby luminescence peak value. A pulse at $\lambda = 532$ nm is applied to the ruby plate via a light guide and the sapphire plate. The luminescence signal is passed at $\lambda = 700$ nm using the same light guide. The shift of the luminescence peak at this signal, equal to 0.5 nm, corresponds to an increase in pressure by 20 kbar.[3]

As an expendable transducer, high-quality ruby with a chromium content of 0.5–0.7% is used.

Ruby-based pressure transducers are used for hydrostatic pressures on the order of 60 kbar. Such a sensor is shaped as a disk, with a diameter of 0.6 mm and a thickness of 0.1 mm, and contains $4 \cdot 10^{14}$ spins. The intensity of the R_1 line is proportional to pressure.

In the CRESST program (Cryogenic Rare Event Search with Superconductivity Thermometers), a cryogenic detector based on sapphire is applied in the physical experiments searching for hypothetical dark matter particles [35]. Such a detector is efficient owing to its low radioactivity (Al and O have no radioactive isotopes).

[3]Dr. A.J. Savvatimsky. Private communication.

Moreover, it is characterized by a low heat capacity, so the signal (ΔT) is quite large. The calorimeter is made of pure, structurally perfect sapphire with a mass of 262 g and is supplied with a strip of superconductor deposited onto one surface to serve as a sensitive thermometer. It is important to eliminate the effect of any vibrations that may deliver energy to the crystal. So, in addition to special isolation suspensions for the apparatus, the crystal is fixed very tightly into its holder to prevent any frictional effects (even microscopic ones). The calorimeter rests on sapphire spheres with a diameter of 1 mm (Fig. 1.18).

This underground system detects single events in sapphire, the energies of these events ranging from about 1 keV to several hundred keV. The system has good energy resolution (0.5 keV) and good time resolution (40 or 100 μs for the onset of a pulse). Small-size cryogenic devices are sensitive to energies even in the region of 1 keV [36].

Laser elements are made of extrahomogeneous crystals of ruby and doped sapphire.

Ruby was the first solid, active medium, introduced into practice as far back as 1960. The content of Cr^{3+} is preset within a rather narrow region (0.018–0.05 mass%). The typical dimensions of the laser elements are diameter equal to 5–10 mm and length equal to 60–200 mm.

Ruby lasers are employed in pulsed holography, medicine, and laser technologies. In pulsed holography, large ruby elements with elevated homogeneity are used. In particular, 8 × 120 mm elements pumped by four flashbulbs are incorporated in a unit for holographic interference, which works in the paired pulse generation mode [37]. The intervals between the pulses are controlled by means of flashes from the pumping bulbs with different time periods, and this allows completion of laser investigations on resonance oscillations of rotating objects.

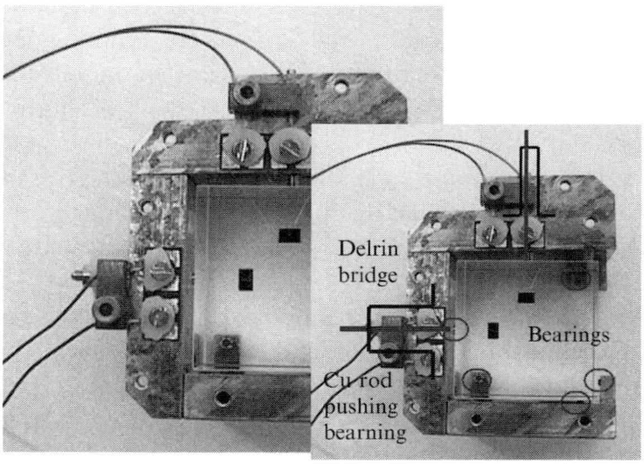

Fig. 1.18 Cryogenic detector based on saphire (Photo: Phillipe Di Stefano)

Production-type facilities are used for the treatment of diamonds, ceramics, ruby, and sapphire plain bearings and so on. For technological ruby lasers the quantity of pulses generated without noticeable laser degradation is in the tens of millions, whereas for high-power lasers this factor is measured in thousands of pulses. The advantages of ruby include high pulsed-radiation power, mechanical strength, and high photostability with respect to pumping bulb emission.

Ruby elements are liquid-cooled. The reliability of the coolant seals increases if located outside the active (ruby) component of the elements. For this purpose, rubies with leucosapphire tips are grown. The standard tip lengths are 5, 20, and 30 mm.

The characteristics of ruby used as an active medium are presented in Table 1.11, the standard types of laser elements and their designations are contained in Fig. 1.34, and ruby optical properties are described in Chap. 2.

Sapphire becomes a promising material for tunable lasers if doped with elements possessing wide absorption bands and wide emission regions. The most interesting among such elements are Ti^{3+}, Cu^{2+}, V^{3+}, Ni^{2+}, Co^{2+}, Co^{3+}, and Mn^{2+}. Concerning three-dimensional elements, the widest luminescence spectrum is characteristic of titanium.

Titanium-doped sapphire (Ti:sapphire, ticor, Ti:S) is one of the key active media of tunable lasers. It has been used since 1982, when the generation of this crystal was reported for the first time [38]. The $Al_2O_3:Ti^{3+}$ crystal is unique due to the fact that, in addition to possessing wideband amplification spectrum that allows pulses of femtosecond duration to be achieved, it also possesses a large cross-section of induced emission. These crystals have been actively investigated in different laser systems using active and passive mode synchronization. Under the conditions of active-mode synchronization pulses with a duration of 150 fs were obtained, while at passive-mode synchronization the pulse duration was diminished to 6 fs [39]. No pulses of femtosecond duration were obtained with pulsed pumping, although such a regimen is interesting as it allows high-pulse energies to be reached. In order to obtain pumping pulses with a duration corresponding to the lifetime of laser levels in $Al_2O_3:Ti^{3+}$ (2.7 μs), YAG:Nd-lasers with 0.5–5 μs pulse durations [40] were developed. The merits of $Al_2O_3:Ti^{3+}$ include the possibility of tuning the radiation wavelength in a wide spectral range, high efficiency at room temperature, high radiation power, stability of generation parameters, and radiation strength. Moreover, $Al_2O_3:Ti^{3+}$ lasers can be pumped by gas, solid-state, and diode lasers, or by flashbulb, and work in pulsed and continuous regimens. One laser of this type can be used instead of several lasers with fixed radiation wavelengths.

Historically, the crystals meant for laser elements were obtained by the Verneuil method. In the present day, the crystals employed for this purpose are grown from the melt by the Czochralski, Kyropoulos, and HEM methods and possess the highest optical homogeneity (Fig. 1.19). The growth media used in the crystal growth chambers are Ar or N_2 + 1.5% O_2. The substitution of O_2 by H_2 in the latter composition decreases the density of micropores. Depending on the function to be performed by the crystals, the content of titanium may vary within limits from 0.02 to 0.3 mass%.

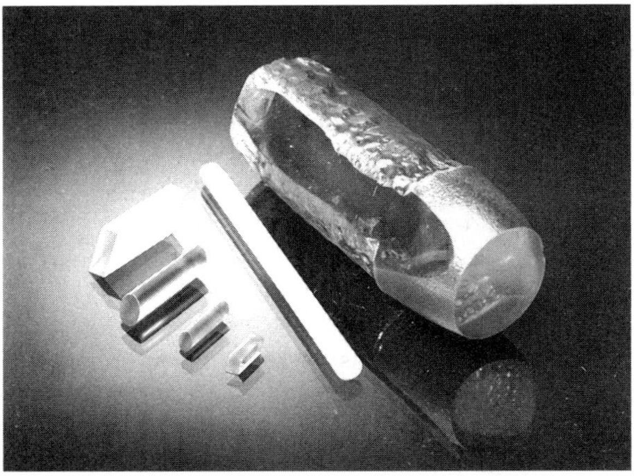

Fig. 1.19 $Al_2O_3:Ti^{3+}$ crystal grown by the Czochralski method and laser elements

The typical dimensions of laser elements are as follows:

- For flashbulb pumping: d= 5–8 mm, l= 60–120 mm
- For laser pumping: d= 4–30 mm or rectangular cross section, l= 4–30 mm
- For powerful short-pulse lasers and laser-pumped amplifiers: d= 80–110 mm, l ~ 30 mm

Pico- and femtosecond lasers and amplifiers are produced. The laser-pumped fs-laser "Mai Tai," manufactured since 1999 (2.5 W, 1 kHz, 710–920 nm tuning range), is one of most widely used apparatuses of this type. All its functions are computer-controlled.

Wide-aperture, large-diameter elements with minimal wave front distortions are utilized in high-power lasers. On the basis of these elements, a compact (6 m^2) 850-TW laser was created with a pulse duration of 32 fs and a spot diameter exceeding 4 cm [41]. Such a power can be achieved if amplifiers or preamplifiers are used together with amplifiers, as well as special adaptive optics (e.g., deformed mirrors, spot brightness correctors, acousto-optical programmed filters, and other facilities). Adaptive optics allow reduction of the thermolens effect, increase in the spot brightness, and improvement of the pulse shape, whereas special compressors can shorten the pulse duration. In "laser-amplifier" systems the power density reaches values of 10^{19} W/cm^2. In lasers with mode self-synchronization pumped by continuous Ar-lasers, 10-GW pulse power has been obtained at pulse durations of 100 fs. The development of petawatt lasers based on $Al_2O_3:Ti^{3+}$ is close to completion. In the opinion of some specialists, a power density of 10^{28} W/cm^2 can be achieved for this crystal [42].

The value of optical loss, or figure of merit (FOM), is an indirect characteristic of Ti:S as an active laser medium. The crystals used in lasers have FOM > 100.

The Ti:sapphire crystals are anisotropic; therefore, one should maintain a certain order of action for the determination of FOM. Otherwise, the values of FOM will be irreproducible and inauthentic.

The order of action for the determination of FOM is as follows:

- Adjust the sample on the adjusting table according to the beam polarization in such a way that the *C* axis of the crystal is parallel to the beam polarization vector (π-polarization). In this position, at the maximum value of the pumping energy absorption, the generation threshold and optical losses at the generation wavelength are the lowest, whereas the bulk laser strength (damage threshold) is the highest.
- Determine the value of optical loss within the 780- to 800-nm range using a polarized source (with 1:100 degree of polarization), the beam divergence should not be more than 1 mrad.
- Shutter the beam to $d \leq 2$ mm.
- Adjust the sample on the adjusting table coaxially to the sounding beam until the maximum reading with a photosensor is received.
- In this position, use the same photoreceiver to baseline the intensity of the sounding beam without the crystal (I_0) and with the crystal (I). This preferably should be realized using a Ti:S laser (10 Hz, 20–30 mJ) operating in the quasi-continuous regimen. It is necessary to control the beam to hit the same point of the photoreceiver.
- Calculate the absorption coefficient $\alpha = \ln I_0{:}I$ and divide the result by the length of the crystal (cm).
- Divide the average titanium concentration (α_{500}, cm^{-1}) by the value obtained in the previous step.

Among the application fields of Ti:S lasers, the following should be mentioned: medicine (ophthalmology, cancer therapy), atmospheric sounding, satellite communication, lidar, photochemical processing, and pumping of other laser types. Femtosecond lasers based on $Al_2O_3{:}Ti^{3+}$ are employed to make holes (with a diameter of 50–200 μm) in diamond and sapphire. The use of the fs-range in medicine is a result of the fact that it does not give rise to pain, as thermal effects are practically absent. The laser characteristics of $Al_2O_3{:}Ti^{3+}$ are presented in 1.12 and its optical characteristics are discussed in Chap. 2.

The $Al_2O_3{:}V^{4+}$ also is considered to be applicable for tunable lasers. The V^{4+} ion has the electronic configuration *3d^1*, which is similar to that of the Ti^{3+} ion. Formed from the d^1 configuration is one ^2D term, which undergoes fission into two multiplets ^2E and $^2T_2$ in an octahedron field. The transitions between these states produce a bimodal absorption band in the visible region of the spectrum and a wide luminescence band in the red and near-IR regions (Fig. 1.20). The luminescence decay time of 2.4 μs is unchanged within the temperature range from 77 to 300 K [43]. The heterocharged ion V^{4+} enters the matrix due to a charge-compensating impurity or defects such as cationic vacancies. The formation of V^{4+} centers in monovalent states (without V^{3+}–V^{2+} centers) is a more complicated problem compared to the case of Ti^{3+}.

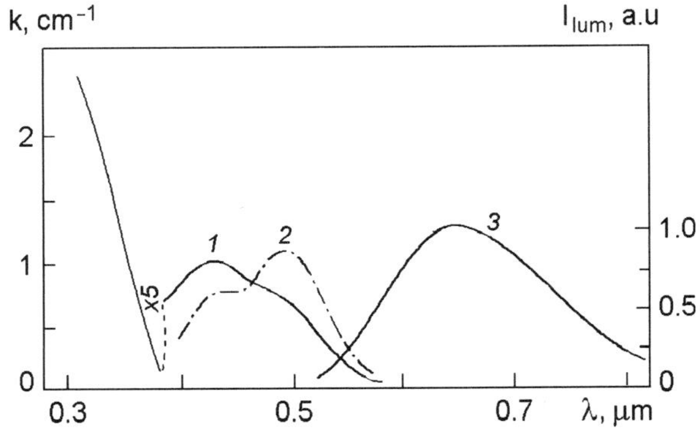

Fig. 1.20 (*1*) Absorption, (*2*) excitation, and (*3*) luminescence spectra of Al_2O_3:V^{4+} crystals at 300 K

Tunable generation also is observed when sapphire is doped with other elements. During the pumping process in nonselective Al_2O_3:Mg, Eu resonator, generation was obtained in the region of 505–565 nm with a maximum at 535 nm and an efficiency of 3%. However, the generated centers did not possess photostability [35]. The Al_2O_3:Cu may be tuned within the 300- to 400-nm range.

Sapphire color center (CC) lasers. Under neutron irradiation and annealing, sapphire exhibits a system of absorption bands with wavelength maxima at 460, 570, 680, 840, and 875 nm [44–46]. The type and nature of the intrinsic CCs responsible for these bands has not yet been reliably determined. However, their spectral parameters – high oscillator strength of the luminescence transition, high quantum efficiency, and possibility of frequency tuning in an important spectral region – have stimulated interest in them as laser-active centers [47]. Efficient, room-temperature-stimulated emission was realized in the optical spectrum regions of 540–620, 770–930, and 950–1,150 nm under pulsed laser excitation of the absorption bands at 460, 680, and 840 nm, respectively. The conversion efficiency of the nanosecond pumping energy amounted to several tens of percent. The maximum output energy under ruby laser excitation was 0.17 J. The active elements exhibited high thermal stability and could withstand heating up to 570 K for 15 min while preserving their characteristics [48]. To achieve a wide practical application of a CC laser, it is necessary to solve a number of problems associated with insufficient photostability of the active centers.

Active laser medium has been reported based on additively colored Al_2O_3:Mg crystals [49]. Under pulsed laser pumping, tunable laser action was obtained in the spectral range from 500 to 590 nm on the CC with an absorption band maximum at 440 nm (due to a perturbation by Mg ions). But a laser-pump-induced degradation of the active CC also was observed in this case.

Passive sapphire gates were proposed by specialists from the Irkutsk State University (Russia) for modulation of the Q factor of laser resonators working in the 0.82- to 0.91-μm wavelength range. Such locks are distinguished by the presence of dichroism of the active centers, which allows change in the initial transmission of linearly polarized light within wide limits (15–50%) by rotating the lock about one of its axes. High uniformity of the optical density over the cross section of the lock remains unchanged. The efficiency of $10 \times 10 \times 15$ mm^3 locks is 25%.

Ruby-based masers (quantum paramagnetic amplifiers) used in radioastronomy are meant for amplification of weak radio signals in the millimeter to decimeter wavelength range. To reduce the level of noise, ruby elements are made of homogeneous crystal and cooled down to the temperature of liquid helium.

Ruby-based phasers have been proposed for acoustic ranges. Acoustic waves were found to circulate in ruby elements at low temperatures for rather long periods of time without dampening. Phasers can be used as frequency standards.

1.4 Use in Medicine

The ancient Indians believed ruby cured bleeding and inflammation, while large ruby stones were believed to make those who carried them invulnerable. The Burmese developed this idea further. In their opinion, it was not sufficient to carry ruby to become invulnerable; the stone had to be inserted into human tissue to become a part of the body. Paracels, a famous doctor of the Middle Ages, used ruby to cure cancer ulcers.

Indeed, sapphire and ruby can be inserted (implanted) in human tissue, as these stones do not react with organic acids and tissues; they are superior to all known constructional materials in inertness. Unlike gold and platinum, corundum crystals are dielectrics. Therefore, in the humid medium of humans no electrochemical potential arises between the implanted crystals; galvanic pairs with other materials (such as metallic crowns) are not formed. Unique inertness, including electrolytic passiveness, biocompatibility, corrosion resistance, and hardness characteristic of sapphire, define its main fields of application in medicine: implants, surgery, and medical instrument making.

Medical-biological investigations carried out during 1977–1983 have shown that sapphire is not toxic for humans and does not cause changes in the functions of the central nervous system, liver, kidneys, protein and fat metabolism, and general reactivity. It does not possess carcinogenic, mutagenic, embryotrophic, or other types of remote effects. Collagen fibrillar capsule growing on the implant passes into bone and muscular tissue, which preserves the normal structure. In contrast to metals, electrically neutral sapphire is not carried by electrochemical reactions into lymph nodes and other parts of the body, does not cause immunodepressions and other changes in the immune system, and does not lead to demineralization of adjacent bone tissue.

Sapphire implants (SIs). The chemical composition of the bones has been intensively studied. However, implants interact with bone tissue not just chemically. Analysis of the first series of implantations testified that, despite the above-mentioned unique inertness of SIs, they showed osteogenic activity in vivo. Investigation of this phenomenon led the authors to conclude the necessity of a crystallographic approach to the problem of the boundary formation between sapphire and bone tissue [50].

Crystallographic aspects. Beyond other conditions, the rate of implant adaptation in the patient depends on the crystallographic conformity between the structures of the implant and the mineral component of the bone tissue (crystalline fibers contained in microfibrils). The mineral component is essential (for example, the content of the mineral component of lower jaw bone tissue is 0.91–1.46 g/cm^3 at a bone density of 1.72–1.77 g/cm^3), and so it is necessary to take into account the crystallographic interaction between the crystals of the bone tissue and the lattice of the implant. Such considerations are similar to heteroepitaxy and intergrowth of crystals.

Biochemical and biomechanical testing of implants made from crystals with different syngony and lattice parameters showed the advantages of sapphire. The values of linear and angular discrepancies between the crystal lattices of sapphire and hydroxylapatite were estimated. It was assumed that the atoms belonged to the same chain if the distances between their centers and the straight line did not exceed the ionic radius (1 Å for Ca ions and 0.6 Å for Al ions). The periodicity and the angles between the chain pairs were calculated, and the crystallographic indexes of the mutually conjugate chains and planes were determined (Table 1.6). The total value of the linear and angular discrepancy between the chain pairs in $Ca_5(PO_4)_3OH$ (the substance A) and sapphire (the substance B) was determined from the formula:

$$\Delta = 2(a_{A1} - a_{B1}):(a_{A2} + a_{B2}) + 2(a_{A2} - a_{B2}):(a_{A2} + a_{B2}) + 2(I_A - I_B):(I_A - I_B)$$

In this expression, a_{A1} and a_{A2} denote the periodicity of the location of metal atoms along the first and the second chain; I_A is the angle these chains make in the substance A; a_{B1}, a_{B2}, I_B are the parameters of the corresponding chain pair in the substance B.

According to the criteria of mutual orientation used while estimating the oriented growth of one substance on another substance [51, 52], such a growth may take place only when the considered crystalline structures contain at least one pair of atomic chains with small linear (<15%) and angular (10–15°) discrepancies.

Table 1.6 Conformity between closely packed chains of Ca and Al atoms in $Ca_5(PO_4)_3OH$ (A) and Al_2O_3 (B) structures

Indexes of planes		Indexes of chain directions				Periodicity along chains (Å)				Angles between chains (deg)		Nonconformity
A	B	A		B		A		B		A	B	Δ
$(\bar{1}100)$	$(1\bar{2}10)$	[0001]	$[11\bar{2}3]$	[0001]	$[20\bar{2}1]$	6.88	3.89	6.50	3.50	53.9	51.8	0.202
$(\bar{2}111)$	$(\bar{1}012)$	$[11\bar{2}3]$	$[1\bar{2}13]$	$[2\bar{2}01]$	$[0\bar{2}21]$	3.89	3.89	3.50	3.50	88.8	85.7	0.246

As follows from Table 1.6, this condition is satisfied. Thus, the main distinctive feature of single crystal implants is their ability to form strong junctions with the mineral component of microfibrils at the crystallographic compliance with their structures.

The biocompatibility of sapphire does not guarantee functional adaptation of the SI in the patient. The service life of implants depends on the structure of the "implant–bone tissue" interface, specifically on the system – fibrobone or bone – in which the implants are integrated. Osseointegration of the SI is provided by an optimum tension value, which should be defined by taking into account the statistical average difference in strength and elastic moduli. For dense bone (e.g., lower jaw), the stretch is 0.15 mm. For the bone of upper jaw, which is less dense, the value of stretch increases up to 0.2 mm. Less close fit leads to fibro-osseous integration. To stimulate osteogenesis, it is expedient to not only provide a close fit of the implant in the bone bed and to create the retention points, but also to activate the surface. The presence of aggregates of point defects, the exit points of dislocations, and microcracks on the SI surface sharply diminishes the energy barrier of crystal nucleus formation and speeds up the process of joining. It is desirable that the implant surface adjacent to the bone tissue have pores measuring 100–300 μm across. This will promote the intergrowth of bone tissue and increase the implant's adaptive potential. A porous, surface-adjacent layer is formed during the growth of profiled single crystals. Capillary holes in the shapers are located in such a manner that the streams of gas-saturated melt flowing out of them collide with the wave reflected by the meniscus, thus shedding vortices only in the vicinity of the preset surface. The pores then are opened up by mechanical treatment.

Functional merits of the SI clearly are seen in the comparison of their basic, functional properties with those of the widely used titanium analogs (Table 1.7).

Mucous membrane tightly encloses the SI neck, thus forming a cuff and preventing the penetration of bacterial infection into the gingival pocket over the interface between implant and bone tissue. The frequency of the adhesion events of gingival epithelium cells to the sapphire surface is three times higher than with titanium, even in the case when the sapphire has a coating applied by the plasma method. The loss of cortical layer (the bone tissue component that is most important for implant fixation) at the neck of the SI is considerably lower. This fundamentally influences the service life of the implants.

It is known that immunologic disturbances are defined by the implant material. The resistance of sapphire to any acid and alkali is immensely higher than that of

Table 1.7 Functional properties of sapphire and titanium dental implants

Implant material	Transfer of implant material to other parts of body	Loss of bone tissue at implant neck in 2 years after implantation (mm)	Change of immunologic status of patients	Formation of tight cuff around implant neck
Titanium	+	0.5	+	–
Sapphire	–	<0.2	–	+

metals and even of polycrystalline aluminum oxide (see Chap. 2). This likely explains the fact that sapphire does not change the immunologic status of patients. At the same time, 73% of the cases involving metallic implant insertion are followed by immunodepression, changes of immunoreactivity, and other immunologic disturbances (especially for male patients) [53]. All these manifestations raise the probability of postoperative complications.

As a rule, implants create a suitable medium for the growth of bacteria. In vivo comparative investigation of the adhesion of *S. mutants* to the surface of sapphire, titanium, Co–Cr–Mo alloy, hydroxylapatite, and polymethyl-methacrylate showed that the adhesion to sapphire was minimal [54]. Histological investigations carried out at different terms of SI adaptation in patients lead to the conclusion that the service life of a correctly inserted implant is unlimited.

It is well-known that the nickel and chromium contained in some implants are carcinogenic, possess cytotoxic effects, and may affect allergies. Approximately 15% of people are especially sensitive to nickel and 8% to chromium. Survival tests for L-132 cells [relative plating efficiency (RPE)] placed in media of different implant materials for a long period of time show that aluminum oxide, platinum, and $TiAl_6V_4$ alloy possess a level of survival close to 100%. For Ni–Cr–Co alloy this characteristic is on the order of 23%, while for Ni–Cr–Mo alloy it is even lower. The SIs are especially suitable for those patients who suffer from intolerance to metals or already have metallic implants inserted.

A number of implants and SI sets have been developed for orthopedics, traumatology, rhinoseptoplasty, spinoplasty, and so forth (Fig. 1.21). The first sapphire vertebra implants of original design were successfully inserted in clinical trials by academician A. A. Korzh [55] in 1980.

Shown in Fig. 1.22 is one example of recently developed implants [56, 57] consisting of two sapphire elements: an implant shaped as a hexahedron and a bar with a groove. The faces of the hexahedron are sharpened to take the shape of milling cutters. The hexahedron is introduced into the gap between vertebrae, moved apart by means of a lever, and then turned around using a wrench. Therein, the sharp edges of the milling cutters impact themselves into the contacting vertebrae, thus fixing position.

Thoracic and lumbar spinoplasty operations, which have no analogs in the rest of the world, are performed at the Kharkov Regional Hospital (Ukraine). Specially developed for such operations are punched sapphire shells, which allow control of the treatment process by means of nuclear magnetic resonance.

A set of implants has been created for dentistry compossed of one-stage and two-stage implants; those with straight and inclined heads; and flat, cylindrical, and thread-type implants (Fig. 1.23). The use of dental SI allows stimulative regeneration of the tissues adjacent to the SI. This can be achieved by exploiting the high optical transparency of sapphire, which enables activation of the immune system and biological processes through transmission of He–Ne laser radiation to the local, cellular level [58]. The radiation finds its way into the polished head and endosal part of the implant via a flexible light guide and connector, then scatters at the boundary between the implant and the bone tissue. In this case, the implant functions as the

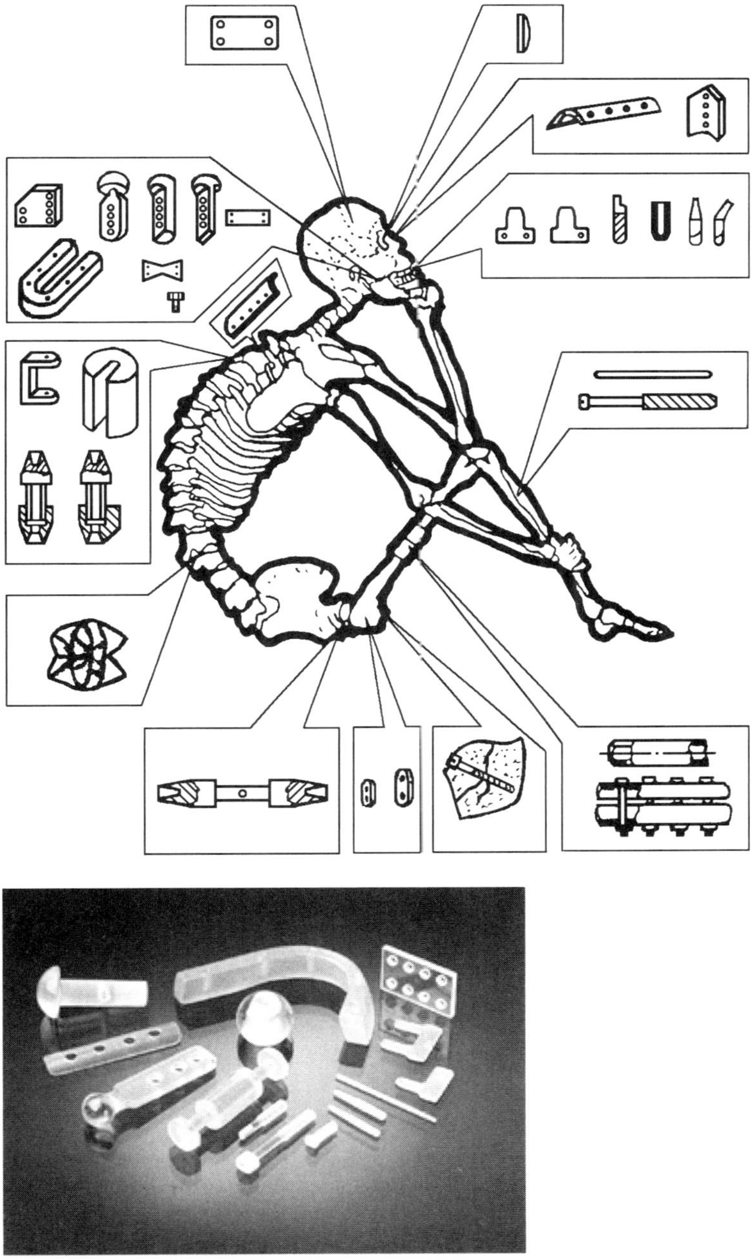

Fig. 1.21 Sapphire implants

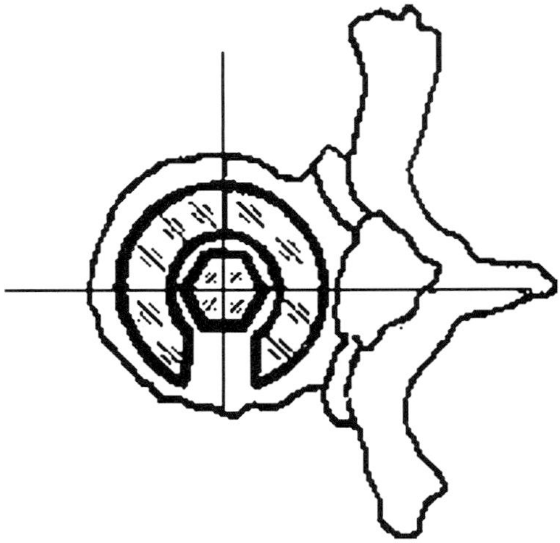

Fig. 1.22 Vertebra endoprosthesis

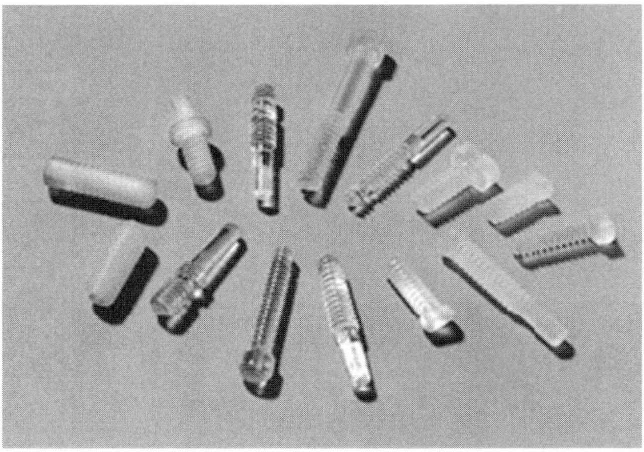

Fig. 1.23 Thread-type implants

final light guide. Several sessions of laser therapy arrest inflammation and reduce hyperemia, while subsequent sessions stimulate osteogenesis. This results in the formation of dense connective tissue around the endosal part of the implant with a tissue structure similar to that of normal, sound tissue. Only sapphire provides the possibility tof utilizing laser therapy for speeding up the adaptation of the implant.

At least 15 standard types of SI have been developed for maxillofacial surgery, such as lower jaw joints for children and adults, fixatives, jaw bone elongators, screws, and so forth (Fig. 1.35). The implants are used at ankylosis, secondary deforming osteoarthritis, distal defects, and fractures of the condyle appendix of the lower jaw [59, 60].

The use of sapphire in friction pairs is of particular interest. As a rule, orthopedic implants are of large size, and therefore, an essential amount of metal is introduced into the patient during traditional implantation. For instance, a hip joint endoprosthesis contains 300–350 g of titanium and alloyed steel (the latter being even less physiological). The metal corrosion products penetrate into the bone and may lead to traumatic osteomyelitis. The service life of the widely used metal-plastic hinges on hip joint prostheses is insufficient (5–7 years) due to the high friction coefficient of the metal-polyethylene pair, which increases over time with use. The service life of joint prostheses depends on the quality of spherical surface polishing. Metallic and ceramic surfaces cannot maintain a high finish class because of the presence of disoriented grains and intergranular boundaries with different physical and mechanical characteristics than those of the grains. Different rates of wear of this microstructure raise the friction coefficient of the pair and lead to elevated wear of the mating component. The rough surface, which is good for conventional intraosteal implants, is undesirable for friction pairs. The probability of adherence of organic molecules to such a surface increases, thus deteriorating the performance of the friction pairs. Sapphire not containing block (grain) boundaries allows surfaces with a high polish quality to be obtained (possessing a slight roughness and containing practically no scratches and pits). Moreover, sapphire is one of the most wear-resistant materials (this fact led to its wide employment in the friction pairs of watches and other devices). The friction coefficient of polished sapphire pairs actually decreases over the life of their operation; the wear index approaches the corresponding value of natural joints (Fig. 1.24).

The wear of the polyethylene component of polyethylene – alumina ceramic pairs (even if the latter is of low quality) is less than with polyethylene – zirconium dioxide pairs [61]. It is natural that for polyethylene coupled with sapphire this characteristic is much better compared to the two ceramic materials. Joint endoprostheses either can be entirely sapphire (e.g., those for lower jaw and shoulder joints) or combined metal – sapphire structures (in particular, for hip joints) (Fig. 1.25).

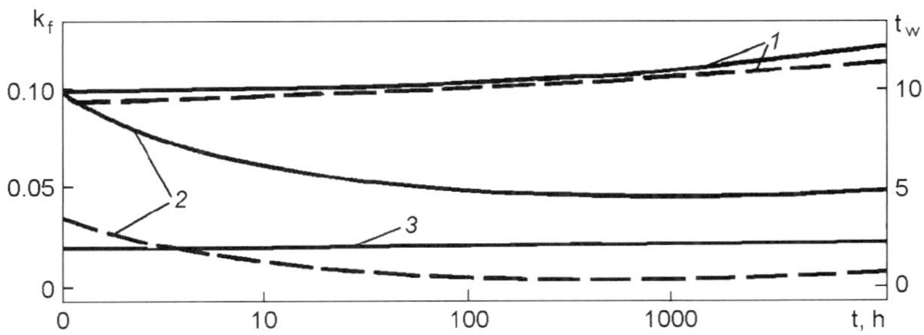

Fig. 1.24 Time dependence of friction coefficient (*solid curves*) and wear resistance (t_W)(*dashed curves*) for natural and artificial hip joints: (*1*) metal-polyethylene, (*2*) Al_2O_3–Al_2O_3, (*3*) natural joint

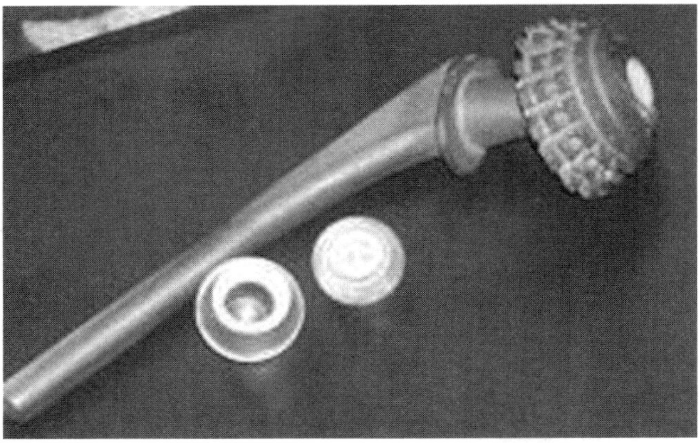

Fig. 1.25 Hip joint and sapphire heads

Sapphire surpasses metals in measures of hardness, wear resistance, and compression strength; however, its bending strength is lower. Therefore, for joint endoprostheses with sapphire heads the geometric parameters of the head and pedicle junction vary.

Nowadays more complicated SIs are being developed. In particular, there are structures that consist of several mobile and one-piece elements (e.g., cardiac valves), intraocular lenses made of special sapphire that absorbs UV light, and so forth.

Clinical results. A check of maxillofacial SI carried out 2 years after implantation (Dr. Ryabokon, Prof. Kutsevlyak) showed that positive results (i.e., complete rehabilitation of the kinematics and chewing function, the absence of complaints, asymmetry, and disturbance of occlusion) were observed in 94.7% of all cases, while in the other 5.3% the results were satisfactory. For metallo-osteosynthesis the share of positive results was equal to only 69.5%, the satisfactory and unsatisfactory results corresponded to 22.5 and 8% of the cases, respectively. Ultrasonic osteometry testified to a higher intensity of bone tissue growth on the implant for sapphire osteosynthesis [62].

Sapphire tools. To a considerable extent, achievements in microsurgery are defined by the sharpness of the cutting tools used. For most serious operations, ancient surgeons manufactured blades of obsidian chips, later substituted by the steel used for an extended period of time.

Maximum achievable blade sharpness depends on the hardness of the material. Diamond is very expensive, and the size of diamond articles is limited by nature itself. Sapphire, second after diamond in hardness, allows fabrication of tools of any size. The production of sapphire, including profiled growth, is well-developed, so sapphire cutting tools of practically any size and shape can be manufactured.

Due to their high hardness and chemical inertness in any media (including those of the patient), sapphire microscalpels are used in all the branches of microsurgery;

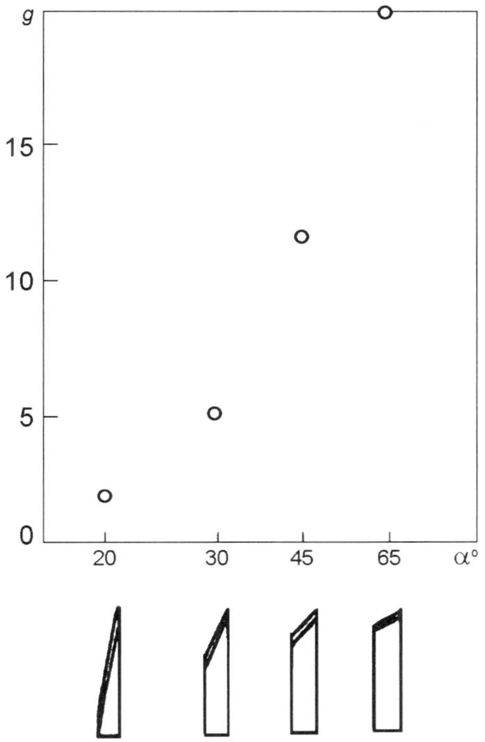

Fig. 1.26 Dependence of the effort (g) for puncturing the tissue on the angle sapphire microscalpel sharpening

large-size scalpels are employed in general surgery. Since sapphire scalpels possess a high sharpness, they are less traumatic than metallic scalpels, require a lower pressure for puncturing tissue (Fig. 1.26), and require less effort for cutting. Owing to these facts, the loss of endothelial cells is reduced. Moreover, the blade sharpness influences the probability of forming cheloid and hypertrophic postoperative scars, as well as the probability of remission at their excision. Sapphire tools can be sterilized by any method and the blades remain sharp for a longer period of time than with metallic analogs. The achieved sharpness depends on the crystallographic peculiarities of the sharpened planes and runs the range of 400 to 500 Å, which corresponds with the cross section of tissue fibers.

For some operations, especially for those conducted under microscope, the microscalpels made of blue sapphire are more efficient; such tools contrast better against the background of tissues. The Institute for Single Crystals has developed 65°, 45°, and 30° microscalpels made of colorless and blue sapphire (Fig. 1.27, Fig. 1.36).

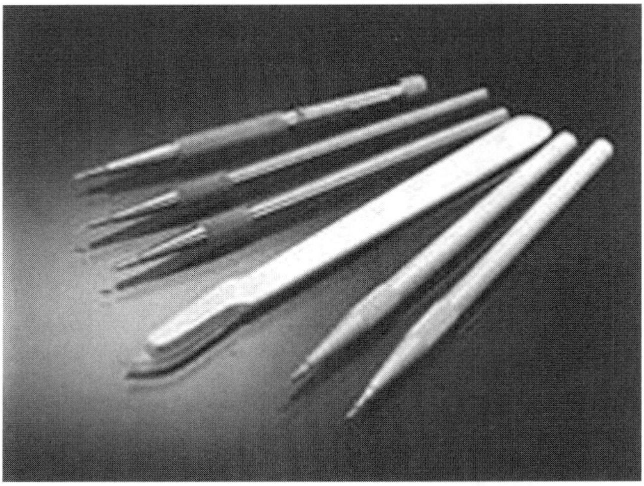

Fig. 1.27 Sapphire microscalpels

Also, 20-deg microscalpels that have no analog among nonmetallic scalpels are produced. The record blade thickness achieved for nonmetallic scalpels is 0.15 mm; superthin blades are made of strengthened sapphire. For abdominal operations knives are produced with a length of 10–35 mm. Such instruments cannot be manufactured of diamond for economic reasons. The cost of sapphire scalpels is an order lower compared to analogous diamond scalpels.

Nowadays sapphire microscalpels are produced in Switzerland, China, Ukraine, and the United States. General operating knives and blue sapphire microscalpels are manufactured in the Ukraine.

The transparency of sapphire scalpel blades allows realization of a unique procedure consisting of the introduction of a single-mode laser beam through a flexible light guide, through the blade, and directly into the cutting zone (Fig. 1.28).

Besides the therapeutic effects (bactericidal, analgesic, anti-inflammatory, antiedemic, immunocorrecting, and desensitizing actions) and better visualization of the blade edge, such scalpels have another significant advantage. They make it possible to simultaneously illuminate and determine the exact location of capillary vessels, nerves, and other anatomical formations both in the cutting zone and during the advance of the blade, as well as more effectively control the cutting depth. The effect of inner illumination can be created. If a microdefect is formed on the cutting edge, it is instantly revealed to the surgeon due to the specificity of light reflection at this point.

a

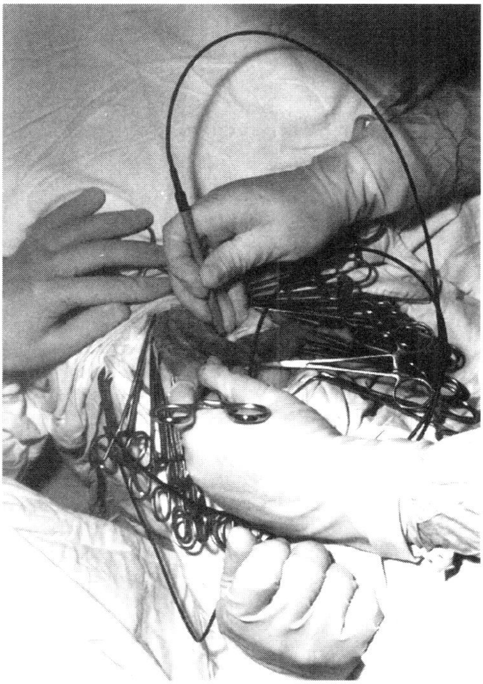

b

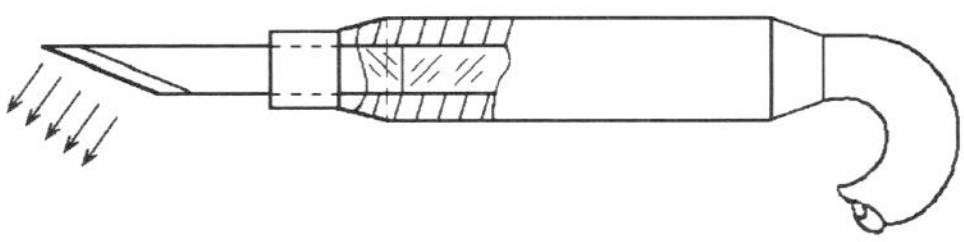

Fig. 1.28 (**a**) Neurosurgical operation which includes introduction of laser beam into the cutting through sapphire microscalpel; (**b**) scheme of microscalpel with light guide

Clinical results testify that, in comparison with traditional surgical instruments, sapphire tools shorten the duration of operations, speed up reparative processes, reduce the period of rehabilitation, and limit the probability of infection. This provides a better clinical outcome of recovery of lost functions of nerves, plexuses, and so forth. The sapphire tools are used for neurosurgical, ophthalmological, cosmetological operations, and surgical treatment of benign tumors (papillomas, cutaneous horn, epulises) of the maxillofacial region [63, 64].

Sapphire in medical equipment. Sapphire possesses higher chemical purity and inertness than most constructional materials. Blood components are not damaged while in contact with sapphire surfaces polished to a high finish class. Sapphire withstands considerable laser loads while maintaining good transmission in the IR-region of the spectrum. For instance, at the wavelength of Er:YAG-laser radiation (2.94 μm) the absorption of sapphire is far less than that of glass, and sapphire withstands loads higher than 1 kJ/cm^2 without damage [65]. These properties define the use of sapphire in medical equipment, for example:

- Pumps for the transfer of blood and other substances (tubes, plungers, valves, pump casings)
- Endoscopes (tips of different shapes)
- Laser surgical Nd:YAG- and Er:YAG-based apparatuses, laser IR-scalpels (conical sapphire contact tips for dissection, hemispherical tips for evaporation, and cylindrical tips for coagulation of tissues in contact with the light guide window)
- Equipment for obtaining high-purity medical preparations (crucibles, cuvettes, tubes, reactors).

Application fields for sapphire continue to become ever wider. As an example, the "sapphire room" concept should be mentioned. A medical project recently has been proposed (but not implemented at present) devoted to making a special room clad with artificial or natural sapphire. Medical chambers based on the generation of electromagnetic fields of different configurations inside closed space are well-known. In the opinion of the authors of Russian Federation Patent No. 2030909, the crystalline field of sapphire can interact with the human biological field, thus stimulating the reserves of the patient for overcoming certain diseases or favoring relaxation.

References

1. Dobrovinskaya E., Litvinov L., Pishchik V. Single Crystals of Corundum. Kiev: Naukova dumka, 1994, 256p [in Russian].
2. Dobrovinskaya E., Litvinov L., Pishchik V. Sapphire and Other Corundum Crystals. Kharkiv: Folio. Institute for Single Crystals, 2002, 349pp.
3. Aristoteles Stones, commented by Lucas ben Serapion. Codex Parisiensis 2772, Ch. III.
4. Konevsky V., Lytvynov L., Nguen Hoang Ngy. Funct. Mater. 7, 2000, 552.
5. Schalzried H. J. Phys. Chem. 28, 1961, 203.
6. Krylova G.I., Repina O.V., Munchaev A.I. Abstract of. Ninth National Conference on Crystal Growth, Moscow, 2000, 436 [in Russian].
7. Litvinov L.A. All About a Ruby. Kharkiv: Prapor, 1991, 155p [in Russian].
9. Bykova S.V., Gonnik M.A., Gonnik M.M. Proceedings of Sixth International Conference on Crystal Growth and Mass Transfer. Obninsk. 2005. 180–181 [in Russian].
10. Egorov L.P., Kotelyanskii I.M., Kravchenko V.B. et al. Proceedings of 14th International Conference on Crystal Growth. Moscow. 2004, 182 [in Russian].
11. Wallace J. Laser Focus World. January, 2003, 19–20.

12. Kurlov V.N., Kiiko V.M., Kolchin A.A., Mileikc S.T.J. Cryst. Growth. 204, 1999, 499.
13. Zhukov L.F., Litvinov L.A., Chugunnyi E.G. Patent USSR. 766237, 1980.
14. Zhukov L.F., Litvinov L.A., Shumikhin V.S. Patent USSR. 1256572, 1984.
15. Ivanina B.M., Litvinov L.A., Priimach B.S., Globus M.E. Patent USSR 1316467, 1985.
16. Ivanina V.M., Konevskii V.S., Litvinov L.A. Patent USSR. 1649859, 1991.
17. Gromova S.N., Dobrovinskaya E.R., Litvinov L.A., Pishchik V.V. Svetotekhnika. 4, 1979, C.8–9.
18. Brailovskii V.B., Gaidukov E.N., Kobzar' A.I et al. Proc. of All-Union Conference on Manufacturing of Profiled Crystals. Leningrad, 1989, 113 [in Russian].
19. Nechiporenko E.P., Litvinov L.A., Pavlenko Yu.B. Patent USSR. 1361197, 1987.
20. Shvid'ko Y. et al. Phys. Rev. Lett. 2003, 10.
21. Schewe P., Riordon J., Stein B. Phys. News Update. 1, 2003, 619.
22. Belt R. Synthetic Corundum as X-Ray Monochromator. X-Ray spectrum. 1(N2), 1972, 55–58.
23. Valbis Ya.A., Litvinov L.A., Ryzhikov V.D. Patent USSR. 1410678, 1988.
24. Ryzhikov V.D., Litvinov A.L., Danshin E.A. et al. Materials for Electron Engineering Moscow. 2001, 61–64 [in Russian].
25. Globus M.E., Grinyov B.V. Inorganic Scintillation Crystals. Kharkiv: Akta, 2000, 72 [in Russian].
26. Ryzhikov V.D. et al. Voprosy atomnoi nauki i tekhniki. Ser. fizika rad. povrezhd. i rad. materialovedenie. [Problems of Nuclear Science and Engineering: Physics of Radiation Damage and Radiation Materials Science.] 3 (81), 2002, 130–132 [in Russian].
27. Litvinov L.A., Krivonosov E.V., Ryzhikov V.D. et al. Voprosy atomnoi nauki i tekhniki. Ser. fizika rad. povrezhd. i rad. materialovedenie. [Problems of Nuclear Science and Engineering: Physics of Radiation Damage and Radiation Materials Science.] 6 (82), 2002, 154–155 [in Russian].
28. Borovkov S.D., Konevskii V.S., Litvinov L.A. et al. Screens for Beam Diagnostics. Preprint IFVE 90–34, 1990 [in Russian].
29. Avakumova L.A. et al. Atomnaya Energiya. 69(5), 1990, 309–313 [in Russian].
30. Kortov V.S., Mil'man I.I., Nikiforov S.V., Moiseikin E.V. Fizika tverdogo tela. 48(3), 2006, 421–426 [in Russian].
31. Bessonova T.S., Gimadova T.I., Litvinov L.A. Zh prikladn. spektrosk. 54, 1991, 433–437 [in Russian].
31. Kortov V.S. et al. Problemy spektroskopii i spektrometrii [Problems of Spectroscopy and Spectrometry] Ekaterinburg, UGTU, 1999, 3–6 [in Russian].
33. Galanciak D., Legowski S., Meczynska., Grinberg M. Funct. Mater. 10, 2003, 212–215.
34. Imbusch G.F. J. Lumin. 53, 1992, 465.
35. Astrom P.C.F., Di Stefano F., Probst et al. Phys. Lett. A356, 2006, P.262–266.
36. Booth N., Cabrera B., Fiorini E. Annu. Rev. Nucl. Part. Sci. 46, 1996, 471.
37. Dashkevich V.I., Tyushkevich B.N. Prib. i tekhn. eksp. 5, 2001, 117–120 [in Russian].
38. Moulton P.F. Optic News. 6, 1982, 9.
39. Sutter D.H., Steinmeyer G., Gallmann L. et al. Optics Lett. 24, 1999, 631.
40. Pashinin V.P. et al. Kvantovaya Elektronika. 32(2), 2002, 121–123 [in Russian].
41. Koishi Yamakawa. Workshop on Technological Bottlenecks in Compact High-Intensity Short Puls Lasers. Paris, April 2003.
42. Morou G. Workshop on Technological Bottlenecks in Compact High-Intensity Short Puls Lasers. Paris. April 2003, 7.
43. Boiko B.B. et al. Dokl. AN BSSR. 31, 1987, 982–984 [in Russian].
44. Kaminskii A.A. et al. Fizika i spektroskopiya lazernykh kristallov [Physics and Spectroscopy of Laser Crystals]. Moscow: Nauka, 1986, 261 [in Russian].
45. Martynovich E.F., Tokarev A.G., Grigorov V.A. Optics Commun. 53, 1989, 254–256.
46. Martynovich E.F., Baryshnikov V.I., Grigorov V.A. Optics Commun.. 53, 1989, 257–258.
47. Voytovich A.P. et al. Sov. J. Quantum Electron. 17, 1987, 561–562.

48. Osiko V.V. Laser Materials. Moscow: Nauka, 2002 [in Russian].
49. Martynovich E.F., Baryshnikov V.I., Grigorov V.A. Pis'ma v JTF. 11, 1985, P.200–202 [in Russian].
50. Litvinov L.A. Izv. AN SSSR. Ser. Fiz. 31, 1988, 1911–1913 [in Russian].
51. Dickens B., Schroeder L.W. J. Res. Nat. Bur.Stand. 85, 1980, 347.
52. Chaplygin G.V. Comprehensive Abstracts of International Conference on Crystal Growth. 1980. Moscow: AN SSSR, 135 [in Russian].
53. Matveeva A.I., Vigdorovich V.A. Stomatologiya. 1, 1992, C.38–40 [in Russian].
54. Streicher R.M. et al. Proc. Int. Congr Fenza. 1991, C.2–5.
55. Korzh A.A., Litvinov L.A., Timchenko I.B. et al. A.s. 1114412, 1983 (SSSR);Patent 16604, 1997 (Ukraine).
56. Korzh A.A., Litvinov L.A., Timchenko I.B. et al. Patent 16596, 1997 (Ukraine).
57. Korzh A.A., Litvinov L.A., Timchenko I.B. et al. Patent 16717, 1997 (Ukraine).
58. Grechko N.B., Litvinov L. A., Kutsevlyak V.I. Application 97126214, favorable action 24.12.97.
59. Kutsevlyak V.I., Ryabokon' E.N., Litvinov L.A. Bulletin on Problems of Modern Medicine. 9, 1994, C.28–31 [in Russian].
60. Kutsevlyak V.I., Ryabokon' E.N., Litvinov L.A. Bull. Stomat. 1, 1994, 71–73 [in Russian].
61. Fudjioka-Hirai J. et al. Biomed. Mater. Res. 21(7), 1987, 913–920.
62. Ryabokon' E.N. Abstract of Thesis. Ukrainian Med. Stomatol. Academia. Poltava, 1995 [in Ukraine].
63. Ryabokon' E.N., Bozhko K.V., Polyakova S.V., Kolpakov S.N., Litvinov L.A. Proceedings of 12th International Conference on Application of Lasers in Medicine and Biology, Kharkiv, 1998. 108–109 [in Ukraine].
64. Kolpakov S.N., Litvinov L.A., Sofienko G.G. Proceedings of 12th International Conference on Application of Lasers in Medicine and Biology, Kharkiv, 1998, C.110–111.

Appendix

Appendix 1.1 Corundum bearings and bushes of devices

The crystal color	Impurity component
Red, pink	Cr_2O_3
Blue	TiO_2+Fe_2O_3
Orange	NiO+Fe_2O_3
Orange (padparajah type)	NiO+Cr_2O_3
Yellow	NiO
Green (tourmaline type)	Co_2O_3+V_2O_3
Dark-red (garnet type)	Cr_2O_3+TiO_2+Fe_2O_3
Violet	TiO_2+Fe_2O_3+Cr_2O_3

Appendix 1.2 (**a**) Sapphire cylindrical bush (**b**) Sapphire noncylindrical bush with two lubricators (**c**) Sapphire bush with noncylindrical hole (**d**) Sapphire bush with one lubricator and cylindrical hole (**e**) Sapphire bush with one lubricator and noncylindrical hole

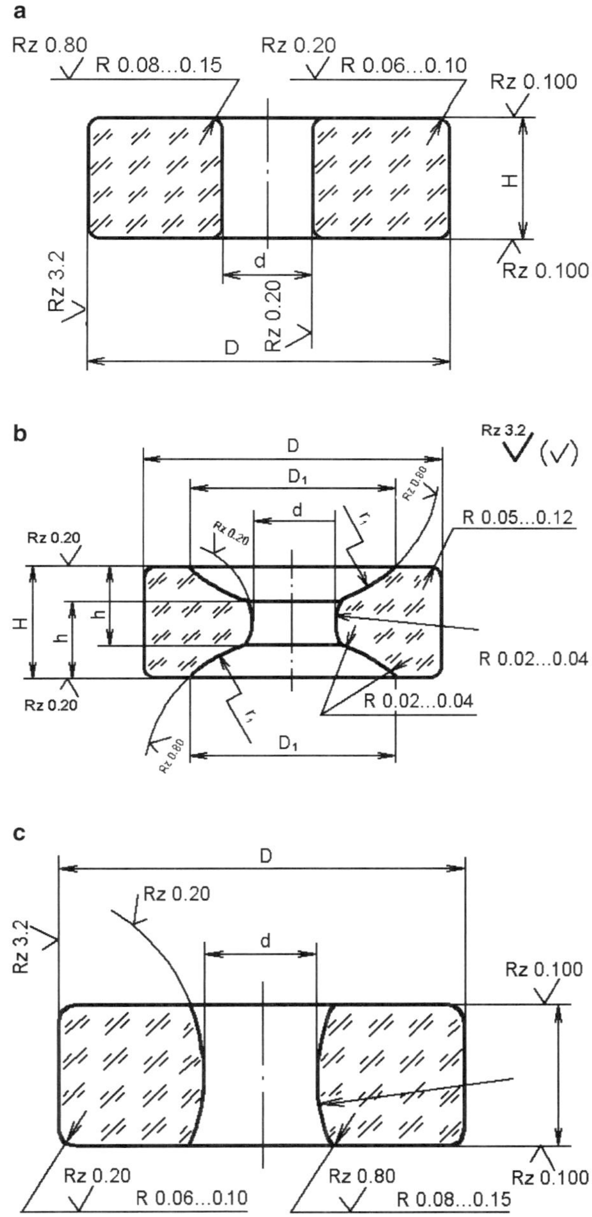

(continued)

Appendix 1.2 (continued)

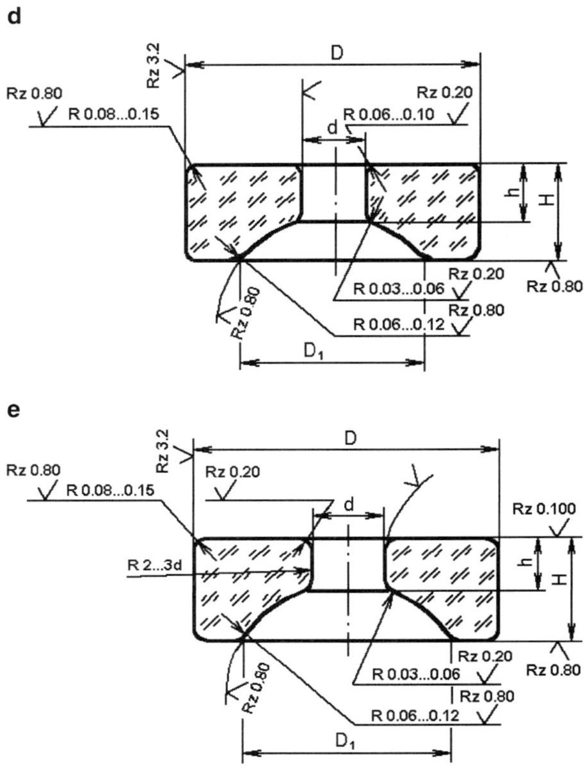

Appendix 1.3 Sharpening angles of sapphire cutter

re-sharpening angles

γ	λ	φ	$φ_1$	α	$α_1$	r mm
+5°	+10°	60°	8°	8°	8°	2,4

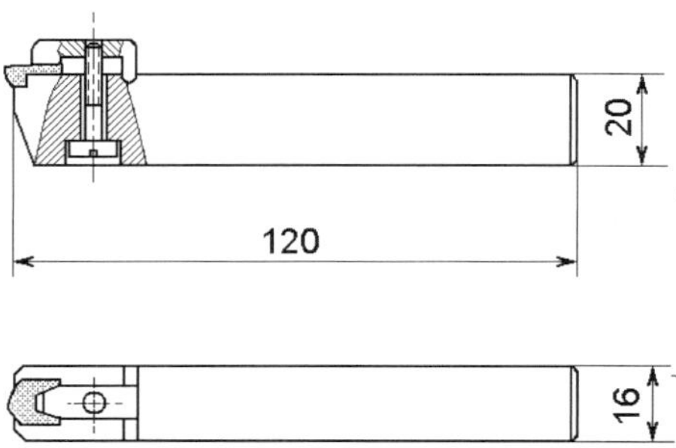

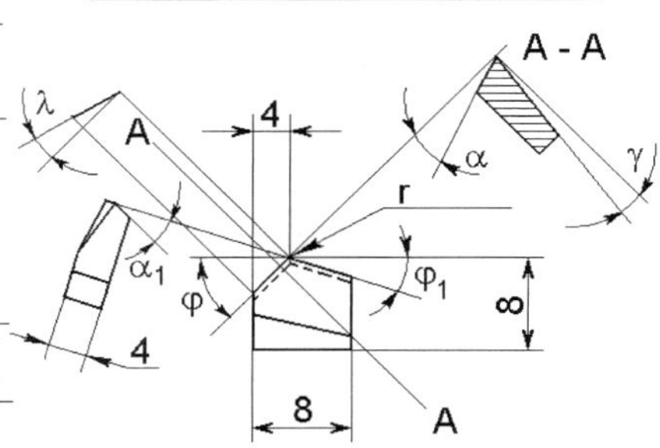

Appendix 1.4

Optical Grade

1. Insertions, lineage, twins, microbubbles, scattering centers are nonexistent.
2. Insertions, lineage, twins are nonexistent. Some scattering centers (microbubbles up to 10 μm) are permitted.
3. Insertions, lineage, twins are nonexistent. Uniformly distributed bubbles not larger then 50 μm and not closer in distance then 500 μm are permitted.
4. Insertions, lineage, twins are nonexistent. Uniformly distributed bubbles not larger then 50 μm and not closer in distance then 500 μm and bubble concentrations not larger 200 μm in diameter and not closer 20 mm are permitted.

Technical Grade

5. Insertions, lineage, twins are nonexistent. Uniformly distributed bubbles not larger then 50 μm and not closer in distance then 500 μm and bubble concentrations not larger 500 μm in diameter and not closer 20 mm are permitted.
6. Insertions, twins, and lineage with angle of disorientation larger then 15 min are nonexistent. Bubbles and bubble concentrations not larger 500 μm in diameter and not closer 15 mm are permitted.

Mechanical Grade

7. Insertions, twins, and lineage and bubble concentrations not larger 200 μm in diameter are permitted.
8. All types of defects, other then cracks, are permitted.

Appendix 1.5 Flange with sapphire window. Dimensions of flange with sapphire window (inch)

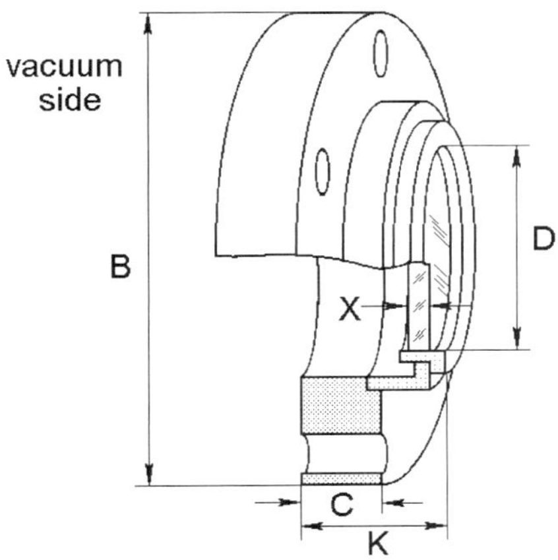

Diameter (D)	B	C	K	X	Pressure (psi) at 293 K
0.46	2.75	0.50	0.52	0.06	250
0.69	2.75	0.50	0.65	0.06	200
0.94	2.75	0.50	0.65	0.06	100
1.44	4.50	0.69	0.86	0.08	75
1.94	4.50	0.69	0.91	0.13	50

Appendix 1.6 The place occupied by sapphire windows

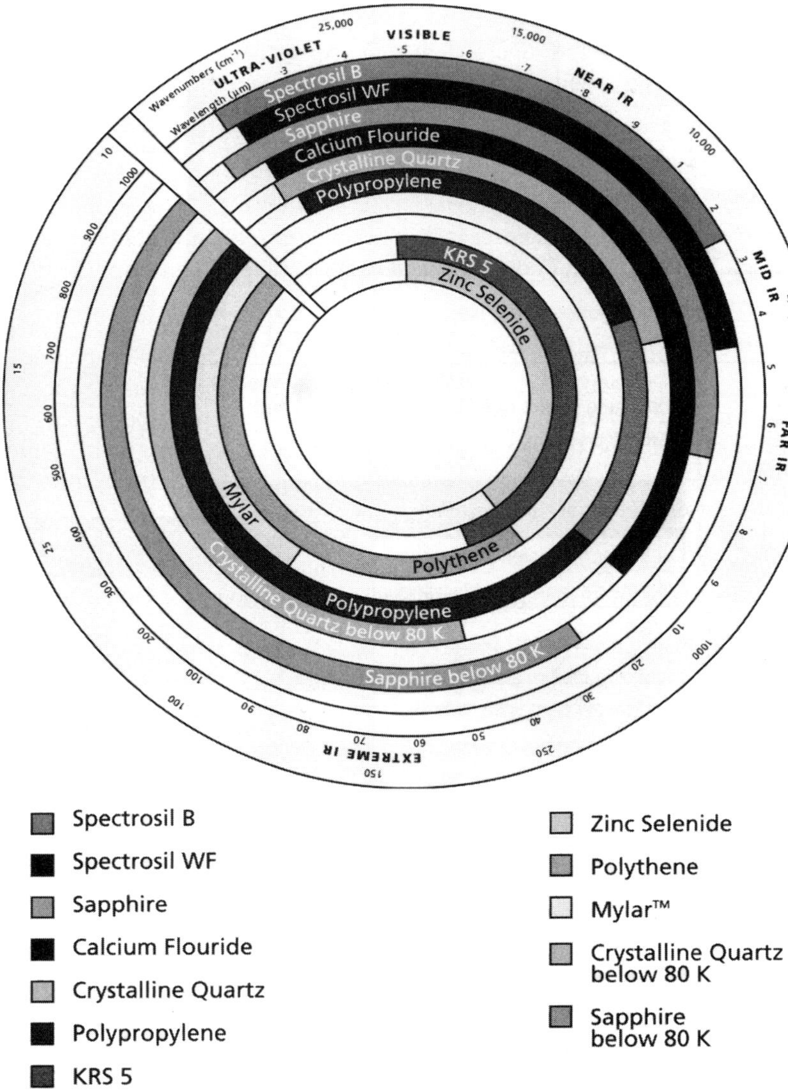

Appendix 1.7 Window with wedge

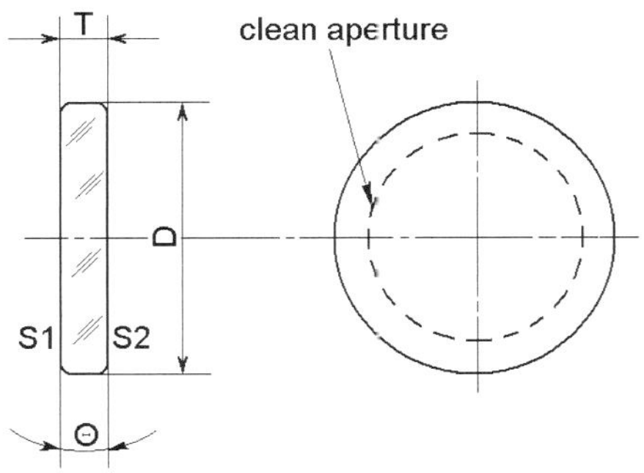

Surface quality	20–10 (both sides)
Flatness	$\lambda/10$ (both sides)
Wedge	1, 3 ± 5 arcmin
Diameter	25, 30, 60 mm, −0.15 mm
Thickness	6, 8 mm, ±0.2 mm
Clean aperture	90% of diameter

Appendix 1.8 Characteristics of TDL-500K sapphire detector

Sensitivity to γ-irradiation (quantum/Gy)	$(1–2) \cdot 10^{11}$
Spread of sensitivity of a series of detectors (%)	5
Fedding during a year (%)	5
Range of constant sensitivity doses (Gy)	10^{-6}–1
Range of registered doses (Gy)	10^{-6}–10
Dose equivalent of own background (Gy)	$<5 \cdot 10^{-7}$
Repeated use (times)	500
De-excitating action of light at an illuminance of 300 lx during 1 min (%)	Not exceeding 3
Energy dependence of sensitivity at 37 keV with respect to 1,250 keV energy (%)	Not exceeding 3

Appendix 1.9 Line peak positions, concentration shifts, and pressure

Line peak position (nm)	Line peak position (cm^{-1})	Concentration shifts (cm^{-1}/mass%)	Concentration shifts (cm^{-1}/mass%)[a]	Pressure shifts (cm^{-1}/kbar)
692.8	14,432.1	1.14	1.01	–
694.2	14,402.8	1.10	0.96	–
697.7	14,335.6	3.63	–	−0.8
698.2	14,330.4	−4.03	–	1.3
698.4	14,312.8	1.61	–	−0.7
701.6	14,256.3	1.34	1.26	−0.8
704.8	14,185.1	1.66	1.76	−1.0
705.5	14,171.9	0.81	–	−1.0
706.9	14,158.6	−4.89	–	1.4
714.5	14,007.4	−4.43	–	2.0
730.1	13,742.2	−29.34	–	4.2
754.5	13,274.9	−11.6	–	1.1
759.8	13,147.5	7.47	–	−2.7
767.2	13,043.4	−3.76	–	1.5
771.3	12,968.3	−3.70	–	–

[a] A. A. Kaplianski, R. B. Rozenblaum. Fiz. Tv. Tala. (1971) **13** p. 2623 [In Russian]

Appendix 1.10 Characteristics of ruby as the active medium

Parameter	Value
Cr^{3+} concentration (cm^{-3})	$(0.7–1.5) \cdot 10^{19}$
Number of radiation-absorbing electrons (cm^{-3})	$1.2 \cdot 10^{22}$
Energy of formation (eV):	
Interstitial oxygen	100
Interstitial aluminum	50
Al_i–V_0 pair	About 3
Light transmission region (μm)	0.2–4
Beam refraction index (n = 6,943 Å)	
Ordinary	1.7634
Extraordinary	1.7556
Temperature coefficient of the refraction index variation (K^{-1}), Angle (deg):	
Brewster	60.4
Full internal reflection	34.5
Cr^{3+} pumping bands (cm^{-1})	25,000; 18,000
Velocity of upper level population at pumping (mol/(cm^2 s))	$3.5 \cdot 10^{21}$
Lifetime of the upper level$\bar{E}$ (s):	
$T = 77$ K	$3 \cdot 10^{-3}$
$T = 77$ K	$4 \cdot 10^{-3}$
Probability of spontaneous transition (s^{-1})	$3 \cdot 10^5$, $3 \cdot 10^7$
Velocity of radiative transition (s^{-1}):	
$^2E \rightarrow A_2$	$1.4 \cdot 10^2$
$^2E \rightarrow {}^4A_2$	$2.3 \cdot 10^4$
Tame of relaxation between the levels $\bar{E}$ and $\overline{2A}$ (s)	up to 10^{-7}
Quantum energy (eV)	1.6
Quantum yield of luminescence in R lines, $\lambda_{ex} = 0.55$ μm, $T = 300$ K	0.54
Energy yield of luminescence in R lines, $T = 300$ K	0.41
Quantum energy yield of luminescence at resonance excitation in R lines, 0.05% Cr_2O_3:	
$T = 300$ K	0.98
$T = 140–200$ K	1
Width of luminescence line (cm^{-1}):	
$T = 77$ K	0.20
$T = 300$ K	20
Character of broadening:	
$T = 77$ K	Nonuniform
$T = 300$ K	Uniform
Temperature shift of the luminescence line maximum (cm^{-1} K^{-1})	$-\Delta T \cdot 0.13$
Polarization of radiation at ruby rod orientation:	
60–90°	$E \perp c$
0°	No
Temperature shift of generation spectrum, $T = 30–300$ K	0.06 Å/K; 12–15 Å
Variation of the intrinsic frequency of the resonator cavity from peak to peak in the free generation mode (cm^{-1}):	
$D \times L = 15 \times 120$ mm	0.0017
$D \times L = 16 \times 240$ mm	0.0031
Time of free oscillation damping in the resonator cavity (s)	$10^{-7}–10^{-8}$
Maximal working temperature of ruby rod (K)	520
Theoretical divergence of radiation:	
Crystal diameter 1 cm, $\lambda = 6{,}943$ Å, angular s	14
Standard of ruby rod, angular min	10–20
Based amplifiers of high uniformity ruby rod	Increase by 20%

(continued)

Appendix 1.10 (continued)

Parameter	Value	
Velocity of ruby rod heating at pulse pumping (K/s)	10^5–10^4	
Coefficient of conversion of the second harmonics of emissions	0.2–0.3	
Amplification per unit of length of ruby rod, 0.05% Cr_2O_3 at a share of the excited ions Cr^{3+}:		
0.1	−1.87	
0.2	−1.38	
0.4	−0.40	
0.5	+0.082	
0.6	+0.57	
0.7	+1.06	
0.8	+1.55	
Generation losses (cm^{-1})	0.025–0.05	
Photoelastic constants, 0.05% Cr_2O_3:		
P_{11}	−0.23	
P_{13}	−0.02	
P_{31}	−0.04	
P_{41}	0.01	
P_{12}	−0.03	
P_{14}	0.00	
P_{33}	−0.20	
P_{44}	−0.01	
Laser transitions	$^2E/\bar{E}\to{}^4A_2$ R_1-line	$^2E(2\bar{A})\to{}^4A_2$ R_2-line
Emission wavelength, 0.05% Cr_2O_3 (Å):	—	—
$T = 77$ K	6,934	6,920
$T = 300$ K	6,943	6,929
$T = 550$ K	6,952	—
Position of Wannier-Stark levels (cm^{-1}), $T = 4.2$ K	14,418	14,447
Beam absorption coefficient, (cm wt% Cr_2O_3)$^{-1}$:	—	
Ordinary	5.1	3.4
Extraordinary	0.35	0.76
γ-ray-induced absorption (10^4 rad), at 0.05% Cr_2O_3 (cm^{-1})	0.002	—
Dichroism of absorption	0.6–0.98	—
Cross sections of ordinary beam absorption (cm^2)	$(2.13 \pm 0.07) \cdot 10^{-20}$	—
R-line width (Å) of crystals:	—	—
Grown by the Czochralski method, $T = 4.2$ K	0.34	—
Verneuil-grown at *T* (K):		—
4.2	0.053–0.071	—
77	0.05	0.05
	0.10	—
300	5	4
Relation intensity of *R* lines, $T = 300$ K:	—	—
In the fluorescence spectrum	2	1
At generation	50	1
Interval of continuous tuning of generation frequency (cm^{-1})	—	16
Emission wavelength, Å (Cr^{3+}–Cr^{3+})	N_1 line	N_2 line
0.5–0.7% Cr_2O_3, 77 K	7,041	7,009
	Free generation	Modulation
Emission pulse energy (J)	1–100	0.1–10
Emission pulse power (W)	10^2–10^5	10^6–10^8
Emission pulse duration (s)	10^{-3}	10^8–10^{-9}
Frequency of pulse respectition (Hz)	Up to 5	Up to 10

Appendix 1.11 Types of ruby laser elements

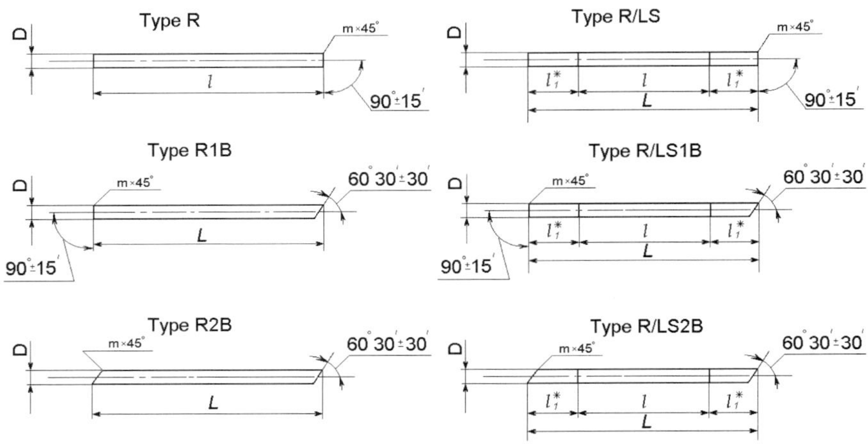

Appendix 1.12 Generation characteristics of Al_2O_3–Ti^{3+} lasers

Source of pumping	Tuning range (μm)	E_{ps} (J)	P_{ps} (W)	E_{th}	Efficiency (%) Full	Efficiency (%) Differential
Continuous coherent pumping – Ar-laser	0.715–1.14	—	1.6	2.6 W	13.3	19.1
Pulse-coherent pumping						
Quasicontinuous laser on YAG with frequency doubling	0.675–0.950		0.38	25 mJ	46–50	68–72
Laser on rhodomyn dye 6G with lamp pumping	0.660–0.986	0.0065	—	2 mJ	44	67
Monopulse laser on YAG with frequency doubling	0.65–1.20	0.03	1.2	0.5–2.5	42–44	70
Pulse nonselective pumping	0.72–0.92	0.3	0.15	11 J	0.2	0.5

Minimum width of generation line (pumping of the amplifier and generator by YAG lasers with frequency doubling) = 0.08 Å

Luminescence lifetime at temperature:

3.9 μs ($T < 150$ K)

2.9 μs ($T = 300$ K)

Saturation energy density in the region of pumping by pulse emission (532 nm, 10 ns, π-polarization) = 7.6 J/cm^2

Absorption cross section near 532 nm = $4.9 \cdot 10^{-20}$ cm^2

Appendix 1.13 Sapphire maxillofacial implants

a

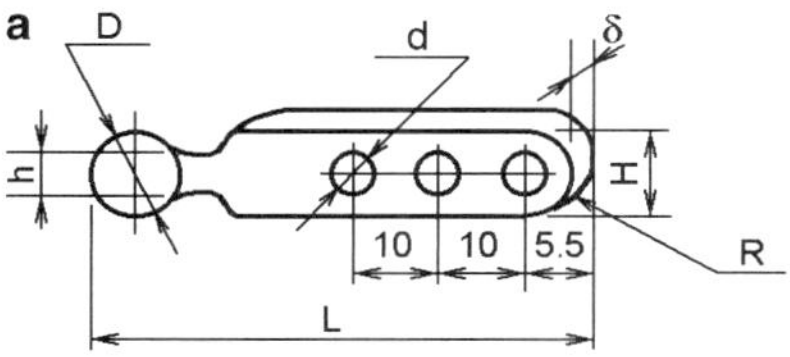

ISJ	11	12	13	14	15
L	55	60	55	50	45
D	12	12	10	10	9
d	4	4	3	3	3
R	6	6	5	4	4
H	12	12	10	8	8
H	8	8	5	5	5
δ	4	4	4	3	3

b

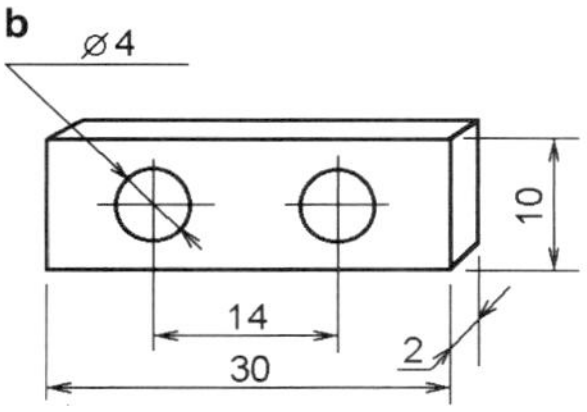

ISJ	41	42	43	44	45	46
L	9	11	13	9	11	9
d	M 4×1	M 4×1	M 4×1	M 3.5×1	M 3.5×1	M 3×1

c

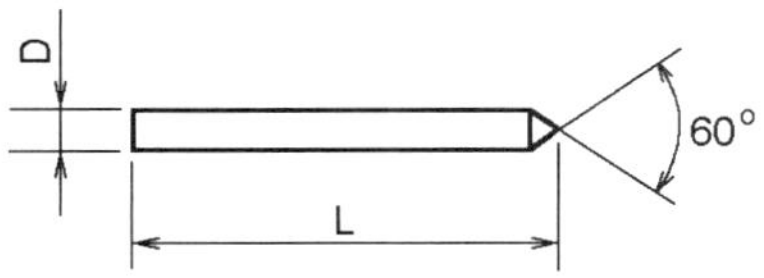

ISJ	51	52	53
L	30	30	35
D	2.3	3	3

d

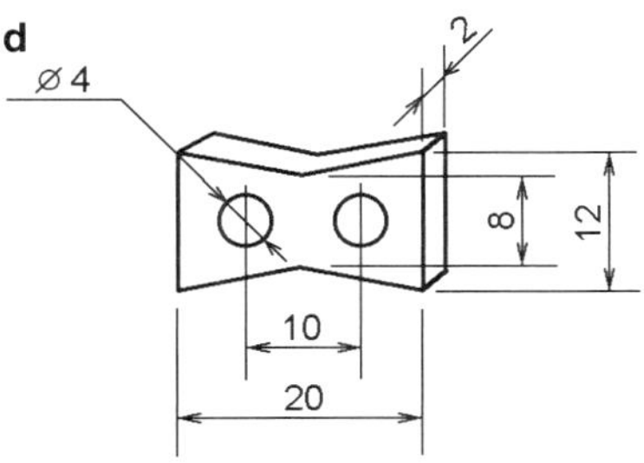

ISJ	61

Appendix 1.14 Sapphire scalpels.

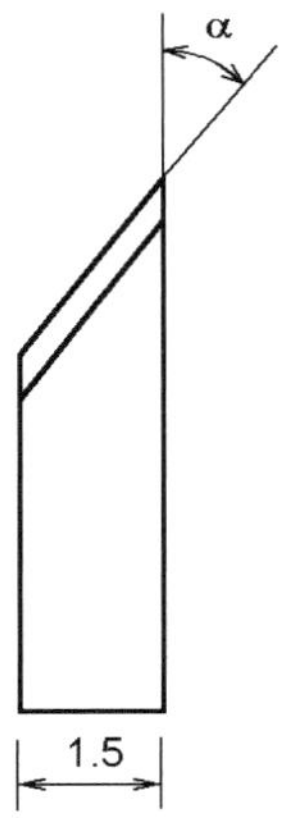

Denotations	α, deg
KLS.01.04	20
KLS.01.02	30
KLS.01.01	45
KLS.01.05	65

Denotations	α, deg
KLS.01.04 blue	20
KLS.01.02 blue	30
KLS.01.01 blue	45
KLS.01.05 blue	65

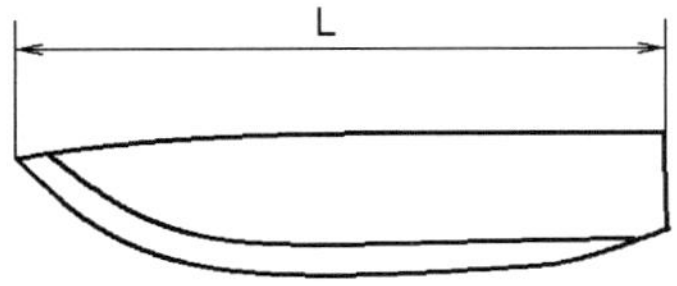

Blades for general surgery
$L_1 = 11, L_2 = 25, L_3 = 30$ mm

Chapter 2
Properties of Sapphire

2.1 Physical Properties

2.1.1 Crystal Structure and Morphology of Sapphire

Corundum crystals belong to the ditrigonal-scalenohedral class of the trigonal symmetry $D^{6}_{3d} - R\,\bar{3}C(L_3 3L_2 3PC)$with symmetry elements:

- Mirror-turn axis of the sixth order (ternary inversion axis)
- Three axes of the second order normal to it
- Three symmetry planes normal to the axes of the second order and intercrossing along the axis of the highest order
- Symmetry center

There are seven simple forms (facets) in this class of symmetry. According to Bravais classification they have the following symbols: pinacoid (0001), hexagonal prisms $\{10\bar{1}0\}$and $\{11\bar{2}0\}$, dihexagonal prism $\{hki0\}$, rhombohedron $\{h0\bar{h}l\}$, hexagonal bipyramid $\{hh2\bar{h}l\}$, and ditrigonal scalenohedron. Depending on its position with respect to the hexagonal crystallographic axes, the rhombohedron may be "positive" $\{10\bar{1}1\}$or "negative" $\{01\bar{1}1\}$[1].

The crystal lattice of α-Al_2O_3 is formed by Al^{3+} and O^{2-} ions. If the O^{2-} anions are depicted as balls, the crystal lattice takes the form of their closest hexagonal packing (Fig. 2.1). The Al^{3+} cations are located in a crystalline field that has no symmetry center (due to crystal lattice distortions). These cations lie in the octahedral hollows between the closely packed O^{2-} ions, filling two thirds of these hollows. The octahedron hollow is surrounded by six balls. If the radius of each is taken as a unit, then the hollow contains a ball with a relative radius of 0.41. Due to the ratio of the ionic radii of O^{2-} and Al^{3+} (equal to 1.40 and 0.57 Å, respectively), the cations are located within the hollows of the anion packing. They slightly distort the lattice, but do not fall outside the stability limits of the octahedron position.

The coordination numbers for Al^{3+} and O^{2-} are 6 and 4, respectively. The three upper O^{2-} ions in the octahedron are turned about 64.3° with respect to the three lower O^{2-} ions and lie in parallel planes (Fig. 2.2a) at a distance of 2.164 Å from each other. The closest packing distortion bound up with the discrepancy between

E.R. Dobrovinskaya et al., *Sapphire: Material, Manufacturing, Applications*,
DOI: 10.1007/978-0-387-85695-7_2, © Springer Science+Business Media, LLC 2009

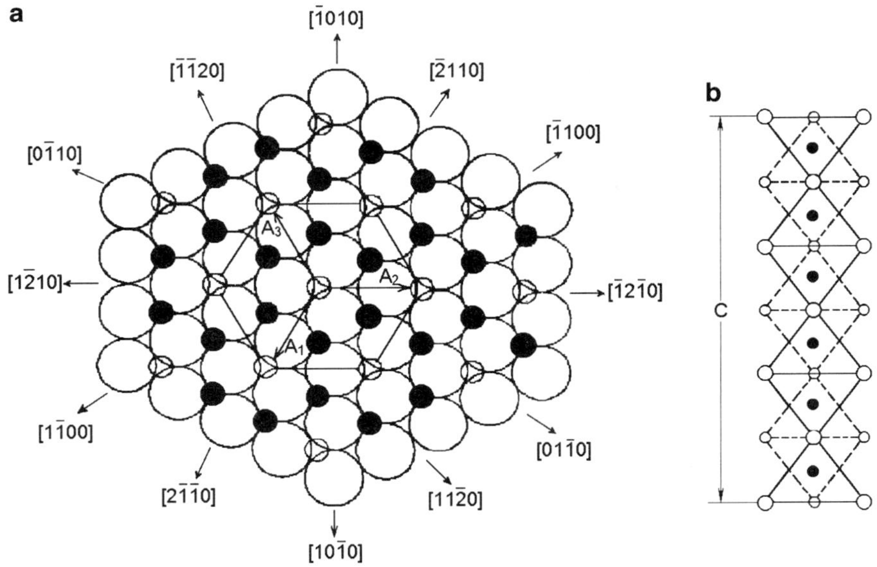

Fig. 2.1 (**a**) Schematic of the arrangement of Al^{3+} (*black circles*) and octahedral hollows (*small light circles*) between two layers of O^{2-} (*large light circles*) in the basal plane (the upper O^{2-} layer is not shown). A_1, A_2, and A_3 are $\langle 11\bar{2}0\rangle$ translational vectors of the hexagonal sapphire cell for the basal plane.(**b**) Schematic of the packing of O^{2-} ions (*light circles*) and Al^{3+} in the direction of the axis C_3 [1]

the size of Al^{3+} and the octahedron position is manifested by the fact that the octahedron is formed by oxygen triangles of different size (Fig. 2.2c), and their rotation angle of 64.3° exceeds the value characteristic of ideal packing (60°).

The spatial arrangement of O^{2-} ions forms the so-called corundum motive. The interchanging octahedron positions of Al^{3+} are replicated in the structure in every three layers. The arrangement of the structural units along the third-order axis are replicated every 2.97 Å (the period of hexagonal lattice identity), i.e., the structure is completely replicated in six O^{2-} layers and six intermediate Al^{3+} layers (Fig. 2.2). The central pair of Al^{3+} ions (AB), three equivalent neighboring pairs (AC), three equivalent remote pairs (AD), and six equivalent most remote pairs (AE) are shown in this figure. In the direction of the *C*-axis, the three distances Al–O are equal to 1.97 Å and the other three distances are 1.87 Å (Fig. 2.2c). The Al–Al and O–O spacing are 2.65 Å and 2.52–2.87 Å, respectively.

As temperature increases, the lattice anisotropy and the bond length increase. In particular, at 2,265 K the length of the bonds Al–O increases from 1.966 to 2.024 Å and from 1.857 to 1.880 Å [2]. The coordinates of the aluminum atoms change to a greater extent in comparison with those of the oxygen atoms. The closest hexagonal packing becomes more regular [3]. The linear expansion coefficient is the highest for the bonds Al–Al ($17.5 \cdot 10^{-6}\,°C^{-1}$).

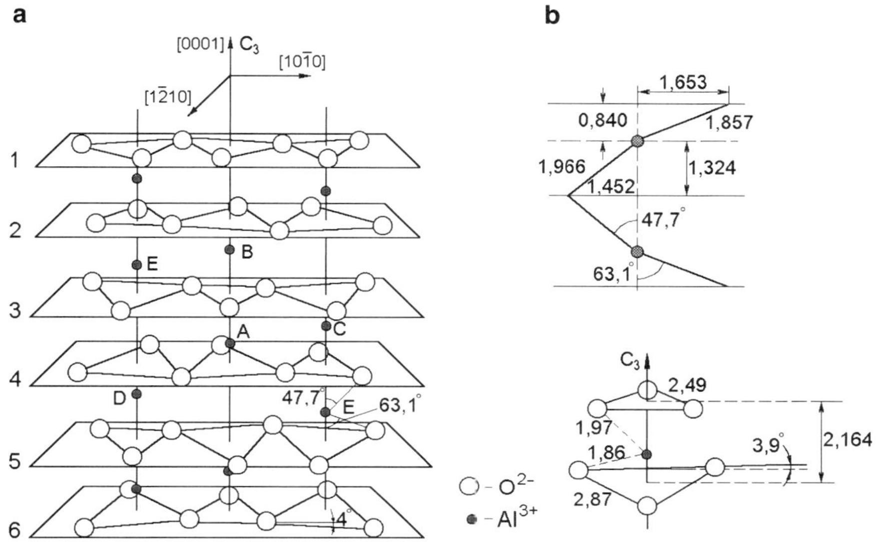

Fig. 2.2 (**a**) Schematic of the arrangement of Al^{3+} ions and O^{2-} layers (*black circles*); (**b**) the distance between Al^{3+} and O^{2-} ions; (**c**) the position of O^{2-} ions surrounding Al^{3+} ion. The planes of the triangle are perpendicular to the *C*-axis

For the ideal closest packing of rigid balls, the ratio of translation vectors is 1.63, while for sapphire it equals 1.58 (the distorted structure has a lower energy in comparison to the ideal).

If Al^{3+} is substituted by another trivalent ion, such as Cr^{3+}, structural variants are possible. In the configurations (Fig. 2.3a–c) chromium is incorporated in the lattice (a) by substituting Al^{3+} next to the anionic vacancy; (b) by substituting Al^{3+} in the regular lattice; and (c) by substituting Al^{3+} next to the cationic vacancy. In the configuration (Fig. 2.3d) Cr^{3+} occupies the position of the cationic vacancy. These configurations are not equivalent from the energetic viewpoint. In position (a) the bond of Cr^{3+} with the lattice is weaker than that of Al^{3+} in the regular lattice (b). The bond with the lattice of the ion Al^{3+} next to the cationic vacancy (c) is stronger than the one in position (b). The configuration (d) leads to diminution of the concentration of the vacancies present in the crystal for charge compensation. This configuration is preferable only at high chromium concentrations.

The structure with the spatial symmetry group $D_{3d}^{6} - R3C$ can be considered to be a slightly distorted closest hexagonal packing of oxygen ions. In the crystal lattice of sapphire, two structural elementary cells — hexagonal (Fig. 2.4) and rhombohedral (Fig. 2.5) — can be distinguished. The bonding between them is shown in Fig. 2.6 [1]. The morphological and structural rhombohedral elementary cells are presented here. The morphological rhombohedral cell along C_3 axis is twice as small as the structural one. The rhombohedral angle of the morphological rhombohedral cell is 85°42 2/3′, whereas that of the structural rhombohedral cell is 55°17′. The rhombohedron vertices are located on the structural vacancies. The base of the

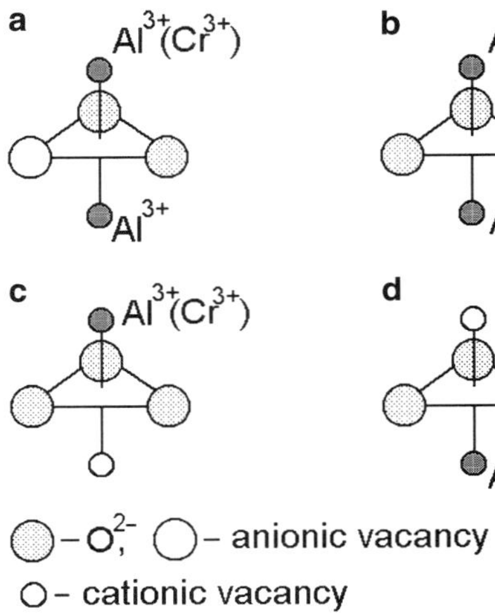

Fig. 2.3 **a–d** Variants on the substitution of aluminum ions by chromium in the sapphire lattice

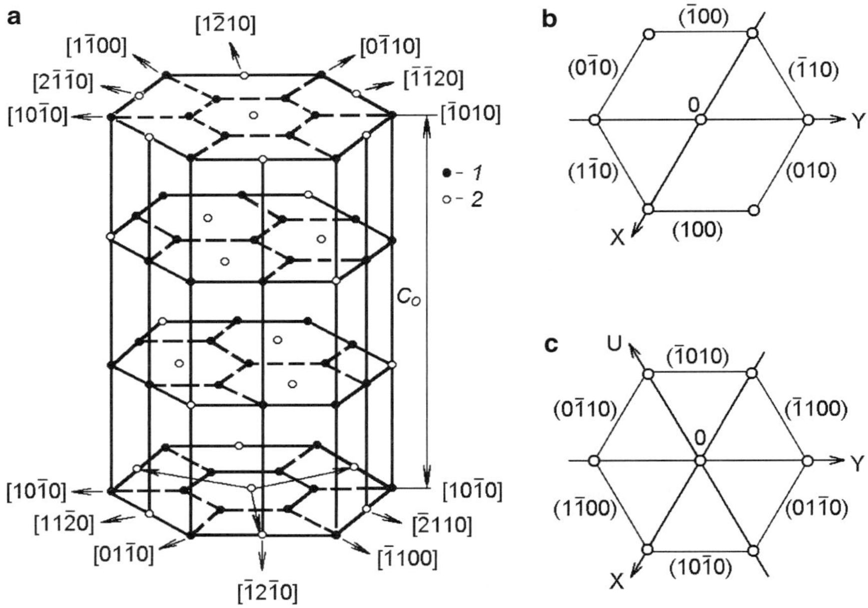

Fig. 2.4 (**a**) Hexagonal cell. (*1*) Aluminum ions; (*2*) octahedral hollows. Indexing of the hexagonal cell facets in a three-axis (**b**) and four-axis (**c**) coordinate system

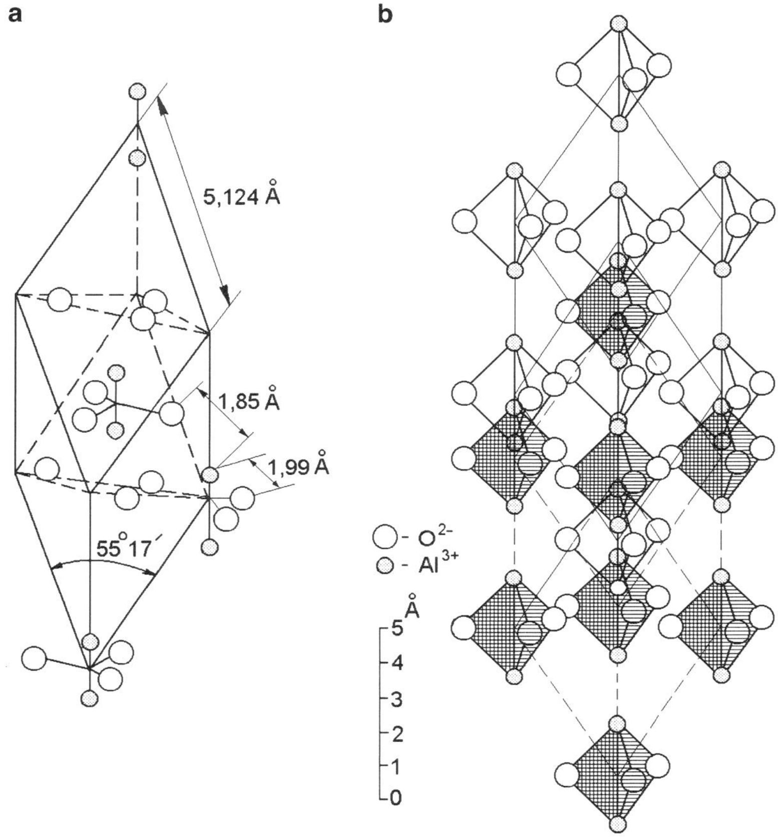

Fig. 2.5 (**a**) Scheme of the rhombohedral elementary cell $\alpha = 55°17'$; (**b**) structural and morphological (*shaded*) rhombohedral elementary cells

structural rhombohedral cell is a group of two molecules centered with respect to the vertex (Fig. 2.5). The coordinates of the ions of this cell are shown in Table 2.34 in the Appendix.

The structure of sapphire also can be depicted by means of coordination polyhedrons. In this scheme, the structural elements are shown not as balls but as polyhedrons formed by straight lines connecting the centers of the anions which surround the cations. The polyhedron vertices touch one another at the centers of the anions. The quantity of the vertices of such a polyhedron is equal to the coordination number of the cation, and the set of polyhedrons presents the mutual location of the cations. The layers formed by the oxygen octahedrons are superimposed one over another in such a way that in the octahedron columns extending along the *C*-axis two filled octahedrons interchange with the unfilled one [4]. The paired octahedrons make a screw axis, which characterizes the corundum motive of the packing along the axis *C* (Fig. 2.7b). The "pure view" of the corundum motive is shown in Fig. 2.7e.

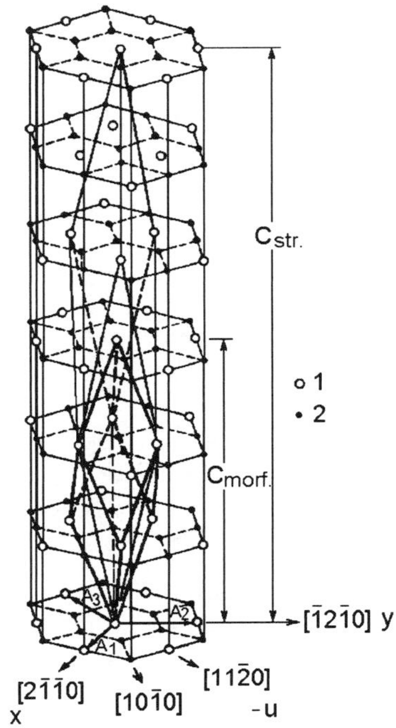

Fig. 2.6 The bonding between the hexagonal and rhombohedral elementary cells: (*1*) octahedral hollows; (*2*) aluminum ions

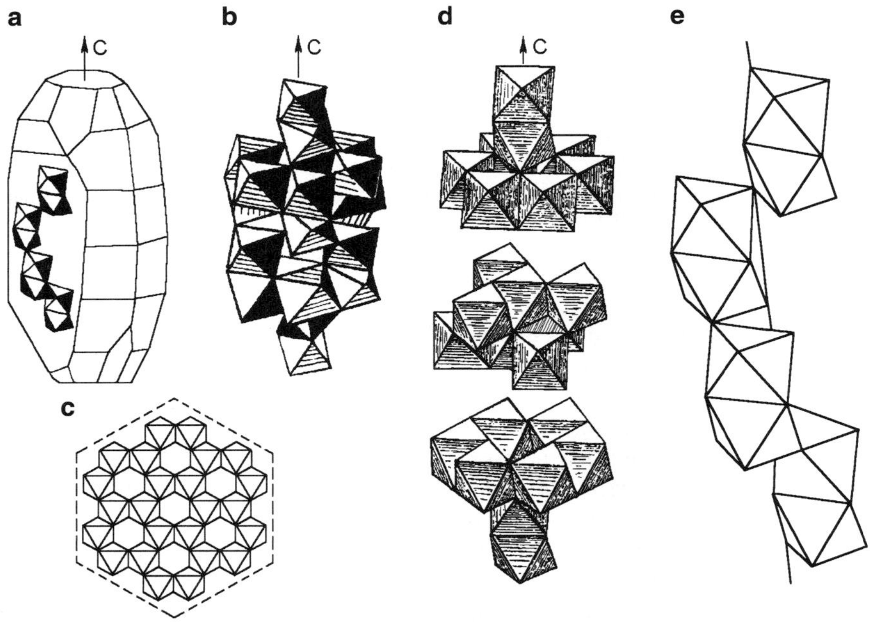

Fig. 2.7 Arrangement of octahedrons in the planes (**a**) parallel and (**c**) perpendicular to the axis *C*; (**b**) the motive of the sapphire structure: screw axis composed of paired octahedrons; (**c**) elementary rhombohedron composed of octahedrons; (**d**) the same rhombohedron in "disassembled" form; (**e**) the "pure view" of the corundum motive

Lattice parameters, or hexagonal lattice constants $a = b$ and c, the interplanar distances are given by the equation

$$d_{hkl} = 1/[(4/3\alpha^2)(h^2 + k^2 + hk) + (1 + 1/c^2)l^2]^{1/2} \quad (2.1)$$

The values of the interplanar distances are presented in Table 2.35.

The lattice parameters increase with increasing temperature. At 295.65 K a = 4.759213 Å, c = 12.991586 Å [5], and the ratio c/a = 2.729776 far exceeds that of a crystal with ideal hexagonal packing $(\sqrt{8/3} \sim 1.633)$. Precise measurements of the temperature dependence of the lattice parameters within 4.5–374 K temperature range (Fig. 2.8, Table 2.36) were carried out using ^{57}Fe Mössbauer radiation [6]. The lattice parameters at high temperatures are presented in Table 2.37.

The parameters of the rhombohedral lattice are shown in Fig. 2.5. The rhombohedral angle equal to 55°17′ is larger than that in the ideal rhombohedron (53°17′) due to the electrostatic interaction of the cations with the anions, which decreases the distance between the oxygen layers. At the isomorphic substitution $^{16}O \rightarrow {}^{18}O$ the length of the bond Al–^{18}O diminishes. The unit cell parameters and the rhombohedron angle also decrease.

Influence of vacancies on the state of the lattice. Electrical neutrality of the lattice is provided by two Al^{3+} (V_{Al}^{3-}) and three O^{2-}(V_O^{2+}) stoichiomentric vacancies. The energy of formation of Schottky defects (E_{Sh}) is E_{Sh} = 20.5 or 4.1 eV per defect, as obtained from the expression characterizing the energy of activation for oxygen

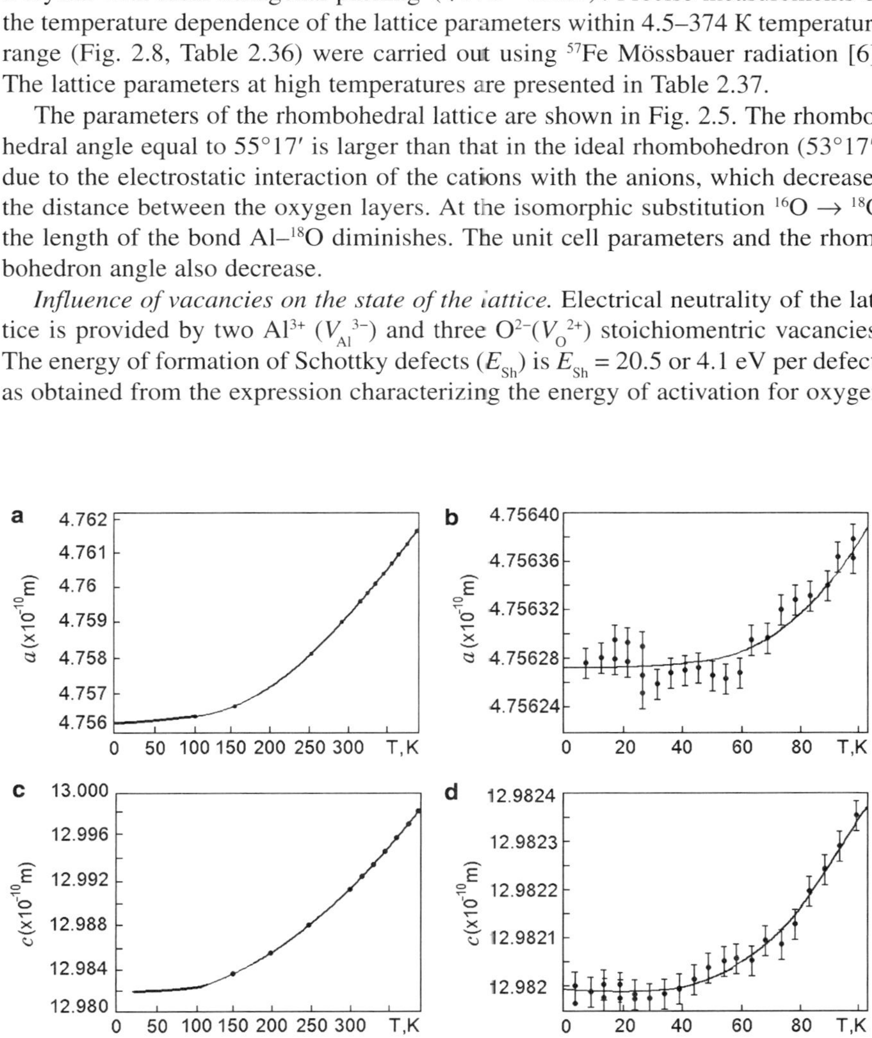

Fig. 2.8 Sapphire lattice parameters in the 4.5–374 K temperature range (**a**,**c**) and more detailed measurements at temperatures below 100 K (**b**,**d**)[5, 6]

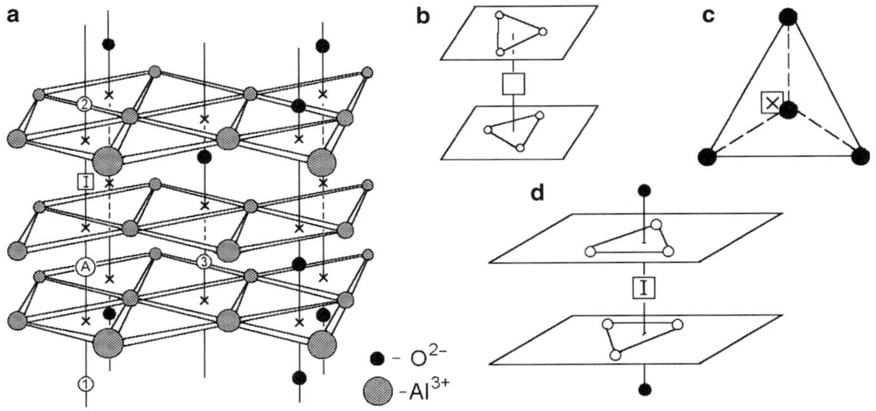

Fig. 2.9 (**a**) Positions of vacancies and interstitial ions *(I)* in the hexagonal α-Al_2O_3 lattice; (×) the places of intersection of the lines where the ions Al^{3+} are located, with the planes in which O^{2-} ions lie; (□) V_{Al}^{3-} position; (☒) V_O^{2+} position; (◫) the interstitial ion; (**b**) the surroundings of a V_{Al}^{3-} vacancy; (**c**) the closest surrounding of a V_O^{2+} vacancy; (**d**) the position of the interstitial ion (◫) in the octahedral bend

diffusion $E_d = 1/5E_{Sh} + E_M$, where $E_M = 2.5$ eV is the energy of oxygen ion migration [7]. The most probable positions of point structure defects are shown in Fig. 2.9b. The vacancy V_{Al}^{3-} is situated between two groups composed of three O^{2-} ions, at a distance of 1.86 and 1.97 Å from the planes in which O^{2-} lie (Fig. 2.9). The vacancy V_O^{2+} is located within an approximately tetrahedral surrounding of Al^{3+} ions at a distance of 1.86 Å from two Al^{3+} ions and 1.97 Å from the other two Al^{3+} ions. The interstitial ion is located between the groups of Al^{3+} ions at a distance of 1.92 and 1.98 Å.

Simulation of sapphire lattice dynamics with vacancies in different charge states makes it possible to obtain data on the frequencies of resonant oscillations induced by the defects, the values of ion charges ($Z_O = -1.98|e|$, $Z_{Al} = 2.97|e|$), and the bonding constants ($k_O = 73.07$ eV/Å^2, $k_{Al} = 192.49$ eV/Å^2)[8]. Using the numerical parameters of interatomic potential and a cluster containing 1,000 atoms, the total density of the phonon states is calculated in theoretically perfect sapphire (Fig. 2.10), which is in good agreement with the corresponding experimental values [9].

Vacancies influence the oscillation spectra of sapphire (Table 2.1). Therefore, the local density of phonons in crystals containing vacancies differs from that of the perfect crystal. The localization of electron density in the center of a vacancy is of the order of 90% for (F^+) centers and 80% for *F* centers in the ground state [10, 11]. The appearance of vacancies in sapphire is accompanied by resonance oscillations at 5.7 and 22.5 THz (Fig. 2.11). The calculated total density of phonon states for all the atoms of a spherical region with a radius of 2.7 Å in the vicinity of oxygen atom (comprising four Al^{3+} and four O^{2-} ions) in nonstoichiometric sapphire is shown in Fig. 2.12.

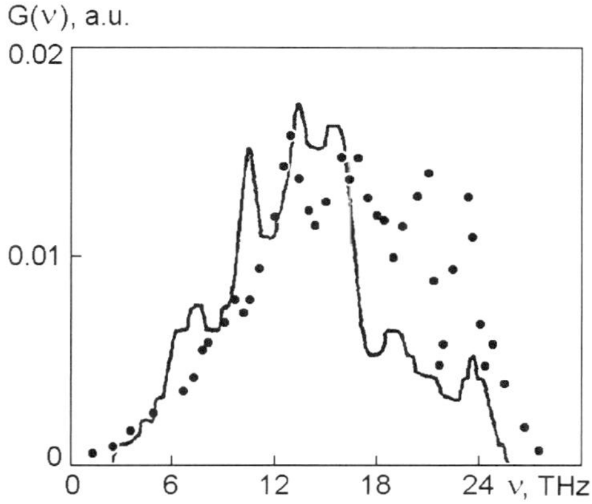

Fig. 2.10 Calculated total density of phonon states in theoretically perfect α-Al_2O_3 crystals. The points denote the experimental values

Table 2.1 Frequencies of localized oscillations induced by defects on the ions nearest to them in the sapphire lattice [8]

			Frequency (THz)		
Defect	Ion	Distance from defect (Å)	*X*	*Y*	*Z*
Anionic vacancy	Al(1)	2.12	5.7; 22.5	6.0; 22.5	5.7; 22.0
	Al(2)	2.24	5.4	8.4	6.6
F^+ center	Al(1)	1.97		3.0; 9.6	8.2
	Al(2)	2.07		2.0; 7.8; 14.0	3.3;11.7
F center	Al(1)	1.89		8.2; 12.0	14.7
	Al(2)	2.02	16.0	14.0	13.0

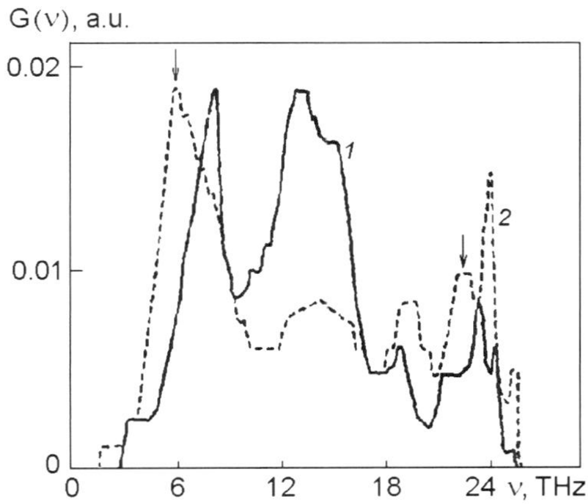

Fig. 2.11 Local density of phonons in the direction *Z* in the Al^{3+} ion position (*1*) nearest to the O^{2-} ion in the perfect sapphire crystal and (*2*) nearest to the anionic vacancy

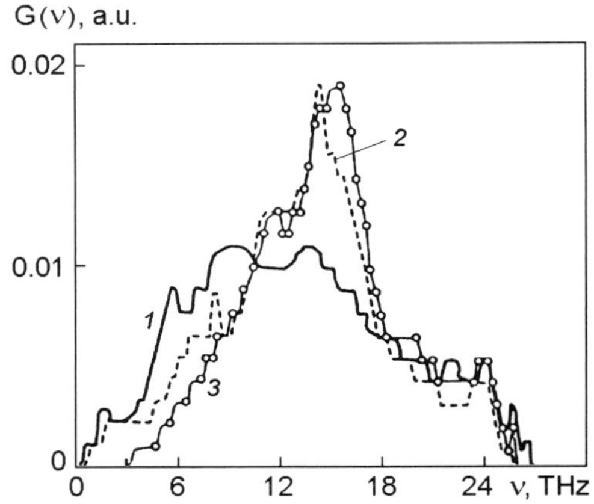

Fig. 2.12 Total density of phonon states for the atoms located in (*1*) a spherical region of radiu s 2.7 Å surrounding the anionic vacancy, (*2*) F^+ center, and (*3*) *F* center

It has been noted [8] that the changes in the oscillation spectra are explained on the basis of effective interaction of a vacancy with its nearest surrounding, which increases the total density of phonon states in the low-frequency region of the spectrum. The coulomb interaction is partly compensated by F^+ centers and the effective interaction increases. When two electrons are captured by the center, the coulomb part is completely restored. This by far increases the effective interaction and shifts the density of states located in the region below 10 THz, to the high-frequency region of the spectrum.

Ionic bond energy influences a number of physiochemical properties of the crystal. For example, the quantity of free bonds per unit of surface area is an approximate measure of surface energy. Bonding in α-Al_2O_3 is mainly ionic, with the share of covalent bonds at ~20%. The large quantity of partly covalent bonds that emanate from each atom of the lattice is a probable cause of the mechanical properties of sapphire. The contribution to the lattice bond energy is made by electrostatic interactions, polarization, repulsion, and Van der Waals forces, with the experimental value of total contributions from these forces at 156.7 eV [12]. Polarization energy arises in the lattice due to the appearance of electric fields bound to lattice defects. The largest contribution belongs to the forces of electrostatic interaction (the calculated value is 189.1 eV), the individual contributions of different forces to the potential ionic energy is presented in Table 2.38. Proceeding from these data, one can calculate the energy necessary for shifting an ion from a lattice site to an octahedron interstitial. Presented below are the calculated values of the energy of formation for vacancies and interstitial atoms in an inelastic, nonpolarized lattice in [7]:

Al^{3+} vacancy	9.1 eV
Interstitial Al^{3+}	10.8 eV
O^{2-} vacancy	3.5 eV
Interstitial O^{2-}	10.5 eV
Five pairs of Schottky defects	5.7 eV
	4.1 eV (Experimental value)
Al-Frenkel pair	10.0 eV
O-Frenkel pair	7.0 eV

The calculated migration energy of V_{Al}^{3-} from the position A (Fig. 2.9a) to the position 1 (i.e., the place of localization of V_{Al}^{3-} displaced into the center of the triangle formed by O^{2-} ions) is 3.8 eV [7]. For the displacement to the positions 2 and 3 there are required energies of 6.6 and 3.8 eV, respectively. The vacancies V_O^{2+} are more mobile in comparison with V_{Al}^{3-} and the interstitial ions, and they can migrate in the planes of the O^{2-} ion locations. The calculated value of the migration energy of V_O^{2+} in the basal plane equal to 2.9 eV [7] is in good agreement with the experimental value (2.5 eV) obtained while measuring the coefficient of oxygen diffusion.

A layer with a content of V_O^{2+} higher than that in the crystal bulk (where V_{Al}^{3-} increases the value of energy per defect) may be formed near the crystal surface where the conditions of electrical neutrality are not so strict. At elevated temperatures, splitting of basal and prismatic dislocations in the process of skipping leads to the formation of packing defects. The energy of such defects (stacking fault energy) is essentially the same for different crystallographic planes at 0.15–0.35 J/m^2 [13].

The main parameters that characterize the lattice of sapphire are given in Table 2.39.

The electronic energy structure of sapphire is typical of that for ionic crystals. Its valence band is formed mainly by the 2p states of oxygen and has a width of 13 eV. The conduction band bottom is formed by the 3s states of aluminum. According to experimental data, the width of the forbidden band is 9.5 eV. It has been shown [14] that the plane (0001) has two unoccupied surface zones in the forbidden band that are located at 2 and 8 eV below the conduction band.

In addition to considering the lattice parameters and nature of the atomic bonds, in the process of epitaxial growth on sapphire one must take into account the electronic structure of the crystallographic planes, which is defined by the type of ions located on the crystal surface. Even parallel chips may end in different ions (Fig. 2.13a) [14].

The planes of the cross section A end in Al and O layers and of the cross section C end in two equivalent Al layers, etc. The densities of charges and their distribution in different cross sections have been calculated [14]. The distribution of charge density around the ions, with a minimum and a maximum of 0.005 and 0.050 e/a^3, respectively, is shown in Fig. 2.13b. The energy of chipping in different cross sections is estimated by studying the surface bonds of the last layer and several previous layers.

The electronic structure of sapphire is responsible for the existence of excitons with an absorption maximum in the region of 9.0–9.25 eV at 300 K.

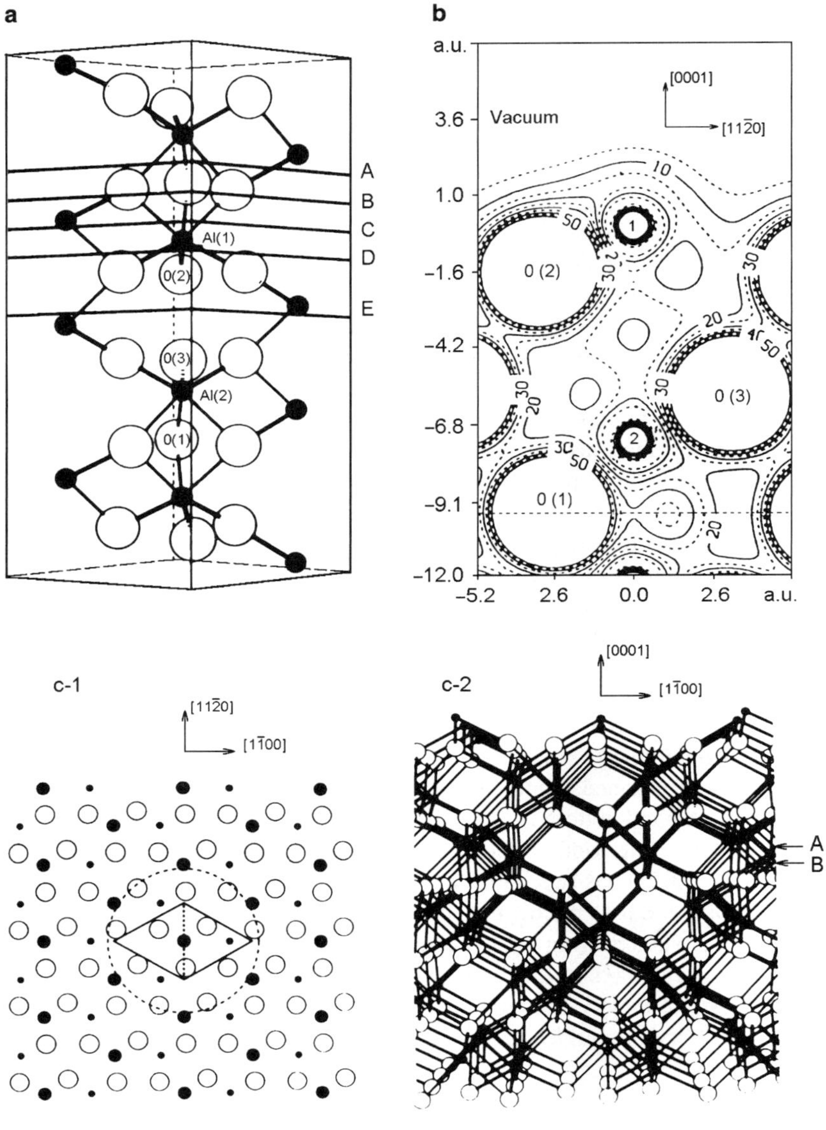

Fig. 2.13 (**a**) Cleavages in different (A–E) cross sections of the sapphire lattice parallel to the plane (0001); (**b**) distribution of the charge density around the ions shown in (**a**). (***c–1***), (***c–2***) the plane (0001), the top view of a block formed by three atomic layers Al–O–Al; O denotes oxygen ions; • and ● are aluminum ions of the upper and lower layers, respectively. Al_2O_3 cell is shown as a rhombohedron, the dashed circle marks the boundary of a cluster

Cleavage in sapphire arises at the intersections of a pair of parallel nets formed by anions. Nets with like charges reduce attractive forces. The larger the distance between the nets, the more vividly the cleavage manifests itself. In a perfect crystal the plane of chipping must pass between these nets. In the basal plane with interchanging O–Al–Al–O–Al–O layers there are no conditions for cleavage, whereas in the plane $(10\bar{1}1)$with interchanging O–O–Al–O–Al–O–O—O–Al–O–Al–O layers the bonds between the layers O–O located at a distance of 1.06 Å are weakened.

Sapphire does not have such a vivid cleavage as diamond and other crystals. For a long period of time, sapphire was considered to exhibit no cleavage at all. Theoretically, it has nine cleavage planes. Six planes are parallel to the facets $\{11\bar{2}0\}$and $\{10\bar{1}1\}$, and to the C-axis; three planes are parallel to the facets $\{10\bar{1}1\}$ and inclined to the C-axis at an angle of 33°, the normal vectors to them make an angle of 57° with the C-axis. Crystals with a small quantity of dislocations and which do not contain blocks may have perfect cleavage in the plane of the morphological rhombohedron $\{10\bar{1}1\}$. However, when block-containing and stressed crystals grown by the Verneuil method are chipped along the prismatic planes, mirror chips with steps of several atomic parameters are often observed (Fig. 2.14). To achieve more desirable chipping of these crystals, the direction of the C-axis is set at an angle 57° to the crystal growth axis at the location of the prism and rhombohedron planes, as shown in Fig. 2.14c [15].

The energy required for destruction along the plane $\{10\bar{1}1\}$is 6 J/m^2, whereas for the basal plane the corresponding value is more than 40 J/m^2 [16]. Shown by the estimation of the surface energy of the planes, considered to be defined by the quantity of free bonds per unit of surface, the minimum quantity of such bonds corresponds to the plane $(10\bar{1}1)$[17].

Slip systems. Nonlinearity of thermal fields leads to plastic deformation of the crystal realized mainly through slipping. Three active slip systems (Table 2.2) and their arrangement with respect to the growth direction make a contribution to the orientation dependence of tangential stresses [18].

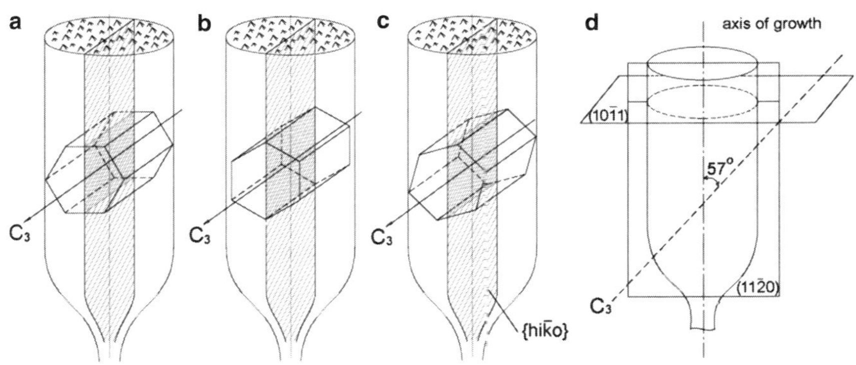

Fig. 2.14 Location of the symmetry elements and the plane of chipping in the crystals grown by the Verneuil method: (**a**) chip along the plane $(11\bar{2}0)$; (**b**) chip along the plane $(10\bar{1}0)$, (**c**) chip along the prism $(hi\bar{k}0)$. The plane of chipping is shaded. (**d**) The location of the planes $(11\bar{2}0)$ and $(10\bar{1}1)$ with respect to the growth axis and to the C-axis for optimization of chipping

Table 2.2 Possible slip systems in sapphire

Type of slip system	Slip plane	Slip direction	Number of equivalent systems
Basal 1	c (0001)	$\langle 1\bar{2}10\rangle$	3
Basal 2	c (0001)	$\langle 10\bar{1}0\rangle$	3
Prismatic 1	$a(1\bar{2}10)$	$\langle 10\bar{1}0\rangle$	3
Prismatic 2	$m(10\bar{1}0)$	$\langle 1\bar{2}10\rangle$	3
Prismatic with inclined slip direction	$a(1\bar{2}10)$	$\langle 10\bar{1}2\rangle$	6
Rhombohedral	$r\{10\bar{1}1\}$	$\langle 10\bar{1}2\rangle$	3

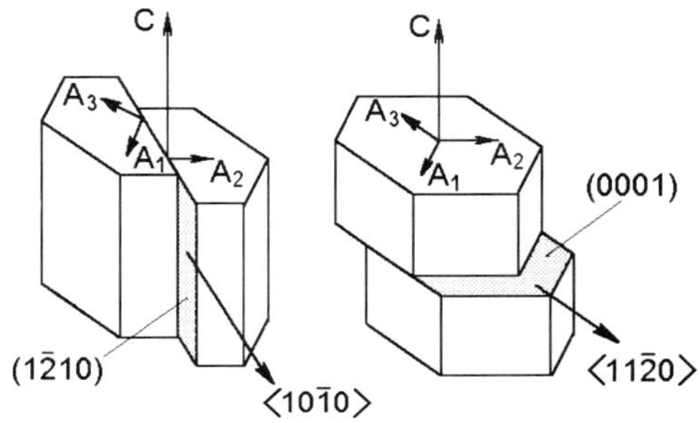

Fig. 2.15 Schematic of prismatic and basal slip systems in sapphire

In the 570–1,170 K range, only easy local slip along the prism planes is observed [19]. The plane of the easiest slip is (0001), with basal slip predominating in the direction (1120). This is observed in sapphire at T > 970–1,170 K, when basal dislocations become mobile. Prismatic slip occurs at T > 1,470 K (or, according to some data [20], at T > 2,270 K). At 1,470–1,670 K the activation energy is 3.8 ± 0.2 eV [21]. The value of basal slip depends on the temperature and the load. The theoretically estimated "ease" of possible slip systems was found to diminish in the following sequence: $(0001)\langle 11\bar{2}0\rangle$, $(0001)\langle 10\bar{1}0\rangle$, $\{10\bar{1}0\}\langle 11\bar{2}0\rangle$, $\{11\bar{2}0\}\langle 10\bar{1}0\rangle$, $\{10\bar{1}0\}\langle 01\bar{1}2\rangle$, $\{11\bar{2}1\}\langle 10\bar{1}0\rangle$, $\{22\bar{4}3\}\langle 10\bar{1}0\rangle$.

The action of the first five systems was revealed experimentally [19, 20, 22]. The first two systems are shown schematically in Fig. 2.15. At extension of whiskers grown along the C-axis, slip on the planes $\{\bar{4}223\}$ in the direction $(01\bar{1}0)$ was established from the orientation of slip traces [23].

Sapphire undergoes deformation under conditions of applied loads P. For plastic deformation, real shearing stresses, τ_c, which act in the given slip system, are important:

$$\tau_c = mP, \tag{2.2}$$

where m is the Schmid-factor (orientation factor); $m = \cos l \cdot \cos c$; l is the angle between the direction of slip and the direction of deformation, c is the angle between the normal axis to the slip plane and the direction of deformation. At $c = l$ the factor m has its maximum value equal to 0.5. In this case the direction of slip lies in the plane which passes through the axis of deformation and the normal axis to the slip plane. Using four-point bend and indentation test data from $1.5 \times 3 \times 25\ \text{mm}^3$ plates (at 1,270 K, 0.5 h, 15 s), the Schmid factor was calculated for several planes and directions of slip (Table 2.3) [24].

The experimental values of shearing stresses at 2,270 K obtained during the growth of sapphire ribbons are $\tau_{cr} = 1$ MPa for the basal slip system and $\tau_{cr} = 10$ MPa for the prismatic [25]. The latter system works only in ribbons in which the deviation of the facet (0001) from the surface does not exceed 3°. At larger deviations the basal system — with the easiest slip (removal of stresses) — starts working. Therefore, in thin basal ribbons only the prismatic system acts; the basal system corresponds to thick ribbons (with a thickness exceeding 5 mm). Chromium impurities which strengthen the crystal increase the onset temperature of slip.

During the growth of sapphire by the Stepanov method in the direction of the *C*-axis, prismatic and rhombohedron slip systems manifest themselves. As calculations show [19], in these crystals basal slip takes place only during the growth in a direction different from ⟨0001⟩ (Fig. 2.16).

Anisotropy of thermoelastic stresses is bound to that of the lattice and with the acting slip systems. The relationship between temperature nonlinearity in sapphire tapes and the vertical component of thermoelastic stresses σ_{zz} is shown in Fig. 2.17 [22]. The first upper magnitudes correspond to σ_{zz} for the ribbon obtained upon the action of the $\{10\bar{1}0\}\langle 2\bar{1}\bar{1}0\rangle$ slip system, the second upper magnitudes relate to the case of the action of the system $\{\bar{1}101\}\langle\bar{1}10\bar{2}\rangle$. These systems give the maximum and the minimum values of σ_{zz}. The lower magnitudes are the calculated values of σ_{zz} in the isotropic approximation. The factor of anisotropy for these orientations is 1.28 and 0.65, respectively.

Twinning may arise in sapphire upon cooling or deformation. Plastic deformation which causes twinning occurs at $T < 1{,}770$ K, a temperature lower in comparison with translational slip, and manifests itself to a temperature of 79 K. Twins in sapphire are defined as the regions limited by plane-parallel coherent boundaries or

Table 2.3 Schmid-factor for the plane (A) $(11\bar{2}3)$ and the loading direction $\langle\bar{2}111\rangle$ and (B) $(11\bar{2}0)$ and the loading direction $\langle\bar{8}80\bar{5}\rangle$

Slip plane	Slip direction	Schmid-factor
(A)		
(0001)	$1/3\langle 2\bar{1}\bar{1}0\rangle$	0.5
(0001)	$1/3\langle\bar{1}2\bar{1}0\rangle$	0.25
(0001)	$1/3\langle\bar{1}\bar{1}20\rangle$	0.25
(B)		
(0001)	$1/3\langle\bar{2}110\rangle$	0.43
(0001)	$1/3\langle 1\bar{2}10\rangle$	0.43
(0001)	$1/3\langle 11\bar{2}0\rangle$	0.25

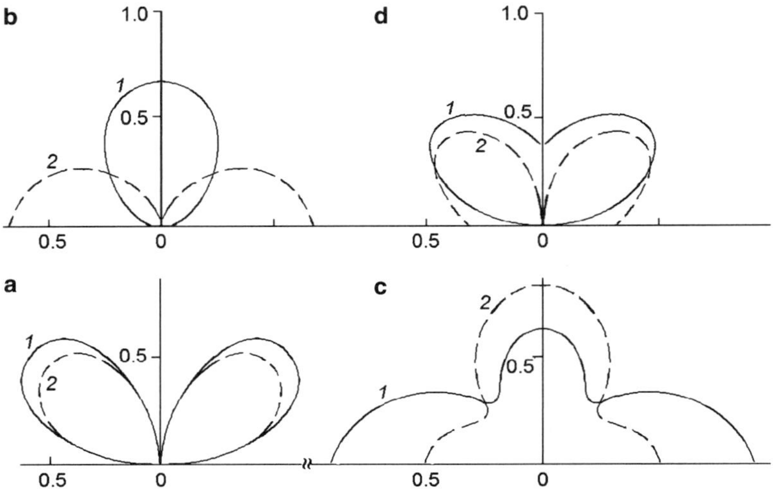

Fig. 2.16 Orientation dependences of tangential stresses (in rel. units) (*1*) in the middle part of a sapphire tube and (*2*) in the vicinity of the crystallization front. Slip systems: (**a**) (0001) $\langle 1\bar{2}10\rangle$; (**b**) $\{1\bar{2}10\}\langle 10\bar{1}0\rangle$; (**c**) $\{10\bar{1}1\}\langle 10\bar{1}2\rangle$; (**d**) $\{1\bar{2}10\}\langle 10\bar{1}2\rangle$

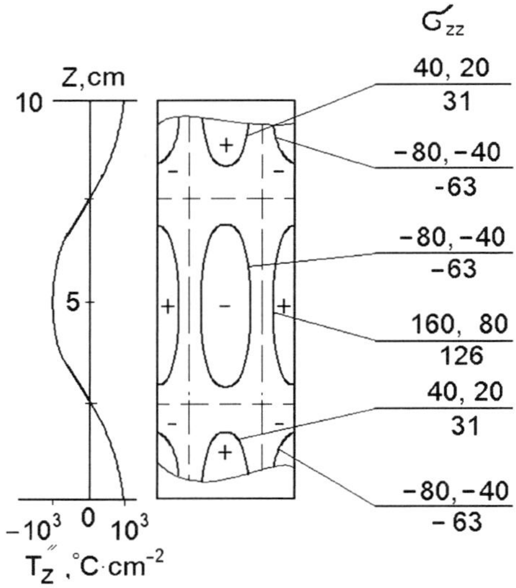

Fig. 2.17 The second derivative of temperature along the length of sapphire ribbon and the distribution of stresses at the action of different slip systems

wedges with noncoherent boundaries. Geometric characteristics of a twin may be visualized proceeding from the scheme of twinning ellipsoids (Fig. 2.18) [26]. As a result of twinning along the plane passing through the center of the circle, the cross section of the circle above this plane transforms into an ellipse. The ellipse and the circle intersect at two points. In the planes K_1 and K_2 perpendicular to the circular

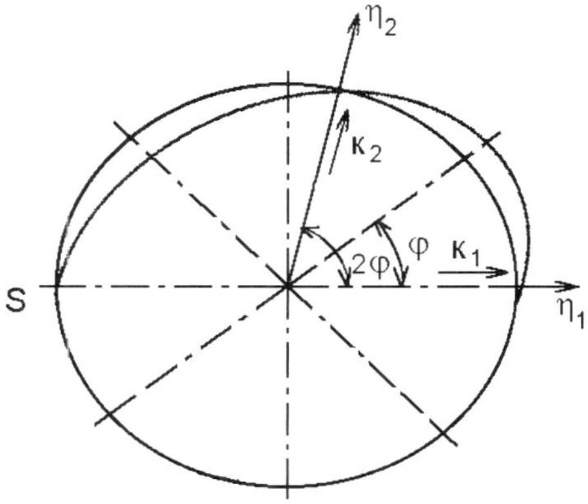

Fig. 2.18 Twinning ellipsoid

plane and passing through these points, the positions of the crystal structure units before and after twinning coincide. The plane K_1 coincides with the twinning plane and contains the direction of shearing (twinning) η_1. The plane K_2 contains the direction η_2 and makes the angle 2φ with the twinning plane. The value of the angle is defined by the specific shear $s = 2/\mathrm{tg}\ 2\varphi$.

In sapphire there are mainly reflection twins and turn twins with the elements of twinning:

K_1 $\{10\bar{1}1\}$; η_1; $\langle 10\bar{1}2\rangle$; K_2 $\{10\bar{1}2\}$; $\eta_2 <$ $\{\bar{2}021\}$; $s = 0{,}635a$

K_1 $\{0001\}$; η_1; $\langle 1\bar{2}10\rangle$; K_2 $\{\bar{2}021\}$; $\eta_2 <$ $\{\bar{2}021\}$; $s = 0{,}202a$

Twinning in the planes $\{10\bar{1}1\}$is characterized by a lesser specific shear and is observed even at cryogenic temperatures. The formation of a twin at compression along the *C*-axis (Fig. 2.19) shows shear along the rhombohedronal plane [27, 28]. The twin boundary is slightly terraced (Fig. 2.19b), with the interval between the individual terraces ~0.7 μm [29].

Joint twins are observed in crystals grown from the solution. The joints shown in Fig. 2.20 are reflection twins with a $\{10\bar{1}1\}$ twinning plane and 64°52′ twin disorientation.

Microtwins measuring 30–75 μm are formed in sapphire with abrasive treatment and indentation. The depth of deformation and the dimensions of twin interlayers depend on the crystallographic orientations and the loads applied.

Periodic bond chains (PBC). According to crystal morphology theory [30], the lattice of sapphire has facets parallel to two or more chains of strong bonds (the so-called *F*-type facets). Hartman found a number of chains with "intense bonds," denoted (in rhombohedral symbols) as $\langle 01\bar{1}\rangle$, $\langle 001\rangle$, $\langle 11\bar{1}\rangle$, $\langle 11\bar{2}\rangle$, $\langle 01\bar{2}\rangle$, $\langle 12\bar{2}\rangle$, $\langle 11\bar{3}\rangle$.

The forms of *F*-type facets are $\{111\}$, $\{001\}$, $\{011\}$, $\{01\bar{1}\}$, $\{112\}$, $\{012\}$, and $\{022\}$ [30]. Other facets are of *S*-type (parallel to one chain of strong bonds) and *K*-type (parallel to no such chains). The gnomonic projection of corundum built by Hartman shows all the *F*-type facets and all the zones parallel to the chains with intense bonds (Fig. 2.108). According to Hartman, the basal plane has 6.6 free

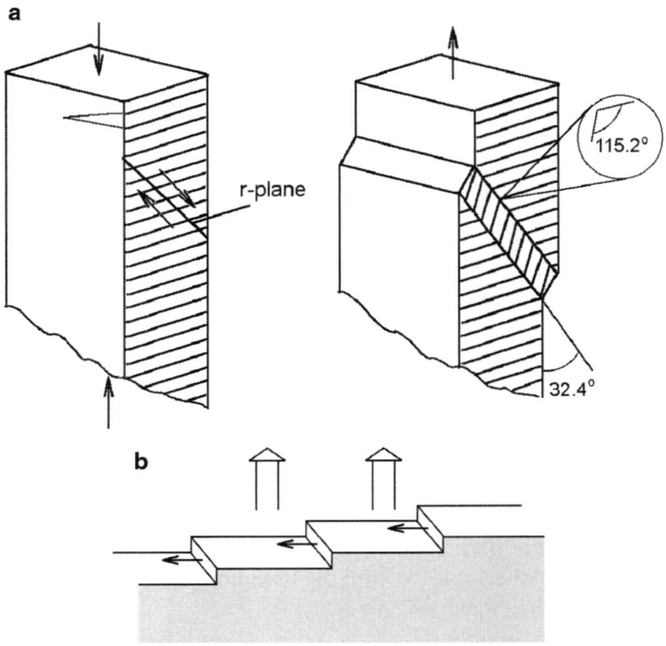

Fig. 2.19 (**a**) Scheme of sapphire twinning in the rhombohedronal plane upon compression. (**b**) Schematic of twin boundary

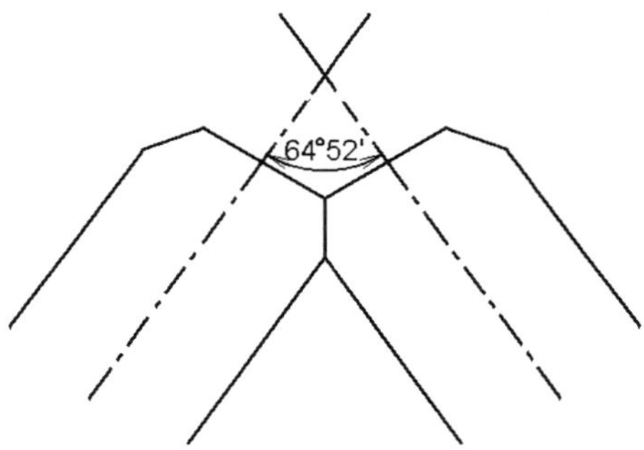

Fig. 2.20 Joint twin of corundum crystals

bonds per unit of plane; for the planes $(10\bar{1}0)$ and $(10\bar{1}1)$ the mentioned quantities are 5.2 and 3.5, respectively. Based on the analysis of chipping planes of crystals grown by the Verneuil method, the directions of $\langle 211\rangle$, $\langle 210\rangle$, $\langle \overline{42}1\rangle$ bonds parallel to the *F*-type face $(1\bar{2}0)$ were established [31].

Morphology of sapphire is defined by the growth conditions and the peculiarities of the structure, including periodic bond chains. The direction of PBC chains defines the stable *F*-faces which will be the slowest growing, and hence of highest morphological importance in the crystal. In order to determine the relative importance of various *F*-faces it is assumed that the growth rate of a face (*hkl*) is directly proportional to its attachment energy, E^{att}_{hkl} [32]. This is the energy released when a layer of thickness d_{hkl} is added to the growing surface. The reverse process in which the energy E^{att}_{hkl} is consumed to produce a certain area of new surface by crystal cleavage suggests a relation between E^{att}_{hkl} and surface energy [33]:

$$\gamma \approx Z E^{att} d_{hkl} / 2V, \tag{2.3}$$

in which Z is the number of formula units in a unit cell of volume V and d_{hkl} is the interplanar spacing of the lattice plane (*hkl*).

Based on the known γ and E^{att} values one can predict the crystal morphology. Relaxation of the lattice is illustrated in Fig. 2.21 [32]. There are two regions, of which the ions in the outermost region (I) are allowed to relax to equilibrium while the inner region (II) remains static. Writing the force and displacement of the ith ion per unit cell of the αth plane as $F_\alpha{}^i$ and $\varepsilon_\alpha{}^i$, respectively, the equilibrium condition is that

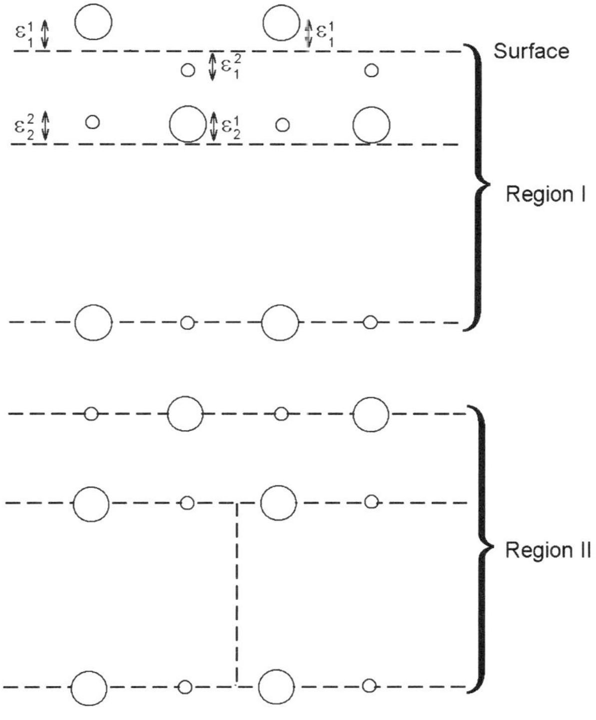

Fig. 2.21 Definition of surface lattice variables

$\partial F_{\alpha}^{\ i} / \partial \varepsilon_{\alpha}^{\ i} = 0$ for all i and α in region I

This can be shown to be equivalent to minimizing the energy of the outer region, E_{I}, with respect to the displacement, $\varepsilon_{\alpha}^{\ i}$, provided the inner region is at equilibrium.

Writing the energy per unit area of given plane (*hkl*) in the bulk as E^{B}_{hkl}, and that in the surface as E^{S}_{hkl}, γ_{hkl} is given by:

$$\gamma_{hkl} = 1/2(E^{B}_{hkl} - 2E^{S}_{hkl}) \tag{2.4}$$

Calculations have been reported [32] in which 20 surface planes were relaxed to equilibrium, while region II contained up to 60 planes of ions, also at equilibrium. The calculated energy of five lowest index surfaces of sapphire are given in Table 2.4, based on both unrelaxed and relaxed surface structures.

Four points emerge from these results. The first is that surface energies corresponding to relaxed structures are appreciably lower than those for unrelaxed surfaces. The second point is that relaxation energies vary from plane to plane. The basal plane energy is reduced by 3.92 J m^{-2}, or 66%, whereas the energy of the rhombohedral surface is reduced by only 1.34 J m^{-2} or 37%. The third point is that lattice relaxation changes the relative stability of the surfaces. The order of unrelaxed surface energies is

$$\{011\}<\{01\bar{1}\}<\{100\}<\{211\}$$

whereas lattice relaxation changes this to

$$\{111\} < \{211\} \approx \{011\} < \{01\bar{1}\}$$

while according to Hartman the predicted order of surface energies is

$$\{011\}<\{01\bar{1}\}<\{121\},\{100\},\{111\},\{021\}$$

Finally, lattice relaxation reduces the difference in energy between the various surfaces. The difference between the highest and lowest unrelaxed energies is 2.93 J m^{-2}, whereas it is only 0.49 J m^{-2} for relaxed surfaces.

Crystal morphologies corresponding to relaxed and unrelaxed surfaces have been drawn using data given in Table 2.4 (Fig. 2.22). The simple rhombohedral

Table 2.4 Calculated energies of the low-index surfaces of sapphire, γ^{0}_{hkl} (J · m^{2})

Surface	Unrelaxed energy	Relaxed energy	Hartman [34]
(111)	5.95	2.03	4.83
$(\bar{2}11)$	6.46	2.23	–
$(01\bar{1})$	3.63	2.29	2.55
$(01\bar{1})$	4.37	2.50	3.80
(100)	5.58	2.52	4.80

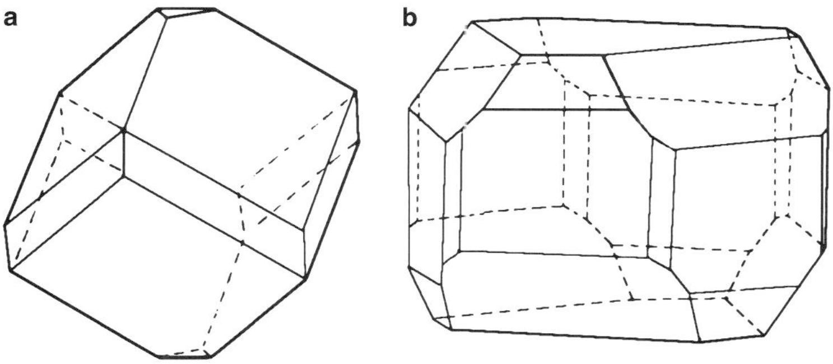

Fig. 2.22 The predicted morphology of sapphire based on (**a**) unrelaxed and (**b**) relaxed surfaces

Table 2.5 Comparison of relative surface energies with the theoretical estimates [35, 36]

	Experimental estimate of facet surface energy		Theoretical estimates	
Surface	Sapphire $\gamma_{(hkjl)}/\gamma_{(0001)}$	Ruby 0.54% Cr_2O_3 $\gamma_{(hkjl)}/\gamma_{(0001)}$	First Principles	Interaction model
$\{\bar{1}012\}$	1.05	1.05	1.12	1.13
$\{11\bar{2}0\}$	1.12	1.00	1.06	1.23
$\{11\bar{2}3\}$	1.06			
$\{10\bar{1}1\}$	1.07	1.02	1.45	1.24
$\{10\bar{1}1\}$			0.795	1.10

habit, predicted assuming no surface relaxation, is replaced by more complex morphology in which the basal planes {111} play an increasingly dominant role.

In a number of papers the value of surface energy of the crystallographic planes was established from the analysis of faceted cavities. The equilibrium shape of internal cavities in sapphire was determined through the study of submicron internal cavities in sapphire and ruby (Table 2.5). Cavities were formed from indentation cracks during annealing at 1,870 K. Equilibrium could be reached only for cavities smaller than 100 nm [35]. Controlled-geometry cavities were introduced in sapphire substrates using photolithographic methods and annealing for prolonged periods [37].

The difference between the largest and smallest relative surface energies for sapphire is about 20%. Cavities in ruby are much more isotropic than in sapphire.

The shape of natural crystals usually has the following facets (denoted by the most often used literal symbols): $c(0001)$, $a\{11\bar{2}0\}$, $r\{10\bar{1}1\}$, $n\{22\bar{4}3\}$, $m\{10\bar{1}0\}$, $s\{02\bar{2}1\}$, $s\{22\bar{4}3\}$, $R\{01\bar{1}2\}$, $p\{11\bar{2}3\}$(Fig. 2.23) and others.[1] The symbols of crystallographic planes in the morphological and the structural classification systems,

[1] In some countries capital letters A, C, M are used.

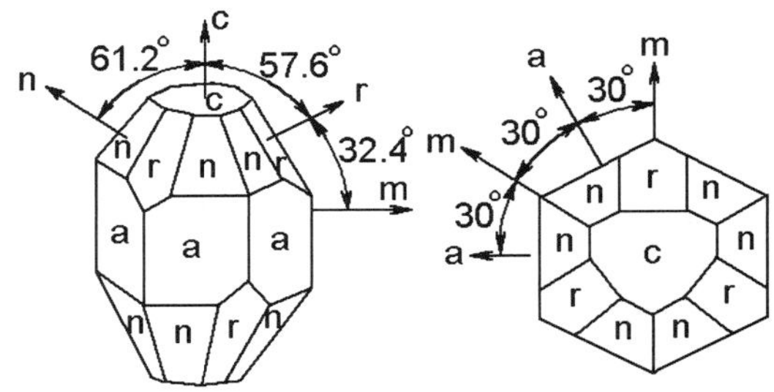

Fig. 2.23 Crystallographic diagram of sapphire

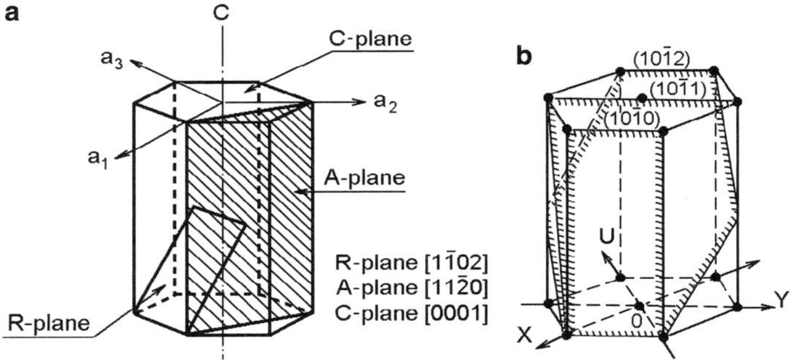

Fig. 2.24 (**a**) Location of the crystallographic planes of sapphire often met in practice. (**b**) Comparison of the symbols of facets with common trace. The trace is parallel to *OY*

the stereographic projections of sapphire, and the Wulff–Bragg angles for some planes are given in Tables 2.40, 2.41 and Fig. 2.109.

Besides the main crystallographic planes, tens of others exist which have their own symbols. The main forms with their spherical coordinates are presented in Table 2.42, as well as others rarely met [38].

The planes that are most often encountered in the practice of sapphire usage[2] are shown in Fig. 2.24. The plane $(10\bar{1}2)$ is inclined to (0001) plane at an angle of 57°36′, while with the plane $(11\bar{2}0)$ it makes an angle of 32°24′. The angles between the normals to the facets of sapphire are given in Table 2.43 [1].

Equilibrium forms. The form of a crystal in which free surface energy is minimal at constant temperature, pressure, and volume is considered to be the equilibrium form. The facets of the equilibrium form may appear on the surface of growing crystals. This is why, by taking into account these facets when orienting

[2] On occasion the most exotic planes find use, too. For instance, the crystallographic planes $(13\bar{4}28)$, $(32\bar{5}43)$, $(65\overline{11}40)$, etc. are utilized in X-ray optics for backward scattering of photons with energies close to that of Mossbaurer ^{57}Fe radiation [5–6].

the crystal, it is possible to influence the character of roughness of the crystallization front (CF) [39].

There are ways to form equilibrium-shaped crystals. One is to heat small crystals in a closed system at high temperatures so that transport processes are hastened. This technique is difficult because of the necessity of controlling the external environment to prevent a volume change of the crystal by evaporation, corrosion, or condensation. The other method is to use internal cavities, which are, in effect, "negative" crystals with very slow lattice diffusion and constant volume in contact with a fixed atmosphere. The conditions of fixed volume in contact with its vapor are easier to maintain with the latter case than with the former case. The deviation of a crystal's shape from its Wulff shape at any time depends on the starting shape and size of the crystal and the rates of kinetic processes that change its shape (Fig. 2.25) [36]. Segregating dopants have a considerable influence on the Wulff shape, even in very dilute concentrations (ppm) (Fig. 2.26) [39].

Experimentally, equilibrium forms are obtained by crystallization or dissolution of spherical crystals, but solvents, impurities, and other factors often change the equilibrium form. Conclusions concerning the equilibrium form thus are more reliable if made on the basis of the shape of a given crystal in its own melt, where the influence of solvents is excluded. However, equilibrium crystal forms in the melt are observed rather rarely, since the form usually is influenced by the temperature field. Under metastable conditions the equilibrium crystal form in its own melt transforms first into cellular and then into skeleton and dendritic forms [40, 41]. Depending on the growth conditions, the form of a crystal that is in equilibrium

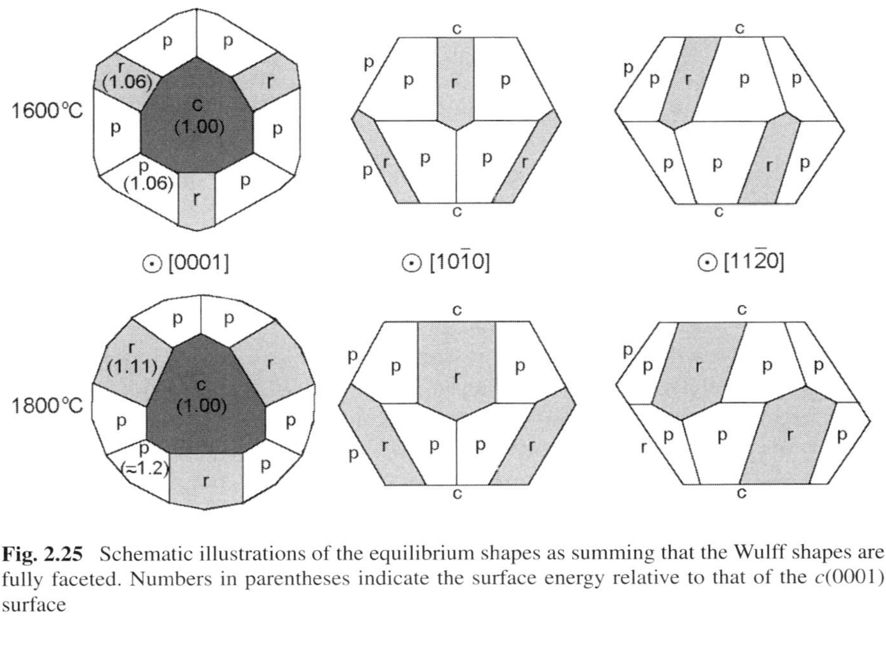

Fig. 2.25 Schematic illustrations of the equilibrium shapes as summing that the Wulff shapes are fully faceted. Numbers in parentheses indicate the surface energy relative to that of the $c(0001)$ surface

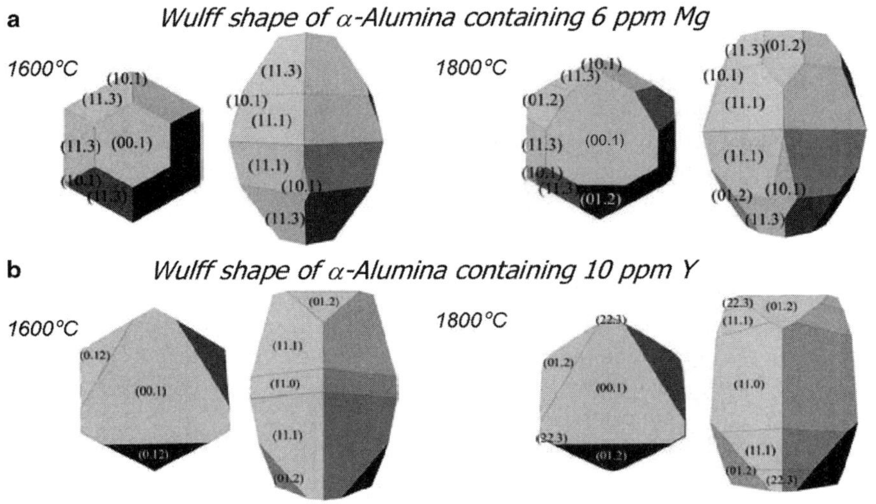

Fig. 2.26 Wulff shape of sapphire containing 6 ppm Mg (**a**) and 10 ppm of Y (**b**)

with the medium may change. However, the form of a skeleton crystal is constant, as it is caused by the properties of the crystal lattice. Therefore, the equilibrium form can be established from the analysis of skeleton and dendrite crystals.

The elementary morphological cell of sapphire corresponding to the positive orientation of the rhombohedron $\{10\bar{1}0\}$ is built on the lattice sites unoccupied by aluminum ions, and describes the arrangement of aluminum ions (Fig. 2.27). The arrangement of these constitutional hollows may be a criterion for estimating the equilibrium form of a crystal. Located inside the elementary morphological rhombohedron are two aluminum ions and three oxygen ions. For the complex Al_2O_3, the growth rate is the highest in the direction of the nearest hollows, which are the rhombohedron vertices. Proceeding from this fact, the morphological rhombohedron $\{10\bar{1}1\}$is to be considered the equilibrium form of sapphire in its own melt [42]. Thus, skeleton and dendrite crystals demonstrate the predominating growth directions that also allow conclusions about the growth rate anisotropy.

Morphology of crystallization front. As is known, in the process of growth the crystal facet does not move uniformly forward; steps or teeth on it appear (Fig. 2.28) [43]. A cellular CF is observed during the growth of sapphire by the horizontally directed crystallization [44] and Stepanov methods [44, 45].

As a rule, the morphology of the CF is investigated on decantation surfaces [43]. In particular, the decantation surface of the crystal grown in the direction $\{10\bar{1}0\}$ consists of vertices of the elementary morphological rhombohedron $\{10\bar{1}1\}$. During growth of sapphire in the direction $[11\bar{2}0]$, the rhombohedron edge lies in the growth plane, so the CF and the decantation surface have a ribbed structure. During growth in the direction [0001] the decantation surface consists of convex and

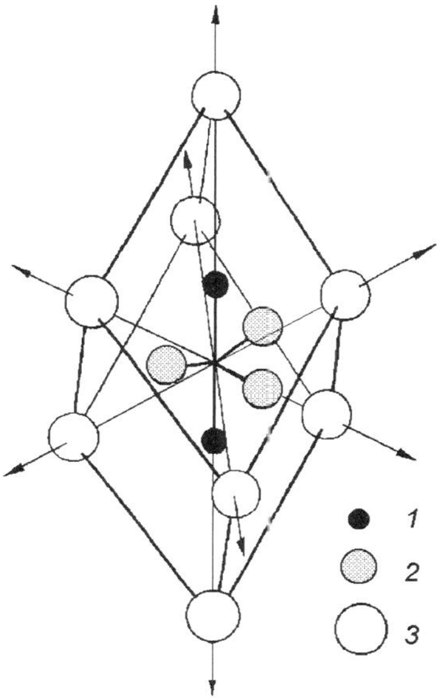

Fig. 2.27 Arrangement of the crystal lattice hollows surrounding the complex Al_2O_3 and the directions of the maximum growth rate; (*1*) aluminum ion, (*2*) oxygen ion, (*3*) octahedral hollows

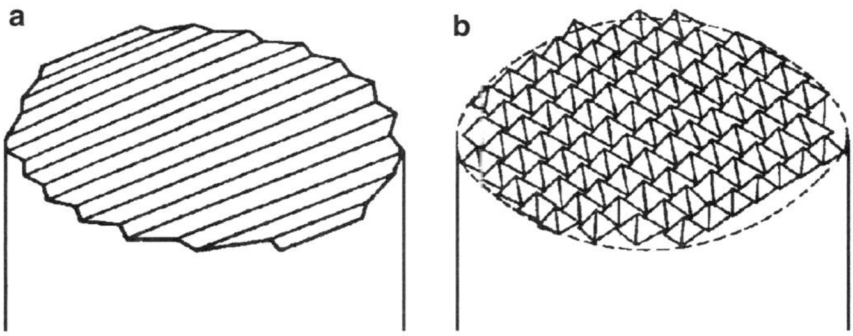

Fig. 2.28 Steps (**a**) and teeth (**b**) on the growing surface of corundum crystals (the Verneuil method; the *C*-axis makes an angle up to 55° with the growth axis)

concave pyramids that correspond to the vertices of the morphological rhombohedron. Thus, the CF of sapphire in its own melt usually is split by the elements of the morphological rhombohedron. The curved surface of the CF is split by those elements of the morphological rhombohedron that coincide with the surface curvature (Fig. 2.29) [46].

Under the conditions of overcooling, skeleton cones are formed on the facets $\{10\bar{1}1\}$due to the priority growth of the vertices and the edges (Fig. 2.30b), which

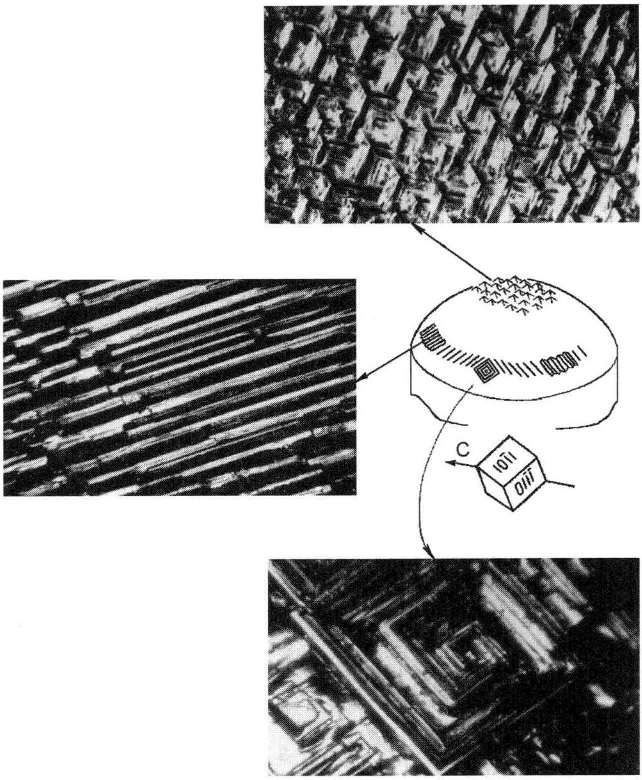

Fig. 2.29 Morphology of the decantation surface of sapphire grown in the direction $[10\bar{1}0]$ (× 50)

may lead further to skeleton crystal growth. Overcooling of the melt favors predominating growth of the rhombohedron vertices and dendritic growth. Characteristically, the branches of the dendrite crystals are oriented in the directions of the rhombohedron vertices and the angles between them correspond to the angles between the rhombohedron diagonals (Fig. 2.30a).

2.1.2 *Optical Properties*

Refraction. In sapphire the refractive index of the ordinary ray is higher than in a great number of optical materials. This is caused by close packing of oxygen ions in the lattice and by ionic polarization. Malitson presented the refractive indexes for

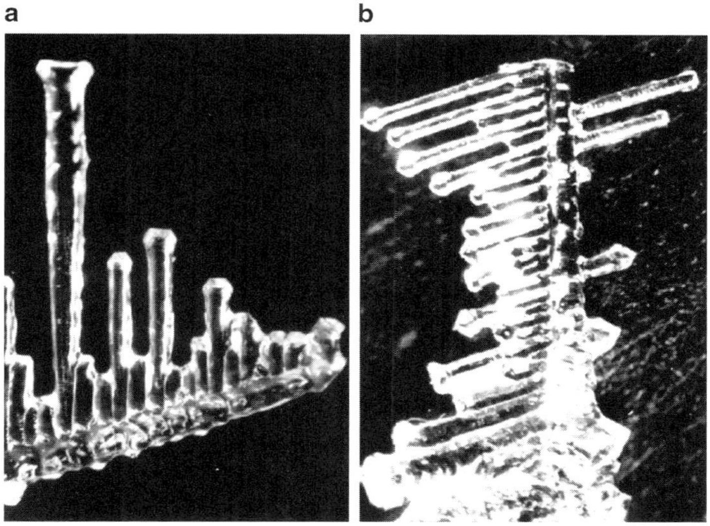

Fig. 2.30 (**a**) Dendritic sapphire branch (×15); (**b**) skeleton cone of sapphire crystal on the plane $(10\bar{1}1)$(×15)

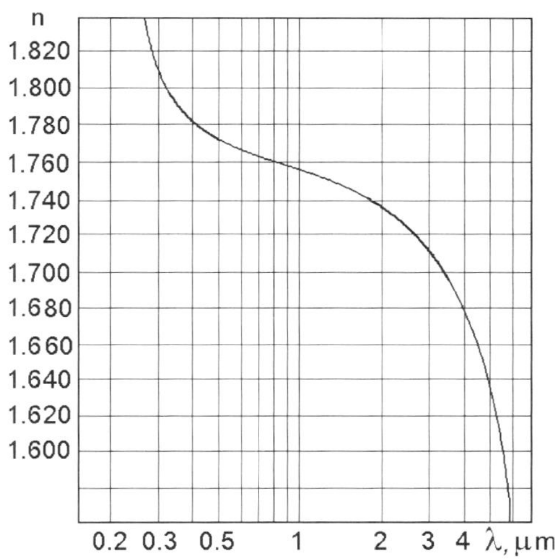

Fig. 2.31 Dependence of the refractive index on wavelength

different wavelengths in the interval 0.2–6 μm [47]. They increase with decreasing wavelength (Fig. 2.31) and reach a maxima of 2.5 in the vacuum ultraviolet (λ = 0.1425 μm). As a rule, the refractive indexes of sapphire in the IR region of the spectra are given without reference to Freneuil loss.

Sapphire is a uniaxial optically negative crystal. It has one direction along the optical axis in which birefringence does not occur and the refractive index of the extraordinary ray is less than that of the ordinary ray. The difference between these values in the visible region of the spectrum is ~0.008. In the UV region, in the vicinity of the maxima point, it grows to 0.012 and drops to zero at the maxima.

Refractive index is temperature-dependent [48, 49]. Figures 2.32 and 2.33 show the temperature-dependencies of the refractive index at 3.39 μm wavelength for the ordinary and extraordinary rays.

Values for the refractive index of sapphire in the region of partial transparency and nontransparency at different temperatures are given in Tables 2.44 and 2.45. The thermo-optical coefficient of the refractive index d*n*/d*t* for the ordinary and extraordinary rays is presented in Table 2.6.

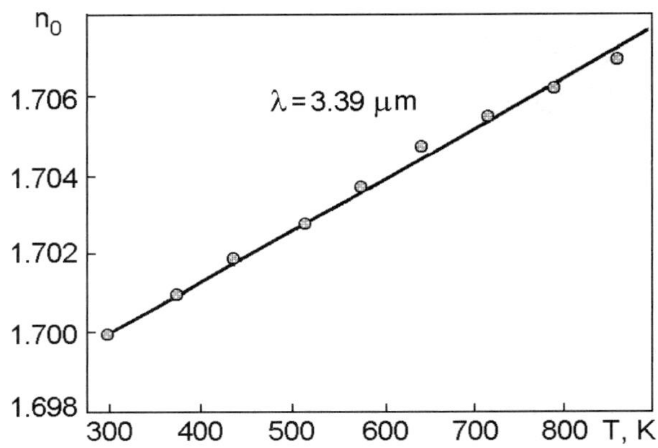

Fig. 2.32 Temperature dependence of the refractive index for the ordinary ray

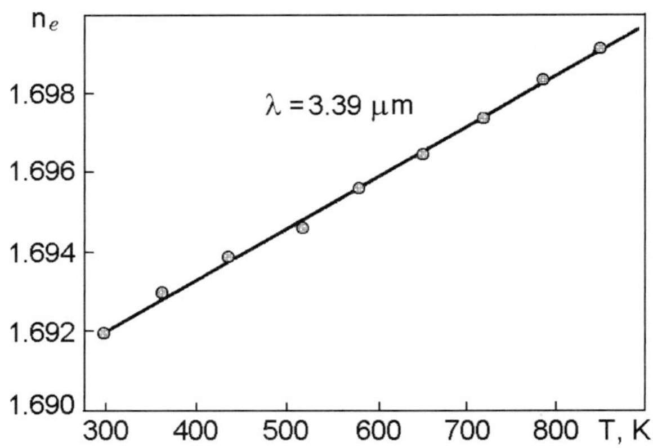

Fig. 2.33 Temperature dependence of the refractive index for the extraordinary ray

Table 2.6 Thermo-optical coefficient for the ordinary (dn_0/dT) and extraordinary rays (dn_e/dT) [49]

Wave-length (μm)	1	1.67	2.5	2.86	3.3	4	5
dn_0/dT (K^{-1})	$8.2 \cdot 10^{-6}$	$8.8 \cdot 10^{-6}$	$9.9 \cdot 10^{-6}$		$1.13 \cdot 10^{-5}$		$1.41 \cdot 10^{-5}$
dn_e/dT (K^{-1})				$1.151 \cdot 10^{-5}$	$1.24 \cdot 10^{-5}$	$1.403 \cdot 10^{-5}$	$1.556 \cdot 10^{-5}$

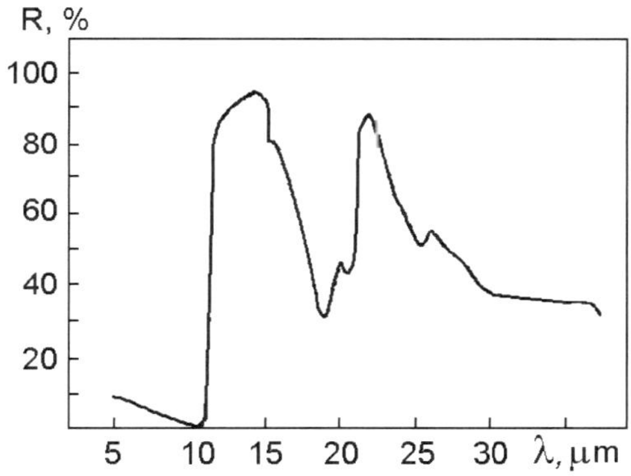

Fig. 2.34 Dependence of the reflection coefficient on wavelength

The equation for refractive index dispersion has the following form:

$$n^2 - 1 = \sum_{i=1}^{3} A_i \lambda^2 / (\lambda^2 - B_i^2), \tag{2.5}$$

where A_i and B_i are constants with $A_1 = 1.023798$, $A_2 = 1.058264$, $A_3 = 5.280792$, $B_1 = 0.06144821$, $B_2 = 0.1106997$, and $B_3 = 17.92656$.

Reflection. The reflection coefficient of a polished sapphire surface ($R_z = 0.1$ μm) at $n = 1.768$ is equal to 7.8%. Reflection essentially is dependent on the wavelength (Fig. 2.34) and the state of the surface, which changes with mechanical and thermal treatment. In the IR region of the spectrum, the reflection coefficient sharply increases at $\lambda = 11$ μm and its maxima are observed at 13.5 and 22 μm. The values of effective coefficient of mirror reflection of a sapphire surface subjected to different types of mechanical treatment (R_0) and thermal treatment in aluminum oxide vapor at 2,250 K for 1.5 h (R_1) and 8 h (R_2) are presented in Table 2.7.

Table 2.7 Coefficients of effective mirror reflection of the surface depending on its roughness and the angle of ray incidence (in %)

θ (deg.)	R_z (μm) 0.1–0.2			1.1–2.5			6.3–10		
	R_0	R_1	R_2	R_0	R_1	R_2	R_0	R_1	R_2
10	8.0	7.8	7.8	0.06	0.06	0.58	0.02	0.04	0.18
20	8.0	7.8	7.8	0.06	0.068	0.60	0.02	0.05	0.18
30	8.0	8.3	8.3	0.10	0.80	0.70	0.04	0.10	0.20
40	10.3	9.5	9.5	0.12	1.00	0.80	0.06	0.18	0.30
50	12.0	11.0	12.0	0.18	1.30	1.10	0.14	0.26	0.50
60	17.5	17.0	17.5	0.30	1.70	1.40	0.34	0.40	0.80
70	27.5	22.5	27.0	0.66	3.7	2.00	0.7	1.00	1.38
80	48.0	45.0	48.0	1.86	15.00	10.40	1.60	1.70	1.90

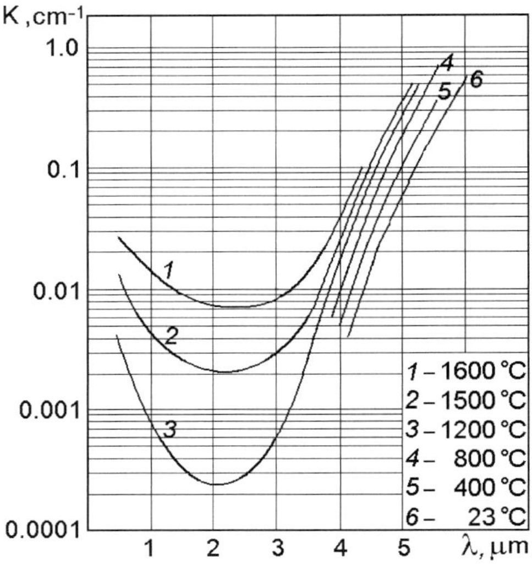

Fig. 2.35 Temperature dependence of the absorption coefficient

Absorption. The absorption coefficient of sapphire is temperature-dependent (Fig. 2.35) [50]. In the IR region, the value of absorption rises with temperature and reaches 0.28 cm^{-1} in the region of premelting temperature.

Structural perfection and impurity composition of crystals essentially affect their absorption. In particular, a considerable value of the absorption coefficient of ruby in the visible region of the spectrum is caused by Cr^{3+} ions. The absorption coefficients at the absorption band maximum for the ordinary and extraordinary waves are presented in Table 2.46.

Ruby is pleochroic, i.e., absorbs the ordinary and extraordinary rays in a different way. In the direction perpendicular to the optical axis the crystal has an intense violet-red color. In the direction parallel to the optical axis its color is yellowish-red and less intense. The absorption spectra of crystals with different impurities are presented in Fig. 2.110.

The absorption coefficient of sapphire is investigated in many papers. Despite disagreement in the results, they have been generalized [51] (Table 2.47). The general scheme for determination of the recommended values is the following. Based on the available literature data on the temperature dependence of the absorption coefficient and taking into account the error of the measurement method, the averaged dependence $k_\lambda(T)$ is built for each wavelength with a step of 0.1 μm. This is used to build the absorption spectra $k_\lambda(\lambda)$.

The absorption coefficients of sapphire are proportional to the following powers of temperature $(T^n)/n = 2.2$ for $\lambda = 4.2$ μm, $n = 2$ for $\lambda = 4.5$–5.5 μm, $n = 1.5$ for $\lambda = 6$ μm, $n = 1.2$ for $\lambda = 6.5$ μm, and $n = 1$ for $\lambda = 7$ μm.

The recommended values of absorption coefficient obtained from an analysis of literature data are to be considered as reference only. For some spectral temperatures, these have been extrapolated.

Transmission. As a rule, transmission in sapphire is described by the curve presented in Fig. 2.36. It was obtained for pure, sufficiently perfect crystals grown by the Czochralski method (Linde Cz UV grade). The samples were 1 mm thick. In the UV region, transmission of thin plates is essentially limited by the Fresnel loss. The presented values do not take into account this loss. Recent data on the transmission of pure perfect crystals in the UV region (UV grade) are presented in Fig. 2.37.

Transmission in the UV region is increased by high-quality polish of the surfaces (20/10 scratch-dig). Low-quality polish may diminish the value of transmission by ~10%. The quality of the surface polish is especially significant for $\lambda < 250$ μm. The presence of point defects and impurities gives rise to absorption at 204, 230, and 400 μm wavelengths. For real samples the value of transmission also depends

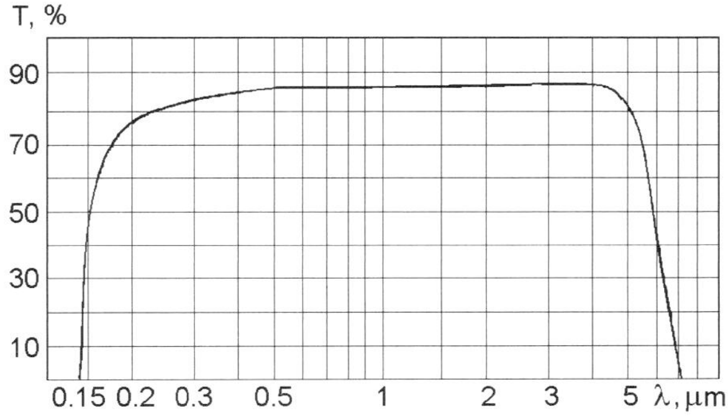

Fig. 2.36 Transmission of sapphire

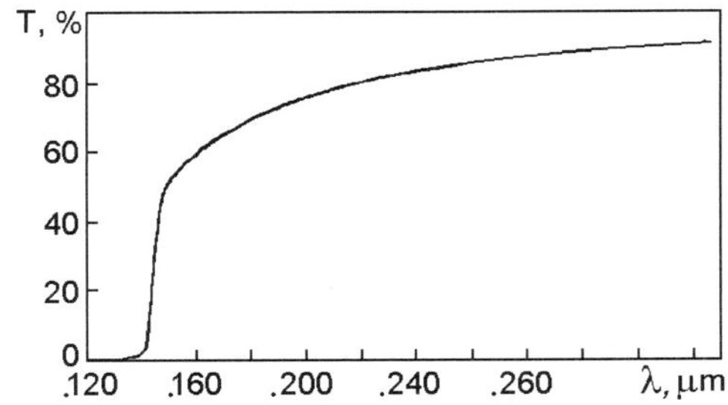

Fig. 2.37 Transmission of sapphire in the UV region

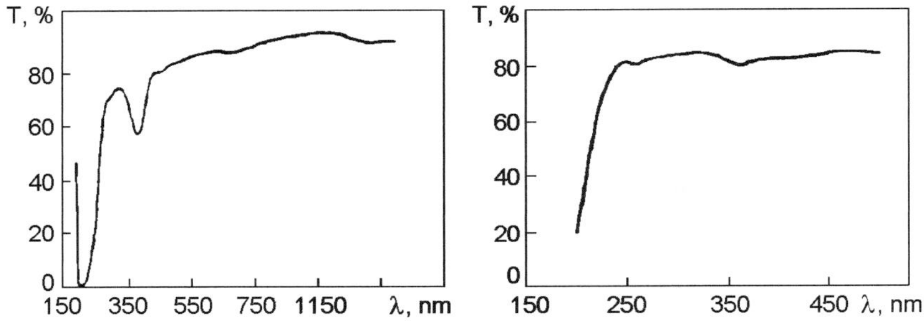

Fig. 2.38 Transmission spectra of 10-mm-thick sapphire grown using (**a**) molybdenum and (**b**) tungsten components

on the medium in the growth chamber and even the equipment components used (Fig. 2.38, [52]).

The X- and γ-irradiation of sapphire increases optical absorption; *n*- and γ-irradiation give rise to the formation of color centers and consequently to the appearance of absorption bands, which can be eliminated with high-temperature annealing.

In the IR region of the spectrum, sapphire possesses high transmission at wavelengths ranging from 1.0 to 5.5 μm. In the 1.0- to 4.0-μm interval, the value of transmission is ~85–86% (Fig. 2.39). Starting from $\lambda > 3$ μm the value of transmission decreases as temperature grows (Fig. 2.40). At $T < 80$ K, the transmission of sapphire increases in the extreme IR region (Fig. 2.40). Due to such an effect, sapphire can be used in cold inner windows for IR measurements.

As sapphire possesses high transmission without taking into account Fresnel loss, it can be applied in those cases when other optical materials require antireflection coatings. These coatings are able to decrease the loss by reflection and essentially increase transmission (Fig. 2.41). Starting from $\lambda > 3$ μm, the value of transmission decreases with the growth of temperature (Fig. 2.42).

Scattering in sapphire depends on the wavelength. Relative scattering diminishes with increasing wavelength and reaches its maximum at $\lambda = 1.3$ μm.

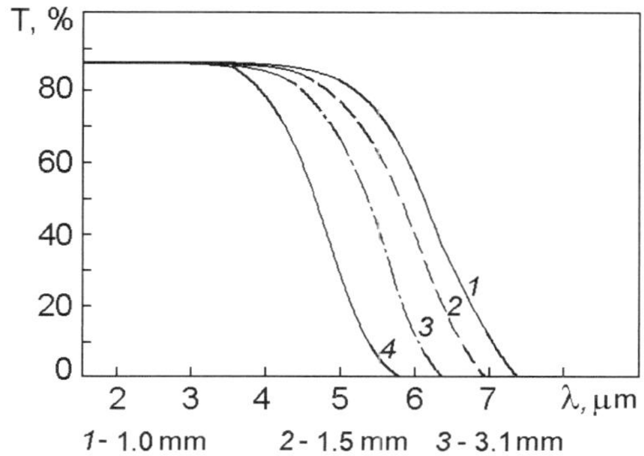

Fig. 2.39 Transmission in IR region. Thick (*1*) 1 mm, 293 K; (*2*) 1.5 mm, 293 K; (*3*) 3.1 mm, 293 K; (*4*) 4 mm, 793 K

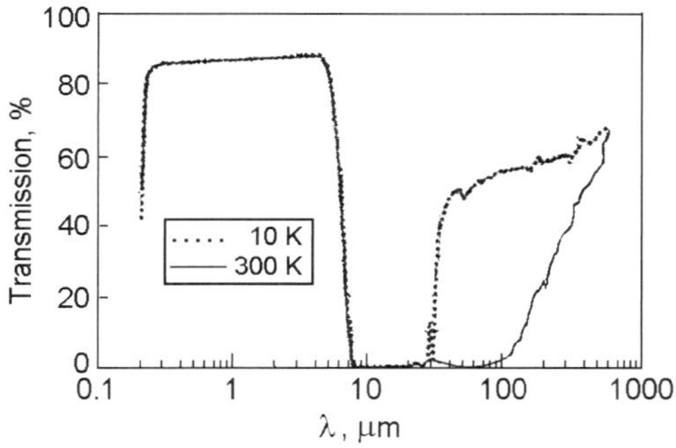

Fig. 2.40 Transmission of 2-mm-thick sapphire in the IR region at 10 and 300 K

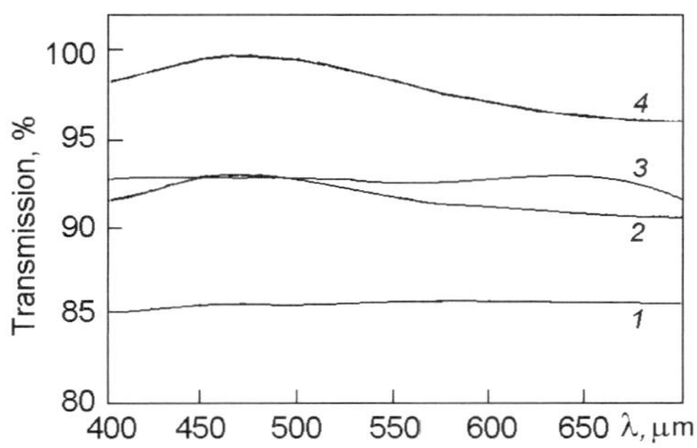

Fig. 2.41 Antireflection coating external transmission: (*1*) no coating; (*2*) one side AR coating; (*3*) one side multicoating; (*4*) both sides AR coating

Photoelastic constant	$2.1 \cdot 10^{-7}$ cm^2/kg [1]
Brewster angle	60.4°
Total internal reflection angle	34.5°
Abbe number ($n_0 - 1/n_f - n_c$)	72.20 (parallel to *C*)
Dispersive power ($n_f - n_c$)	0.011

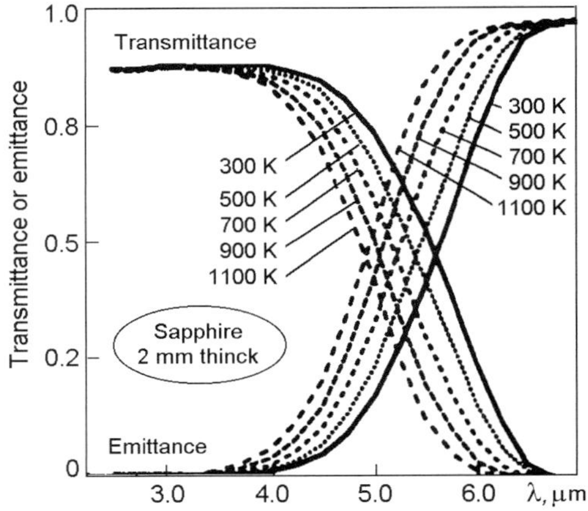

Fig. 2.42 Transmittance and emittance normal to surface (|| *C*-axis of the crystal)

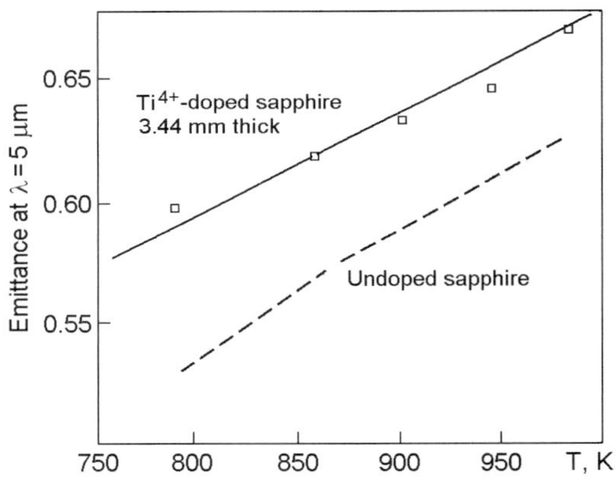

Fig. 2.43 Emittance of Ti^{4+}-doped sapphire

Emittance. Emittance is the ratio of the radiative energy of a body to the corresponding energy of black body radiation at the same temperature. Emission remains low up to $\lambda\sim4$ μm (Fig. 2.42). It increases with wavelength and temperature. Doping of sapphire increases emittance (Figs. 2.43 and 2.44) [53].

Optical properties in the region of phase transition undergo changes due to a considerable distinction in the properties of the liquid and solid phases. For instance, at $\lambda = 0.6$ μm in the visible region of the spectrum the scattering coefficient (K) of the melt is ~25 cm^{-1} at 2,400 K, whereas at 2,300 K (solid phase) $K = 0.36$ cm^{-1}. For the same wavelength the refractive index at the phase transition changes from 1.78 to 1.81 in a narrow temperature interval (60–70°). In the 0.3- to 30-μm region the values of K differ by 2–2.5 orders. As seen from the optical property behavior of sapphire crystal and its melt in the region of the phase transition

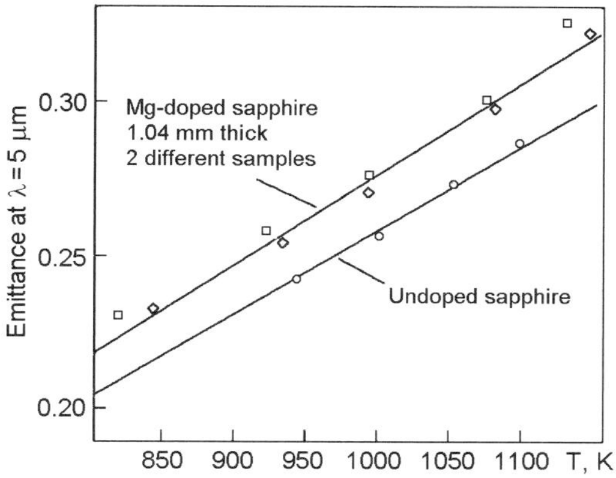

Fig. 2.44 Emittance of Mg-doped sapphire

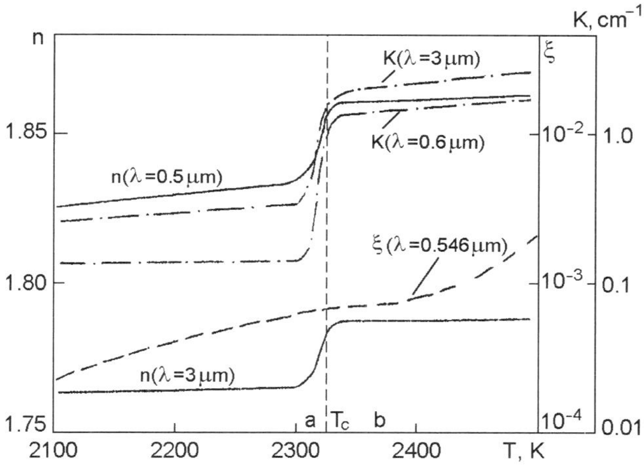

Fig. 2.45 Temperature-dependent behavior of the refractive index (*n*), absorption index (ξ), and absorption coefficient (K) for (**a**) sapphire and (**b**) its melt in the region of the phase transition (for 25-mm-thick melt-crystal layer)

(Fig. 2.45), it is only the absorption coefficient that smoothly changes, but it has no essential influence on the radiation-conductive heat transfer in this region [51].

Luminescence of sapphire is caused by impurities or lattice defects. Several weak luminescence bands are observed in sapphire. Their intensity depends on the type of impurity and the crystal's history. Luminescence in the region of 290–335 and 650–774 nm is related to the lattice defects and impurity ions, respectively. In particular, chromium ions emit in the region of 650–774 nm; manganese ions emit in the region of 680 nm; the 620-nm line is attributed to vanadium ions; and the

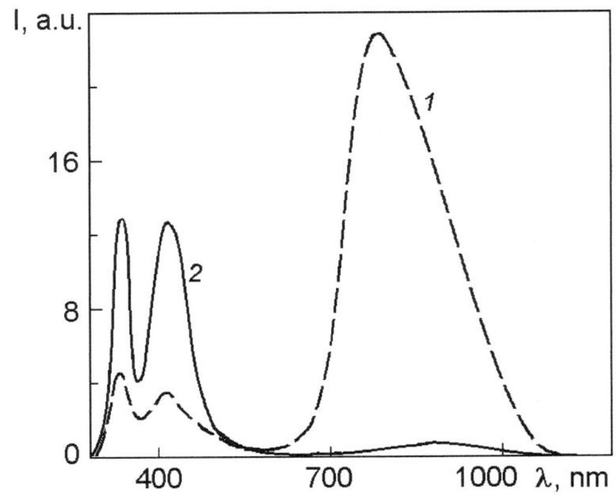

Fig. 2.46 Cathodoluminescence spectrum of (*1*) nonirradiated sapphire and (*2*) sapphire preliminarily irradiated on a linear accelerator

480-, 530-, and 600-nm lines correspond to divalent cobalt ions. The most widespread luminescent impurities are chromium and titanium. The Cr:Al_2O_3 crystals possess intense RL with a dominating band at 695 nm (1.78 eV), prolonged afterglow at room temperature (exceeding 10 h), considerable accumulated light sum, and typical TSL peak at 300°C, the intensity of which correlates with the content of chromium.

The comparison of sapphire grown by different methods [54] shows that crystals obtained by hydrothermal methods possess the minimum accumulated light sum, weak glow at laser, X-ray, electronic excitation, and fast afterglow decay. Meanwhile, these crystals have the lowest radiation resistance. As a rule, irradiation of the crystal and increasing temperature increase the glow intensity of some bands or change the ratio of their intensities. Only in Ti^{3+}:Al_2O_3 does the luminescence intensity at nitrogen temperatures turn out to exceed that at room temperature [55]). Figure 2.46 gives a characteristic example of the behavior of the glow band intensities ratio [56].

The intensity of the low-energy part of the luminescence spectrum sharply diminishes at helium temperatures, as the phonon component of luminescence is frozen out. Many types of luminescence (e.g., cathodoluminescence and photoluminescence) in nonactivated crystals have the same nature. Therefore, the absence of photoluminescence at cathodoluminescence reported in some papers can be referred to the distinction in the density of excitation.

TSL of nonactivated sapphire is distinguished by the presence of a thermopeak at 408 K. In UV-irradiated Ti:Al_2O_3, TSL is characterized by thermopeaks with the maxima at 423 and 495–500 K. The TSL spectrum consists of 290- and 410-nm bands. The intensity and spectrum of TSL depend on the titanium content and on the oxidation–reduction potential of the crystal growth and annealing media. To explain the dependence, three models of activation–vacancy centers are proposed [57].

The TSL spectra of sapphire irradiated with neutrons and excited with X-rays contain bands at 2.4, 3.7, and 2.8–3.1 eV, which coincide with the luminescence bands at photoexcitation. After proton irradiation the luminescence bands at 260, 330, 390, 410, and 510 nm were registered [58]. Several luminescence (emission) bands (3.80, 3.18, 2.82, 2.43, 2.23 eV) appeared, irrespective of thee sample's prehistory, at the excitation of sapphire irradiated by neutrons [59].

Besides luminescence in thermal equilibrium with the lattice, nonequilibrium luminescence or emission upon relaxation of electrons on electron and electron-oscillating levels also are observed on intense pumping with a Cr:Al_2O_3 laser. This luminescence possesses a continuous spectrum, and its intensity is defined by the wavelength of the exciting laser and is by several orders weaker than equilibrium luminescence.

The luminescence caused by lattice defects of sapphire grown under thermodynamically nonequilibrium conditions often is observed in different regions of the spectrum. The 330 nm band is caused by the main radiative transition IB → IA in F^+ centers. Emission in the 450- to 660-nm region is related to *R* centers (three unified *F* centers, e.g., three anion vacancies that captured three electrons) [60]. Glow in the 420-nm region also is related to aggregate centers. According to data reported [61], this is a P^- center (the neighboring anion and cation vacancies). The capture of an electron by this pair leads to the formation of a *P* center, whereas the capture of a hole leads to the appearance of a P^{2-} center.

Emission in the 510-nm band is observed in neutron-irradiated samples, attributed to interstitial Al^+ [62]. This ion can be stabilized by electron capture. It has been identified [63] according to the absorption band at 4.1 eV, the luminescence band at 2.45 eV, and a narrow luminescence band at 3.8 eV. Figure 2.47 shows the proposed scheme of its energy levels.

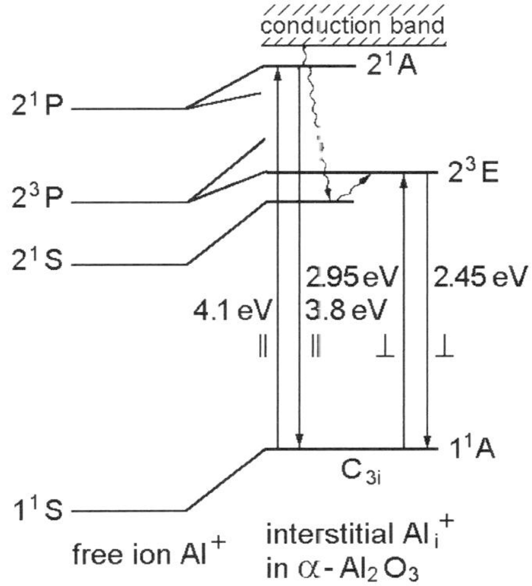

Fig. 2.47 Scheme of energy levels of Al^+ centers

Emission in the 550- and 595-nm bands observed in nonirradiated crystals are related to vacancy aggregates [64]. This is accompanied with glow in the 329- to 335- and 388- to 420-nm bands also related to vacancy centers. During γ-irradiation, the charge is transferred from centers that emit in the region of 550–595 nm to centers with emission in the 510-nm band.

It is difficult to compare quantitative luminescence results obtained by different researchers, as these results may reflect random features of the studied samples. The influence of anionic and cationic nonstoichiometry on luminescence was investigated [65] on samples with the same composition and initial structural perfection (Figs. 2.48–2.50).

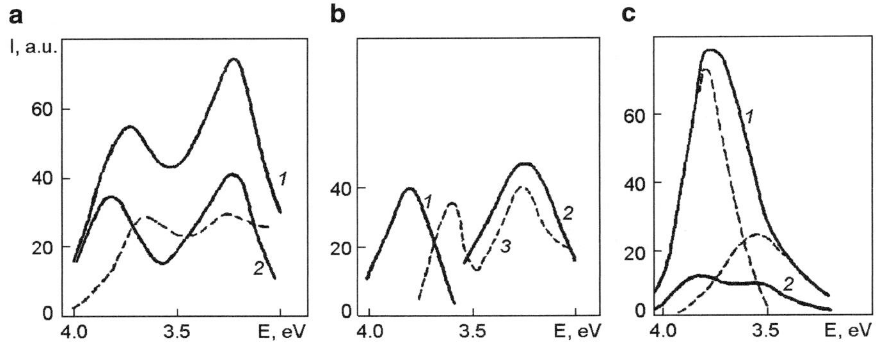

Fig. 2.48 Spectrum of fast-decaying luminescence for α-Al_2O_3 with anionic nonstoichiometry. (**a**) T = 80 K. (*1*) spectrum measured at the instant of termination of excitation pulse (5 ns); (*2*) since 50 ns. *Dashed line* shows the difference spectrum (*1–2*). (**b**) T = 80 K. Resolution of the luminescence spectrum into components with different decay times τ, ns: (*1*) 50, (*2*) 30, (*3*) short component with $\tau \leq 5$ ns. (**c**) T = 295 K. (*1*) at the instant of termination of excitation pulse, (*2*) since 200 ns. *Dashed line* corresponds to resolution of the spectrum into the components

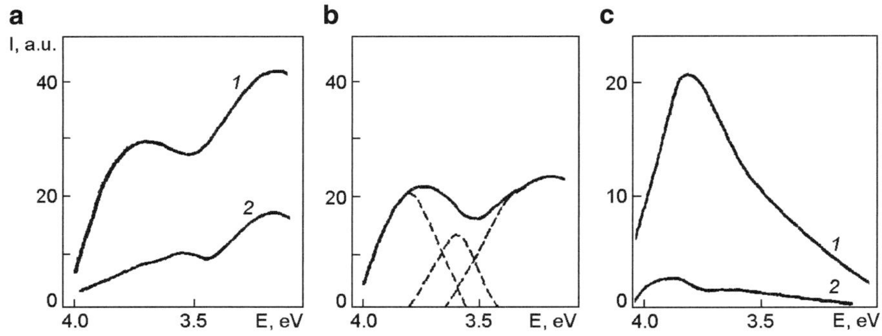

Fig. 2.49 Spectrum of fast-decaying luminescence for α-Al_2O_3 with cationic nonstoichiometry. (**a**) T = 80 K; (*1*) spectrum measured at the instant of termination of excitation pulse; (*2*) since 25 ns. (**b**) T = 80 K. Resolution of the difference spectrum (*1–2*) into components. (**c**) T= 295 K; (*1*) at the instant of termination of excitation pulse; (*2*) since 400 ns

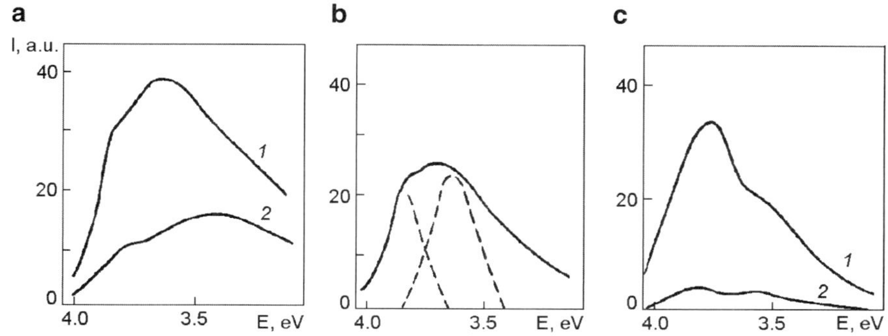

Fig. 2.50 Spectrum of fast-decaying luminescence for α-Al_2O_3 with anionic nonstoichiometry (normalized sample). (**a**) T= 80 K; (*1*) spectrum measured at the instant of termination of excitation pulse; (*2*) since 25 ns. (**b**) T= 80 K. Resolution of the difference spectrum (*1–2*) into components. (**c**) T= 295 K; (*1*) at the instant of termination of excitation pulse; (*2*) since 400 ns

Table 2.8 Spectral and kinetic characteristics of centers responsible for luminescence bands 3.8 eV in sapphire

Type of center	Luminescence band (eV)	W (eV)		t (s)	
		80 K	300 K	80 K	300 K
F^+	3.8	0.34	0.41	$7 \cdot 10^{-9}$	$7 \cdot 10^{-9}$
Al^+	3.82	0.18	0.22	–	–
	2.45	–	–	–	$5 \cdot 10^{-2}$
ALE	3.8	0.35	0.46	$2.8 \cdot 10^{-8}$	$2.8 \cdot 10^{-8}$

As seen from the figure, the luminescence intensity is highest in the crystals with anionic nonstoichiometry, while in the crystals with cationic nonstoichiometry it is the lowest. The luminescence decay times differ significantly, too. It is interesting to investigate the luminescence with E_m = 3.8 eV. The luminescence band may be caused by F^+ centers, interstitial Al^+ ions, and autolocalized excitons (ALE) in the anionic sublattice. The glow of these defects has different band half-widths (W) and intercenter decay times (Table. 2.8).

Clarification of the nature of the defects responsible for the observed glow is possible through comparison of Table 2.8 data and that of Figs. 2.40–2.42 with the spectral and kinetic parameters of the 3.8-eV band luminescence in the time interval 10^{-9}–10^{-7} s for the crystals with different nonstoichiometry (Table 2.9).

The fast-decaying 3.8-eV luminescence probably is caused by interstitial aluminum. The half-width of the observed glow band is significantly smaller and closer to that of the Al^+ center glow band than to other centers (Table 2.8). The half-width of the luminescence band (0.25 eV), which is larger than for Al^+ center luminescence (0.2 eV), is explained by the fact that the defect is not isolated, but forms the Frenkel pair Al^+–V_{Al}^{3-}.

Table 2.9 Spectral and kinetic parameters of 3.8 eV luminescence band [65]

	W (eV)		*t* (s)		*I* (rel.unit)	
Type of sapphire	80 K	300 K	80 K	300 K	80 K	300 K
Stoichiometric	0.25	0.38	30	200	25	30
Anionic	0.25	0.25	50	35	40	40
nonstoichiometry	–	0.4	–	200	–	40
Cationic	0.3	0.4	35	200	20	20
nonstoichiometry						

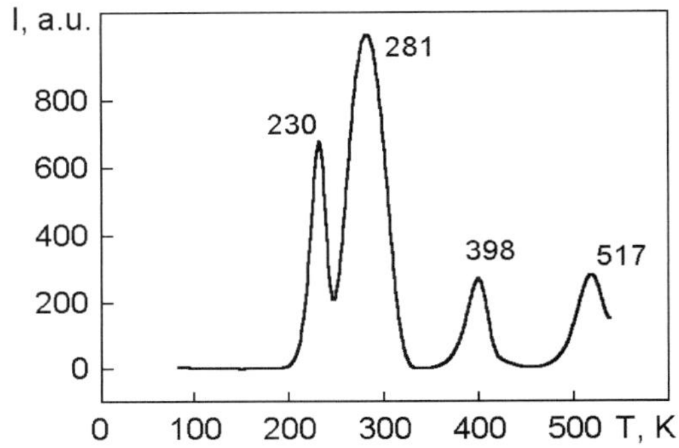

Fig. 2.51 Thermoluminescence curve for sapphire. The heating rate is 0.15 K/s

The threshold energy of the aluminum ion shift into the interstice is 18 eV [66]. Such a shift can be realized by a 300-keV electron. The concentration of these defects is less than ~10^{12} cm^{-3} per pulse. As evident from general considerations, the movement of the cation to the interstice is promoted by the presence of free space formed by anion vacancies in the nearest coordination sphere. Therefore, at higher V_O^{2+} concentrations the shift of the aluminum into the interstice leads to the formation of a pair of defects with an increased distance between pair components. This conclusion agrees with the fact that the lifetime of a Frenkel pair in samples with an ionic nonstoichiometry at 80 K (50 ns) exceeds that of such a pair in stoichiometric samples (~30 ns). Besides active laser media, the luminescent (thermoluminescent) properties of sapphire and ruby are used in radiation dosimeters (see Chap. 1), pressure gauges, luminescence receivers, and ionizing radiation converters.

The thermoluminescence spectrum of undoped sapphire is characterized by several bands (Fig. 2.51). In the low-temperature region, a peak at 100 K is observed in all samples. Its intensity is very low and can be recorded only at considerable amplification. Two other peaks have maxima within the 223–240 and 261–281 K ranges depending on the individual sample. They overlap in part, so the maximum temperature shifts. The intensity ratio between those peaks depends on the excitation duration. Several partially overlapping peaks are observed in the 350–600 K range [67].

Studies of the 280-K thermoluminescence peak decay kinetics show it has a complex structure and is a superposition of at least two neighboring elementary peaks. As the light sum increases, the intensity of the low-temperature component rises faster than that of the main component. The studied trapping centers (TC) are shown to be defects present in the samples prior to excitation. The values of TC activation energy in undoped sapphire form a single oscillatory series, the oscillatory quantum energy $h\omega_{TL}$ being 0.079 eV (642 cm^{-1}).

2.1.3 Mechanical Characteristics

Density, measured by the method of hydrostatic weighing to an accuracy of 0.05%, varies from 3.97 to 3.99 g/cm^3 for sapphire and to 4.013 g/cm^3 for dark-colored ruby containing 2.97% of Cr_2O_3. Its dependence on the content of chromium is linear. Some impurities, such as titanium, calcium, and so forth, diminish the value of density. Calculated values of density somewhat exceed experimental values, and this is connected to the presence of micropores, microcracks, and so forth in the real crystals.

Hardness. Mohs' hardness is equal to 9; the value of hardness measured according to the 15-point Ridgway scale is 12. Knoop hardness measured parallel and perpendicular to the *C*-axis is 1,900 and 2,200 kg/mm^2, respectively. Knoop hardness values range from 1,525 to 2,200 kg/mm^2. The values of microhardness measured by microindentation of the surface are 19.40 GPa (parallel to the optical axis), 23.15 GPa (at an angle of 60° with the axis), and 22.0 GPa (perpendicular to the axis). The microhardness of $\alpha\text{-}Al_2{}^{18}O$ is noticeably higher (the bond $Al_2\text{–}{}^{18}O$ is shorter than the bond $Al_2\text{–}{}^{16}O$).

The friction coefficient has been investigated under a load applied perpendicularly to polished end surfaces. The lower component of the friction pair was rotating, whereas the upper remained immovable. The sliding friction coefficient of such sapphire pairs depends on orientation (Fig. 2.52) [68]. For friction against steel it is dependent on the pressure and the lubricant used (Table 2.10). The abrasion resistance (according to Mackensen) is 0.12 mm.

The elastic constants of sapphire measured at room temperature [69, 70] and the temperature dependence of pliability are presented in Tables 2.48 and 2.49.

The values of ultimate strength from different types of sapphire testing are contained in Table 2.57.

Tensile strength is 275–400 MPa ($4\text{–}6 \cdot 10^4$ psi);

Bending strength is 450–895 MPa. According to other sources, the bending strength in the direction parallel to the *C*-axis is 1.03 GPa ($1.5 \cdot 10^5$ psi) and in the direction perpendicular to the *C*-axis is 758 MPa ($1.1 \cdot 10^5$ psi) (catalogues of leading sapphire producers).

Compression strength is 2 GPa ($3 \cdot 10^5$ psi).

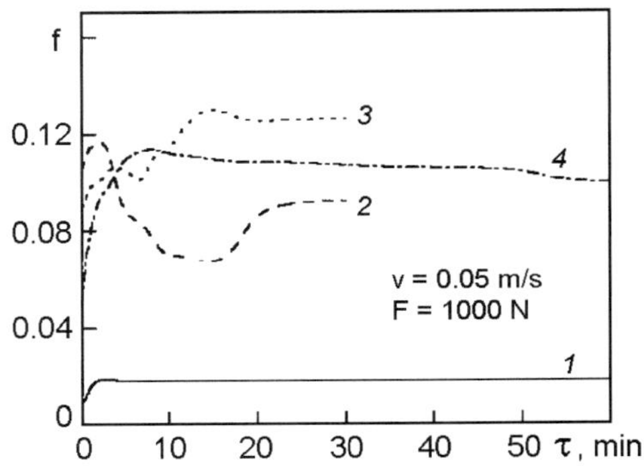

Fig. 2.52 Influence of the crystallographic orientation on the friction coefficient for sapphire pairs. (*1*) (0001), (*2*) (10$\bar{1}$0), (*3*) (11$\bar{2}$0), (*4*) (10$\bar{1}$2). For all cases the second component was a sample with a surface parallel to the plane (0001)

Table 2.10 Friction coefficients at different pressures

P	*P/S* (GPa)	Sapphire–steel	Ruby–steel
5	1.7	0.33/0.13	0.40/0.12
10	2.6	0.31/0.15	0.37/0.15
15	3.0	0.28/0.17	0.30/0.12
20	3.3	0.25/0.15	0.28/0.11
25	3.5	0.23/0.15	0.26/0.12
30	3.7	0.22/0.14	0.25/0.12

The magnitudes in front of the slash correspond to the dry friction coefficients; those behind the slash are the coefficients after clock mechanism oiling

Young's modulus (elasticity modulus) is the value opposite to pliability: $E = 1/S_{33}$. For different crystallographic directions this value is the following $E_{[0001]} = 1/S_{33}$, $E_{[11\bar{2}0]} = 1/S_{11}$. Presented below are the Young's moduli (GPa) measured by different authors [1]:

Crystallographic direction	Young's Modulus			
[0001]	469.7	467.0	461.2	328.0
[11$\bar{2}$0]	494.0	447.1	358.9	324.0
[10$\bar{1}$0]	–	–	–	322.0

The value of the Young's modulus is temperature-dependent: $E_{[0001]} = 435$ GPa ($63 \cdot 10^6$ psi) at 323 K, $E_{[0001]} = 386$ GPa ($56 \cdot 10^6$ psi) at 1,273 K. At present, the value of the Young's modulus most often used in calculations is 345 GPa ($50 \cdot 10^6$ psi).

Compression modulus is 250 GPa ($36 \cdot 10^6$ psi).

Shear modulus (rigidity modulus) is 145 GPa ($21 \cdot 10^6$ psi)–175 GPa ($26 \cdot 10^6$ psi).

Rupture modulus is 350–690 MPa (5–10 $\cdot$ 10^4 psi).

Weibull modulus is the probability of brittle failure for a material under a stress exceeding its ultimate strength σ_S. If P_0 is the probability of failure per unit volume of material, then the probability of failure for a sample with volume V is found from the equation

$$(1 - P_0)^V = 1 - P \tag{2.6}$$

The expression $1-P_0$ is the probability of survival per unit volume of material under the stress σ_s; $1-P$ is the probability of survival for the entire sample with volume V under the same stress.

Weibull's theory describes the probability of cumulative failure P as a function of the applied tensile stress σ and predicates on tensile-stress initiated fracture at strength-limiting surface flaws. Depending on the state of the surface (and even of the ends) and the sample's history, the Weibull modulus may vary from 5.38 (as grown) and 6.35 (annealed) [71] to 6.07 upon chemical–mechanical polishing [72] and even up to 9.79 [73].

As shown [73], chemical–mechanical polishing can improve the high-temperature strength characteristic of *C*-plane sapphire by 150% and the room-temperature Weibull modulus by 100% (Figs. 2.53 and 2.54). The results were obtained on samples polished in different companies.

Poisson coefficient is 0.27–0.30, depending on the crystallographic orientation. The strength characteristics of sapphire are structure-sensitive, and this fact explains essential distinctions in the results of measurements performed by different authors. The temperature-dependence [1] of reduced critical chipping stress and the ratio of chipping yield stresses for ruby of 90- and 60-degree orientation are presented in Figs. 2.55–2.57.

Activation energies E and activation bulk V^* depend on the stress τ^* (Fig. 2.58). The temperature dependence of activation energy has two regions: below 1,800 K, where the value of $E^* = 335–500$ kJ/mol, and above 1,800 K, where the value of activation energy grows to 3,350 kJ/mol. Based on the value of activation energy, the character of its dependence on temperature and stresses, one can expect the

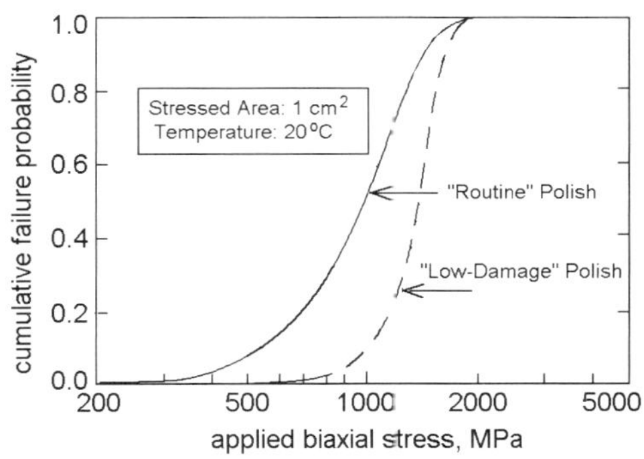

Fig. 2.53 Cumulative failure probability of routinely polished and "low-damage" polished *C*-plane sapphire

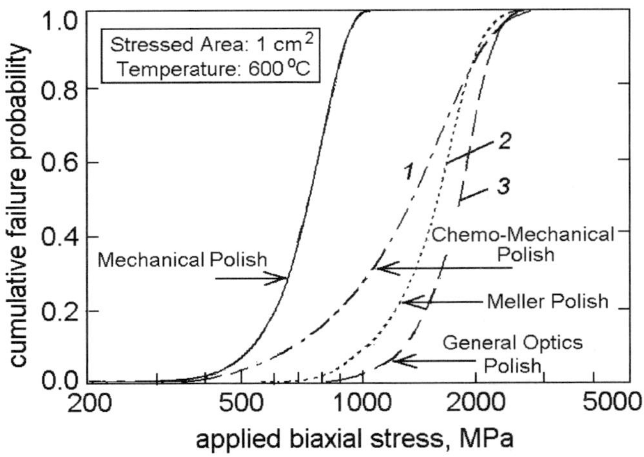

Fig. 2.54 Cumulative failure probability of mechanically and chemical–mechanical polished *C*-plane sapphire. (*1*) Roughness (rms nm) ≥1, measured strength (MPa) 1,082 ± 425, Weibull modulus 2.80 ± 0.42. (*2*) Roughness (rms nm) ≥2.5, measured strength (MPa) 1,330 ± 313, Weibull modulus 4.860 ± 0.31. (*3*) Roughness (rms nm) ≤1, measured strength (MPa) 1,534 ± 18, Weibull modulus 6.07 ± 0.66

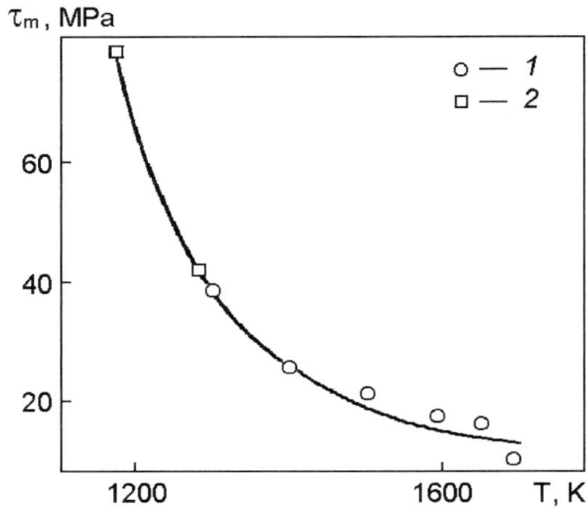

Fig. 2.55 Temperature dependence of reduced yield stress for sapphire: (*1*) tension; (*2*) bending

existence of two mechanisms of plastic deformation in sapphire. At temperatures up to 1,800 K, the thermally activated process of moving over the Peierls-Navarro barriers may take place; at higher temperatures, dislocation creep seems to be more probable.

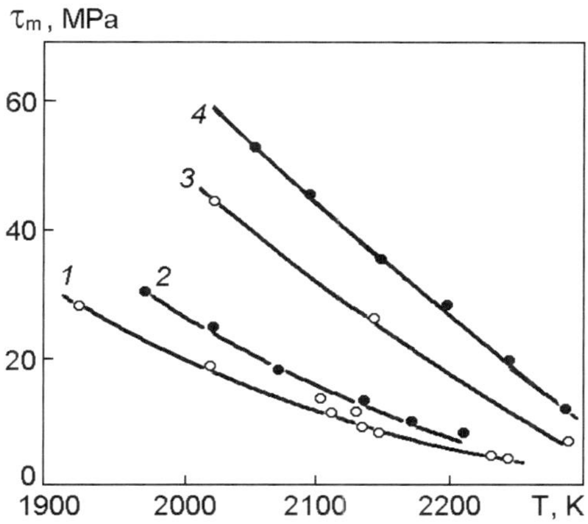

Fig. 2.56 Temperature dependence of critical chipping yield stress (extension rate is $8.7 \cdot 10^{-5}$ c^{-1}): (*1*) sapphire of 60-degree orientation; (*2*) the same, ruby; (*3*) sapphire of 90-degree orientation; (*4*) the same, ruby

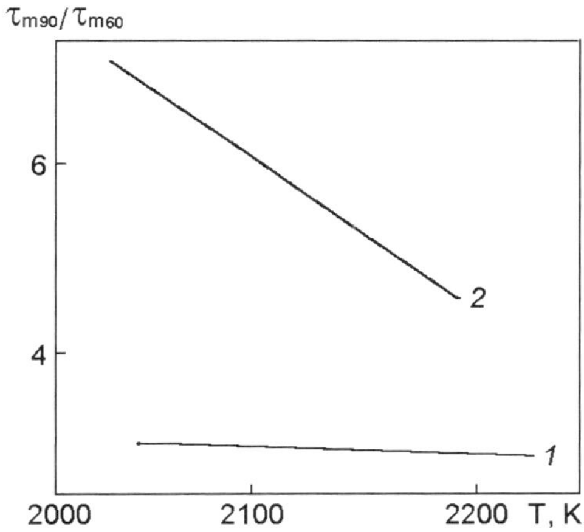

Fig. 2.57 Dependence of the ratio of chipping yield stresses for 90- and 60-degree orientation ruby on temperature and on deformation rate: (*1*) extension rate $8.7 \cdot 10^{-5}$ s^{-1}; (*2*) $5.3 \cdot 10^{-6}$ s^{-1}

2.1.4 Dynamic Strength of Sapphire

Understanding the processes of material destruction under high-intensity pulsed loads is of great importance for military, space, and explosion technologies and other primary pulsed energy sources. Collision of high-velocity microparticles (~10 km/s or more) with surfaces gives rise to instantaneous (~10^{-10} s) release of energy

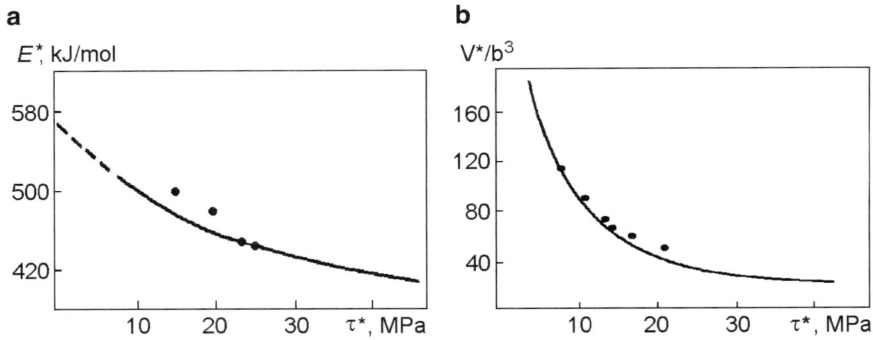

Fig. 2.58 Dependence of the activation energy (**a**) and of the activation bulk (**b**) on stress

on the order of ~5 · 10^7–5 · 10^9 J/kg, or 10–500 eV/atom, within a small volume at the point of impact. There, pressure sharply increases up to ~10–100 Mbar, the temperature reaches ~10^5–10^6 K, and shockwaves arise. The response of solids to shock loading has been studied for more than a century. However, to date the processes that occur under such conditions have not been unambiguously interpreted in terms of physics.

Investigations show that, due to a short period of loading, interaction between separate areas of the surface is practically absent. Within some areas the rate of deformation may be extremely high and the substance may be heated. Under the influence of high-energy action with a velocity exceeding 3 km/s, the transition of metals into the liquid state may occur in the shock zone. The mechanism of plastic deformation also changes. It is found that in metallic single crystals additional slip planes arise, and the contribution of twinning is revealed even in those materials in which deformation twins are not formed under normal conditions [74, 75].

For normal and moderately high deformation rates, the values of ultimate strength and flow limit in solids diminish with heating. Under conditions of submicrosecond shockwave loading, certain metals show athermic behavior of strength characteristics; in some cases these characteristics increase with heating up to T_m. Numerical simulation of high-rate deformation of metals by the molecular dynamics method indicates formation of a large number of nonequilibrium packing defects. This obviously results in instability and high mobility of the crystalline structure. In experiments, such an effect manifests itself as temporary loss of material strength caused by shockwaves.

Measurement of crystal lattice deformation at the moment shockwaves pass through is possible. The crystal lattice parameters are determined by the angle of X-ray diffraction. These parameters change as shockwaves pass through the crystal, indicated by oscillation of the Bragg angle.

With shockwave loads (10–100 MPa), the deformation state of a crystal changes from elastic to plastic deformation, i.e., crystals start "flowing." As the period of shockwave action is very short (approximately a nanosecond), crystal lattice ordering is not violated by the flow of material. However, under such conditions the lattice undergoes isotropic uniform compression, i.e., compression in the direction of wave propagation and in the perpendicular plane. But if the material has acquired the properties of a liquid, Pascal's law comes into force. In actuality, X-ray diffraction experiments usually reveal uniform compression of the crystal lattice in all directions.

Measurements of shockwave-induced crystal lattice deformation in silicon and copper show that these materials have fundamentally different response mechanisms [76]. In particular, at a load of 10–100 MPa, the state of elastic deformation in copper crystals transforms into plastic deformation, while the crystal lattice remains unchanged. Only its isotropic compression in the direction of wave propagation and in the perpendicular direction arises. However, the crystal lattice of silicon is compressed by ~11% in the direction of wave propagation, while no changes occur in the perpendicular direction. Despite pressures exceeding the static flow limit and strong deformation in the longitudinal direction, silicon does not acquire a state of plastic deformation; its response remains completely elastic.

It is known that under the action of increasing loads dislocations in crystals start to move, interact, and generate new dislocations. Macroscopically, dislocation motion is just plasticity of the material. However, transition into the state of plasticity requires a certain period of time that depends on the initial concentration of dislocations and their mobility. As with other covalent crystals, silicon is characterized by extremely low dislocation mobility. Therefore, during short-term action of forces no transition into a state of plastic deformation is observed. For copper crystals the time of transition into the state of plasticity is short (10–100 ps), but nevertheless, under the action of an equivalent shock load copper acquires a hydrodynamic state. For ionic crystals, the density and mobility of dislocations are several orders higher than in covalent crystals, which provide the conditions necessary for crystal transition into the plastic state during the interaction of forces. In general, it is assumed that structural rearrangements in solids under shock compression may last 10^{-9}–10^{-7} s or less.

Shockwave loading is accompanied by an increase in temperature, which depends on the shockwave amplitude. At amplitudes of several tens of GPa, the temperature increment of homogeneous heating is hundreds of degrees. Local heating on the slip lines may exceed the temperature of homogeneous heating. Inhomogeneous heating leads to an essential short-duration loss of strength. Subsequent temperatures decrease through diffusion heat conduction in areas of intense heating and results in recovery of strength properties.

The behavior of elastoplastic materials is characterized by the splitting of shockwaves into elastic (elastic Gugonio precursor) and plastic components. Some experimental methods for studying shear strength are based on the elastoplastic structure of expansion wave and nonhydrodynamic shockwave damping. In modern experimental methods, the space–time profiles of loading and unloading waves are registered by means of fast piezoelectric, piezocapacitance, and piezoresistive transducers.

Among the mentioned experimental methods, one should distinguish the technique for registration of the main normal stresses behind the wavefront. In this case, additional calculations of the medium flow are not required, since the dynamic flow limit is determined as a stress difference.

Destruction is induced by interference of unloading waves. While passing through a material, shockwaves change its microstructure, hardening the material. The processes of hardening and destruction occur simultaneously. Shockwave damping in the material leads to breakage of the interface between the striker and the material, halting the introduction of the striker. Shockwave damping in the striker, caused by an interaction with the side unloading wave, speeds up mass flow by creating a new compression pulse. Destruction of the sample begins with the formation of a channel chip-off crack oriented along the symmetry axis. Such a crack appears due to the focusing of side unloading waves. With subsequent loading the crack itself becomes a source of unloading waves, which leads to the formation of circular coaxial chip-off cracks. The material starts fragmenting, carried out of the contact zone by the striker, and a hollow is formed below this zone. The latent stage of surface destruction ends when, due to geometry, the shockwave reflected by the bottom of the hollow starts passing through the side walls. The reflected pulse, which reaches the face surface near the contact zone, creates a cross crack (face chip-off).

During the past decade, the behavior of brittle materials has been thoroughly investigated. In particular, complete deformation diagrams have been obtained and the different behaviors of ceramics under the influence of shockwaves have been considered. The most significant results achieved in this field are connected with the discovery of destruction waves during shock compression of glasses. The formation of destruction waves is one of the mechanisms of catastrophic strength loss in materials with high hardness, and demonstrates nonlocal response to loading. In ceramics, interatomic bonds may break under shock. The highest resistance to destruction is characteristic of materials with the highest values of dynamic compressibility and dissociation energy.

Systematic investigations of sapphire's response to shock loading started a few years ago. Sapphire products may undergo shock loading in a number of application fields, such as protective windows of environmental detectors, aerospace applications, and so forth. Individual data obtained on randomly chosen samples vary in result, but comparative analysis of the mechanical properties of different glasses and ceramics used as protective materials (Table 2.11) shows that sapphire is superior to all known transparent materials and is on par with ceramic materials employed for protection against bullets and shell splinters.

Presented below are results of an experimental check of sapphire's response to shock loads taking into account the samples' history, which includes their structural perfection (dependent on the growth method) and prior chemical-mechanical or thermal treatments. An attempt was made to establish a relationship between the behavior of the material and its grain structure. The shock load values were chosen based on practical requirements of sapphire screens and windows to withstand shocks with an energy of 6–8 J. These investigations were carried out on sapphire

Table 2.11 Mechanical characteristics of protective materials

Material	Density (g/cm^3)	Flexural strength (MPa)	Microhardness (GPa)	Fracture toughness (MN/m$^{3/2}$)
B_4C (opaque)	2.5	410	32	2.5
Si_3N_4 (opaque)	2.35	800	15.8	6
SiC (opaque)	2.35	400	28	3–4
ALON (clear)	3.69	380	18	2
Sapphire (clear)	3.97	760	20	4

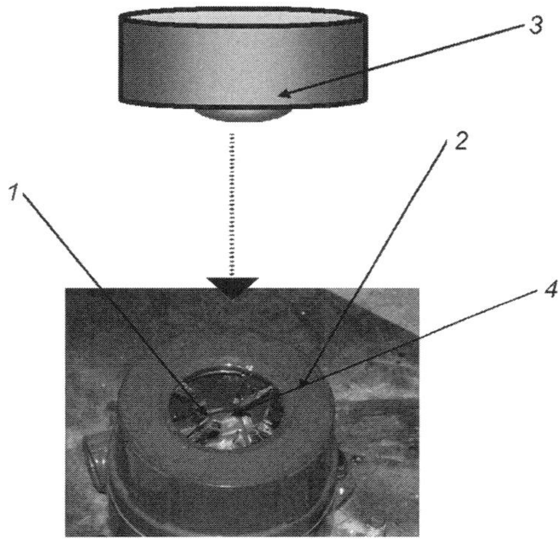

Fig. 2.59 Schematic of shock testing (*1*) sapphire sample; (*2*) fixing of the sample; (*3*) falling ball; (*4*) cracks formed at testing

cylinders with a diameter of 75 mm and thickness of 4.6 mm, and on 78 × 78 × 4.6 mm plates, made from crystals grown by the HDS, Kyropoulos, and Stepanov methods. The surfaces C, A, and R were polished to 120/100, 60/40, and 40/20. The tests were performed in accordance with standard requirements with the shock energy varied from 4 to 10 J (Fig. 2.59).

It is known that the maximum rate of loading transfer through any material is equal to the velocity of sound in this material, specifically $(E/\gamma)^{1/2}$, where E is the Young's modulus and γ is the density of the material. The velocity of sound in sapphire, the rate of shock loading transfer, may reach $(1.04–1.12) \cdot 10^4$ m/s, which considerably exceeds the velocity of a bullet (300–850 m/s). As a general rule, the action of shock loading lasts approximately a hundredth of a second, and the period of energy withdrawal from the point of shock is by several orders faster. At the instant of shock, a series of stress waves emanate from the point of shock. The waves reach the boundaries of the sample or blocks, twin interlayers, or so on, and

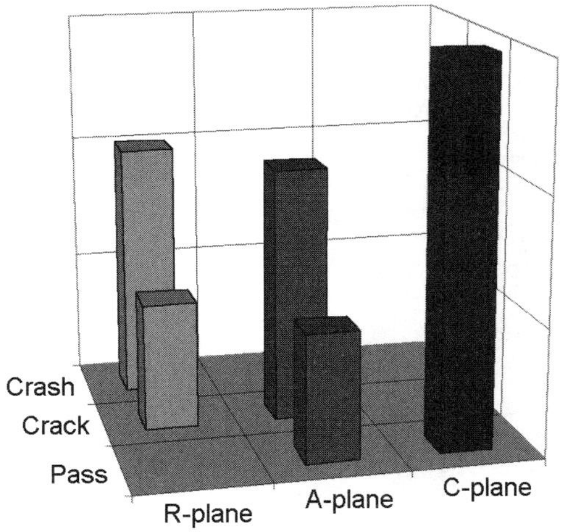

Fig. 2.60 Behavior of sapphire with different orientations during shock testing. The maximum shock energy is 7 J

are reflected, diminishing their intensity slightly. Superimposition of the direct and reflected waves results in the formation of standing waves, which may give rise to a sharp increase of stresses in the antinodes and to destruction of the sample. So, the progress of events is defined by processes that depend on the shape and size of the sample, the place of shock, and the structural perfection and orientation of the crystal.

Figure 2.60 shows the strength comparison of sapphire samples of different crystallographic orientations grown by the same method. Practically all samples with the (0001) orientation subjected to a special thermal treatment withstand three to five shocks of up to 7 J. The *R*-oriented samples are destroyed after the first shock, and only a third of the $(11\bar{2}0)$ orientation samples survive tests carried out under the same conditions.

It is known [77] that the size of grain structure elements in sapphire, a_e, and the properties connected with them depend on the growth method (see Fig. 2.61). Samples grown by different methods and subjected to different annealing regimens have fundamentally different values of a_e. A positive correlation exists between strength and a_e, and the previously obtained dependence $k_c(a_e)$ [77] has the same tendency, i.e., the larger the value of a_e, the greater the cracking resistance. It may be assumed that due to large values of a_e and the "density" of their boundaries, the waves emanating from a point of shock do not propagate over the whole of the sample's bulk, but are partly reflected from the grain boundaries. In this case, a part of the shock energy is scattered, and the standing waves formed in the antinodes will not have the energy sufficient for destruction of sapphire. With other conditions being equal, samples with a high-quality polish exhibit the highest strength (Fig. 2.62).

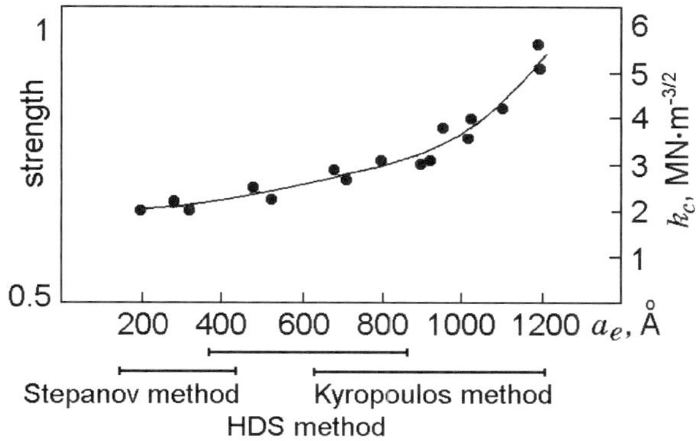

Fig. 2.61 Dependence of the relative strength and cracking resistance of sapphire crystals grown by different methods on the size of their grain structure elements

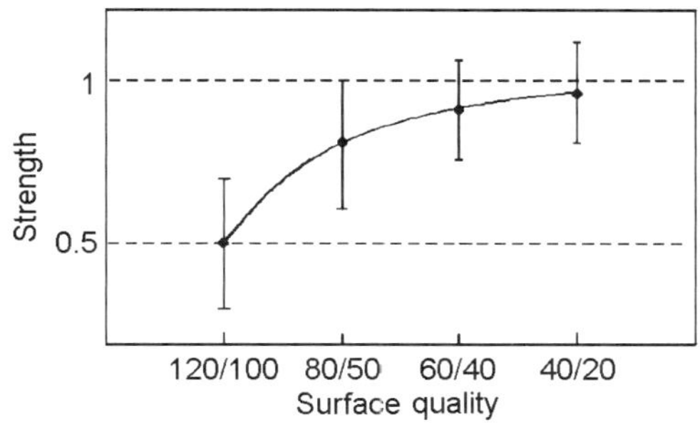

Fig. 2.62 Dependence of sapphire strength on surface quality

The depth of the defective layer formed on the surface of sapphire during mechanical treatment depends on the crystal growth method and on the grain structure, with other conditions being equal. For instance, crystals grown by the Verneuil and Stepanov methods have small grain structure elements, the depth of their defective layer is the largest, and their relative strength and cracking resistance during shock loading are the lowest (Fig. 2.63a).

In sapphire crystals, which have ~20% covalent bonds, all the destruction types are observed: brittle failure, plastic flow (preceding destruction), and chip-off. At centric impact, brittle failure may be observed in the center of the sample, but destruction of the periphery zones is not excluded (Fig. 2.63b).

The value of failing stress at chip-off is shown to depend on the shape and duration of tensile stress pulses, the stress-strain state of the medium, and different physical factors such as temperature, initial microstructure, and so forth. Thus, chip-off strength is a function of many variables. The energy criterion for chip-off destruction is based on the work required for chip-off in the region of dilatation

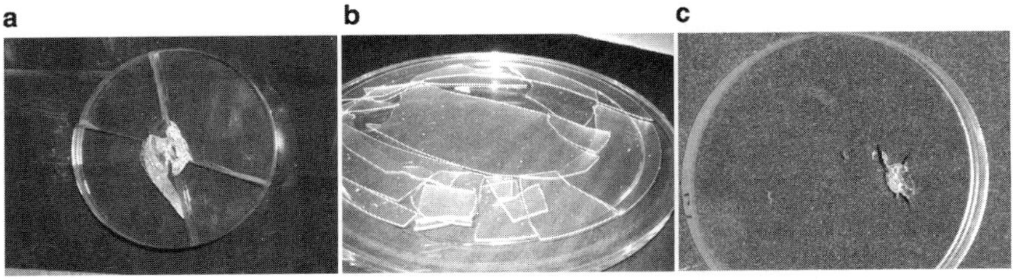

Fig. 2.63 (**a**) Centric impact. Brittle failure in the center of the sample. (**b**) Centric impact. Destruction of the periphery areas of the crystal. (**c**) Centric impact. Chip-off in the periphery of the sample

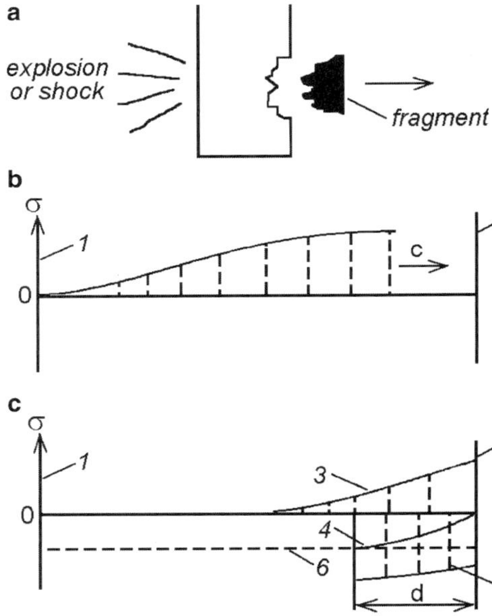

Fig. 2.64 Schematic of (**a**) chip-off, (**b**) pulse propagation, and (**c**) profiles of stresses arising with a shock: (*1*) sample's surface subjected to shock; (*2*) back side of the sample; (*3*) compression stress profile; (*4*) resulting profile; (*5*) stretching stress profile (reflected wave); (*6*) critical breaking stress; *d* (↔) distance between the surface and the place of breaking

wave interaction and is accomplished at the expense of elastic energy in the stretching pulse. Chip-off will occur if the elastic energy is sufficient for realization of this process.

To understand the mechanism of chip-off [75], one should consider the compression stress pulse that passes through a tested sample in consequence of shock (Fig. 2.64). The compression wave (Fig. 2.64b) reaches the free surface of the sampleand is reflected as a stretching wave. The reflected stretching wave then interacts with the incident compression wave (Fig. 2.64c). At a distance *d* from the back surface of the sample, the resulting stretching stress becomes equal to the critical normal breaking stress, the material breaks, and a piece of it separates from the surface. Thus, chip-off occurs exclusively due to wave effects.

The described progression of destruction also is observed in rectangular plates. As a rule, the angular zones are the first to be broken.

Plastic flow of sapphire under shock loading has been revealed. On the sapphire surface, a sharp deformation relief is formed at the point of shock (Fig. 2.65).

Earlier, similar patterns of deformation relief caused by plastic flow were observed only with diffusion welding of sapphire at $T > 2{,}200$ K and high specific pressures. The size and character of the deformation relief depend on the orientation of the sample (Fig. 2.66). Samples of (0001) orientation have a symmetrical trace of shock on the surface and their deformation relief is sharp. The shock energy is spent mainly on the formation of such a relief, i.e., on plastic flow of the material. Therefore, crystals with the (0001) orientation usually withstand shocks better (Fig. 2.66a). In *R*-oriented crystals plastic flow is extremely faint (Fig. 2.66b). In many cases it is not revealed at all, and the applied shock energy leads to destruction of the crystal. The behavior of orientation *A* is intermediate between the *C* and *R* orientations. An essential share of shock energy may be spent on plastic flow of the surface at the point of shock and such crystals may withstand shock loading.

Such an analysis of experimental facts yields the following conclusions:

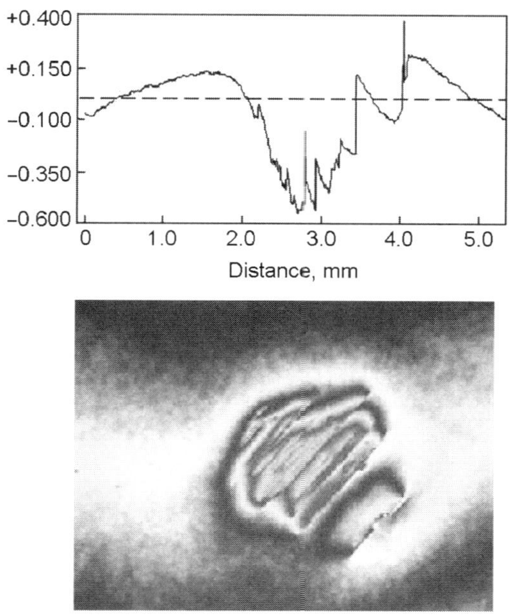

Fig. 2.65 Deformation relief formed on the surface at the point of shock. Centric impact. Brittle failure at the place of shock. The study was performed by means of the Zago apparatus

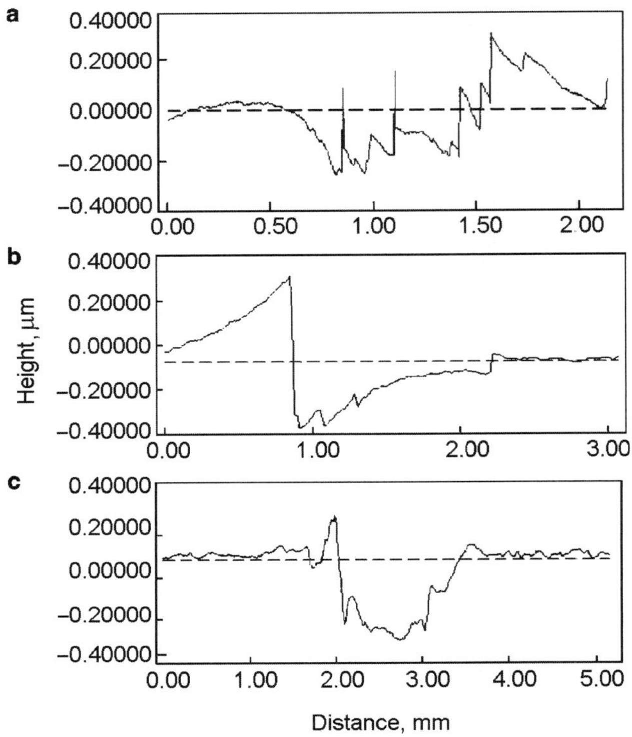

Fig. 2.66 Deformation relief formed on the surface of the crystals with the orientation (**a**) *C* (0001), (**b**) $R(10\bar{1}2)$, (**c**) $A(11\bar{2}0)$

- The behavior of sapphire can be predicted only by comparing the kinetic shock energy with the interatomic and intermolecular bond energies
- The physical–mechanical parameters of sapphire, such as Young's modulus, compressive and tensile strengths, Poisson's ratio, shock viscosity, and so forth, are only structural characteristics that lose their physical sense during the material's transition into the plastic state under shock loading
- The strength of sapphire under shock loading depends on the method of crystal growth and annealing, the crystallographic orientation, and the surface quality
- The dependence of sapphire strength on the crystal growth method is defined by the substructure of the crystals
- The values of cracking resistance and shock strength of sapphire are correlated
- At the point of shock plastic flow of sapphire occurs, and a sharp deformation relief is formed
- Shock loading of sapphire gives rise to all types of destruction: brittle failure, plastic flow which precedes destruction, and chip-off
- Special, high-temperature treatment fundamentally raises the shock strength of sapphire

2.1.5 Thermal Properties

Melting temperature is 2,323 K.

Boiling temperature is 3,253 K.

Linear expansion coefficient of sapphire (α) depends on temperature and orientation (Fig. 2.67) [78]. Like other thermal parameters established by different authors and presented in catalogues, the values of α differ significantly.

The most often used values of the linear expansion coefficient:

$\perp C$	293–323 K	$5.0 \cdot 10^{-6}$ K^{-1}
$\parallel C$		$6.6 \cdot 10^{-6}$ K^{-1}
60° orientation	293–323 K	$5.8 \cdot 10^{-6}$ K^{-1}
$\perp C$	1,273 K	$(7.9–8.3) \cdot 10^{-6}$ K^{-1}
$\parallel C$		$(8.8–9.0) \cdot 10^{-6}$ K^{-1}
60° orientation	1,273 K	$7.7 \cdot 10^{-6}$ K^{-1}

At room temperature (295.65 K) the linear extension coefficients for the lattice constants are: $\alpha_a = 5.22 \cdot 10^{-6}$K^{-1}; $\alpha_c = 5.92 \cdot 10^{-6}$K^{-1}.

The investigation of thermal extension of sapphire in the region of low temperatures, which appears to be topical in view of this materials use in superconducting resonators, shows that below 50 K α is lower than 10^{-7} K^{-1} [79], and this value is negligible in most cases. However, impurities may give rise to low-temperature anomalies. For example, ruby with the magnetic impurity Cr^{2+} has an anomaly of temperature expansion caused by the Yan-Teller effect. Anomalies also discovered in sapphire with deviations from the stoichiometric content [80]. The results of the thermal expansion study of a sample with an anionic vacancy content of 10^{18} cm^{-3} (10^{-3}%) are shown in Fig. 2.68. The measurements were carried out using a capacitance dilatometer on a sample with 16 mm diameter and 100 mm width.

The values of *a* are determined in 17–115 K temperature interval with $\pm 2 \cdot 10^{-8}$ K^{-1} error at $T > 50$ K and with $\pm 5 \cdot 10^{-9}$ K^{-1} error at $T < 50$ K. Thermal expansion of

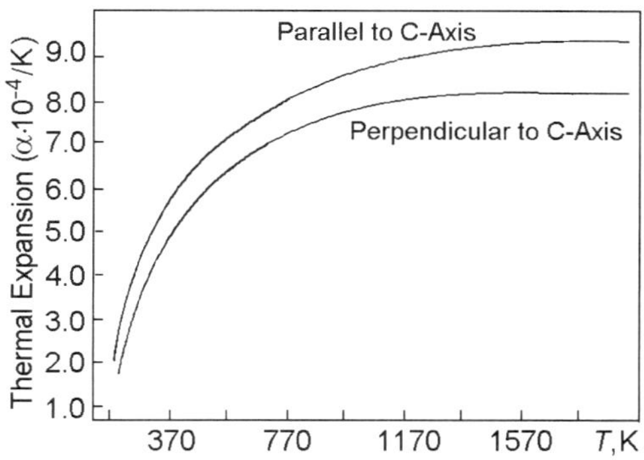

Fig. 2.67 Temperature dependence of the linear expansion

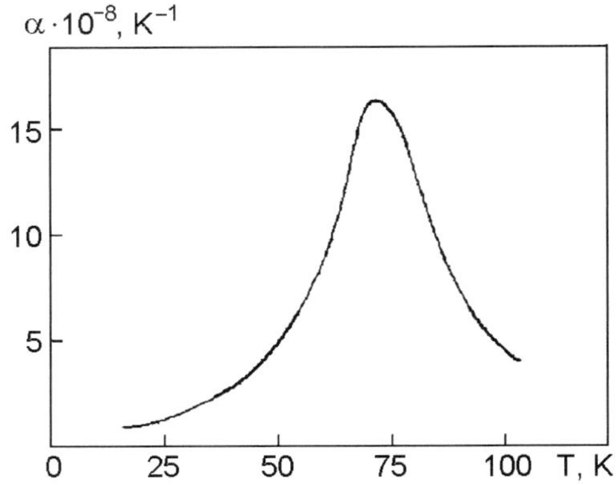

Fig. 2.68 Expansion of sapphire with anionic vacancies in the low temperature region. The sample is oriented along the axis $[10\bar{1}0]$

sapphire with anionic vacancies at $T < 100$ K turns out to considerably exceed that of the stoichiometric crystal. The value of α reaches its maximum of $16 \cdot 10^{-8}$ K^{-1} at 70 K.

Sapphire is a universal reference standard material for thermal expansion. The sapphire reference standards used in high-temperature comparative dilatometers with the kinematic polycrystalline corundum system must have thermal expansion equal to that of corundum polycrystals. Such reference standards are made from sapphire rods with $d = 6$ mm and having the rod axis oriented at an angle of $59 \pm 2°$ with the C-axis. Accurate values of the relative extension and the differential temperature coefficient of linear expansion are contained in Table 2.50 [81].

Thermal conductivity coefficient (λ_T) of sapphire also depends on the temperature and the orientation. The temperature dependences of sapphire and ruby λ_T are presented in Figs. 2.69 and 2.70, the temperature-dependence of thermal conductivity λ_T, W/(m K) (q) and of thermal resistance (W) are shown in Figs. 2.71 and 2.72.

The most often used values of the thermal conductivity coefficient are the following:

- At 298 K $\perp C$: 30.3 W/(m K)
  $\| C$: 32.5 W/(m K)
- At 273 K 60° orientation 46.06 W/(m K)
- At 373 K 60° orientation 25.12 W/(m K)
- At 673 K 60° orientation 12.56 W/(m K)

Thermal conductivity of sapphire at T_m, obtained by extrapolation is equal to 3.4 W/(m K), the ratio of thermal conductivity of sapphire to that of its melt $\lambda_{sol}/\lambda_{liq} = 1.65$.

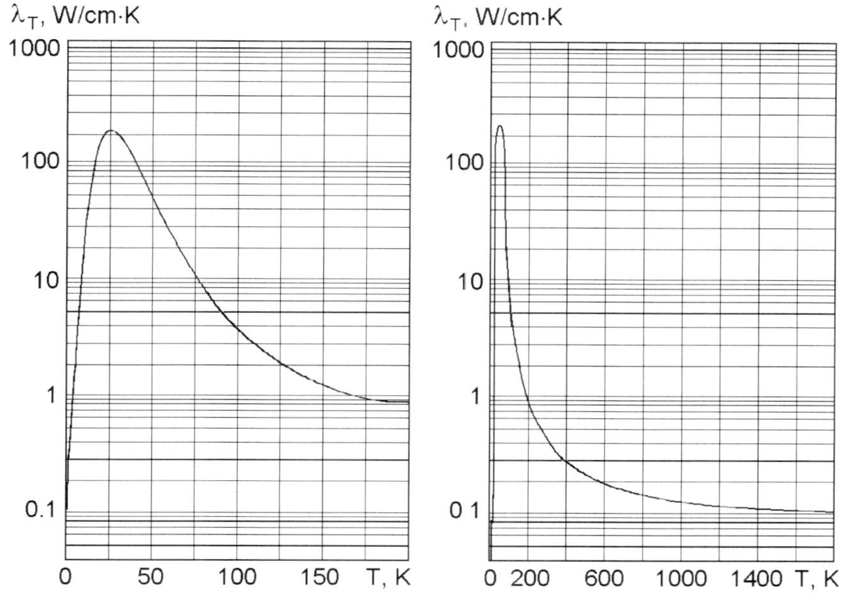

Fig. 2.69 Temperature dependence of the thermal conductivity coefficient for sapphire: (**a**) low temperatures; (**b**) high temperatures

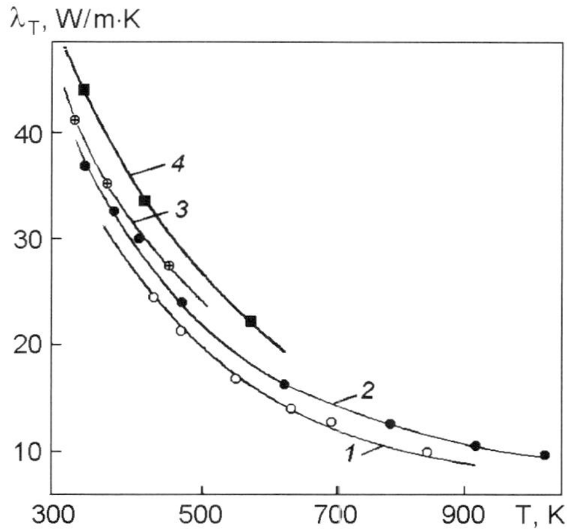

Fig. 2.70 Temperature dependence of the thermal conductivity coefficient for ruby with varying chromium content: (*1*) 1.1%; (*2*) 0.75%; (*3*) 0.16%; (*4*) 0.003%

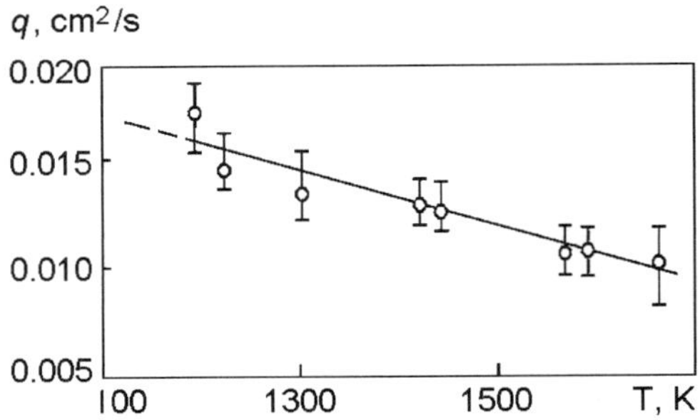

Fig. 2.71 Temperature dependence of thermal conductivity

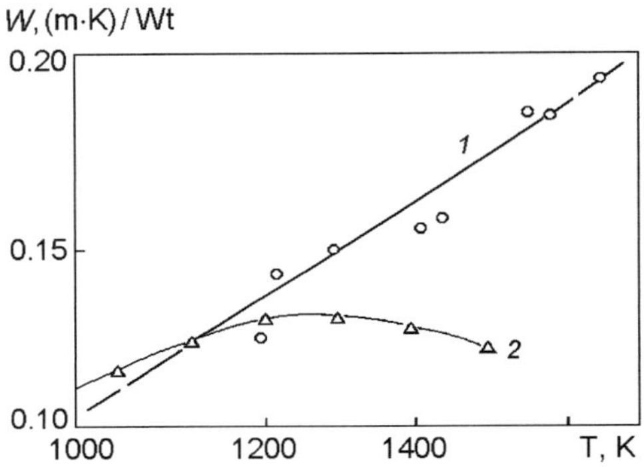

Fig. 2.72 Temperature dependence of thermal resistance: (*1*) calculated data; (*2*) experimental data

Specific heat of sapphire (c_p) at 293 K: 0.181–0.187 cal/(g K), at 1,273 K: 0.300 cal/(g K).

Specific heat of sapphire at low temperatures:

T (K)	10	25	50	100	150	200	298.1
c_p (kJ/(kmol K))	0.04	0.20	1.9	13.4	33.5	51.92	79.13

Molar heat capacity ($c_{p\lambda}$) at 293 K: 18.63 cal/(mol K), at 1,273 K: 29.86 cal/(mol K).

The temperature dependences of specific heat and molar heat capacity are shown in Fig. 2.73 and 2.74. The numerical values for thermal conductivity and heat capacity of ruby versus temperature are given in Table 2.51.

Molar heat capacity (kJ/(mol K) can be estimated in the 298–2,300 K temperature interval by means of the following dependence:

$$c_{p\mu} = 4.2(26.12 + 4.388 \cdot 10^{-3} T - 7.269 \cdot 10^{5} T^{-2}). \tag{2.7}$$

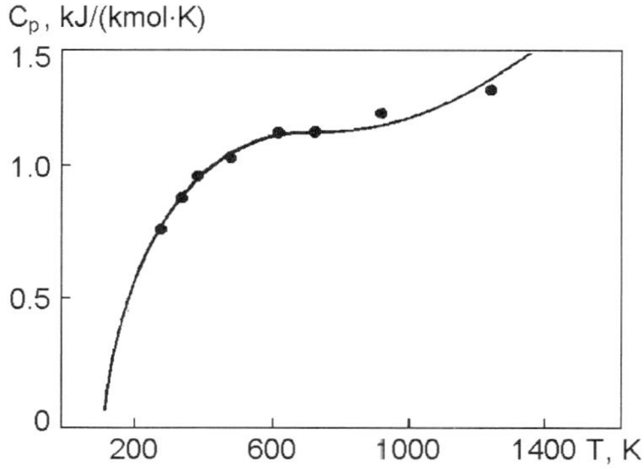

Fig. 2.73 Temperature dependence of specific heat

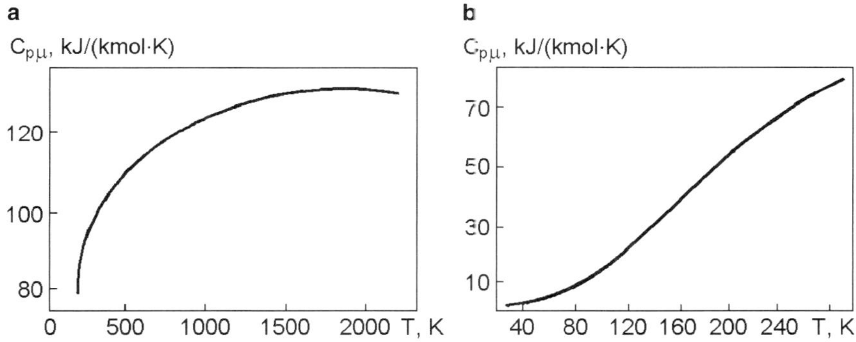

Fig. 2.74 Temperature dependence of molar thermal capacity: (**a**) high temperatures; (**b**) low temperatures

Presented below are the values of some thermodynamic parameters of Al_2O_3 at room temperature and melting temperature:

$$\Delta H_{298} = 1{,}671{,}000\,\text{kJ/mol}; \quad \Delta F_{298} = 157{,}800\,\text{kJ/kmol};$$
$$\Delta S_{298} = 51\,\text{kJ/(kmol K)}; \quad \Delta H_{\text{m}} = 108{,}900\,\text{kJ/kmol}.$$

The temperature dependence of the thermodynamic parameters of sapphire is given in Tables 2.52, 2.54, and 2.56. Since the processes that occur at attenuation of sound waves have the same character as those defining thermal resistance, Table 2.53 presents the temperature dependence of the sound absorption coefficient for ruby.

Coefficients of diffusion of Cr^{3+} and Fe^{3+} in sapphire and in ruby (0.5 mass% of Cr) are temperature-dependent (Table 2.12) [32]. The coefficient of oxygen diffusion in sapphire is $4.2 \cdot 10^{-20}$ cm^2 s^{-1} at 1,473 K [83]. That is, the diffusion path of

Table 2.12 Cr^{3+} and Fe^{3+} diffusion coefficients in sapphire and in ruby

	Diffusion coefficients ($cm^2\ s^{-1}$)			
	Sapphire		Ruby	
T (K)	Cr^{3+}	Fe^{3+}	Cr^{3+}	Fe^{3+}
1,473	$3.64 \cdot 10^{-14}$	$2.54 \cdot 10^{-14}$	$5.81 \cdot 10^{-14}$	$3.73 \cdot 10^{-14}$
1,523	$7.97 \cdot 10^{-14}$	$6.77 \cdot 10^{-14}$	$1.11 \cdot 10^{-13}$	$1.09 \cdot 10^{-13}$
1,573	$2.94 \cdot 10^{-13}$	$2.05 \cdot 10^{-13}$	$3.24 \cdot 10^{-13}$	$3.16 \cdot 10^{-13}$
1,623	$4.43 \cdot 10^{-13}$	$3.70 \cdot 10^{-13}$	$6.94 \cdot 10^{-13}$	$5.17 \cdot 10^{-13}$

oxygen does not exceed 2.5 Å/h. In the temperature interval from 1,853 to 2,113 K for ^{16}O the diffusion coefficient is $1.56 \cdot 10^{-10}\ m^2\ s^{-1}$ [84], the activation energy being equal to 788 ± 29 kJ/mol.

The dependence of diffusion on crystallographic direction has been studied insufficiently. It has been reported [85] that ^{22}Na diffusion along the C-axis is almost twice as high as along the perpendicular direction.

Implantation of ions into a sapphire surface and evolution of the surface-adjacent layer upon thermal treatment. Ions of metals and gases, including those which implantation into the surface-adjacent sapphire layer during the growth process is either impossible or difficult, can be introduced by ionic bombardment. Ions of Mn, Ni, Cr, Xe, and so forth are implanted at temperatures from −170 up to 250 K. The doses vary from $3 \cdot 10^{16}$ to $5 \cdot 10^{17}$ ions/cm^2, the density of ionic current is lower than 2.5 mA/cm^2 for doses not exceeding $1 \cdot 10^{17}$ ions/cm^2, and the ion energy is 300 keV to 1 MeV. The implantation depth reaches several thousand Å. Upon deep implantation the surface layer becomes amorphous. The degree and depth of surface damage depend on the implantation energy. In particular, the critical dose for sapphire amorphic-phase transition is ~$2 \cdot 10^{15}$ ion/cm^2 for 150 keV Cr ions implanted at ~77 K or 300 keV Ni ions implanted at ~100 K. For 400 keV Xe ions the critical dose is $3 \cdot 10^{16}$ ions/cm^2 [86]. During postimplantation annealing at 873–1,873 K the damaged surface-adjacent layer recrystallizes. The process of annealing is accompanied with migration and redistribution of the implanted ions.

Unlike Mn, Ni, or Xe ions, trivalent Cr, Ga, and Fe ions can be incorporated into the aluminum sublattice upon annealing in air. At 1,473 K all Cr^{3+} ions enter the sublattice and further migration of ions is not observed.

2.1.6 Electrical Properties of Sapphire

At room temperature sapphire is one of the best dielectrics. The character of electric conduction (χ) is described by the equation:

$$\chi = \chi_1 \exp(-E_1^* / kT) + \chi_2 \exp(-E_2^* / kT) \tag{2.8}$$

where E_1* and E_2* are the activation energies of charge carriers. The activation energies of low- and high-temperature conduction are 2.25 and 5.50 eV, respectively. Electrical resistance (ω) is temperature-dependent (Table 2.13).

Table 2.13 Temperature dependence of electrical resistance

T (K)	ω (Ω m)	T (K)	ω (Ω m)
293	$1 \cdot 10^{14}$	973	$2.5 \cdot 10^{7}$
373	$2 \cdot 10^{13}$	1,073	$3.5 \cdot 10^{6}$
473	$4 \cdot 10^{12}$	1,273	$1 \cdot 10^{6}$
573	$1.1 \cdot 10^{11}$	1,373	$4 \cdot 10^{5}$
673	$1.6 \cdot 10^{10}$	1,473	$4 \cdot 10^{4}$
773	$1.3 \cdot 10^{9}$	1,773	$2 \cdot 10^{3}$
873	$1.9 \cdot 10^{8}$		

In catalogues of different sapphire producers other data can be found. For instance, the bulk resistance at 298 K has been reported to be equal to 10^{14} Ω cm, which coincides with the magnitude contained in Table 2.13, whereas in other catalogues the said value is 10^{16} Ω cm. For 773 and 1,273 K temperatures values of ω equal to 10^{11} and 10^{6}Ω cm, respectively, are reported. It should be noted, however, that at present one cannot persuasively claim that these data are correct.

Resistivity of sapphire (Ω cm)	
293 K	$\perp C$: $5.0 \cdot 10^{18}$
	$\parallel C$: $1.3–2.9 \cdot 10^{19}$
773 K	$>10^{12}$
1,273 K	$>10^{9}$

The value of resistivity diminishes with increasing temperature, as with other oxides (Fig. 2.75) [87]. The electrical conduction value abruptly changes from 10 to 0.03 Ω^{-1} cm^{-1} at the liquid to solid phase transition.

At 1,573–2,023 K the behavior of sapphire and polycrystalline aluminum oxide is similar to that of p-type semiconductors at high pressures of oxygen and of n-type semiconductors at low pressures [88]. At high temperatures, the type of electrical conduction in sapphire varies depending on the partial pressure of oxygen in the surrounding atmosphere (Fig. 2.76) [89].

The change in mechanisms of electrical conduction is explained by the theory of high temperature thermochemical processes, when the crystal is in equilibrium with the vapors of crystal-forming components. In the state of high-temperature equilibrium with low partial pressures of oxygen, donor-type defects are predominantly formed, whereas at high pressures acceptor-type defects arise. In both cases a

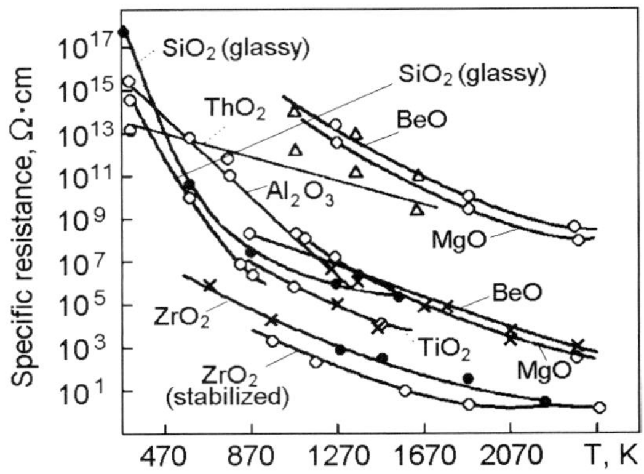

Fig. 2.75 Temperature dependence of resistivity for Al_2O_3 and some other oxides

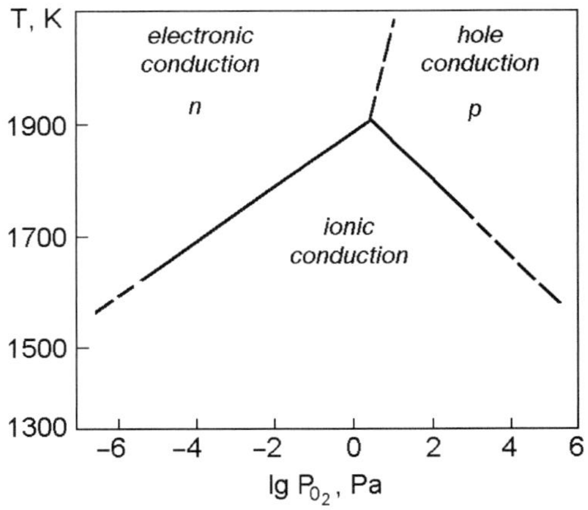

Fig. 2.76 Mechanisms of electrical conduction of sapphire

deviation from stoichiometric content of Al_2O_3 occurs. At medium partial pressures of oxygen no deviation from stoichiometry should be expected, perhaps in addition to the formation of donors, acceptors, and associates.

Irradiation of sapphire changes its electrical conduction [90]. Upon irradiation with doses up to 0.1 MGy the value of electrical conductivity decreases, whereas at higher doses (>1 MGy) the primary value of electrical conduction is almost restored (Fig. 2.77) [91, 92].

Dielectric constant of sapphire at 298 K in 10^3–10^9 Hz interval is

$\parallel C$	11.5
$\perp C$	9.3

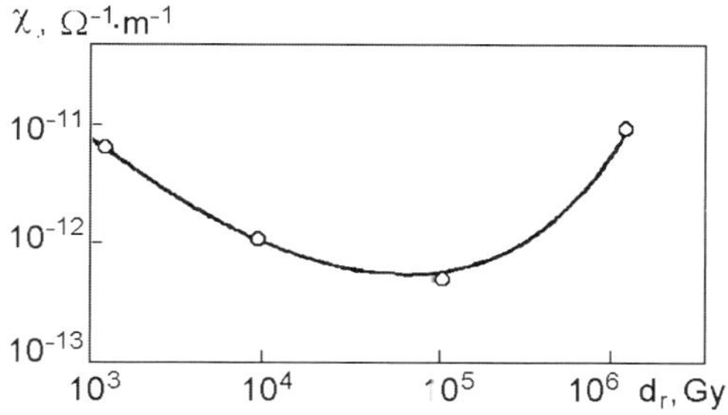

Fig. 2.77 Dose dependence of the value of electrical conduction for sapphire

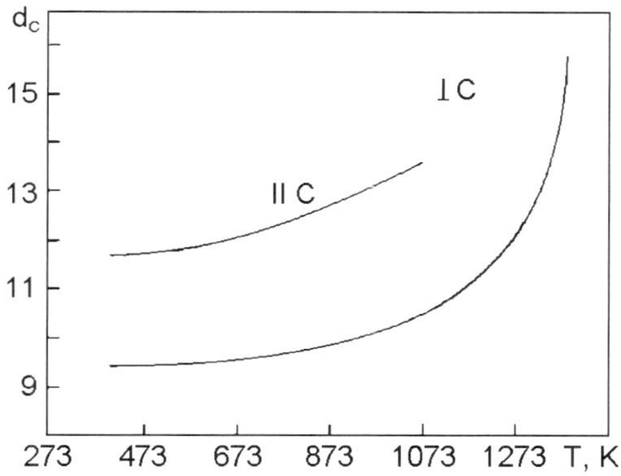

Fig. 2.78 Temperature dependences of the dielectric constant in the directions parallel and perpendicular to the optical axis

(In some catalogues the value of the dielectric constant at $1 \cdot 10^2$–$3 \cdot 10^8$ Hz frequencies in the direction $\| C$ is 10.55, in the direction $\perp C$ it is equal to 8.6.) The value of the dielectric constant increases with temperature irrespective of the crystallographic orientation (Fig. 2.78).

*Tangent of dielectric loss angle (*tg δ) is the same in the directions parallel and perpendicular to the optical axis. For a frequency of 1.0 MHz it is equal to $1 \cdot 10^{-4}$; for higher frequencies tg $\delta < 1 \cdot 10^{-4}$. At 298 K and frequencies up to 10^{10} Hz, tg δ in the direction parallel and perpendicular to the C-axis it essentially differs:

$$\perp C: \ 3.0 \cdot 10^{-5}$$
$$\| C: \ 8.6 \cdot 10^{-5}$$

Dielectric losses in sapphire are temperature-dependent (Figs. 2.79 and 2.80). For comparison, this figure also presents the temperature-dependence of tg δ in γ-irradiated sapphire (the irradiation dose is 10^4 Gy).

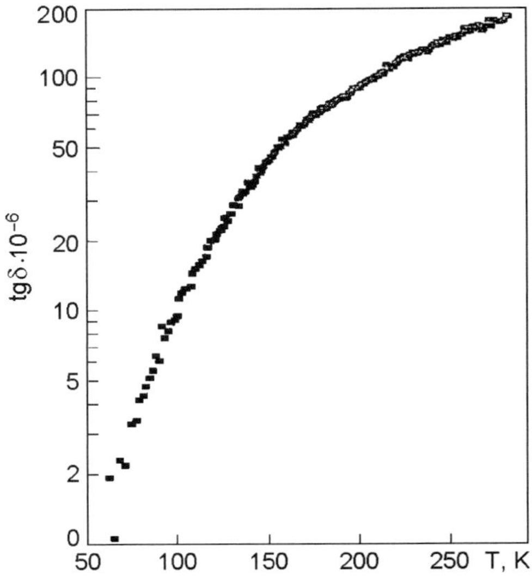

Fig. 2.79 Dielectric losses at low temperatures and 145 GHz

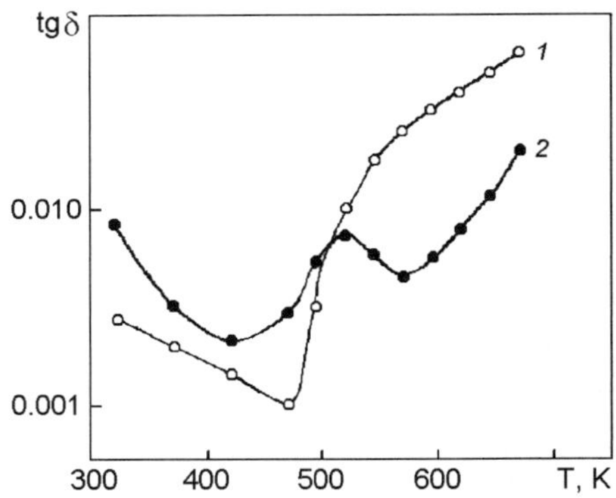

Fig. 2.80 Temperature dependences of tg δ in the initial (*1*) and irradiated crystals (*2*)

As is seen, the value of dielectric loss increases in the region of low temperatures and diminishes as temperature rises. In the 293- to 430-K interval, the frequency dependence of tg δ is more pronounced than its temperature dependence. Figure 2.81 presents the frequency dependences of tg δ at different temperatures for nonirradiated sapphire (a) and for sapphire irradiated by a 10^4 Gy dose (b). The observed nonlinear behavior of the frequency dependence of tg δ is supposed to be caused by polarization processes in sapphire.

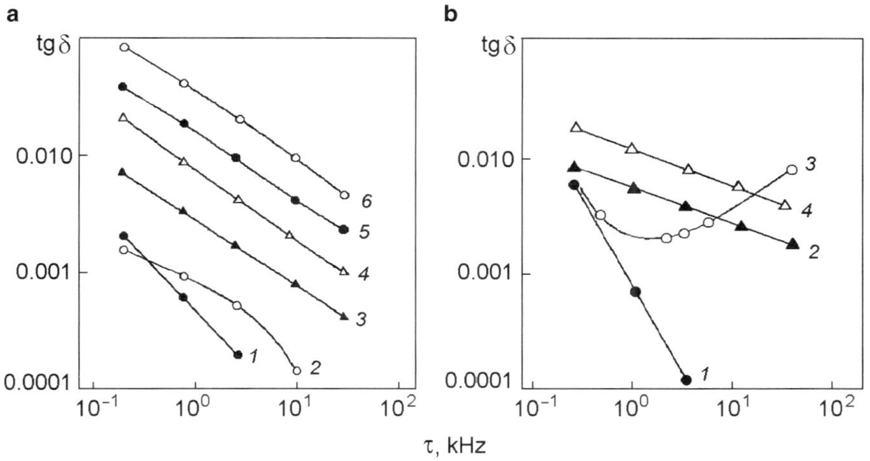

Fig. 2.81 Frequency dependences of tg δ at temperatures for (**a**) nonirradiated and (**b**) irradiated sapphire: (*1*) $T = 323$ K; (*2*) 432 K; (*3*) 523 K; (*4*) 573 K; (*5*) 623 K; (*6*) 673 K

Magnetic susceptibility of sapphire:

$$\| C: \quad 0.21 \cdot 10^{-6}$$
$$\perp C: \quad 0.25 \cdot 10^{-6}$$

2.1.7 Laser Properties

Sapphire acquires laser properties after the introductio n of activating additives. It should be noted that the crystal can be used as a laser medium only in the case of isomorphic substitution of aluminum.

The coefficient of activator distribution, K_d, is dependent on the thermodynamic parameters of the matrix-activator system and the growth conditions (Table 2.14). It is difficult to calculate K_d taking into account thermodynamic and kinetic factors. This value can be estimated within the framework of the energy theory of isomorphism by the relation [93]:

$$\ln K_d = \Delta H_m / R_g (T_c^{-1} - T_m^{-1}) - Q_s / R_g (1.2 T_c^{-1} - t_i) \tag{2.9}$$

where ΔH_m is the melting heat, R_g is the gas constant, T_c is the crystallization temperature of the mixture, T_m is the melting temperature of the activator, t_i is an empirical constant, and Q_s is the energy of mixing for the solid solution. The introduction of large-size cations (Sc, Ln) into sapphire considerably diminishes the distribution coefficient.

Table 2.14 Presents estimates of K_d for different activators, as well as a comparison of K_d with experimental values K_{de}

Ion	T_m of oxide (K)	r/r_1*	Q_s (kcal/mol)	Q_l (kcal/mol)	K_d	K_{de}
Cr^{3+}	2,538	0.042	15.9	−3.2	0.30	0.4
Fe^{3+}	–	0.063	35.7	−7.1	0.05	0.01
Ga^{3+}	2,013	0.048	20.6	−4.1	0.18	0.05
Sc^{3+}	2,750	0.106	101	−20.2	0.0002	0.005

*r/r_1 is relative difference of interatomic distances; Q_l is the shift energy for oxides with Al_2O_3 in liquid phase

Among Al_2O_3-based laser media, the most widespread are $Cr:Al_2O_3$ and $Ti:Al_2O_3$. The application fields of these materials are considered in Chap. 1.

$Cr:Al_2O_3$. Ruby has the following advantages as an active medium [94, 95]: high pulse power; possibility of operation at room temperature; the highest mechanical strength compared with other laser media; good photochemical resistance to pump source radiation; and high threshold of surface and bulk damage under the influence of its own radiation. The characteristics of ruby as a laser medium are given in Table 1.11.

Ruby efficiently absorbs pump lamp emission in the blue and green regions of the spectrum. Ruby laser emission can be registered by low-inertia vacuum photoreceivers. In the region of ruby element radiation, such photoreceivers possess high sensitivity and stability. In this connection, ruby lasers compare favorably with more efficient lasers in the threshold location of objects upon laser illumination. At the same time, the threshold of excitation and generation of ruby is higher than with four-level active media.

Optical strength. Surface damage of sapphire (chips or cracks) and bulk damage (stars or tracks) may occur immediately at the onset of laser irradiation. The threshold of surface damage depends on the quality of the surface and usually is lower than the threshold of bulk damage. Microcracks in the surface layer and even edge bevels can focus laser irradiation. Another cause of damage is radiation absorption by abrasive particles contained in the surface-adjacent layer.

For short pulses the threshold damage power is inversely proportional to the duration of irradiation, t_p. For long-duration pulses the damage threshold does not depend on t_p. The location of the transitional region depends on surface structure. For quasiamorphous and crystalline surfaces, the dependence behavior changes at $t_p = 5 \cdot 10^{-6}$ s and $t_p = 2 \cdot 10^{-6}$ s, respectively. The surface-adjacent layer absorbs ~10% of the laser pulse energy. In a 1-μm thick layer, a pressure of ~200 N/m² is created when the radiation power density is 0.5 GW/cm² and $t_p = 30$ ns, leading to "microexplosion."

The strength of sapphire crystal ends can be increased by diminishing the depth of the defective surface-adjacent layer. In particular, for thicknesses of 5, 2, 0.03, and 0 μm, the surface damage threshold (in relative units) is 0.47; 1.0; 2.5, and 4, respectively.

To a lesser extent, the surface damage threshold depends on crystal orientation. Rotation of the crystal end plane by 90° with respect to the vector of laser beam polarization does not influence the damage threshold for the basal plane, but for the prism plane this value changes by 1.3–1.5 times. Such a difference is explained by redistribution of polarized radiation between microcracks, developing mainly in the direction of the *C*-axis, and by other defects of the surface-adjacent layer.

When irradiation intensity is close to the threshold density of the surface evaporation energy, the main damage mechanism for the output surface becomes evaporation in the regions of local energy maxima. Estimation of the threshold evaporation energy of an output end, when irradiated by a monopulse ruby laser with a power density of 0.5 GW/cm^2 and t_p= 30 ns, indicates an energy density threshold equal to 20 J/cm^2. The output end usually is destroyed at lower powers versus the input end.

As a rule, volume damage in ruby takes place at irradiation energy densities higher than 0.7 J/cm^2, due to absorption of the radiation by inclusions and other absorbing defects, or to self-focusing of the radiation. A significant part of the energy is absorbed by Cr^{4+} ions and impurities, some of which change their phase state under irradiation. The volume of the phase also changes, leading to additional volumetric stresses. Besides passive absorption, some impurities (e.g., Mg^{2+}) diminish the output energy as the number of pulses increase. Upon strong excitation, titanium impurities lead to the formation of stable color center (CC) and decrease the bulk strength.

Bulk strength also may be reduced due to the initial or pumping-induced optical inhomogeneity of RE, leading to nonuniform energy distribution over the element's cross section and to damage in zones subjected to the most intense irradiation. High-power lasers use either highly homogeneous rubies (grown by the Czochralski method) or an optical design in which the beam repeatedly passes through different regions of the active body during generation.

The influence of pores on the bulk strength is not significant. Obviously, regions containing pores do not participate in the process of generation and have low local energy densities.

Bulk damage may give rise to destruction of the crystal ends. For a pulse energy of 0.2 J and t_p= 30–70 ns, the first Fresnel diffraction maximum from the edges of the bulk damage region, located 8–10 mm from the end, may favor damage of the surface.

In the process of RE operation, individual damaged regions similar in size to pores accumulate in the bulk. Their quantity grows with time up to several tens per cubic centimeter. Damaged regions shaped as stars grow to several tenths of a millimeter, while the track type may reach tens of centimeters. However, total damaged volume usually does not exceed 1–8% of the element's volume. Exclusion of such a volume from generation does not lead to a noticeable change in the pulse energy or radiation direction. Even elements with numerous damaged regions withstand hundreds of thousands of pulses. Large damaged areas at the ends and in the bulk can be detrimental, if they overlap a considerable part of the RE cross section. Experimental data on the surface and bulk strength of ruby and sapphire, obtained in the regimen of one-mode generation, are presented in Table 2.15.

Radiation resistance of ruby. Some equilibrium number of CC is observed even in nonirradiated ruby. Radiation resistance of RE is characterized by the change in pulse energy, E_p, and the threshold energy, E_t (Tables 2.16–2.18).

A correlation exists between the generation parameters and the total distortions of the electron paramagnetic resonance spectra (the parameter β in Tables 2.17 and 2.18). The spread in the data presented is related to the impurity content of the RE and other individual peculiarities of the crystals.

Valence transitions of impurity ions lead to changes in laser radiation parameters during the process of irradiation. The quantity of emitting ions diminishes and the processes of energy migration and dissipation change.

The probability of change in the number of dopant electrons during the irradiation process is defined by the ratio of the energy of electron ionization (capture) to the Madelung constant α_M. The state of the ion, I_{Me}^{+n}, is stable if the ionization energy is higher than the absolute magnitude of α_M [96]:

Table 2.15 Surface and bulk radiation resistance of ruby and sapphire

	Light damage threshold (W/cm^2)		
Crystal	Surface	Bulk	t_p (s)
Ruby (0.05% of Cr)	10^8–10^9	–	$2 \cdot 10^{-9}$–$2 \cdot 10^{-6}$
	~10^6	–	$5 \cdot 10^{-6}$–$1 \cdot 10^{-4}$
		$(0.6–1.0) \cdot 10^{10}$	$(2–3) \cdot 10^{-8}$
Sapphire	$(1.0–2.0) \cdot 10^{10}$	–	$2 \cdot 10^{-8}$

Table 2.16 Change of characteristics of ruby elements after γ-irradiation

Ruby element	Irradiation conditions	Change of characteristics
0.04–0.05% of Cr^{3+}	$0.8 \cdot 10^3$ rad	$E'_p : E_p = 1.7$
	$1.7 \cdot 10^3$ rad	$E'_p : E_p = 3$
	$1.5 \cdot 10^4$ rad	$E'_p : E_p = 1$
0.04% of Cr^{3+}	$2.7 \cdot 10^4$ rad	
	20 min	$E'_p : E_p = 1.3$
	30 min	$E'_p : E_p = 1.35$, recovery of E_p after 15–20 pulses
0.1% of Cr^{3+}	$1 \cdot 10^6$ rad	$E'_p \geq E_p, E'_t \geq E_t$ $E'_p : E_p = 0.8:1.8 = 0.44$ $E'_t : E_t = 375:145 = 2.6$
0.05% of Cr^{3+}	$1 \cdot 10^7$ rad	Rise of delay time of generation onset from 260 to 427 ms

E'_p, E'_t are value after radiation

Table 2.17 Change of characteristics of ruby elements irradiated with fast electrons

Irradiation dose (electrons/cm²)	$E'_t:E_t$	$E'_p:E_p$	β^a Before irradiation	After irradiation
$7.4 \cdot 10^9$	1.05	1.75	0.47	0.08
$7.4 \cdot 10^{10}$	0.96	3.8	0.16	0.1
$7.4 \cdot 10^{11}$	1.07	0.96	0.7	0.8
10^{14}	0.80–0.75	1.25–1.3	–	–
10^{15}	1.03	1.67	0.26	0.16
10^{15}	1	1	0.1	0.16
10^{15}	0.86	1.35	0.24	0.06
10^{15}	0.87	2.3	0.3	0.02
10^{15}	0.86	1.03	0.12	0.08
10^{15}	0.95	1.14	0.17	0.05

Energy: 10 MeV; flux density: $1.8 \cdot 10^{11}$ electrons/(cm² s)

[a] β is total distortion of EPR spectra

Table 2.18 Change of characteristics of ruby elements irradiated with fast protons (100 MeV)

Irradiation dose (electrons/cm²)	$E'_t : E_t$	$E'_p : E_p$	β Before irradiation	After irradiation
10^{12}	1.65	0.43	0.12	0.4
10^{15}	1.56	0.57	0.2	0.4
10^{15}	1.09	1.2	0.13	0.12
10^{15}	0.78	2.85	0.1	0.07

$$I_{M_e} + n > |\alpha_{\mathrm{M}}| \tag{2.10}$$

In this formula, distortions of the lattice by the impurity ions are not taken into account. For the cationic site of ideal sapphire, $\alpha_{\mathrm{M}} = -35.2$ eV and the potentials I_{Me} of Al ions have the following values: $I_{\mathrm{Al}^{+}} = 18.83$ eV; $I_{\mathrm{Al}^{2+}} = 28.45$ eV; [97] and $I_{\mathrm{Al}^{3+}} = 119.98$ eV. The difference $I_{\mathrm{Al}^{3+}} - |\alpha_{\mathrm{M}}| = 84.8$ eV indicates high stability of the Al^{3+} state in corundum. For the impurity ions, $I_{\mathrm{Cr}^{2+}} = 33.2$; $I_{\mathrm{Cr}^{3+}} = 52$; $I_{\mathrm{Fe}^{2+}} = 30.6$; $I_{\mathrm{Fe}^{3+}} = 57.1$; and $I_{\mathrm{Mn}^{2+}} = 34$ eV.

The formation of a charge-deficient or charge-excessive cation in the second coordination sphere does not change the impurity valence. Therefore, the pairs Mn^{2+}–Mn^{4+} are not formed in the ideal lattice of sapphire. The appearance of an electronic state that does not satisfy the relation (II.9) testifies to a considerable content of structural defects stabilizing this state.

Annealing of irradiated RE leads to recombination of radiation defects. Even after short-term light treatment E_p recovers.

The intensity of induced absorption can be reduced by the introduction of vanadium or titanium. Titanium works as a protector only in crystals annealed in vacuum or in a reducing medium.

Ti:Al$_2$O$_3$ is one of the best active media for tunable lasers.

Characteristic of *d*-elements is a strong interaction of the unfilled d-shell of the ion with the crystalline field, which results in widening of the absorption bands and in luminescence. Wide luminescence bands are used for tuning laser radiation frequency and the spectrum of titanium is one of the widest. Unlike with ruby, the splitting scheme of titanium levels does not contain absorption in the region of pumping and generation.

For Ti:Al$_2$O$_3$-based lasers, efficiencies comparable to or exceeding those of dye lasers in the 0.7- to 0.9-μm range have been achieved. The crystals possess high laser resistance. Breakdown upon irradiation with 0.532-nm wavelength and $t_p = 10$ ns occurs at 4–5 J/cm^{2}energy density. The generation characteristics of Ti:Al$_2$O$_3$ are presented in Table 1.12.

For pumping, the absorption band in the blue-green region of the spectrum corresponding to the main transition $^2T_{2g} \rightarrow {}^2E_g$ is used. The most intense absorption band is located in the UV region. This region contains the first excited ^{2}S level of free Ti^{3+} ions, which corresponds to the configuration 4S^1 where the absorption is overlapped by the fundamental absorption of the matrix. The transition d $\rightarrow$ S, resolved with respect to parity and spin, corresponds to charge transfer.

Absorption in the region of generation depends on the content of titanium and reaches 0.15 cm^{-1}, ~10% of the absorption at the main line maximum. Excitation of Ti:Al$_2$O$_3$ in the UV region is accompanied by luminescence, with maxima at 0.25 and 0.315 μm. The crystal has a wide luminescence band in the 0.6- to 1.1-μm region (Fig. 2.82).

In comparison with the absorption spectra, the luminescence spectra show a noticeable distinction between the radiation intensities of different polarizations: the intensity of π-polarization ($E||C$) is almost three times higher than with σ-polarization. The luminescence band is asymmetric and has a maximum at 0.79–0.80 μm. The quantum luminescence yield is 0.81–1.0 at 300 K, the lifetime 3.9–2.6 ms.

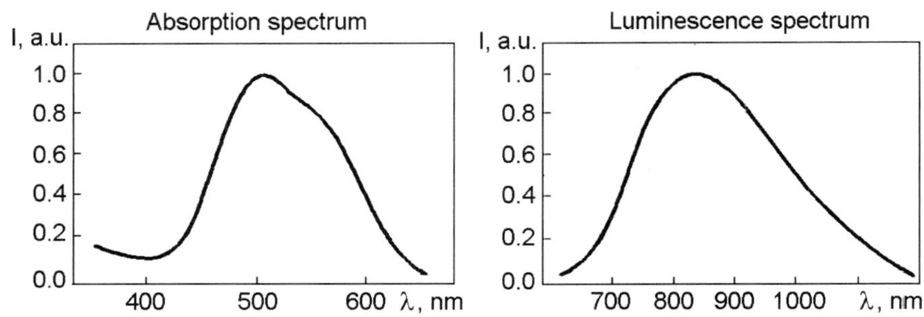

Fig. 2.82 Absorption and luminescence spectra of Ti:Al$_2$O$_3$

Besides the luminescence band in the IR-region, bands that overlap the visible region also are observed. The total luminescence spectrum of Ti:Al_2O_3 contains an IR-luminescence band of Ti^{3+} with a peak at ~0.78 μm and an additional band in the short-wavelength region with a peak at about 0.42 μm. Upon pumping with high-power radiation of 0.532-μm wavelength, the luminescence lifetime is 4.0 ms for the 0.78-μm band. Its increase is related to luminescence pumping in the blue-green region (~0.42 μm). The lifetime of the short-wavelength band is 7.4 ms, the lifetime of luminescence in the 0.42-μm band is 2.3 ms.

Upon UV pumping two additional luminescence lines are observed: a polarized line with a peak at ~0.425 μm and a π-polarized line with a peak at about 0.575 μm. The lifetime of the latter is 2.1 ms.

The spectral characteristics of both laser media depend on the oxidation–reduction potential of the annealing medium. For high-power (pico- and terawatt) femtosecond Ti:Al_2O_3-based lasers, acousto-optically programmed filters, pulse compressors, and other means for improvement of the wave front and pulse shape have been elaborated.

2.1.8 Wettability

The effectivity of sapphire use in chemical engineering and in medical implants is defined, among other factors, by the wetting of its surface with liquid substances in contact. The wettability is among the "bioactivity" factors of the implant surface. Depending on the surface destination, both the maximum and minimum wetting may be required. At the same time, there are substantially no literature data on the sapphire wetting. In ref. [98], which deals with the contact wetting angle, the surface-free energy of sapphire has been determined to be 52.95 erg/cm^2, but the corresponding crystallographic plane has been identified. In the mineralogical literature, the wettability characteristics of 92–80° are indicated most often; these data are not associated with crystallographic characteristics, too. That is why the angular range is so broad. Sapphire is anisotropic; thus, the surface-free energy values differ considerably for various crystallographic planes [99].

The wettability is characterized quantitatively by the wetting angle Q^0 formed at a solid surface along the phase interface. The value, according to Foxe equation, is related to the material surface energy and depends on the surface-free energy (SFE) values of three phase interfaces being in contact at the wetting line (s–1, s–g, l–g) as follows:

$$\cos Q = (\sigma_{s-g} - \sigma_{s-1}) / \sigma_{1-g}. \tag{2.11}$$

It is the number of free bonds per unit surface area that can be considered to be an approximated measure of that energy.

In real crystals the SFE depends on the number, type, and distribution of structure defects. To unify these parameters, the crystals were grown at the speed of

10 mm/h and annealed in vacuum at 1,950°C. Then 10-mm-thick disks of 18 mm in diameter were cut out of the crystals; the orientation error was less than 1°. The sample surfaces polished to $R_z = 0.05$ μm were wiped with cotton wetted with alcohol and then with water-wetted cloth and dried in air. Measuring rest drop wetting angle Q^0 using optical microscopy, the wetting was studied for pure sapphire at crystallographic planes (0001); {10$\bar{1}$0}; {11$\bar{2}$0}; {10$\bar{1}$2}; and for crystals doped with 0.1% Cr at (0001) plane and with 0.05% Ti at {11$\bar{2}$0}one. The surface was wetted with distilled water, glycerin, and isotonic solution (0.9% NaCl) in air at 25°C. The wetting angle measurement error was 0.5–1.3%.

The wetting angles for the main crystallographic planes of sapphire are within 68 to 85° range both in polar liquid (distilled water) and a nonpolar one (glycerin) (Figs. 2.83 and 2.84). The nonpolar glycerin wets the surfaces of all samples studied better

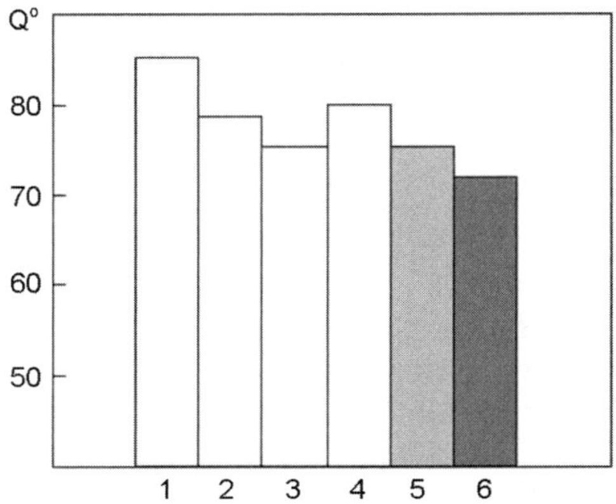

Fig. 2.83 Wetting angle of sapphire planes with water: (*1*) (0001), (*2*) {10$\bar{1}$0}, (*3*) {11$\bar{2}$0}, (*4*) {1012}, (*5*) {1120}(0.05 Ti), (*6*) (0001) (0.1 Cr)

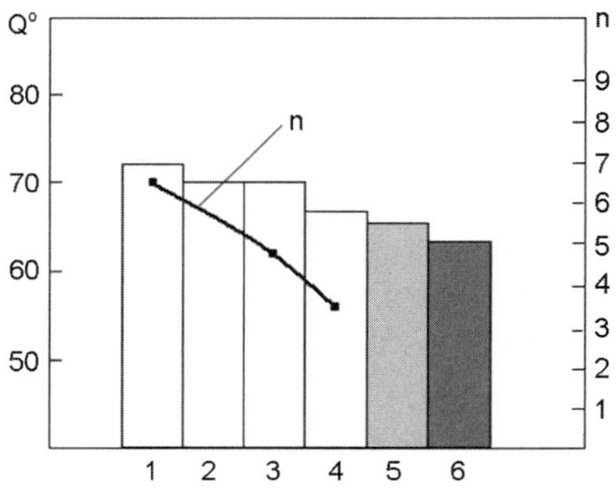

Fig. 2.84 Wetting angle of sapphire planes with glycerin: (*1*) (0001), (*2*) {10$\bar{1}$0}, (*3*) {11$\bar{2}$0}, (*4*) {1012}, (*5*) {1120}(0.05 Ti), (*6*) (0001) (0.1 Cr)

than water. The crystallographic orientation and dopants more weakly affect the Q^0 value, although the general variation character remains unchanged. The crystal doping is a factor of importance enhancing the wettability. Doping with Cr ($C = 0.1\%$) causes $Q^0_{(0001)}$ decrease by 12° and 8° for water and glycerin, respectively. The same effect is revealed at Ti doping ($C \sim 0.05\%$), the $Q^0_{(11\bar{2}0)}$ decrease amounting 3°. In isotonic NaCl solution ($C = 0.9\%$), the crystal composition and crystallographic parameters are observed to influence the wettability, too, but the variation trend does not correlate with the data on other liquids in all the cases. This can be explained by differences in effects of the solution ionic components (Na^+ and Cl^-) on the physicochemical sorption processes at the planes studied. The values of surface energy for the sapphire crystallographic planes estimated from the wetting angles are in correlation with the results of wear resistance tests carried out before. The maximum wear rate with free abrasives is seen to correspond to minimum Q^0 values (maximum energy) in water for $\{10\bar{1}0\}$and $\{11\bar{2}0\}$(see Fig. 2.83). This demonstrates that water is involved actively in the wearing process as a surface-active substrate. The maximum biological inertness of a sample corresponds to minimum surface energy and respectively to the maximum possible wetting angle. These data are in correlation also with corrosion resistance of various crystallographic planes of sapphire [100].

The physicochemical processes in the treatment area in the presence of water can be described briefly as follows. At the crystal–liquid interface, considerable adhesion forces occur between the liquid boundary layer molecules and the crystal surface. These forces hinder the sliding at the interface. The wetting phenomenon is explained usually by such an interaction of force fields. The surfaces in friction are separated as a rule by a thin lubricant layer that is connected so strongly that their direct contact is excluded, and the relative displacement occurs along the intermediate liquid layer. The surface activity of a liquid depends on its molecular structure, the activity being the stronger the higher the molecular polarity. As the liquid molecules adsorbed at the crystal surface are mobile, the molecular interaction is propagated into the crystal near-surface layer along the microcracks. The wedging effect of the liquid molecules (the Rebinder effect) develops the microcracks toward macrocracks, and thus causes the shearing of particles from the crystal surface, so the dispersion process is facilitated. Since water wets the $\{10\bar{1}0\}$and $\{11\bar{2}0\}$ planes better during the grinding, it is just these planes where the maximum wear rate is observed.

Thus, the dependence of sapphire wetting with water and glycerin on the crystallographic orientation has been studied. The wetting angle for both liquids is maximum at the (0001) plane. The minimum wetting angle values are observed at $\{10\bar{1}0\}$and $\{11\bar{2}0\}$for water and at $\{10\bar{1}2\}$for glycerin. A correlation has been found between the water wettability of the main crystallographic planes of sapphire and the wear rate of those planes under grinding with free abrasive using water as the suspension medium.

2.2 Chemical Properties

Molecular mass	101.9612
Electronegativeness	2.58
Dissociation energy of molecule, $D_{Al_2O_3}$	740 kcal/mol
Formation heat for Al_2O_3 molecule	377 kcal

2.2.1 *Dissolution*

Dissolution of crystals is studied to determine the symmetry, facets of equilibrium form, traces of radioactive decay (tracks of decay), dislocation structure, as well as to increase the strength by removing the defective surface-adjacent layer formed at mechanical treatment. The rate of dissolution V_s is a function of the surface energy and the bond strength and to a certain extent depends on the defectiveness of the surface-adjacent layer. The process of dissolution (etching) of crystals usually begins at the most active points, such as the places where dislocations reach the surface. The shape of etch pits (Fig. 2.85) depends on the location of the chains of the strongest bonds [101] which also define the appearance of crystals.

The dependence of the rate of dissolution on the orientation of crystals was revealed as far back as the nineteenth century [102]. The rate grows as the density of ions on the facet increases and the energy of bonds between the ions diminishes.

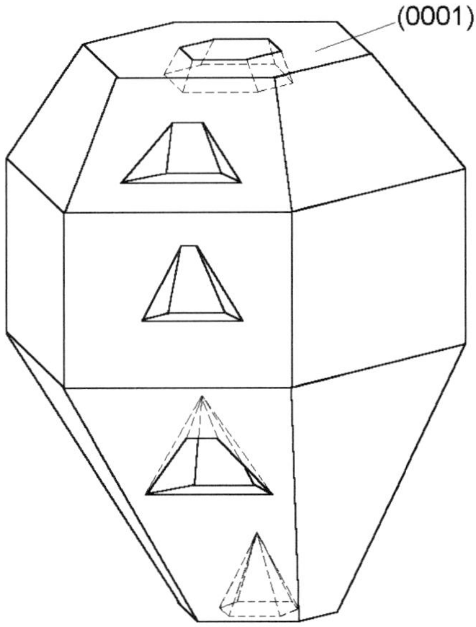

Fig. 2.85 Etch pits shown as an intersection of the figures of the growth form with fictitious (negative) etching polyhedrons

The bond type also matters. With other conditions being equal, the activation energy of covalently bonded ions is higher due to the fact that at oscillation shifts higher energy is required to overcome the bonding force.

Crystal facets, edges, and vertices dissolve at different rates. In the process of dissolution, slopes of the surfaces of the corresponding orientations move along the isoclines in the direction of trajectory of the peak of the dissolution body (Fig. 2.86) [103].

The isoclines are contiguous to the normals to the facets and coincide with them at least at one point, e.g., in the directions [$\bar{1}\bar{1}2$]and [$1\bar{2}1$]. In the zones where the vertices and the edges are located, V_s has the maximum value. The vertex motion rate is defined by the relation

$$V_{[0\bar{1}1]} = \frac{V_{[\bar{1}\bar{1}2]}\sqrt{3}}{2} = \frac{V_{[1\bar{2}1]}\sqrt{3}}{2} \qquad (2.12)$$

At dissolution of sapphire in H_3PO_4 at 573 K the dominating final equilibrium form (*L*-form) is the rhombohedron [$10\bar{1}2$]; dissolution of other planes obviously suppressed by adsorption. The dissolution stages of sapphire ball (the view along [$20\bar{2}1$]) are shown in Fig. 2.87.

The facet ($10\bar{1}2$) moves at the constant movement d_r. The final form of dissolution of a spherical sample consists of the facets of equilibrium form.

The equilibrium form of Al_2O_3 crystals was theoretically determined by Heimann [104] using the PBC method. The facets {$10\bar{1}2$}, {$11\bar{2}0$}, {$10\bar{1}4$}, {0001} belong to the dissolution form of concave hemisphere (*G*-form). As in the case of *L*-form, the rhombohedron {$10\bar{1}2$} is a significant facet of *G*-form. The{$11\bar{2}0$},

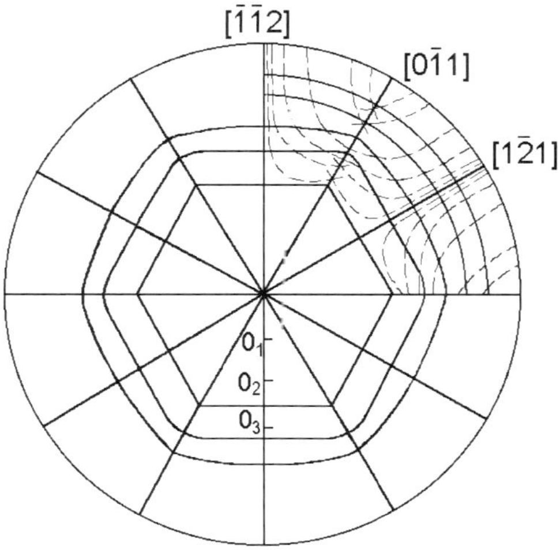

Fig. 2.86 Isoclines at dissolution of a hexagonal crystal

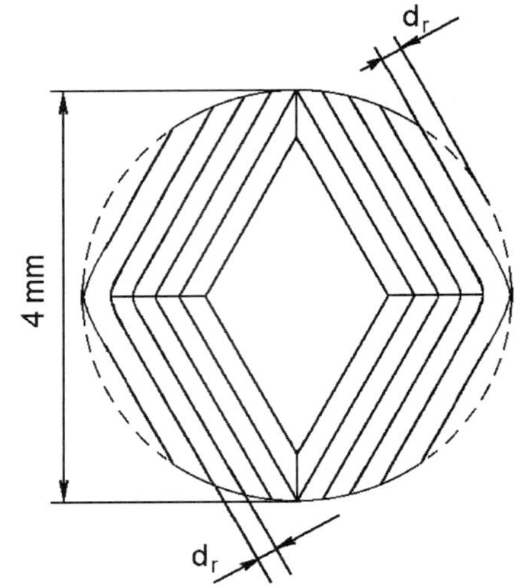

Fig. 2.87 Stages of dissolution of sapphire ball in H_3PO_4 at 573 K

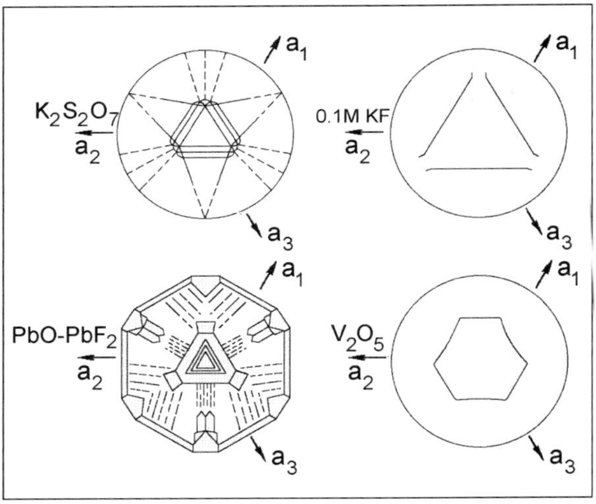

Fig. 2.88 Idealized view of etched hollow hemispheres with a diameter of 4 mm on (0001) facet of sapphire

{0001}, and the rhombohedron $\{10\bar{1}4\}$ follow. As follows from Fig. 2.88, Table 2.19 and the forms of dissolution of sapphire in solvents (Figs. 2.89–2.94), all the observed *L*-forms can be explained by the participation of the facets $\{11\bar{2}0\}$, $\{10\bar{1}4\}$, {0001} on *G*-form polyhedron [105–107].

By cutting off the vertices of this polyhedron the *L*-form is obtained. It consists of two rhombohedrons shifted by 60° and of two scalenohedrons (Fig. 2.93).

Table 2.19 *G* and *L* forms of sapphire

Solvent	Theoretical *L* form	Experimental *L* form	Type (Fig. 2.93)	Theoretical *G*-form (Fig. 2.89)	Experimental[a] *G* form (Fig. 2.88)
$K_2S_2O_7$	$[0\bar{h}h1]$, $[\bar{h}0h1]$, 2[*hkil*]		I	{0001}+ $\{11\bar{2}0\}$ + $\{10\bar{1}2\}$ + $\{\bar{1}014\}$	(0001), 3$(\bar{h}0hl)$, $(\bar{h}0hl)$, (*hkil*)
<800°C		$[10\bar{1}3]$>$[14.\bar{5}.\bar{9}.18]$			
>800°C		$[14.\bar{5}.9.18]$> $[10\bar{1}3]$			
$PbO–PbF_2$ (1:1), 950°C	$[0\bar{h}h1]$, $[\bar{h}0h1]$, 2[*hkil*]	$[12\bar{3}]$>$[30\bar{3}5]$ $[\bar{5}054]$> $[1.\bar{1}8.14.15]$	I	{0001} + $\{11\bar{2}0\}$ + $\{10\bar{1}2\}$ + $\{\bar{1}014\}$	(0001), 3$(\bar{h}0hl)$, $(\bar{h}0hl)$, 2 (*hkil*), $(hh\overline{2h}l)$
0.1 M KF, 600°C, 3 kbar	$[\bar{h}0h1]$, $[2[\bar{h}0h1]$	$[\bar{5}054]$>$[\bar{1}015]$>$[10\bar{1}3]$, $[60\bar{6}1]$	II	{0001} + $\{11\bar{2}0\}$ + $\{10\bar{1}2\}$	(0001), $(\bar{h}0hl)$, $(\bar{h}0hl)$
V_2O_5, 950°C	$[\bar{h}0h1]$	$[\bar{5}054]$	III	{0001} + $\{10\bar{1}2\}$	(0001), $(\bar{h}0hl)$, $(\bar{h}0hl)$

[a] Due to unsatisfactory geometry of hollow hemisphere the prism facets $(hh\overline{2h}0)$ can be observed only in exceptional cases ($PbO–PbF_2$)

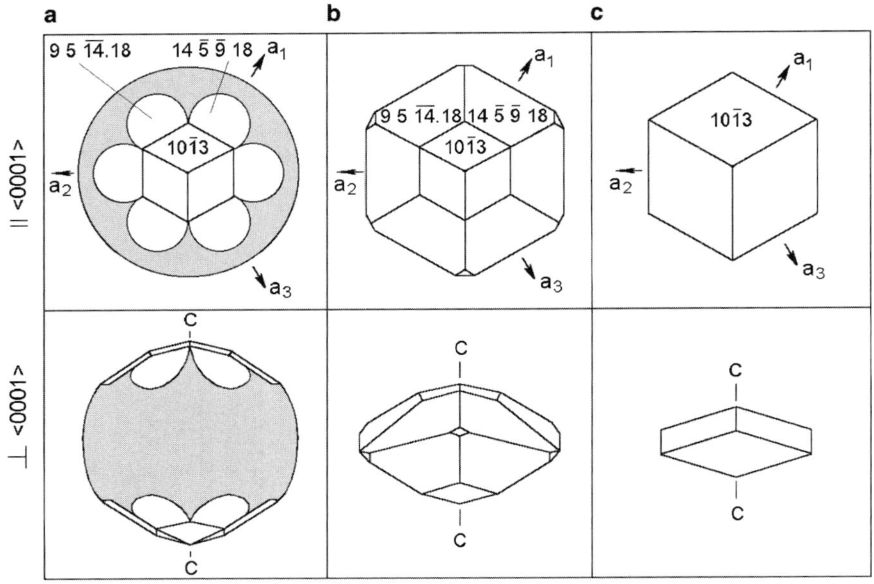

Fig. 2.89 Forms of dissolution of sapphire (*L* forms) in $K_2S_2O_7$ melt at $T < 1{,}073$ K. Shaded regions show primary sphere surfaces

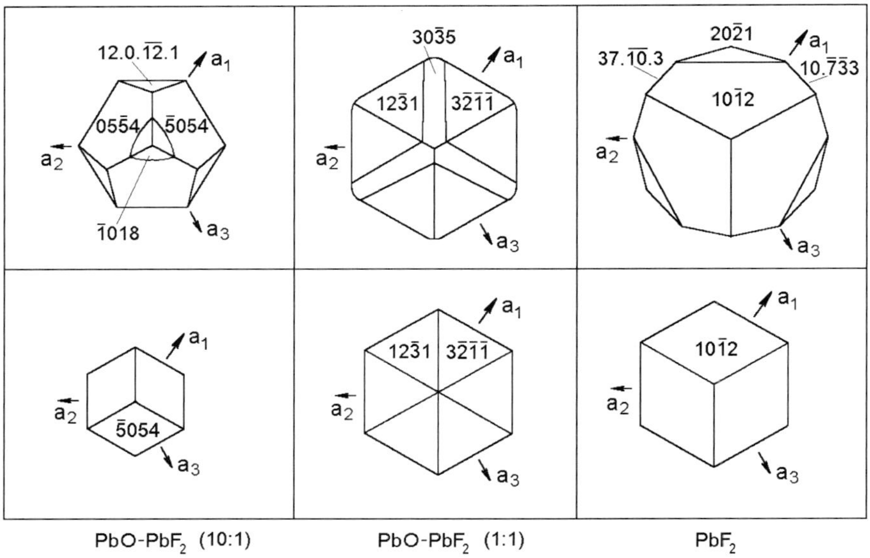

Fig. 2.90 Forms of dissolution of sapphire in pure PbF_2 melt and in PbO–PbF_2 (with 1:1 and 10:1 molar ratios) at 1,223 K

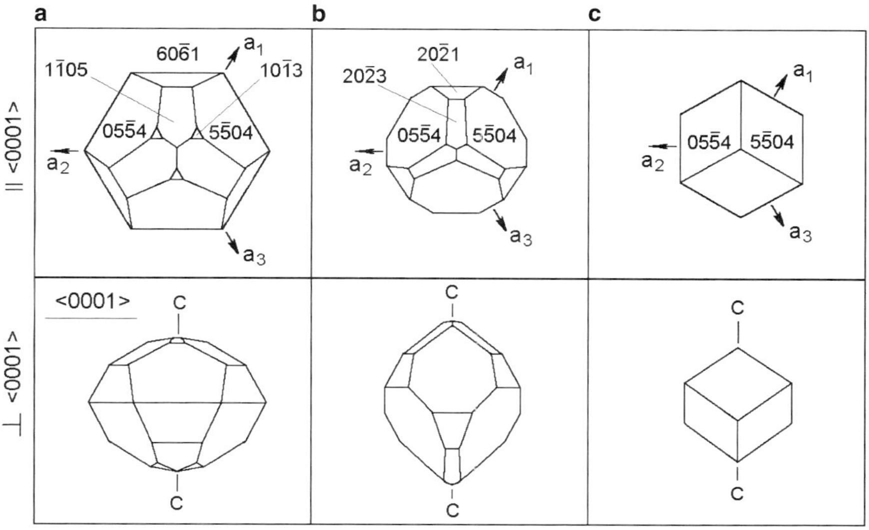

Fig. 2.91 Forms of dissolution of sapphire in 0.1 M KF and K_2CO_3 at 873 K and 3 kbar

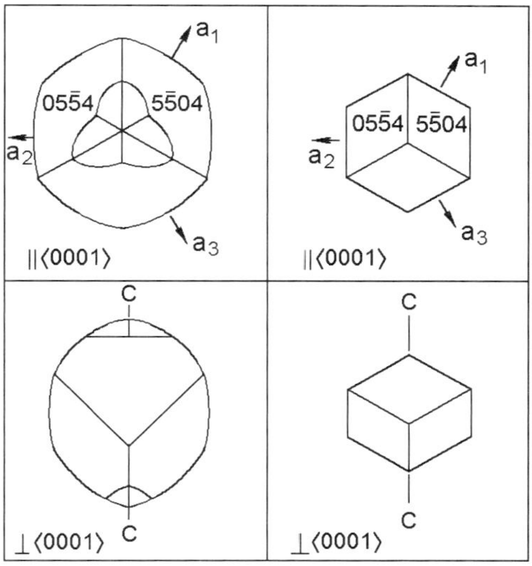

Fig. 2.92 Forms of dissolution of sapphire in V_2O_5 melt at 1,223 K

Experimentally these facets were found at dissolution of sapphire in $K_2S_2O_7$ melt. The cut of the final form is temperature-dependent (see Table 2.19).

The polar diagram V_s (Fig. 2.94) obtained at dissolution of sapphire spheres at their interaction with carbon in a gaseous medium vividly shows the presence of dissolution anisotropy [108]. To build this diagram, the segments proportional to V_s

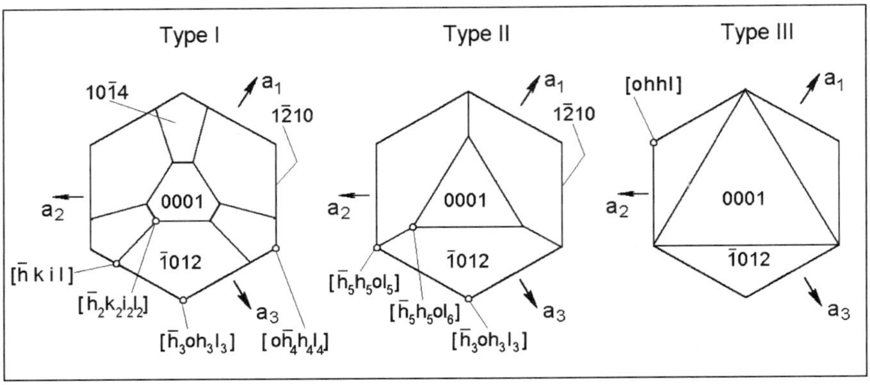

Fig. 2.93 *G*-polyhedron form of sapphire (in [0001] direction) with *L* facets formed at cutting off the vertices

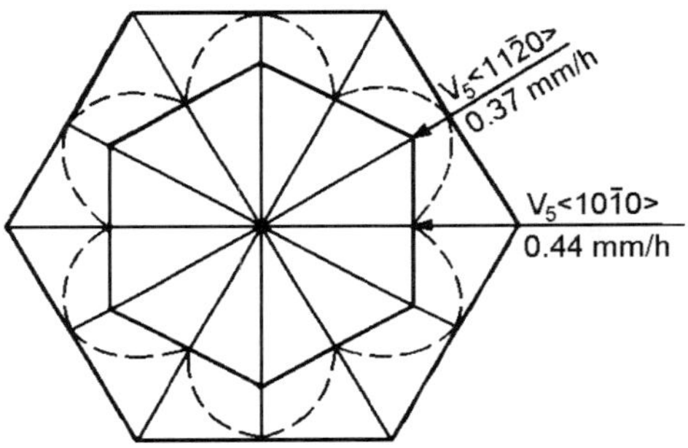

Fig. 2.94 Polar diagram of the rate of dissolution

in ⟨11$\bar{2}$0⟩ and ⟨10$\bar{1}$0⟩ directions were laid off along radius vectors. The internal envelope of the diagram presents the form of dissolution and the external one shows the stationary growth form for the basal plane.

The ideal dissolution form of sapphire that depends on the lattice parameters and is independent of the conditions of dissolution is a parallelogram dodecahedron with rhombohedral symmetry group. The real form depends on the impurities, nature of the solvent, and the conditions of dissolution (Fig. 2.95, Table 2.20) [110].

On all types of facets and at all the temperatures the dissolution rate of ruby is higher. On the surface of ruby spheres a rougher layer is formed.

The role of water in the processes of hydrothermal dissolution of sapphire is investigated in connection with crystallization under hydrothermal conditions. Solubility of sapphire in pure water is very low even at thermobaric parameters.[3]

[3]Solubility of 8 mg of sapphire in 100 ml of water at 1,070 K and a pressure of 6 kbar is 0.0035%.

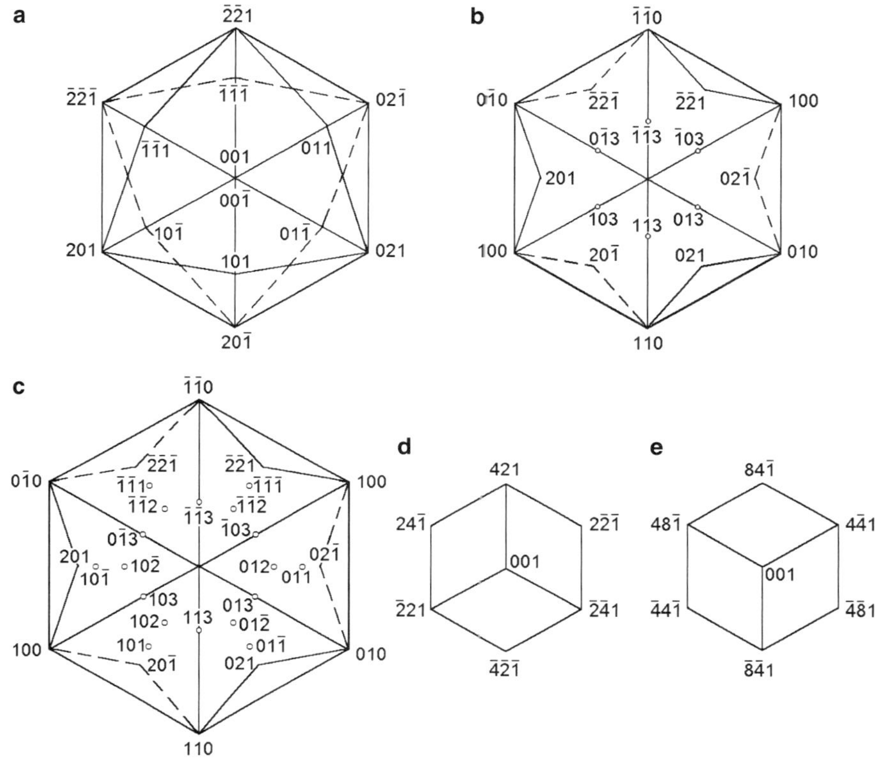

Fig. 2.95 Stereographic projections of dissolution forms of sapphire and ruby: (**a**) ideal dissolution form. (**b–e**) Real dissolution forms obtained under the conditions: (**b**) thermochemical etching of sapphire and ruby spheres; (**c**) dissolution of ruby in carbon oxide; (**d**) dissolution of sapphire spheres in the melts of $PbO–PbF_2$ (1:1); $PbO–PbF_2$ (10:1), 273 K; 0.1 M KF or 0.1 M K_2CO_3, 873 K, 3 kbar; Y_2O_5, 1,223 K [109]; (**e**) dissolution of sapphire spheres in pure PbF_2, 1,223 K

However, in the presence of mineralizers, in particular, alkalis, the solubility sharply increases. The results of experimental investigations of dissolution kinetics of quartz and sapphire in water solutions in autoclaves at 373–723 K are presented in Fig. 2.96 [111]. The numeric results are given in Table 2.21.

The data contained in the table data make it possible to calculate the dissolution rate constant according to the formula:

$$dC / dt = kS(C_k - C_\tau), \tag{2.13}$$

where C_k is the content of Al_2O_3 in saturated solution, C_τ is the concentration of the dissolving substance at the moment of time τ, and S is the surface of the dissolving material.

Table 2.20 Anisotropy of dissolution rates of sapphire and ruby spheres and polar diagrams V_s in basal plane

		V_s (mm/h)		Polar diagrams	
T (K)	*hkl*	Sapphire	Ruby	Sapphire	Ruby
1,773	001	0.033	0.396	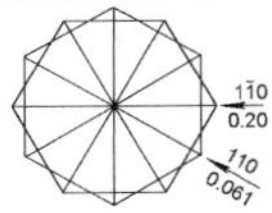	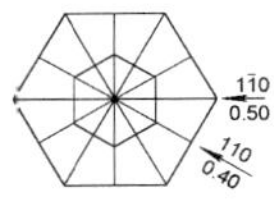
	110	0061	0.445		
	021	0.085	0.530		
	113	0.100	0.538		
1,873	001	0.085	0.475	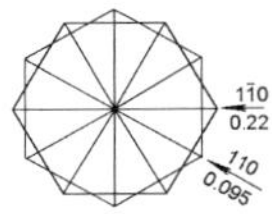	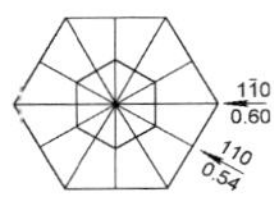
	110	0.095	0.535		
	021	0.110	0.685		
	113	0.210	0.695		
1,973	001	0.330	0.860	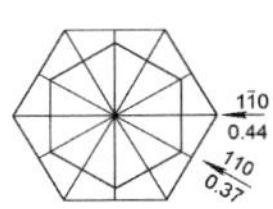	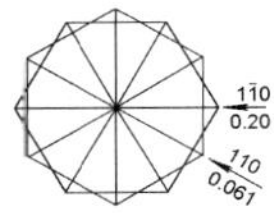
	110	0.370	0.870		
	021	0.380	0.900		
	113	0.400	0.960		

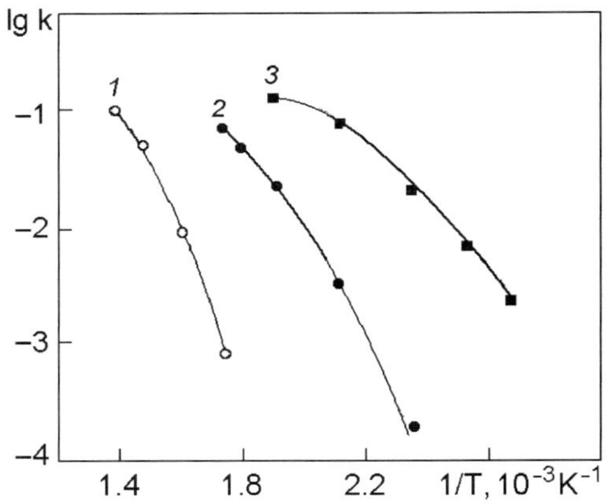

Fig. 2.96 Logarithm of the constants of quartz and corundum dissolution rates in NaOH and NH_4F water solutions as a function of reciprocal temperature (500 atm. pressure): (*1*) corundum, 0.5 M NaOH, (*2*) quartz, 1 M NaOH solution, (*3*) quartz, 1 M NH_4F solution

The lower temperature boundary is limited by 573 K, as at temperatures below the said magnitude it requires too long a time to reach equilibrium: 150 days at 523 K, 838 years at 423 K, 175,000 years at 373 K. Continuous decrease of the value of apparent activation energy – 33.5, 31.2, and 15.4 kcal/mol – with the growth of temperature in 573–623, 623–673, and 673–723 K intervals is explained by the fact that in the process of dissolution the resistance of the surface-adjacent layer to

Table 2.21 Dissolution of sapphire in 0.5 M NaOH water solution at different temperatures and a pressure of 500 atm

τ (h)	V_s (g/l)	τ (h)	V_s (g/l)	τ (h)	V_s (g/l)	τ (h)	V_s (g/l)
573 K		623 K		673 K		723 K	
14	1.3	5	6.2	0.5	5.0	0.167	4
39	3.5	10	7.7	1	7.9	0.5	8.6
70	5.1	15	12.1	1.5	11.0	1	12.4
108	8.7	17.5	14.2	2	12.2	3	20.2
144	11.1	20	13.2	3	14.9	5	20.1
180	14.3	25	15.9	4	16.7	7.5	20.9
234	16.5	30	17.6	5	18.6	10	20.6
360	16.5	36	18.7	7	21.1	15	20.6
		48	17.7	9	20.0		
		60	17.8	12	20.1		
				18	19.6		
				36	20.7		
				48	20.2		

The experiments in which equilibrium takes place are separated by bold line. The average values for calculating the dissolution rate constants are the following: 573 K – 16.5; 623 K – 17.9; 673 K – 20.1; 723 K – 20.5 g/l.

Table 2.22 Contamination of acid solutions by aluminum after treatment of sapphire by boiling acid solutions, (mk g)/mol

Acid solution		Time of boiling (h) 3	6	Acid solution		Time of boiling (h) 3	6
HCl	5 mol	0.04	–	HNO_3	5 mol	1.5	–
HCl	12 mol	0.2	2.6	HNO_3	14 mol	14	35
H_3PO_4	5 mol	6.5	–	H_2SO_4	5 mol	1.8	–
H_3PO_4	12 mol	11	11	H_2SO_4	18 mol	90	150

diffusion through it diminishes. In ref. [111] such an effect is attributed to the decreasing thickness of the layer of adsorbed water molecules. The components dissolved in aqueous medium have a destructive influence on the surface-adjacent layer of water, the intensity of which depends on hydration properties of the ions.

Due to low solubility of sapphire in acids, one can estimate their interaction from the contamination of acids by aluminum (Table 2.22) [112].

In a number of analytical methods for controlling stoichiometry of crystals and the distribution of impurities in them (atomic-emission and atomicsorption spectrometry with flame atomization of substances), the analyzed material is introduced into a flame atomizer in the form of solution. The most suitable solvent for the analysis of sapphire is concentrated orthophosphoric acid.

Orthophosphoric acid is one of a few low-temperature solvents used for corundum crystals. It has the form of reaction mixture containing $H_4PO_4^+$ cations and

$H_3P_2O_7^-$, and $H_4P_3O_{10}^-$ anions formed as a result of H_3PO_4 condensation, and a change of its acid–base function. The dissolving action of condensed phosphoric acids at increased elevated temperatures is caused by loosening of the anionic part of corundum lattice and by passing of cations into solution due to the complexation with $H_3P_2O_7^-$ and $H_4P_3O_{10}^-$.

A noticeable dissolution of sapphire in concentrated orthophosphoric acid starts at 483 K, the optimum temperature being 543–573 K. The conditions of dissolution are shown in Table 2.23 and in Fig. 2.97.

Further growth of temperature is unreasonable, as it leads to condensation of phosphoric acids, and as a consequence to the formation of low-soluble glasslike products. (This phenomenon gave rise to the error of some investigators who reported the polishing action of H_3PO_4 at $T > 673$ K. In fact, it was the effect of pseudopolishing caused by the formation of a glasslike aluminum phosphate film on the crystal surface.)

The mixture of H_3PO_4 with H_2SO_4 possesses more stable properties. The ratio of H_2SO_4 to H_3PO_4 ranges from 1:1 to 1:3 (Table 2.24).

Impurities and dispersivity of samples increase their solubility (Fig. 2.98). The action of stronger solvents (Figs. 2.99 and 2.100) was studied. The dissolution heat of Al_2O_3 in lead fluoride is 9.8 kcal/mol [114].

Table 2.23 Conditions of dissolution of aluminum oxide in condensed phosphoric acid [113] (the volume of initial 70% orthophosphoric acid was 10 ml)

Substance	Weight (g)	T (K)	Dissolution time (min)	Conventional solubility (g/ml)[a]
α-Al_2O_3 powder	0.3...0.4	540	16	0.050
α-Al_2O_3 crystal	0.15	560...570	30	0.16
α-Al_2O_3:Ti powder	0.3	540	13	0.090
α-Al_2O_3:Ti crystal	0.2	570	25	0.020

[a] Mass of the substance (g) passing into solution at 540 K, in reference to 1 ml of initial 70% phosphoric acid

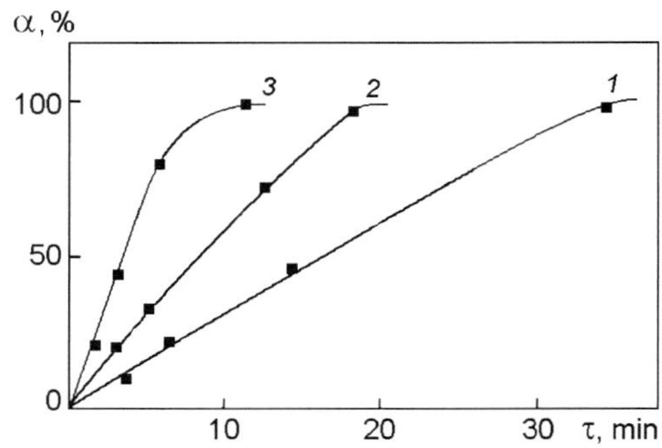

Fig. 2.97 Time dependence of the share of dissolved Al_2O_3. A weight of 0.3 g. (*1*) 215°C; (*2*) 270°C; (*3*) 330°C

Table 2.24 Temperature dependence of dissolution rate for (0001) plane (μm/h)

	H_2SO_4: H_3PO_4		
T (K)	3:1	2:1	1:1
523–568	20		
578–583	60	40	25
583–588	65		
618–628	120		

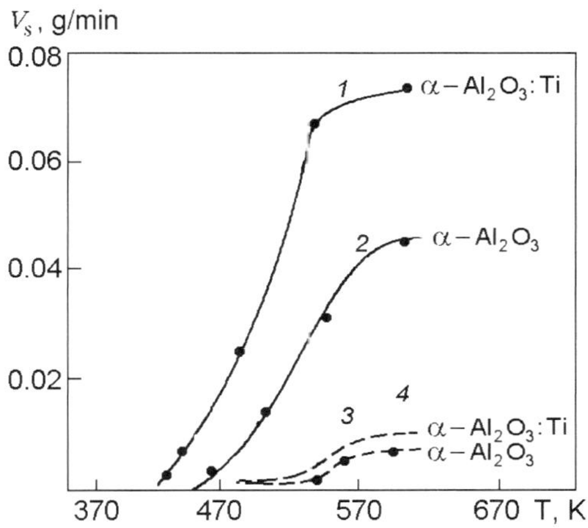

Fig. 2.98 Temperature dependence of sapphire dissolution rate V_s (*1*) Ti:Al_2O_3 with a dispersity up to 40 μm (raw material); (*2*) Al_2O_3 with a dispersity up to 40 μm (raw material); (*3*) Ti:Al_2O_3 crystal; (*4*) Al_2O_3 crystal

The dissolution rate of sapphire also grows after irradiation. As a rule, it is connected with surface damage or with the absorption of bombarding particles, such as krypton, by the surface-adjacent layer.

2.2.2 *Thermochemical Polishing*

In some cases it is necessary to remove the damaged layer that arises on the surface of polished crystals without significant worsening of the polishing quality. This can be realized by thermal treatment of sapphire in vacuum or in gaseous media. However, to obtain the desired effect, high temperatures (over 1,973 K) are required.

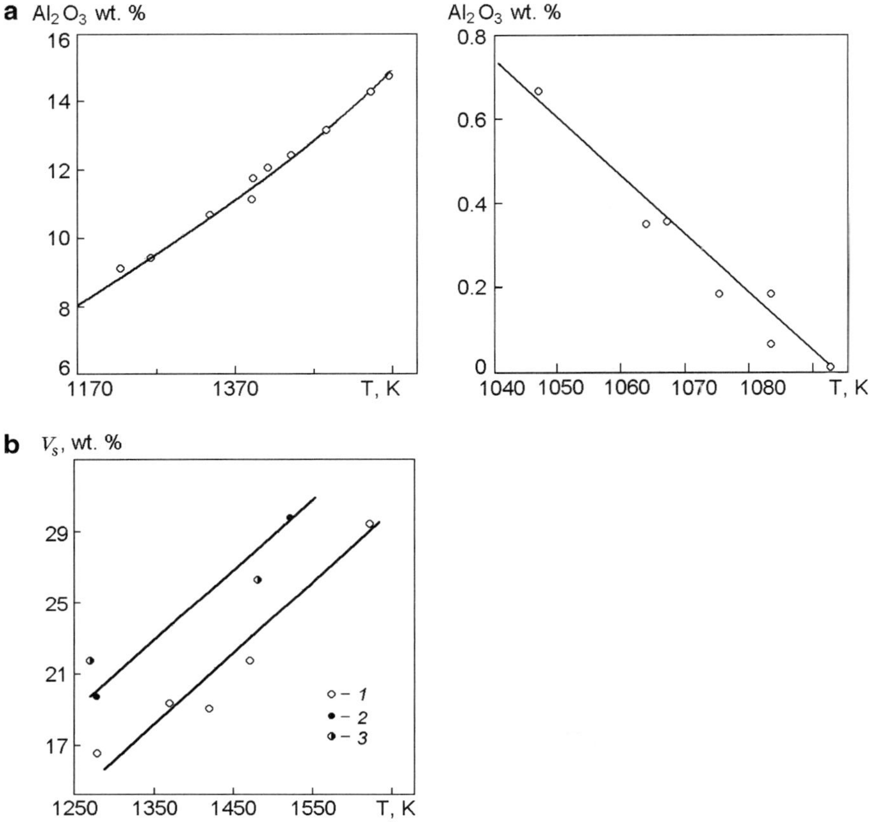

Fig. 2.99 Solubility in PbF_2 melt for (**a**) sapphire and Al_2O_3:Ga, Cr, (*1*, **b**); Al_2O_3: 46% Ga (*2*, **b**); Al_2O_3: 1.32% Ga (*3*, **b**) [114]

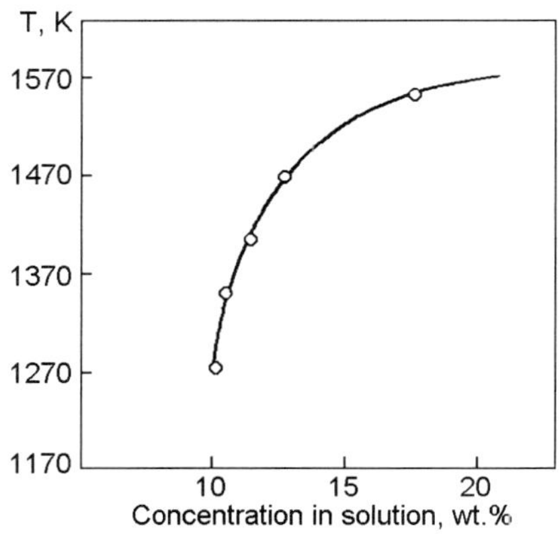

Fig. 2.100 Solubility of sapphire in PbO–0.35 Ba_2O_3–1.2 PbF_2 melt [114]

Thermochemical polishing proceeds at lower temperatures. For instance, in potassium bisulfate melt a noticeable dissolution starts at 723 K. When heated up to temperatures higher than the melting point (419 K), potassium bisulfate transforms into pyrosulfate and then at $T > 673$ K it becomes the normal sulfate:

$$2KHSO_4 = K_2S_2O_7 + H_2O \tag{2.14}$$

Dissolution of aluminum oxide in potassium bisulfate melt proceeds according to the scheme:

$$Al_2O_3 + 3K_2S_2O_7 = Al_2(SO_4)_3 + 3K_2SO_4 \tag{2.15}$$

Stirring increases the dissolution rate. The dissolution rate in $K_2S_2O_7$ is described by the equations [115]:

$$\lg V_s = 4.84 - 4310/T \text{ (at stirring)} \tag{2.16}$$

$$\lg V_s = 5.63 - 5350/T \text{ (without stirring)} \tag{2.17}$$

The activation energies are 23.5 ± 1 and 20.0 ± 0.5 kcal/mol, respectively.

Widespread solvents do not always provide a polishing effect (Table 2.25). As follows from the table, solvents can both polish and etch the surface.

Table 2.25 Dependence of the state of sapphire surface on the type of solvent and regimen

Solvent	T (K)	V_s (μm/min)	Action on surface	Reference
H_3PO_4	570	50	Selective etching	[116]
	690		Polishing	
KOH, concentrated	620		Selective etching	[116]
$Na_2B_4O_7 \cdot 10H_2O$	1,170	1	Weak polishing	
	1,220	2.6	Weak polishing	
	1,270	3	High-quality polishing	
	1,320	4	Breaking of corners/edges	
	1,370	5	Etching	
	1,320	12	Strong etching	
$Na_2B_4O_7 \cdot 10H_2O$ + LiF	1,070	2	Weak polishing	
2% LiF	1,120	3	High-quality polishing	
7% LiF	1,170	5	High-quality polishing	
10% LiF	1,120	5–6	Strong etching	
$Na_3AlF_6/Na_2B_4O_7 \cdot 10H_2O$ (2:1)	1,270	–	Polishing effect	[117]
$KHSO_4/Na_3AlF_6$ (3.5:1)	1,120	–	Polishing effect	[117]
PbF_2/PbO	820–920	–	Polishing effect	
V_2O_3	1,070	1.5	Weak polishing	
	1,170	2	Polishing effect	
	1,270	2.3–2.5	High-quality polishing	
	1,320	3–4	Etching	

In ref. [115] chemical polishing was realized by means of $Na_2B_4O_7$ vapor. The crystal was heated up to 1,273–1,473 K. In ref. [118] the crystal that was heated up to 1,273 K was polished by colloid silicon dioxide under pressure.

The purpose of thermochemical polishing is to remove the surface-adjacent layer at minimal worsening of the quality of polishing. This can be achieved by providing thermodynamic equilibrium of dissolution process, when the rate of crystal dissolution is proportional to the difference between the saturation concentration and the concentration of the dissolved sapphire. The said problem is solved by using V_2O_3-based polishing solvents.

Metavanadates of alkali metals [119, 120] are grown by melting V_2O_3 and, e.g., LiOH. The temperature dependence of the saturation concentration of C_s is described by the relation

$$\lg C_s = A - B/\mathrm{T}, \tag{2.18}$$

where *A*, *B* are constants (Table 2.26). Anhydrous powder of alkali metal metavanadate powder and a calculated amount of aluminum oxide powder are placed into platinum crucible, where this mixture is heated up to 1,123–1,148 K and kept during 2 h. Then the sample is dipped into the crucible. After the treatment the crystal is washed in alkali metal chloride melt.

Colloidal SiO_2solution is one of the best reagents for chemical surface polishing and the medium into which the products of the low-temperature reaction

$$\mathrm{x}Al_2O_3 + ySiO_2 + H_2O \rightarrow Al_{2x}Si_yO_{3x+2y}H_2O \tag{2.19}$$

are removed. Gaseous medium also is effective enough for thermochemical polishing (etching, evaporation). For such purposes hydrogen, fluoride (SF_4, $CCIF_4$, CHF_3), and carbon media are used.

$CClF_3$ is used as a polishing etchant for the planes $(10\bar{1}2)$, $(11\bar{2}0)$, (0001), and $(10\bar{1}4)$.

The etching temperature is 1,670–1,500 K. The maximum etching rate makes 0.6 μm/min at 1,770 K. The treatment in hydrogen flow is most often used.

Hydrogen medium at 1,173–1,973 K activates the processes of healing of surface defects at the expense of the reduction and evaporation of aluminum oxide. Changes in sapphire morphology are observed at temperatures starting from 1,573–1,623 K (Fig. 2.101).

The character of the opening up surface points to the fact that the material is removed layer by layer, parallel to the basal plane. At $T > 1{,}923$ K, the effect of

Table 2.26 Values of constants for sapphire dissolved in metavandates of alkali metals

Solvent	A	B
$LiVO_3$	2.93	3,020
$NaVO_3$	2.64	2,680
KVO_3	2.12	1,940

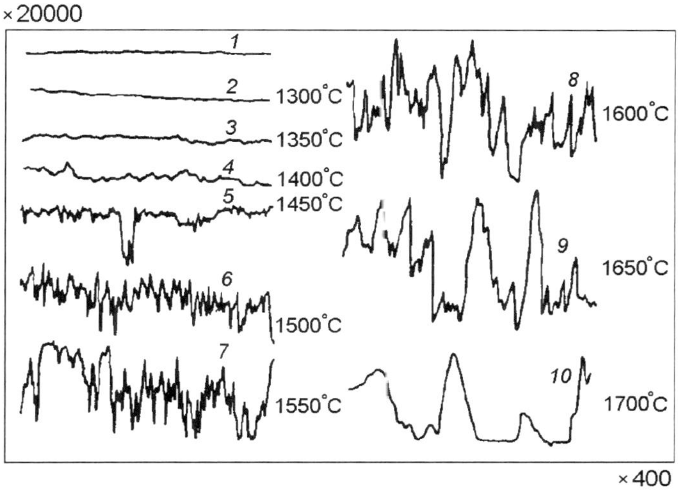

Fig. 2.101 Changes in sapphire surface microrelief at successive thermal treatment in dry hydrogen. The time of holding the sample at each temperature is 3 h. The plane (0001). (*1*) Initial sample; (*2–10*) samples treated at different temperatures; ordinate axis – vertical increase; abscissa axis – horizontal increase

such a selective removal of material vanishes and the process acquires polishing character.

A noticeable loss of the weight of the samples is observed starting from 1,623 K. The evaporation rate is $0.37 \cdot 10^{-2}$ mg/cm^2 h. In dry hydrogen medium at 1,623–1,673 K complete transformation of the damaged surface-adjacent layer without noticeable changes in the surface quality and in the dimensions of the product takes place. The rise of the temperature to 1,973 K leads to a sharp increase in the evaporation rate to 1.8 mg/cm^2 h. As the rate of sapphire evaporation in vacuum is insignificant (at 1,973 K it makes ~0.05–0.1 mg/cm^2 h) and has a dissociative character, the rise of evaporation can be explained by chemical interaction of sapphire with hydrogen and by carryover of the products of the reaction

$$Al_2O_3 + 2H_2 \rightarrow Al_2O + 2H_2O \qquad (2.20)$$

The observed transport of aluminum oxide obviously is connected with Al_2O decomposition and the formation of Al_2O_3 and Al:

$$3Al_2O \rightarrow Al_2O_3 + 4Al \qquad (2.21)$$

The rate of sapphire etching in hydrogen is the logarithmic function of the temperature (Fig. 2.102) [121].

At deposition of silicon on sapphire substrate the latter is annealed in hydrogen in the same reactor where silicon layers are grown. Under the conditions when the rate of hydrogen flow is 1–5 l/min, the annealing time is 15–180 min and removal of material at 1,473–1,932 K is accompanied with polishing. Although the polishing

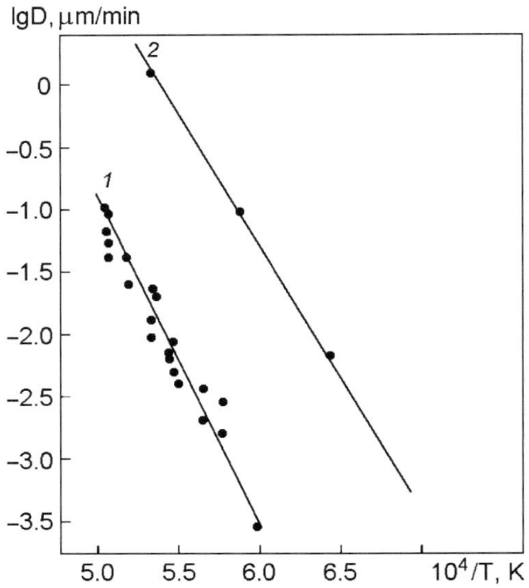

Fig. 2.102 Temperature dependence of the rate of sapphire evaporation in hydrogen; (0001) plane. (*1*) Successive treatment with holding during 3 h at each temperature; (*2*) treatment in the presence of graphite

rate (from 0.003 μm/min at 1,473 K to 0.3 μm/min at 1,923 K) is lower than the rate of polishing in melts, the quality of polishing is higher. As a rule, the silicon films grown on substrates thermochemically polished in $Na_2B_4O_7 \cdot 10H_2O$ and H_3PO_4 melts have polycrystalline structure. Besides the Kekuchi lines, the electron diffraction patterns of the surfaces of these substrates contain diffusion rings. After removal of a 3-μm-thick sapphire layer by etching in hydrogen at 1,773–1,923 K, the electron diffraction patterns clearly show the presence of Kekuchi picture.

Carbon-containing media, for example, CO and CO_2, to a larger extent favor evaporation of the surface-adjacent layer. Therefrom, the processes of recrystallization of the defective layer starting from 1,173 K and reduction of aluminum oxide with its subsequent evaporation starting from 1,623 K at of 10^{-2}–10^{2} Pa pressure take place [116, 117, 121].

Aluminum oxide reduction by carbon oxide in vacuum proceeds according to the scheme:

$$Al_2O_3 + CO = Al_2O_2 + CO_2 \quad (2.22)$$

$$Al_2O_3 + CO = 2AlO + CO_2 \quad (2.23)$$

$$Al_2O_3 + 2CO = Al_2O + 2CO_2 \quad (2.24)$$

$$Al_2O_3 + 3CO = 2Al + 3CO_2 \quad (2.25)$$

The quality of the etched surface depends on the temperature, gas pressure, and crystallographic orientation (Fig. 2.103).

The observed difference between the etching rates of the basal and prism planes is explained by different densities of oxygen atoms in these planes. For nonselective etching (polishing) it is necessary to etch off ~10^{-8}–10^{-7} g/(cm^2 s) layer, depending on the temperature and the crystallographic orientation.

Etching with the help of carbon-containing medium also is used for obtaining figure holes in sapphire. For this purpose special graphite shapers are applied; carbon-containing medium is formed between the shapers and sapphire at sufficiently high temperatures [123].

Ionic plasma etching reduces the temperature of the process to 470–770 K. As etching gases, argon, CF_3, CF_4, SF_6, CHF_3 (with threshold values of the bombarding ion energy of 10–20 eV), CO and CO_2 [at a pressure of 60–80 Pa and $T > 200$ K the etching rate for the plane (0001) is 10^{-5}–$4 \cdot 10^{-4}$ cm/s], as well as BCl_3, BBr_3 (the etching rates are 1.5 nm/s at 310 K and HF power of W/cm^2) are used. The time dependences of the thickness of the etched-off layer are shown in Fig. 2.104.

Magnetic plasma etching through a mask is used for etching bands with a width of about 5 μm in sapphire substrates. For this purpose the gas mixture of 20% of Cl and 80% of BCl is applied.

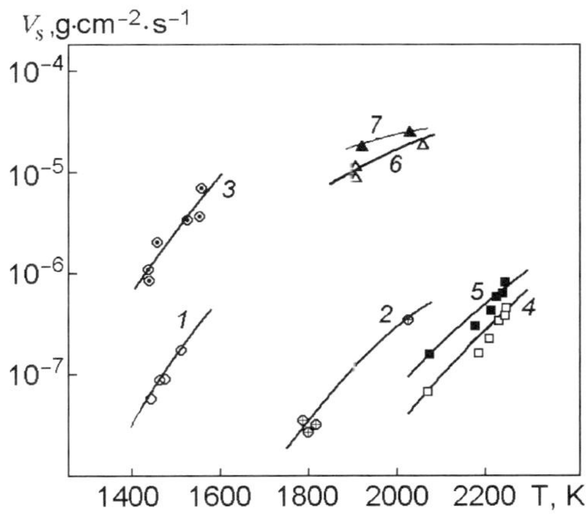

Fig. 2.103 Etching rate for different planes of sapphire depending on the temperature at a constant pressure [122]: (*1*) (0001), *P*= 4–6 · 10^{-2} Pa; (*2*) (0001), *P*= 1–3 · 10^{-1} Pa; (*3*) (0001), *P*= 1–4 Pa; (*4*) (0001), *P*= 10–40 Pa; (*5*) (11$\bar{2}$0), *P* = 1–4 Pa; (*6*) (0001), *P* = 60–80 Pa; (*7*) (1$\bar{1}$02), *P*= 60–80 Pa

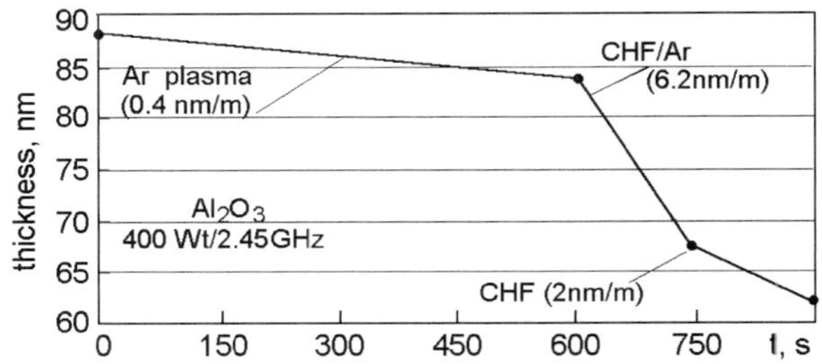

Fig. 2.104 Dependence of the thickness of etched sapphire layer on the time of ion action

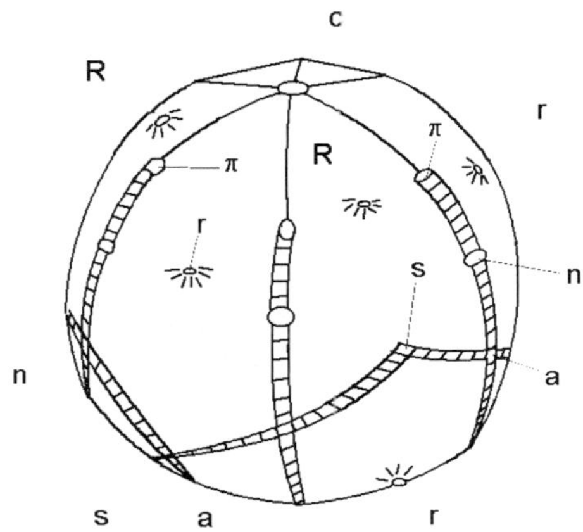

Fig. 2.105 Evaporation forms on annealed sapphire ball. Isolated vertexes: *c*: 0001; *R*: $01\bar{1}2$, *r*: $10\bar{1}1$, *a*: $01\bar{1}2$, π: $1\bar{1}02$, *s*: $11\bar{2}3$, n: $22\bar{4}3$

Evaporation forms of the crystals are similar to the dissolution forms. The mechanisms of evaporation and dissolution seem to be analogous. First, the surface areas with elevated energy are removed. At dissolution in mother liquor medium, the growth and the dissolution forms completely coincide. When the chemical composition of the medium is changed, the dissolution and evaporation forms may change, too. The evaporation form of a sapphire ball in hydrogen or in carbon-containing medium is a combination of isolated vertexes (Fig. 2.105) [124].

The rate of Langmuir evaporation of the plane $(10\bar{1}0)>(\bar{2}110)>(0001)$ [125]. The evaporation morphologies of the planes $(10\bar{1}0)$ and $(\bar{2}110)$ correspond to the layer mechanism; one of the facets (0001) has the monolayer (normal) mechanism. The evaporation energy is 210 ± 20 kcal/mol.

Al_2O_3 molecule has an essentially higher value of formation heat (377 kcal) in comparison with other oxides (e.g., 177 and 192 kcal for Fe_2O_3 and SiO_2, respectively); therefore, it is more difficult to reduce sapphire to the metal.

2.2.3 Corrosion Resistance

Sapphire possesses sufficient resistance to sulfur vapor up to ~1,300 K; to sodium and potassium up to 1,100 K (in Na melt its mass diminishes by 1% during 160 h at 1,200 K); to tin up to 2,200 K; in lead and bismuth up to 1,300 K; in caustic soda up to 1,100 K; in iodine and iodides up to 1,300 K; in hydrogen sulfide and CO_2 up to 1,500 K; and in titanium tetrachloride at 1,300–1,400 K. Constructional sapphire components work in Mg, Al, Cr, Co, Ni, Cs, Pb, Bi, and Zn melts and in their alloys, as well as in rare earth, cast iron, and steel melts. The crystals are resistant to hot water solutions and melts of a number of compounds (Table 2.27, Fig. 2.106).

Corrosion resistance of sapphire surface. As sapphire is widely used in chemical technologies; it is necessary to know the conditions of maximum corrosion resistance of the surface of sapphire products. Corrosion resistance of sapphire to different aggressive media has been studied [112–114], but the obtained data concern the bulk resistance.

The resistance of the surface to aggressive media is influenced by the defectiveness of the surface-adjacent layer formed at mechanical treatment. Moreover, it

Table 2.27 Loss of mass under the action of chemical reagents (%)

Reagents	T (K)	Time of action (h)	Loss of mass	
			Sapphire[a]	Ruby[b]
Boiling in water solutions				
HNO_3 (1:4.6)	400	6.6	0.01	0.03
HNO_3 (1:1)	400	6.0	0.01	0.03
KOH (70%)	400	6.0	0.01	0.03
H_2SO_4 (1:9)	400	6.0	0.01	0.03
HCl (1:1)	400	6.0	0.01	0.03
HF (1:1)	400	6.0	0.01	0.06
Fusing				
Lithium metaborate	1,300	1.0	93.4	–
Lithium megaphosphate	1,300	1.0	7.2	10.0
Potassium hydrosulfate	1,100	0.7	2.4	4.1
Boron anhydrite	1,200	0.7	4.2	0.03
Na_2O_2 + soda 1:5	900	0.25	0.5	0.2
NaF	1,300	1.0	4.1	11.2
H_3PO_4	600	1.0	0.2	–
HCl	1,100	1.0	0.05	0.03
LiF	1,200	1.0	0.4	1.0
Holding in melts				
Copper	1,350	4.0	0.05	–
Steel St3	1,900	4.0	0.02	–
Ni:Cr	1,500	4.0	0.01	–
Carbon steel	1,900	4.0	0.13	–

[a]~7 cm^2 working surface
[b]~2 cm^2 working surface

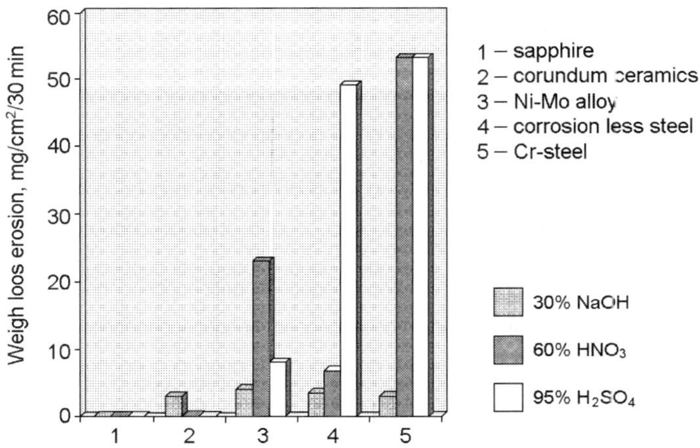

Fig. 2.106 Erosion resistance of sapphire in acids and alkalis

depends on the bulk structure defects that reach the surface; the density and distribution of such defects depend on the crystal growth method. Sapphire is anisotropic; therefore, it is necessary to take into account the nature of atomic bonds of the crystallographic planes, their electronic structure, and energy properties.

To study the influence of the mentioned factors, the crystals were grown by different methods from the same raw material, i.e., the crystals obtained by the Verneuil method. The growth medium was reductive (the Verneuil method), vacuum (the Kyropoulos and HDS methods), and argon (the Stepanov method). The samples of different crystallographic orientations were treated to obtain different degrees of roughness. For comparison, smooth basal and prism planes wedged out in the process of crystal growth were considered. The crystals with the maximum anionic nonstoichiometry were obtained by annealing at 2,253 K in $CO + CO_2$ medium at a residual pressure of 60 Pa. The crystals possessing the maximum cationic nonstoichiometry were produced by annealing in oxidative medium at 2,023 K; the content of O_2 in the annealing space was 12–15 vol%. The influence of point defects was studied by changing the density of cationic and anionic vacancies [125].

The corrosion resistance of the surface was investigated by the method of layer-by-layer etching using 0.3 ml of condensed phosphoric acid spread onto the surface of the sample [100]. After holding the sample at 573 K during 30 min, the film that formed on its surface was dissolved in bidistilled water. The degree of surface destruction was determined from the content of aluminum in the solution. The process of etching was repeated six times. The thickness of the dissolved layer (h, cm) was determined using the ratio:

$$h = \frac{m \cdot 1.89}{\gamma \pi r^2} \tag{2.26}$$

where 1.89 = $Al_2O_3/2Al$ is the ratio of molecular masses; m is the average value of aluminum mass that passed into solution at one etching process; $\gamma = 3.98$ g/cm^3 is the density of Al_2O_3; and r is the radius of the sample, cm. The relative standard deviation from a single result of the measurement of h is 0.15.

The rate of the chemical reaction is a function of the surface energy and of the strength of the crystal lattice bond. It depends on the degree of destruction of the surface and the surface-adjacent layer. Mechanical treatment decreases the strength of the surface more than by two times (Table 2.28), depending on the degree of its destruction.

It is known that dislocations reaching the surface increase the rate of its dissolution. However, at $\rho = 10^3$–10^5 cm^{-2} this dependence is barely noticeable for the same crystallographic plane (Table 2.29). Not only dislocations but also 1.00-mm pores that reach the surface do not have a significant influence on the value of corrosion resistance. The contribution of pores cannot be considered decisive either. An essential increase in the content of chromium apparently raises the strength of the plane $(10\bar{1}0)$ but does not influence the plane (0001) (Table 2.30). Moreover, the value of corrosion resistance is not noticeably influenced by point defects in the form of anionic and cationic vacancies.

At the same treatment, a decisive contribution to the corrosion resistance of the surface belongs to their crystallographic orientation. Irrespective of the methods of crystal growth and thermal treatment, the value of h for the plane (0001) is 0.72–

Table 2.28 Corrosion resistance of sapphire surface depending on the type of treatment [Kyropoulos method, (0001) plane]

Type of treatment	Damaged layer depth (μm)	Thickness (h) of layer removed at one etching act (μm)
Natural facet	0	0.72
Polishing	1–5	0.77
Fine lapping	20–30	0.96
Rough lapping	>40	1.8

Table 2.29 Corrosion resistance of sapphire surface depending on dislocation density

Growth method	ρ (cm^{-2})	Plane	Type of surface treatment	h (μm)
Verneuil	$(1–5) \cdot 10^5$	(0001)	Polishing	0.75
			Natural facet	0.82
Kyropoulos	10^3	(0001)	Polishing	0.77
			Natural facet	0.72
HDS	$5 \cdot 10^3$	(0001)	Polishing	0.83
Stepanov	10^5	(0001)	Polishing	0.73
			Natural facet	0.72
	10^5	$(10\bar{1}0)$	Polishing	0.29
		$(10\bar{1}0)$	Natural facet	0.23
	10^5	$(11\bar{2}0)$	Polishing	0.15
		$(11\bar{2}0)$	Natural facet	0.12

Table 2.30 Corrosion resistance of the surface of doped sapphire grown by Verneuil method

Crystal	Content of doping addition (mass%)	Plane	h (μm)
Al_2O_3	$1 \cdot 10^{-4}$	$(10\bar{1}0)$	0.25
	$1 \cdot 10^{-4}$	(0001)	0.80
Al_2O_3:Cr	$1 \cdot 10^{-1}$	$(10\bar{1}0)$	0.25
	$6 \cdot 10^{-4}$	$(10\bar{1}0)$	0.20
	1.3	$(10\bar{1}0)$	0.12
	1.3	(0001)	0.80
Al_2O_3:Ti	$6 \cdot 10^{-2}$	$(10\bar{1}0)$	0.24
	$6 \cdot 10^{-2}$	(0001)	0.80

Table 2.31 Dependence of V_s of aluminum oxide on crystallographic parameters at etching with $K_2S_2O_7$ (870 K, 1 min)

Crystallographic plane	V_s (μm/min)	Character of etching	Reference
(0001)	2.18 ± 0.2	Selective	[119]
$(10\bar{1}0)$	1.03 ± 0.5	Polishing	[119]
$(11\bar{2}0)$	0.95 ± 0.2	Polishing	[126]
70–80° deviation from *C*-axis	1.33 ± 0.5	Mixed	[126]

0.83 μm; for the planes $(11\bar{2}0)$ and $(10\bar{1}0)$ it equals 0.12–0.15 and 0.12–0.29 μm, respectively. These data agree with the results of etching sapphire surface by potassium bisulfate which is a stronger etchant (Table 2.31).

Corrosion is caused by a number of surface phenomena. The atomic structure in the vicinity of free surface is distinguished by surface relaxation (decrease of the interplanar distances in the surface-adjacent atomic nets) and reconstruction (change of the symmetry in the surface-adjacent layers) characteristic of the crystals with covalent bond. These phenomena raise the surface energy proportionally to the reticular density of the crystallographic planes.

An approximate measure of the surface energy is the quantity of free bonds per unit of the surface. According to some estimations [17], the maximum quantity of such bonds in the plane (0001) is 6.6; in the plane $(10\bar{1}0)$ it is 5.2. Therefore, the plane (0001) has the lowest corrosion resistance. In the direction of the most closely packed row of the lattice, the work necessary for escape of the atoms out of the plane is the least and the process of destruction is the fastest [127]. Another origin of an abnormally low strength of the basal plane is revealed while examining flat nets composed of aluminum and oxygen ions.

In the planes (0001) and $(10\bar{1}0)$ such nets are formed by the atom O–Al–Al–O–Al–Al–O and Al–O–Al–O, respectively. The nets $(10\bar{1}0)$are electrostatically neutral. In the plane $(11\bar{2}0)$ Al– and O– nets alternate with each other and form "packages." The boundaries of these "packages" are located between the flat nets formed by oxygen atoms.

Thus, the nets Al–Al lie only in the basal planes. The bond Al–Al is weaker than Al–O not only chemically, but also mechanically. While examining the surface bonds of the last layer and of some previous layers in the semiendless lattice, it is established [14] that the chipping energy of the plane (0001) is the lowest in the Al layer. The energy of Al_2O_3 lattice is high enough (152 eV). The total contribution of Al^{3+} to the potential ion energy exceeds the one of O^{2-}, but the contribution of the van der Waals energy for O^{2-} (−1.3 eV) essentially exceeds that for Al^{3+} (−0.1 eV) [7]. Therefore, aluminum atoms from the nets Al–Al enter into the reaction more readily.

So, as concerns the technological factors, the most essential influence on the corrosion resistance of sapphire belongs to mechanical treatment. Rough lapping lowers the resistance of the crystal surface more than by two times. Crystallographic factors make a decisive contribution to the corrosion resistance. The resistance to aggressive media of different crystallographic planes differs by several times.

Interaction with tungsten is of great interest, because tungsten is one of a few constructional materials used at the growth and annealing of sapphire. The reaction proceeds according to the equations:

$$W + Al_2O_3 = 1/3(WO_3)_3 + 2Al \tag{2.27}$$

$$W + Al_2O_3 = 1/4(WO_3)_4 + 2Al \tag{2.28}$$

The results of the calculations of the isobar–isothermal potential ΔG_T for these reactions per one gram-atom of W and the temperature dependence of the rate of the interaction between W and Al_2O_3 are shown in Tables 2.32 and 2.33.

Loss of tungsten weight is expressed by the linear time dependence

$$d\Delta P/d\tau = K_i \tag{2.29}$$

where K_i is the interaction rate constant. This constant is practically independent of the mass of Al_2O_3 and the area of contact of the reagents (see Table 2.33). The time dependence of the specific weight loss ΔP is shown in Fig. 2.107.

The constant of the rate of the interaction at 2,473–3,473 K (Fig. 2.107b) is described by the equation:

$$K_i = 9.5 \times 10^2 \exp(-70{,}000/RT) \tag{2.30}$$

Table 2.32 Change of the isobar-isothermal potential of W + Al_2O_3 reaction

	Change in ΔG_T (cal/g at. W)	
T (K)	For reaction (2.25)	For reaction (2.26)
2,100	+157.00	+157.10
2,300	+147.00	+147.10
2,500	+139.00	+139.10
2,700	+131.00	+131.15
2,900	+124.00	+124.15
3,100	+116.00	+116.20
3,300	+107.00	+107.20
3,500	+99.00	+99.25

Table 2.33 Temperature dependence of the rate of interaction between W and Al_2O_3 in vacuum (10^{-4} mHg) [126]

T (K)	Area of contact (cm²)	$K \cdot 10^3$ g/(cm² min) at different masses (g)		
		0.3	0.7	1.0
2,073	2	–	0.056	–
	4	0.055	0.055	0.057
2,273	2	–	0.21	–
	4	0.20	0.21	0.22
2,473	1	–	0.8	–
	2	0.7	0.9	0.8
2,673	1	–	2.1	–
	2	2.4	2.0	2.1
2,873	1	–	5.6	–
	2	5.4	5.5	5.5
3,073	1	9.2	9.1	9.1
3,273	1	25.0	24.9	25.2
3,373	1	47.6	50.1	52.4

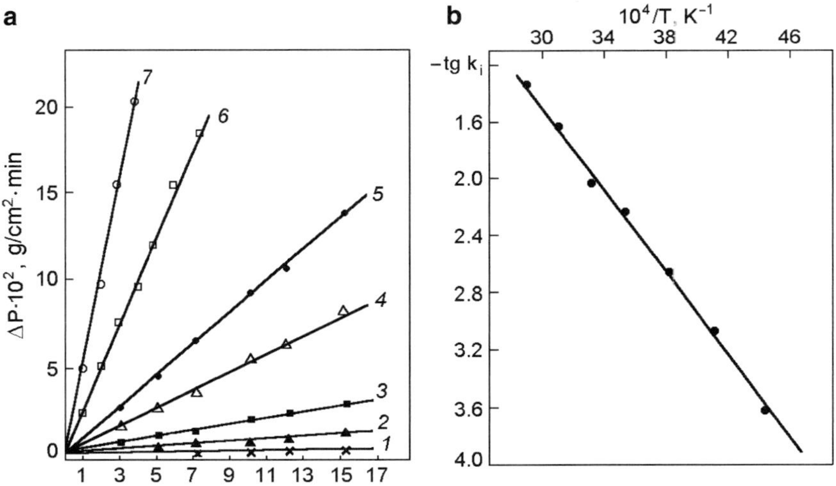

Fig. 2.107 (**a**) Kinetics of the interaction of tungsten with Al_2O_3 (time dependence of the specific weight loss ΔP). (*1*) 2,273 K; (*2*) 2,473 K; (*3*) 2,673 K; (*4*) 2,873 K; (*5*) 3,073 K; (*6*) 3,273 K; (*7*) 3,473 K. (**b**) Temperature dependence of the interaction rate K_i

As follows from Fig. 2.106, for different aggregative states of Al_2O_3, the dependence of lg K_i on $1/T$ is linear. This testifies to the fact that the interaction between W and Al_2O_3 proceeds by the same mechanism within the whole of the temperature interval.

The interaction reaction has the following stages:

- Dissociative evaporation of Al_2O_3

- Chemical adsorptive action
- Desorption and migration of the reaction products

The activation energy at the interaction of a one-atom substance with a two-atom molecule can be calculated using the empirical formula [128]

$$E_a = 1/4D_{xy} \tag{2.31}$$

where D_{xy} is the dissociation energy of XY molecule.

Al_2O_3 molecule has the structure:

```
Al – O – Al
||       ||
O        O
```

As $D_{Al_2O_3}$ = 740 kcal/mol, an average energy of 123 kcal corresponds to one bond. In a first approximation the interaction energy can be presented in the following form:

$$W + Al_2O_2 = O \rightarrow (WO_3)_3 + Al \tag{2.32}$$

where $Al_2O_2 = O$ is the two-atom molecule consisting of an oxygen atom and Al_2O_2 "atom."

For such a molecule D_i ~ 250 kcal/mol. As follows from formula (2.31), E_a ~ 63 kcal/mol, which is in a good agreement with the experimentally defined value E_a ~ 70 kcal/mol [129]. Thus, it can be assumed that the limiting stage of the process is dissociative evaporation of Al_2O_3 with an activation energy of about 70 kcal/mol.

References

1. Klassen-Neklyudova M.V., Bagdasarov Kh.S. (eds.). Ruby & Sapphire: A Coll. Book. Moscow: Nauka, 1974.
2. Aldeberi P., Traverse J.P. J. Am. Ceram. Soc. 65(9), 1982, pp. 460–464.
3. Ishizawa N., Miyata T., Minato I. et al. Acta Crystallogr. 36(2), 1980. pp. 228–230.
4. Belov I.V. Structure of Ionic Crystals and Metallic Phases. Moscow: USSR Academy of Sciences, 1947.
5. Shvid'ko Yu.V., Lucht M., Gerdau E. et al. J. Synchr. Rad. 9, 2002, pp. 17–23.
6. Lucht M., Lerche M., Wille H.C. et al. J. Appl. Cryst. 36, 2003, pp. 1075–1081.
7. Dienes G.I., Welch D.O. Phys. Rev. 7, 1975, pp. 3060–3070.
8. Kislov A.N., Mazurenko V.G., Korzov K.N. et al. Fizika tverdogo tela. 45(9), 2003, pp. 1696–1699.
9. Loong C.-K. J. Eur. Ceram. Soc. 19, 1999, p. 2241.
10. Stashans A., Kotomin E.K., Calais J.L. Phys. Rev. B. 49, 1994, p. 14854.
11. Xu Yong-Nain, Gu Zhong-Quan, Zhong Xue-Fu, Ching W.Y. Phys. Rev. B. 56, 1997, p. 7277.
12. Seitz F. Modern Theory of Solid. New York: McGraw-Hill, 1940, 83pp.
13. Lagerlof K.D.P., Mitchell T.E., Heuer A.H. et al. Acta met. 32, 1984, pp. 197–105.
14. Gue J., Ellis D., Lam D. Phys. Rev. B. 45, 1992, pp. 13647–13656.
15. Gumilevskiy A.A. Watches and Watch Mechanisms. 1(131), 1962, pp. 26–29 [in Russian].
16. Wiederhorn S.M. J. Am. Ceram. Soc. 59, 1969, p. 485.
17. Hartman P. Proceedings of Mineralogical Society. Leningrad: USSR Academy of Sciences. 91(6), 1962, pp. 672–682 [in Russian].

18. Bakholdin S.I., Galaktionov E.V., Krymov V.M., et al. Izv. AN SSSR, Ser. Fiz. 58, 1994, pp. 32–41.
19. Kozlovskaya E.P. Ph.D. Thesis. Moscow: Institute for Crystallography, USSR Academy of Sciences, 1971.
20. Scheuplein R., Gibbs J. J. Am. Ceram. Soc. 47, 1964, p. 81.
21. Seong-Young Yoon, Jong-Yound Lee. J. Mat. Res. 11, 2001, pp. 221–226.
22. Antonov P.I. et al. Production of Profiled Single Crystals and Articles Therefrom. Leningrad: Nauka, 1981. p. 108.
23. Kotchik D., Tresler R., Barber D. J. Mater. Sci. 17(7), 1982, pp. 1977–1987.
24. Young S.Y., LeeJ.Y. J. Korean Association of Crystal Growth. 10(4), 2000, pp. 337–343.
25. Kuandykov L.L. Study of Sapphire Strips Growing by Stepanov Technique at Different Orientations. D. Thesis. St. Petersburg, A.F.Ioffe FTI, 2004 [in Russian].
26. Blistanov A.A. Crystals of Quantum Electronics. Moscow: MISIS, 2000 [in Russian].
27. Harris D.C. Proceedings of SPIE Conference on Window and Dome Technologies and Materials VI, 1999. 3705, 1999, pp. 2–11.
28. Harris D.C., Shmid F., Black D. et al. Proceedings of SPIE Conference on Window and Dome Technologies and Materials VII, 2001. 3050, 2001, pp.226–235.
29. Regan T.M., Harris D.C., Stroud R.M. et al. Proceedings of SPIE Conference on Window and Dome Technologies and Materials VII, 2001. 4375, 2001, pp. 31–40.
30. Hartman R., Perdok D.C. Acta Crystallogr. 8, 1955, p. 521.
31. Renaudo C., Orione P., Causa M. J. Cryst. Growth. 244, 2002, pp. 53–62.
32. Mackrodt W.C., Davey R.J., Black S.N., Docherty R. J. Cryst. Growth. 80, 1987, pp. 441–446.
33. Hartman P., Bennema P. J. Cryst. Growth. 49, 1980, p. 145.
34. Hartman P. J. Cryst. Growth. 49, 1980, p. 166
35. Jung-Hae Choi, Doh-Yeon Kim, Bernard J., Hockey et al. J. Am. Ceram. Soc. 80(1), 1997, pp. 62–68.
36. Kitayama M., Glaeser A.M. J. Am. Ceram. Soc. 88(12), 2005, pp. 3492–3500
37. Jung-Hae Choi, Doh-Yeon Kim, Bernard J., Hockey et al. J. Am. Ceram. Soc. 85(7), 2002, pp. 1841–44.
38. Chuhrov F.V. Minerals: A Reference Book, Vol.2. Moscow: Nauka, 1965, p.65.
39. Galmarini S., Aschauer U., Bowen P., Parker S.C. Poster, Switzerland, 2005. Lausanne, Bath.
40. Kanevsky V.S., Krivonosov E.V., ALitvinov L., Tkachenko S.A. Funct. Mater.. 6, 1999, pp. 370–373.
41. Kanevsky V.S, Krivonosov E.V., Lytvynov L.A.,Tkachenko S.A. Funct. Mater. 7, 2000. p. 156.
42. Belustin A.V. IKAN SSSR Trans. 8, 1953, pp. 247–252.
43. Borodenko Yu.A. Ph.D. Thesis. Kharkov: Institute For Single Crystals. 1986, 148pp.
44. Tkachenko S.A. Ph.D. Thesis. Kharkov: Institute For Single Crystals. 2000, 190pp.
45. Nicoara J., Nicoara D., Sofonea V. J. Cryst. Growth. 104, 1990, pp. 169–174.
46. Dobrovinskaya E.R., Litvinov L.A., Pishchik V.V. Encyclopaedia of Sapphire. Kharkov, 2004, 500pp [in Russian].
47. Malitson I.H. Opt. Soc. Am. 52, 1962, p. 1377.
48. Thomas M.E., Andersson S.K., Sova R.M., Joseph R.I. SPIE, 1997. 3060, 1997, p. 258.
49. Yang D., Thomas M.E., Kaplan S.G. Proceedings of SPIE, 2001. 4375, 2001, p. 53
50. Gravnak D.A., Burch D.E. JOSA. 55, 1965, p. 625.
51. Lingart Yu.K. Sci. Dr. Thesis. Moscow. 1990, 486pp.
52. Es'kov E.V., Sabelnikova M.M., Ignatov Yu.A. VI International Conference on Chemistry of Solids, Kislovodsk, Russia, 2006, p. 510.
53. Harris D. Proceedings of SPIE, 1999. 3705, 1999, p. 786.
54. Flerov V.I., Flerov A.V., Konyaev V.M et al. Kristallografia. 38(1), 1993. pp. 222–228.
55. Sirota N.N., Zhdanov E.F. Dokl. AN BSSR. 11, 1967, pp. 772–773.
56. Muminov V.A., Tursumetov M.A. Izv. AN UzbSSR, Ser. Fiz.-Math. 3, 1977, pp. 60–63.

57. Kviatkovskiy S.F., Konevskiy V.S., Krivonosov E.V., Litvinov L.A. Zh. Prikl. Spektroskopii. 51, 1989, pp. 90–94.
58. Vakhidov Sh.A., Khattamov D.R., Yangibaev M. Izv. AN UzbSSR, Ser. Fiz.-Math. 4, 1976, pp. 65–67.
59. Bayramov B.Kh., Ubaydullaev Sh.B., Khoshkhozhev Z.M. Fiz. Tverdogo Tela. 18, 1976, pp. 2019–2023.
60. Gulchuk P.F., Litvinov L.A., Chernina E.A. Zh. Prikl. Spektroskopii. 43, 1985, pp. 132–134.
61. Pugets A.N., Springis M.I., Valbis J.A. Phys. Stat. Sol. A62, 1980, pp. k85–k87.
62. Evans B.D., Stapelbrock M. Solid State Comm. 33, 1980, pp. 765–770.
63. Springis M.I., Valbis J.A. Phys. Stat. Sol. 123(b), 1984, pp. 335–343.
64. Litvinov L.A., Chernina E.A. Ukr. Fiz. Zhurn. 30, 1985, pp. 811–814.
65. Aluker E.D., Gavrilov V.V., Konevskiy V.S. et al. Optika i Spektroskopia. 70, 1991, pp. 75–81.
66. Hughes A.E. Radiat. Eff. 97, 1986, pp. 161–173.
67. Gumenjuk A.F., Kutovyi S.Y., Grebenovych M.O. Funct. Mater. 12, 2005, pp. 72–77.
68. Voloshin A.V., Litvinov L.A. Funct. Mater. 9, 2002, p. 555.
69. Voronkova E.M., Grechushnikov V.I., Distler G.I., Petrov I.P. Optical Materials for Infrared Engineering. Moscow: Nauka, 1965, p. 268 [in Russian].
70. Butabaev Sh.M., Perelomova N.V., Smyslov I.I. Kristallografia. 14, 1972, pp. 678–679.
71. Borden M.K., Askinazi J. Proceedings of SPIE 1997. 3060, 1997, p. 246.
72. Jacoby K.T., Goodrich S.M. Proceedings of SPIE 2005. 5786, 2005, pp. 188–194.
73. Klein C.A., Schmid F. Proceedings of SPIE 2005. 5786, 2005, pp. 17.
74. Glushak B.L.et al. Study of Material Strength under Dynamic Loadings. Novosibirsk: Nauka. 1992, p. 294 [in Russian].
75. Loveridge-Smith A. et al. Phys. Rev. Lett. 86, 2001, p. 2349.
76. Glushak B.L., Novikov S.A., Ruzanov A.I., Sadyrin A.I. Failure of Strainable Media under Pulse Loadings. Nizhny Novgorod: Nizhegorodsky University, 1992, 192pp [in Russian].
77. Dobrovinskaya E.R., Litvinov L.A., Pishchik V.V. Sapphire and Other Corundum Crystals. Kharkov: Institute of Single Crystals, Folio Publ, 2002, p. 300.
78. Krzhizhanovskiy R.E., Shtern Z.Yu. Thermal-Physic Properties of Non-Metal Materials. Leningrad: Energia, 1973, p. 333.
79. White G.K., Roberts R.B. High Temp. High Pressure. 15, 1983, p. 321.
80. Tolmachev A.M., Kiryanova E.A., Litvinov L.A. Kristallografia. 39, 1994, p. 150.
81. Kompan T.N., Korenev A.S., Lunin A.Ya. et al. Izmeritelnaya Tekhnika, 8, 1999, pp. 38–42.
82. Votinova V.V., Ignatov D.V. Transactions of A.A. Baykov Institute for Metallurgy, Moscow, USSR Acad. of Sciences Publ. 1961. 8th Issue. P. 263–268.
83. Reedy K.P.R., Cooper A.B. J. Am. Ceram. Soc. 65, 1982, p. 634.
84. Reed D., Wuensch B., Bowen H. Annual Report M.J.T. Research in Materials. 1978, p. 241.
85. Frishchat A.N. Ber. Dent. Keram. Aes. 44, 1971, p. 441 [in German].
86. Masataka Ohkubo, Tansumi Hioki, Jun-ichi Kawamoto. J. Appl. Phys. 60(4), 1986, pp. 1325–1335.
87. Compbell N.E. (ed.), High TEmperature Engineering Misscow: IL Publ. 1959, P. 75 [Russian Translated Edition is cited]
88. Papis J., Kingary W.D. J. Am. Ceram. Soc. 44, 1961, p. 459.
89. Kitazawa K.R.L. Coble Am. Ceram. Soc. 57, 1974, pp. 245–250.
90. Abdukadirova I.Kh. The Fifth International Conference on Modern Problems of Nuclear Physics, Samarkand, 2003, pp. 161–162.
91. Abdukadirova I.Kh Pis`ma Zh. Tekhn. Fiz. 26, 2003, p. 17.
92. Abdukadirova I.Kh. Superhard Mater. 1, 2004, p. 12.
93. Urusov V.S., Kravchuk I.F. Geokhimia. 8, 1976, p. 1204.
94. Litvinov L.A. Ruby: Production and New Application Fields. Moscow: NIITEKHIM, 1978, p. 73 [in Russian].
95. Litvinov L.A. Everything About Ruby. Kharkiv: Prapor. 1991, p. 150 [in Russian].

96. Kulagin N.A., Sviridov D.T. Electron Structure Calculation Methods for Free and Impurity Ions. Moscow: Nauka, 1986, p. 202 [in Russian].
97. Bartran R.H., Swenberg C.E., Fournier J.T. Phys. Rev. A139(3), 1965, pp. 941–948.
98. Smite R. J. Biomed. Mater. Res. 21, 1987, p. 917.
99. Voloshin A., Lytvynov L. Funct. Mater. 9, 2002, p. 554.
100. Litvinov L., Druzenko T., Potapova V., Blank A. Crystallogr. Rep. 46, 2001, p. 346.
101. Wolff G.A., Frawley J.J., Heitanen J.R. J. Electrochem. Soc.111, 1964, pp. 22–27.
102. Lavizzari G. Nouveaux phenomenes des corps cristallises, Lugano, 1865.
103. Izmailova V.N., Yampolskaya G.P., Summ B.D. Surface Phenomena in Protein Systems. Moscow: Khimia, 1988 [in Russian].
104. Heimann R.B. Aufloesung von Kristallen. Springer, Wien. New York, 1975.
105. Siesmayer B. Diss. Freie Universitat Berlin. – 1973.
106. Ardamatsky A.L. Diamond Treatment of Optical Elements. Leningrad: Mashinostroenie, 1978 [in Russian].
107. Siesmayer B., Heimann R., Franke W., Lacmann R. J. Cryst. Growth. 28, 1975, pp. 157–161.
108. Berezhkova G.V. et al. IV All-Union Conference on Crystal Growth, Vol. 2 Erevan, 1972, p.39.
109. Liesmayer R., Heimann R., Franke W., Lacmann. J. Cryst. Growth. 28, 1975, p. 157.
110. Govorkov V.R., Smirnova A.E. Kristallografia. 24, 1979, pp. 1095–1097.
111. Rumyantsev V.N. Kristallografia. 20(4), 1975, pp. 870–872.
112. Blank A.B., Litvinov L.A., Pishchik V.V. et al. Zavodskaya Laboratoriya. 8, 1992, pp. 6–8.
113. Blank A.B. Analytical Chemistry in the Studies and Production of Inorganic Functional Materials. Kharkiv: Institute for Single Crystals Publication, 2005, 350pp [in Russian].
114. Timofeeva V.A. Crystal Growth from Solutions—Melts. Moscow: Nauka, 1978, p. 77.
115. USA Patent 3, 753, 775, 1973.
116. Osinskiy V.I., Goncharenko T.I., Lyakhova I.N. PSE. 1, 2003, pp. 93–99 [in Russian].
117. Champion J.A., Clemence M.A. J. Mater. Sci. 2, 1967, p. 153.
118. Japan Patent 4,011,099, 1977.
119. Ryabov A.N., Kiseleva T.I., Kulikova L.B. Zh. Prakt. Khimii. 48, 1975, pp. 407–408.
120. Krasheninnikov A.A. Thesis. Leningrad: LTI, 1973.
121. Batygin V.I., Kulikov V.I., Kulikova S.V. et al. Fizika i Khimia Obrabotki Materialov. 2, 1978, pp.126–130.
122. Sidelnikova N.S., Adonkin G.T., Budnikov A.T. et al. Funct. mater. 4, 1997, pp. 92–96.
123. Beletskiy A.S., Litvinov L.A., Chernina E.A. USSR Inventor's Certificate 1522805, 1989 [in Russian].
124. Voytsekhovskiy V.N. Kristallografia. 13, 1968, pp. 563–565.
125. Konevskiy V.S., Krivonosov E.V., Litvinov L.A. Izv AN SSSR, Neorganicheskie Materialy. 25, 1989, p. 1486.
126. Goryachkovskiy Yu.G., Kostikov V.I., Mitin B.S. et al. Fizika i Khimia Obrabotki Materialov. 4, 1978, pp. 70–74.
127. Tamman G., Botschwar A.A. Z. Anorg. Allg. Chem. 146, 1925, p. 420.
128. Glesston C., Leidler K., Eyring G. Theory of Absolute Reaction Rates. Moscow: IL, 1948, 583pp [Rusisan Translated Edition is Cited].
129. Hart P.B. Br. J. Appl. Phys. 18, 1967, p. 1389.
130. Kingary W.D., Bowen H.K., Uhlmann D.R. Introduction to Ceramics, 2nd ed. 1976, New York: Willey.
131. Bond W.L.. Crystal Technology. Wiley, New York,1980.
132. Handbook of Optics, Vol 2, Ch. 33, McGraw-Hill, 1995.

Appendix

Table 2.34 Coordinates of ions of two molecules in the rhombohedral unit cell α-Al_2O_3, α = 5,124 Å. Origin of coordinates is in the rhombohedron vertex [7]

Ion	Type	*x*	*y*	*z*
1	Al	0	0	0.3750
2	O	−0.2457	−0.1418	0.6334
3	O	0.2457	−0.1418	0.6334
4	O	0	0.2837	0.6334
5	Al	0	0	0.8918
6	Al	0	0	0.3750
7	O	0.2457	0.1418	−0.6337
8	O	−0.2457	0.1418	0.6337
9	O	0	−0.2837	−0.6337
10	Al	0	–	−0.8918

Table 2.35 Interplanar distances in α-Al_2O_3. Cu-radiation

d	I/I_0	(*hkl*)	(*hkil*)	θ	2θ
3.479	75	012	$01\bar{1}2$	11°32′	23°4′
2.552	90	104	$10\bar{1}4$	15°50′	31°40′
2.379	40	1110	$11\bar{2}0$	17°1′	34°2′
2.165	<1	006	0006	18°45′	37°31′
2.085	100	113	$11\bar{2}3$	19°30′	39°
1.964	2	202	$20\bar{2}3$	20°15.5′	40°31′
1.74	45	024	$02\bar{2}4$	23°35′	47°10′
1.601	80	116	$11\bar{2}6$	25°46′	51°32′
1.546	4	211	$21\bar{3}1$	26°46′	53°32′
1.514	6	122	$12\bar{3}2$	27°22.5′	54°45′
1.51	8	018	$01\bar{1}8$	27°27′	54°54′
1.404	30	124	$12\bar{3}4$	29°43′	59°26′
1.374	50	030	$03\bar{3}0$	30°26′	60°52′
1.337	2	124	$12\bar{3}5$	31°22′	62°44′
1.276	4	208	$20\bar{2}8$	33°4′	66°8′
1.239	16	1.0.10	$1.0.\bar{1}.10$	34°11′	68°22′
1.2343	8	119	$11\bar{2}9$	34°20′	68°40′
1.1898	8	220	$22\bar{4}0$	35°49′	71°38′
1.16	<1	306	$30\bar{3}6$	36°52′	73°44′
1.147	6	223	$22\bar{4}3$	37°17.5′	74°35′
1.1382	2	311	$31\bar{4}1$	37°42′	75°24′
1.1255	6	312	$31\bar{4}2$	38°12′	76°24′
1.1246	4	128	$12\bar{3}8$	38°14.5′	76°29′
1.0988	8	0.2.10	$0.2.\bar{2}.10$	39°18.5′	78°37′
1.0831	4	0.0.12	0.0.0.12	39°59.5′	79°59′
1.0781	8	134	$13\bar{4}4$	40°13′	80°26′
1.0426	14	226	$22\bar{4}6$	41°53′	83°46′
1.0175	2	402	$40\bar{4}2$	43°10′	86°20′

(continued)

Table 2.35 (continued)

d	I/I_0	(*hkl*)	(*hkil*)	θ	2θ
0.9976	12	1.2.10	$1.2.\bar{3}.10$	44°15′	88°30′
0.9857	<1	1.1.12	$2.2.\bar{4}.12$	44°55.5′	89°51′
0.9819	4	404	$40\bar{4}4$	45°9′	90°18′
0.9431	<1	321	$32\bar{5}1$	47°34′	95°8′
0.9413	<1	1.2.11	$1.2.\bar{3}.11$	47°42′	95°24′
0.9345	4	318	$31\bar{4}8$	48°9′	96°18′
0.9178	4	229	$22\bar{4}9$	49°20′	98°40′
0.9076	14	324	$32\bar{5}4$	50°5′	100°10′
0.9052	4	0.1.11	$0.1.\bar{1}.11$	50°16′	100°32′
0.8991	8	410	$41\bar{5}0$	50°44′	101°28′
0.8884	<1	235	$23\bar{5}5$	51°35′	103°10′

Table 2.36 Parameters of sapphire crystal lattice in 4.5 ... 375 K temperature interval

T (K)	*a* (Å)	*c* (Å)	*T* (K)	*a* (Å)	*c* (Å)
4.5	4.7562786	12.9819485	312.533	4.759636	12.992930
100.0	4.7563553	12.9823018	322.359	4.759884	12.993740
150.0	4.7566469	12.9834100	332.184	4.760158	12.994581
200.0	4.7572561	12.9854463	342.010	4.760427	12.995414
250.0	4.7581322	12.9883093	351.836	4.760705	12.996261
286.143	4.758977	12.990872	361.661	4.760976	12.997130
286.968	4.758998	12.990932	371.339	4.761251	12.998000
287.125	4.759001	12.990946	374.000	4.761310	12.998155
288.108	4.759020	12.991024	375.000	4.761341	12.998241

Table 2.37 Parameters of sapphire crystal lattice in 593–2,170 K temperature interval

T (K)	*a* (Å)	*c* (Å)	*T* (K)	*a* (Å)	*c* (Å)
593	4.767 ± 0.001	13.008 ± 0.003	1043	4.785 ± 0.001	13.054 ± 0.004
713	4.772 ± 0.001	13.021 ± 0.003	1253	4.794 ± 0.001	13.082 ± 0.004
883	4.775 ± 0.001	13.037 ± 0.004	2170	4.844	13.27

Table 2.38 Contributions of ions occupying different positions of sapphire crystal lattice, to the potential energy (eV) [7]

Ion and its position	Madelung energy	Repulsion energy	van der Waals energy	Total energy
Al^{3+} in the lattice	−109.8	19.5	−0.1	−90.4
O^{2-} in the lattice	−52.8	13.0	−1.3	−41.1
Al^{3+} in the octahedron interstice	−8.9	16.0	0.1	7.0
O^{2-} in the octahedron interstice	5.9	9.4	−3.6	11.7

Table 2.39 Parameters characterizing sapphire crystal lattice

Radius of Al^{3+}	0.51 Å
Area of Al^{3+}	$0.9 \cdot 10^{-16}$ cm^2
$2Al^{3+} + 3O^{2+}$ occupy	$19.9 \cdot 10^{-16}$ cm^2
Oxide formation heat	1,680 J/(g mol), 580 J/at.
Lattice energy	3,663–3,708 (3,708–theoretical) kcal/mol
Covalent interaction energy	314 kcal/mol
Effective ion energy	381 kcal/mol
Atomization energy (experimental)	731 kcal/mol
Madelung constant	
Referred to the interatomic distance	$A_R = 25.155$
Referred to the rhombohedral cell parameter	$A_a = 66.92486$
Repulsion coefficient	1,941 Å^{-1}
Compressibility coefficient	$0.32 \cdot 10^{-3}$ kbar^{-1}
Contact angle	
Angle of wetting with water	Sapphire (0001) – 84°
	Sapphire $(10\bar{1}2)$ – 79°
	Sapphire $(10\bar{1}0)$ – 78°
	Sapphire $(11\bar{2}0)$ – 74°
	Ruby 0.1 Cr (0001) – 73°
Angle of wetting with glycerin	Sapphire (0001) – 73°
	Ruby 0.1 Cr (0001) – 63°
Energy of the surface (0001)[a]	4.83 J·m^2
Average surface energy at 2120 K[b]	0.9 J·m2

[a] Calculated by Hartman [34]
[b] Ref [132]

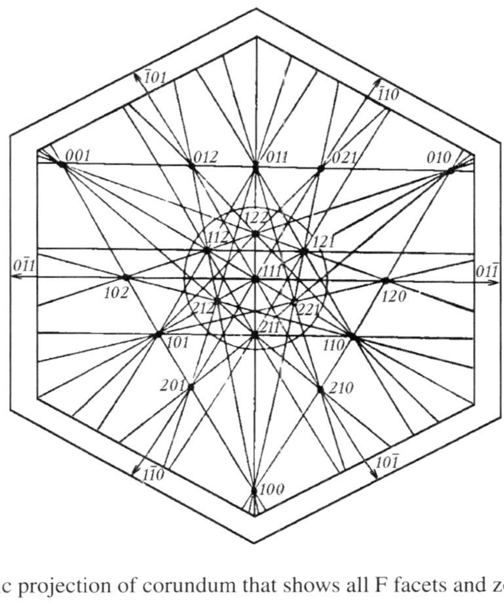

Fig. 2.108 Gnomonic projection of corundum that shows all F facets and zones parallel to chains with intense bonds

Table 2.40 Symbols of crystallographic planes in morphological and structural units

Hexagonal symbols of planes in unit		Rhombohedral symbols of planes in structural unit
Morphological	Structural	
$\{0001\}$	$\{0001\}$	$\{111\}$
$\{11\bar{2}0\}$	$\{11\bar{2}0\}$	$\{10\bar{1}\}$
$\{10\bar{1}0\}$	$\{10\bar{1}0\}$	$\{2\bar{1}1\}$
$\{10\bar{1}1\}$	$\{\bar{1}012\}$	$\{011\}$
$\{02\bar{2}1\}$	$\{10\bar{1}1\}$	$\{100\}$
$\{01\bar{1}2\}$	$\{10\bar{1}4\}$	$\{211\}$
$\{22\bar{4}3\}$	$\{11\bar{2}3\}$	$\{210\}$
$\{11\bar{2}3\}$	$\{11\bar{2}6\}$	$\{321\}$
$\{20\bar{2}5\}$	$\{10\bar{1}5\}$	$\{122\}$
$\{40\bar{4}1\}$	$\{\bar{2}021\}$	$\{\bar{1}11\}$
$\{21\bar{3}1\}$	$\{12\bar{3}2\}$	$\{21\bar{1}\}$

a

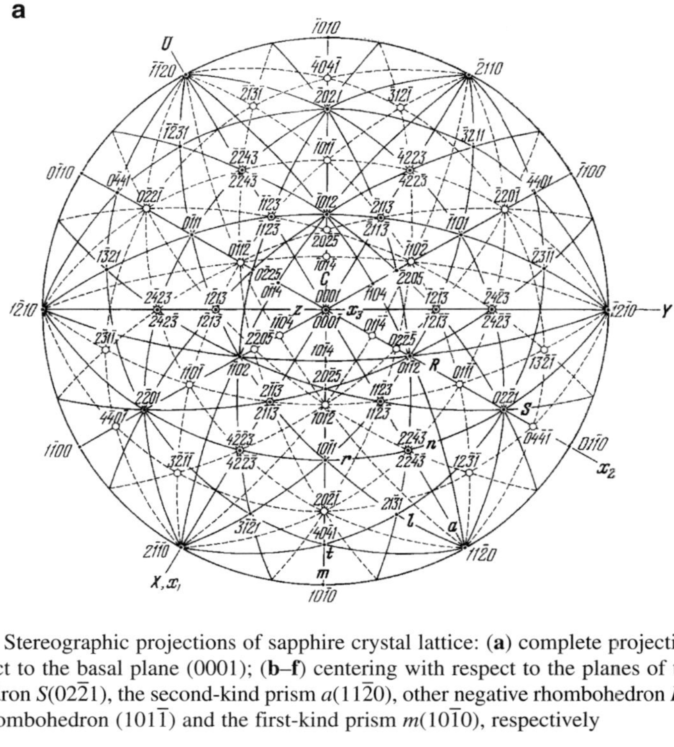

Fig 2.109 Stereographic projections of sapphire crystal lattice: (**a**) complete projection centered with respect to the basal plane (0001); (**b–f**) centering with respect to the planes of the negative rhombohedron $S(02\bar{2}1)$, the second-kind prism $a(11\bar{2}0)$, other negative rhombohedron $R(01\bar{1}2)$, the positive rhombohedron $(10\bar{1}\bar{1})$ and the first-kind prism $m(10\bar{1}0)$, respectively

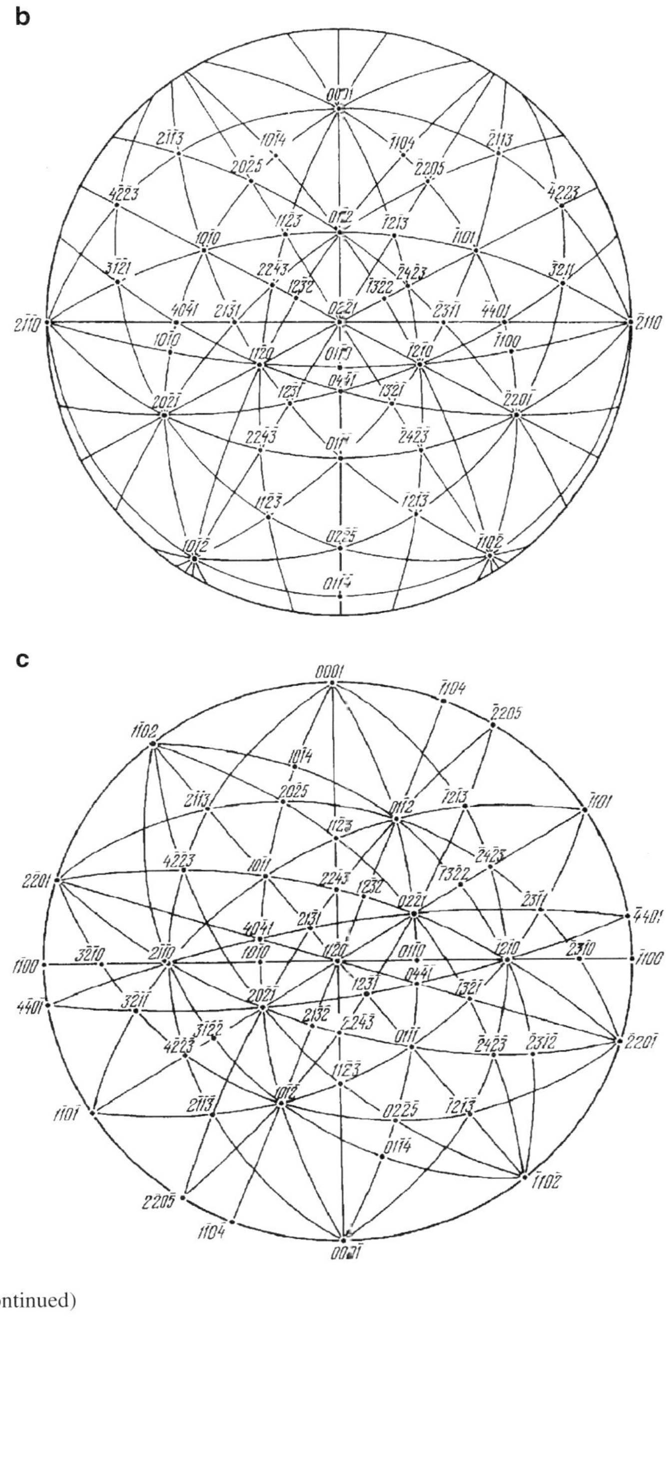

Fig 2.109 (continued)

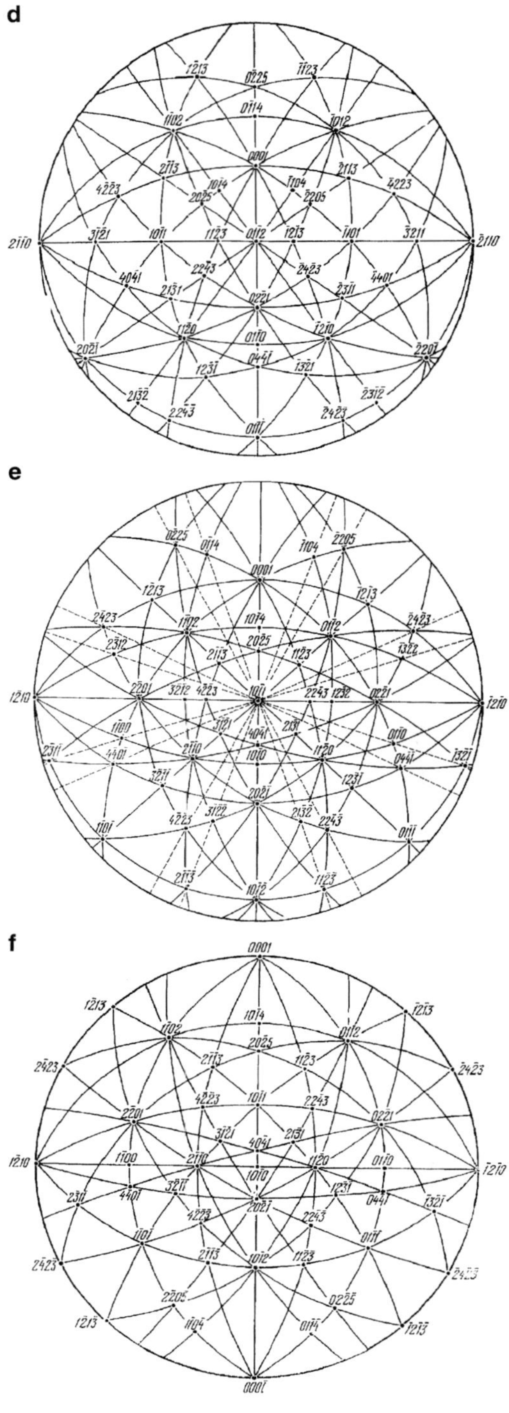

Fig 2.109 (continued)

Table 2.41 Wulff-Bragg's angles for some planes[a]

Symbol of plane	Interface distance (Å)	Order reflections	Angle (θ)
$\{10\bar{1}1\}$	3,479	1	12°48′
		2	26°17′
$\{01\bar{1}2\}$	2,552	1	17°35′
		2	37°09′
$\{11\bar{2}0\}$	2,379	1	18°54′
		2	40°22′
$\{0001\}$	2,165	3	20°51′
		6	45°26′
$\{02\bar{2}1\}$	1,964	1	26°06′
		2	54°42′
$\{11\bar{2}3\}$	1,601	1	28°46′
$\{21\bar{3}1\}$	1,514	1	30°36′
$\{10\bar{1}4\}$	1,510	1	30°41′
$\{10\bar{1}0\}$	1,374	3	34°07′
$\{20\bar{2}5\}$	1,0988	1	44°31′
$\{22\bar{4}3\}$	1,0426	1	47°40′

[a] Symbols of plane correspond to a morphological elementary cell. The angles are calculated by the interplane distances (Nat.Bat. Standarts US 1959.9.#539). Cu-radiation $K_{\alpha 1}$

θ angles for some planes of rhombohedral sapphire cell, $R\bar{3}c$ Cu-radiation

hkl	θ	*I*
01·2	12°48′	74
10·4	17°35′	92
X 11·0	18°55′	42
Z 00·6	20°50′	< 1
11·3	21°43′	100
02·4	26°19′	43
11·6	28°47′	81
21·1	29°56′	3
01·8	30°40′	7
21·4	33°19′	32
Y 03·0	34°11′	48

I is the intensity of X-rays passing through the crystal [133]

Table 2.42 Crystallographic planes of sapphire lattice and their spherical coordinates

Usual forms

		φ	ρ			φ	ρ
c	0001	–	0°00′	*k*	7.7.$\overline{14}$.6	30°00′	72°33′
a	$11\bar{2}0$	30°00′	90°00′	*υ*	$44\bar{8}3$	30°00′	74°38′
m	$10\bar{1}0$	60°00′	90°00′	X	5.5.$\overline{10}$.3	30°00′	77°36′
s	$02\bar{2}1$	0°00′	72°24′	*u*	11.11.$\overline{22}$.3	30°00′	78°41′
δ	$10\bar{1}3$	60°00′	27°41′	*z*	$22\bar{4}1$	30°00′	79°37′
d	$10\bar{1}2$	60°00′	38°12′	*λ*	7.7.$\overline{14}$.3	30°00′	81°05′
r	$10\bar{1}1$	60°00′	57°36′	*υ*	8.8.$\overline{16}$.3	30°00′	82°11′
n	$22\bar{4}3$	30°00′	61°13′	*E*	$33\bar{6}1$	30°00′	83°02′
ξ	7.7.$\overline{14}$.9	30°00′	64°47′	*ν*	$44\bar{8}1$	30°00′	84°46′
w	$11\bar{2}1$	30°00′	69°53′	*ω*	14.14.$\overline{28}$.3	30°00′	85°31′

rr ($10\bar{1}1$):($\bar{1}101$) = 93°58′	*rc* ($10\bar{1}1$):(0001) = 57°36′	*ra* ($10\bar{1}1$):($11\bar{2}0$) = 43°02′
nn ($22\bar{4}3$):($4\bar{2}\bar{2}3$) = 51°58′	*nc* ($22\bar{4}3$):(0001) = 61°13′	*rn* ($10\bar{1}1$):($22\bar{4}3$) = 26°00′
zz ($22\bar{4}1$):($4\bar{2}\bar{2}1$) = 58°55′	*zc* ($22\bar{4}1$):(0001) = 79°37′	rz ($10\bar{1}1$):($22\bar{4}1$) = 35°32′

Rare forms

e $52\bar{7}0$	*p* $50\bar{5}1$	*P* $05\bar{5}1$	*j* 10.10.$\overline{20}$.3	τ $41\bar{5}3$	φ $32\bar{5}1$
f $71\bar{8}0$	*q* $70\bar{7}1$	*o* $22\bar{4}5$	ε 11.11.$\overline{22}$.3	ξ $31\bar{4}2$	*T* 11.8.$\overline{19}$.3
γ $10\bar{1}5$	*R* $01\bar{1}2$	*π* $11\bar{2}3$	ψ 7.4.$\overline{11}$.9	*l* $21\bar{3}1$	*χ* 2.4.$\bar{6}$.1
x $30\bar{3}2$	η $01\bar{1}2$	*M* $11\bar{2}2$	ρ 2.8.$\overline{10}$.9	*σ* 20.5.$\overline{25}$.9	
α $50\bar{5}2$	*A* $05\bar{5}2$	*B* $44\bar{8}5$	*D* 28.4.$\overline{32}$.27	*L* $12\bar{3}1$	
b $70\bar{7}2$	β $07\bar{7}2$	*Y* 10.10.$\overline{20}$.9	*H* 16.4.$\overline{20}$.15	*Σ* 5.20.$\overline{25}$.9	

The spherical coordinates ρ and φ fix the direction and plane; ρ corresponds to the polar distance from the Wulff net center; φ corresponds to the geographic longitude and is measured from the Wulff net circumference

Table 2.43 Angles between the normals to sapphire facets (planes)

c 0001															
90,0	*m* $10\bar{1}0$														
90,0	30,0	*a* $11\bar{2}0$													
32,2	57,8	–	*q* $20\bar{2}5$												
57,6	32,4	43,0	25,4	*r* $10\bar{1}1$											
81,0	9,0	31,2	48,8	23,4	*t* $40\bar{4}1$										
38,2	–	57,6	33,9	47,0	–	$01\bar{1}2$	R								
–	51,8	–	–	84,2	–	–	$10\bar{1}2$								
72,4	–	34,4	59,3	55,6	58,8	34,2	–	$02\bar{2}1$	S						
–	17,6	–	75,4	50,0	–	–	–	–	$20\bar{2}1$						
42,3	–	47,7	20,5	27,3	–	19,6	–	38,8	–	*p* $11\bar{2}3$					
61,2	–	28,8	–	26,0	–	32,0	–	30,5	–	18,9	*n* $22\bar{4}3$				
–	–	17,3	–	25,7	19,2	50,4	–	39,5	–	–	18,4	$21\bar{3}1$		*l*	
–	–	–	–	–	–	–	–	–	–	53,0	–	–	$\bar{3}\bar{1}21$		
–	–	–	–	–	–	–	–	36,4	–	–	–	–	–	$12\bar{3}1$	
69,9	–	–	–	–	–	–	–	49,1	–	21,7	42,1	29,9	–	–	*w* $11\bar{2}1$

Table 2.44 Refractive index in partial transparency region

	T (K)					
λ (μm)	297	1473	1773	1873	1973	2293
0.5	1.774	1.805	1.814	1.817	1.820	1.830
0.75	1.762	1.799	1.802	1.805	1.808	1.818
1.00	1.756	1.785	1.794	1.797	1.800	1.810
1.25	1.751	1.781	1.790	1.793	1.793	1.806
1.50	1.747	1.784	1.787	1.790	1.793	1.803
1.75	1.743	1.777	1.787	1.789	1.792	1.802
2.00	1.738	1.775	1.784	1.787	1.790	1.800
2.25	1.732	1.769	1.778	1.781	1.784	1.794
2.50	1.726	1.761	1.770	1.773	1.776	1.786
2.75	1.718	1.753	1.762	1.765	1.768	1.778
3.00	1.712	1.745	1.754	1.757	1.760	1.770
3.25	1.703	1.738	1.747	1.750	1.753	1.763
3.50	1.696	1.731	1.740	1.743	1.746	1.756
3.75	1.686	1.723	1.732	1.735	1.738	1.748
4.00	1.675	1.715	1.724	1.727	1.730	1.740
4.25	1.664	1.704	1.713	1.716	1.719	1.729
4.50	1.651	1.691	1.700	1.703	1.706	1.716
4.75	1.638	1.680	1.689	1.692	1.695	1.706
5.00	1.624	1.665	1.674	1.677	1.680	1.690
5.25	1.608	1.648	1.657	1.660	1.663	1.673
5.50	1.592	1.627	1.636	1.639	1.642	1.652
5.75	1.566	1.603	1.612	1.615	1.618	1.628
6.00	1.540	1.575	1.584	1.587	1.590	1.600

Refractive index (n_0)

λ (nm)	n_o	λ (nm)	n_o
206.7	1.834	1014.0	1.75547
265.2	1.8336	1033.0	1.755
284	1.82427	1129.0	1.75339
296.7	1.81595	1378.0	1.749
313.0	1.80906	1550.0	1.746
346.6	1.79815	1693.0	1.74368
354.3	1.796	2067.0	1.736
365.0	1.79358	2480.0	1.726
404.7	1.785592	2564.0	1.723
546.0	1.77078	2703.0	1.719
579.1	1.76871	2778.0	1.716
643.9	1.76547	2857.0	1.714
706.5	1.76303	2941.0	1.712
894.4	1.75796	3303.0	1.70231

Sellmeier equation*

$$n_0^2 = 1 + \frac{1.4313493\lambda^2}{\lambda^2 - (0.0726631)^2} + \frac{0.65054713\lambda^2}{\lambda^2 - (0.1193242)^2} + \frac{0.53414021\lambda^2}{\lambda^2 - (18.0228251)^2}$$

$$n_e^2 = 1 + \frac{1.5030759\lambda^2}{\lambda^2 - (0.07402288)^2} + \frac{0.55069141\lambda^2}{\lambda^2 - (0.1216529)^2} + \frac{6.5927379\lambda^2}{\lambda^2 - (20.072248)^2}$$

*Ref [134]

Table 2.45 Refractive index in opaque region

	T (K)			T (K)			T (K)	
λ (μm)	293	1773	λ (μm)	293	1773	λ (μm)	293	1773
6.2	1.51	1.56	10.5	0.6	1.0	21.5	0.3	2.7
6.4	1.48	1.54	11.0	0.25	0.8	22.0	0.1	2.1
6.6	1.46	1.52	11.5	0.05	0.5	22.7	14.5	1.8
6.8	1.44	1.50	12.0	0.05	0.4	23.0	11,0	2.0
7.0	1.41	1.48	13.0	0.05	0.3	23.5	7.3	2.5
7.2	1.39	1.45	14.0	0.05	0.4	24.0	5.8	3.4
7.4	1.36	1.43	15.0	0.1	0.45	24.5	4.9	4.9
7.6	1.33	1.41	15.5	0.2	0.5	25.0	4.6	5.7
7.8	1.31	1.38	15.9	2.1	0.5	25.5	4.8	5.6
8.0	1.28	1.35	16.2	0.4	0.7	26.0	5.3	5.3
8.2	1.24	1.33	16.5	0.4	0.8	26.3	5.8	5.0
8.4	1.21	1.30	17.0	0.6	1.2	27.0	5.0	4.5
8.6	1.18	1.29	17.5	12.2	1.4	27.5	4.5	4.7
8.8	1.15	1.24	18.0	8.0	2.0	28.0	4.2	5.0
9.0	1.12	1.20	18.5	5.2	3.3	28.5	4.0	5.0
9.2	1.08	1.18	19.0	4.0	5.1	29.0	3.9	4.8
9.4	1.05	1.15	19.2	3.8	5.4	30.0	3.8	4.4
9.6	1.00	1.12	20.0	2.5	4.5	31.0	3.6	4.2
9.8	0.96	1.09	20.5	1.3	3.9	32.0	3.5	4.1
10.0	0.90	1.06	21.0	0.6	3.3	33.0	3.4	4.0

Table 2.46 Absorption coefficient for ruby

Wave	λ (μm)	k_λ (cm^{-1})	λ (μm)	k_λ (cm^{-1})	λ (μm)	k_λ (cm^{-1})
Ordinary	0.555	18,000	0.410	24,400	0.260	38,400
Extraordinary	0.550	18,200	0.400	25,000	0.260	38,400

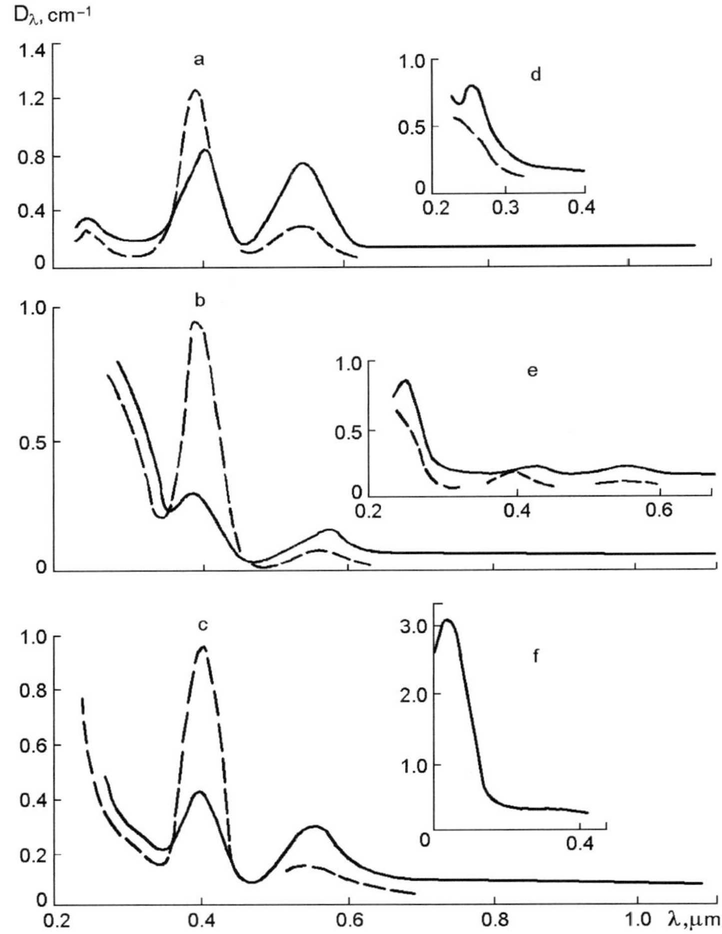

Fig 2.110 Spectral absorption curves for corundum crystals with impurities: (**a**) chromium (ruby); (**b**) vanadium; (**c**) vanadium and chromium; (**d**) iron; (**e**) iron and chromium; (**f**) titanium; (**g**) iron and titanium (blue sapphire); (**h**) cobalt; (**i**) cobalt and chromium; (**j**) nickel; (**k**) manganese; solid lines – ordinary wave, dashed lines – extraordinary wave; in the spectra (**f**) and (**k**) the samples are perpendicular to the optical axis, in the other spectra the samples are parallel to it

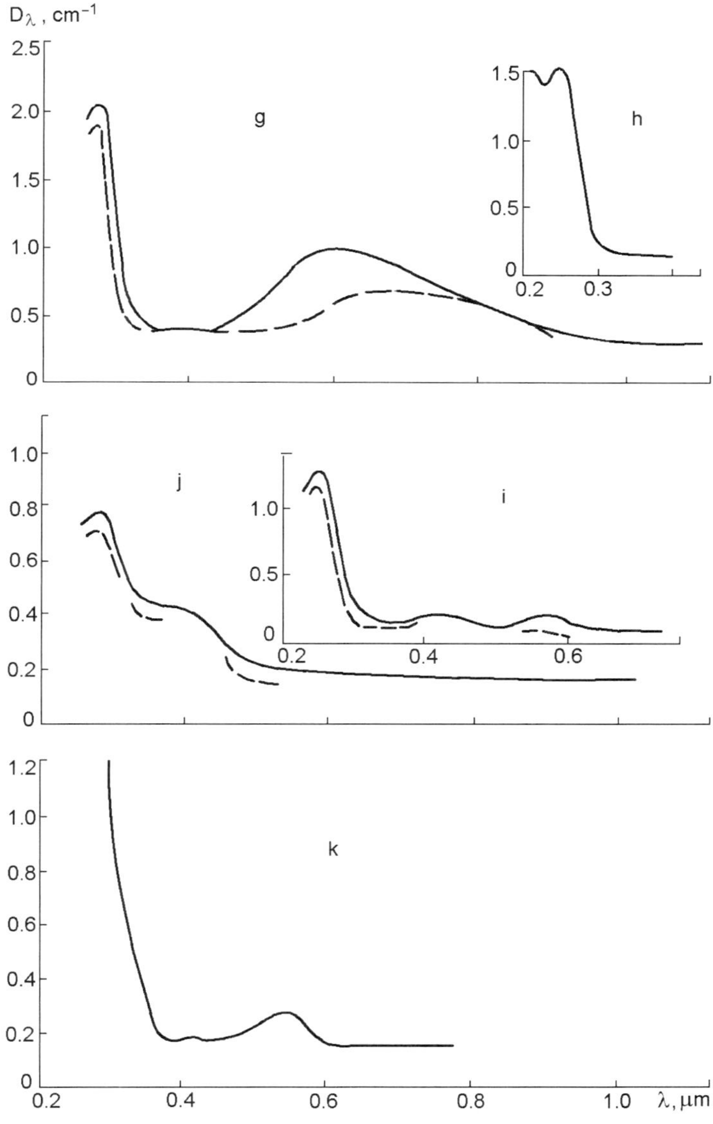

Fig 2.110 (continued)

Table 2.47 Absorption coefficient depending on temperature (cm^{-1})

λ (μm)	*T* (K)										
	300	500	700	900	1100	1300	1500	1700	1900	2100	2300
1	2	3	4	5	6	7	8	9	10	11	12
0.5	0.0049	0.0069	0.0088	0.012	0.014	0.019	0.024	0.039	0.23	0.34	0.39
0.6	0.0041	0.0061	0.0081	0.010	0.013	0.016	0.022	0.035	0.21	0.31	0.36
0.7	0.0035	0.0055	0.0072	0.0090	0.011	0.014	0.020	0.032	0.19	0.29	0.33
0.8	0.0031	0.0049	0.0065	0.082	0.010	0.013	0.018	0.029	0.18	0.27	0.30
0.9	0.0027	0.0045	0.0061	0.0075	0.0092	0.012	0.016	0.026	0.16	0.24	0.27
1.0	0.0025	0.0041	0.0056	0.0068	0.0083	0.011	0.015	0.024	0.15	0.23	0.25
1.1	0.0024	0.0038	0.0051	0.0063	0.0076	0.0097	0.014	0.022	0.14	0.21	0.23
1.2	0.0022	0.0035	0.0046	0.0058	0.0070	0.0088	0.013	0.020	0.13	0.19	0.22
1.3	0.0021	0.0032	0.0042	0.0054	0.0065	0.0082	0.011	0.018	0.12	0.18	0.20
1,4	0.0020	0.0029	0.0037	0.0050	0.0061	0.0076	0.011	0.017	0.11	0.17	0.19
1.5	0.0019	0.0026	0.0035	0.0046	0.0058	0.0071	0.010	0.016	0.10	0.16	0.18
1.6	0.0018	0.0025	0.0032	0.0042	0.0054	0.0066	0.009	0.015	0.099	0.15	0.17
1.7	0.0018	0.0023	0.0030	0.0039	0.0051	0.0062	0.0085	0.014	0.091	0.14	0.16
1.8	0.0017	0.0022	0.0028	0.0037	0.0047	0.0058	0.0079	0.013	0.086	0.13	0.15
1.9	0.0016	0.0021	0.0027	0.0035	0.0044	0.0056	0.0076	0.012	0.081	0.13	0.14
2.0	0.0016	0.0020	0.0026	0.0033	0.0042	0.0053	0.0074	0.012	0.076	0.12	0.13
2.1	0.0015	0.0019	0.0025	0.0032	0.0039	0.0052	0.0071	0.012	0.071	0.12	0.13
2.2	0.0015	0.0019	0.0024	0.0030	0.0037	0.0050	0.0069	0.012	0.068	0.11	0.12
2.3	0.0015	0.0018	0.0023	0.0030	0.0036	0.0049	0.0069	0.012	0.065	0.11	0.12
2.4	0.0014	0.0018	0.0023	0.0029	0.0035	0.0048	0.0069	0.012	0.064	0.11	0.12
2.5	0.0014	0.0018	0.0023	0.0028	0.0035	0.0048	0.0069	0.012	0.063	0.11	0.12
2.6	0.0014	0.0018	0.0023	0.0028	0.0035	0.0049	0.0070	0.013	0.063	0.11	0.12
2.7	0.0014	0.0018	0.0023	0.0029	0.0036	0.0051	0.0074	0.015	0.064	0.11	0.12
2.8	0.0014	0.0018	0.0024	0.0031	0.0039	0.0055	0.0078	0.017	0.067	0.11	0.12
2.9	0.0014	0.0018	0.0025	0.0033	0.0044	0.0061	0.0087	0.020	0.069	0.12	0.13
3.0	0.0015	0.0019	0.0026	0.0037	0.0050	0.0071	0.010	0.024	0.075	0.12	0.14
3.1	0.0016	0.0021	0.0029	0.0042	0.0061	0.0088	0.014	0.030	0.084	0.13	0.15
3.2	0.0017	0.0022	0.0033	0.0050	0.0082	0.013	0.020	0.040	0.097	0.15	0.17
3.3	0.0019	0.0025	0.0039	0.0067	0.012	0.021	0.034	0.057	0.12	0.17	0.20
3.4	0.0021	0.0030	0.0050	0.0096	0.019	0.036	0.057	0.087	0.15	0.21	0.25
3.5	0.0026	0.0040	0.0076	0.016	0.034	0.062	0.090	0.13	0.20	0.27	0.32
3.6	0.0034	0.0061	0.013	0.028	0.061	0.097	0.14	0.18	0.26	0.37	0.43
3.7	0.0052	0.011	0.024	0.049	0.091	0.14	0.20	0.26	0.36	0.49	0.58
3.8	0.010	0.020	0.046	0.081	0.14	0.20	0.28	0.37	0.52	0.68	0.78
3.9	0.021	0.036	0.064	0.12	0.20	0.29	0.40	0.54	0.70	0.91	1.06
4.0	0.034	0.055	0.094	0.17	0.27	0.39	0.55	0.71	0.95	1.18	1.46
4.1	0.051	0.082	0.13	0.23	0.36	0.52	0.74	0.97	1.24	1.68	1.98
4.2	0.076	0.12	0.18	0.32	0.49	0.71	0.97	1.28	1.63	2.04	2.47
4.3	0.12	0.17	0.26	0.43	0.65	0.97	1.28	1.66	2.10	2.63	3.21
4.4	0.16	0.24	0.36	0.50	0.86	1.30	1.66	2.15	2.67	3.34	4.04

(continued)

Table 2.47 (continued)

λ (μm)	T (K)										
	300	500	700	900	1100	1300	1500	1700	1900	2100	2300
1	2	3	4	5	6	7	8	9	10	11	12
4.5	0.23	0.32	0.47	0.77	1.17	1.63	2.15	2.75	3.47	4.23	5.09
4.6	0.30	0.43	0.62	0.97	1.42	2.03	2.71	3.52	0.430	5.25	6.36
4.7	0.40	0.53	0.79	1.20	1.76	2.47	3.31	4.30	5.33	6.51	7.94
4.8	0.51	0.68	1.00	1.49	2.19	2.97	4.14	5.25	6.51	7.82	9.70
4.9	0.64	0.86	1.24	1.82	2.67	3.60	4.97	6.31	7.94	9,70	11.50
5.0	0.78	1.06	1.51	2.19	3.19	4.43	5.98	7.47	9.40	11.50	13.8
5.1	0.96	1.30	1.86	2,65	3.86	5.41	7.24	9.12	11.5	13.8	16.8
5.2	1.17	1.57	2.26	3.26	4.64	6.51	8.71	11.0	13.6	16.3	19.8
5.3	1.44	1.96	2.75	3.98	5.67	7.32	10.4	13.1	16.1	19.3	23.3
5.4	1.74	2.44	3.31	4.86	6.81	9.55	12.4	15.6	19.0	22.9	27.1
5.5	2.12	2.97	4.04	5.84	8.58	11.7	14.9	18.6	22.6	27.1	32.1
5.6	2.51	3.55	4.90	7.02	10.3	14.1	17.8	22.2	26.5	31.6	37.4
5.7	2.94	4.23	5.98	8.58	12.0	16.8	21.2	25.9	30.7	36.9	44.0
5.8	3.41	4.93	7.36	10.3	14.2	20.3	25.1	30.7	35.7	43.0	50.5
5.9	3.86	6.03	8.84	12.4	17.1	24.0	30.2	36.0	42.3	50.1	58.4
6.0	4.79	7.82	11.1	16.3	21.9	28.4	35.2	42.3	50.1	58.4	67.1
6.1	6.81	9.77	13.4	19.9	26.3	33.6	41.4	49.7	57.5	66.6	77.6
6.2	9.62	12.0	16.6	24.2	31.6	39.8	47.9	57.5	66.1	75.9	87.8
6.3	12.6	15.4	20.6	29.3	37.7	47.1	55.8	66.1	75.9	87.1	100
6.4	16.3	19.6	25.9	34.7	44.7	54.9	65.1	75.9	87.1	98.5	112
6.5	20.4	24.4	31.1	41.1	52.5	63.1	75.3	87.8	100	113	123
6.6	25.1	30.2	38.0	48.6	60.7	73.0	87.1	100	113	128	142
6.7	30.4	35.8	44.7	56.7	70.2	83.2	99.2	114	128	142	158
6.8	36.3	43.0	52.5	65.1	80.7	94.0	113	130	144	160	176
6.9	43.0	50.9	62.1	74.7	92.6	103	128	144	161	179	196
7.0	50.1	61.2	73.6	85.8	105	124	142	161	181	199	219

Table 2.48 Elastic constants

Rigidity, $c_{ij} \cdot 10^{10}$ Pa	Compliance, $S_{ij} \cdot 10^{-11}$ Pa^{-1}
$c_{11} = 49.68$	$S_{11} = 0.2353$
$c_{33} = 49.81$	$S_{33} = 0.2170$
$c_{44} = 14.74$	$S_{44} = 0.6940$
$c_{12} = 16.36$	$S_{12} = 0.0716$
$c_{13} = 11.09$	$S_{13} = 0.0364$
$c_{14} = 2.35$	$S_{14} = 0.489$

Table 2.49 Temperature dependence of compliance ($\times 10^{-11}$ Pa^{-1})

T (K)	S_{11}	S_{33}	S_{44}	S_{12}	S_{13}	S_{14}
0	0.2326	0.2151	0.6765	−0.0693	−0.0362	0.0465
373	0.2326	0.2152	0.6776	−0.0695	−0.0363	0.0467
423	0.2330	0.2154	0.6802	−0.0698	−0.0364	0.0471
473	0.2336	0.2158	0.6841	−0.0704	−0.0365	0.0477
523	0.2344	0.2163	0.6889	−0.0710	−0.0366	0.0483
573	0.2353	0.2169	0.6942	−0.0716	−0.0368	0.0491
623	0.2364	0.2176	0.700	−0.0723	−0.0369	0.0500
673	0.2375	0.2183	0.7061	−0.0731	−0.0370	0.0509
723	0.2386	0.2191	0.7126	−0.0740	−0.0370	0.0519
773	0.2399	0.2200	0.7193	−0.0748	−0.0371	0.0530
823	0.2412	0.2209	0.7263	−0.0757	−0.0372	0.0541
873	0.2425	0.2218	0.7334	−0.0766	−0.0372	0.0552
923	0.2438	0.2228	0.7407	−0.0775	−0.0373	0.0564
973	0.2453	0.2237	0.7482	−0.0784	−0.0373	0.0576
1023	0.2467	0.2247	0.7559	−0.0796	−0.0374	0.0587
1073	0.2482	0.2258	0.7638	−0.0808	−0.0374	0.0600
1123	0.2497	0.2268	0.7719	−0.0820	−0.0374	0.0613
1173	0.2512	0.2279	0.7803	−0,0832	−0.0375	0.0626

Table 2.50 Accurate values of relative elongation and differential temperature coefficient of linear expansion (α)

T (K)	Relative elongation, $dL/L \cdot 10^6$	$\alpha = dL/L \cdot 10^6$ K^{-1}	T (K)	Relative elongation, $dL/L \cdot 10^6$	$\alpha = dL/L \cdot 10^6$ K^{-1}
93	−622.162	0.283	573	1887.723	7.752
73	−492.952	2.795	673	2684.375	8.159
273	−99.998	4.911	773	3516.39	8.470
293	0	5.234	1073	6172.19	9.217
373	464.467	6.279	1273	8062.719	9.686
473	1139.9	7.166	1773	13185.68	10.806

Table 2.51 Temperature dependence of λ_T(T) for ruby (W/m K)

T (K)	λ_T	T (K)	λ_T	T (K)	λ_T	T (K)	λ_T	T (K)	λ_T
5.0	170	28.0	920	50	850	100	140	205	108
6.0	206	30.8	950	51.6	700	102	130	209	120
6.6	303	31.0	970	51.7	680	108	120	210	109
7.2	295	34.0	990	55.4	515	116	110	215	130
7.3	299	35.0	1020	55	500	130	130	220	110
8.4	207	35.0	1070	57	450	138	120	225	100
9.0	200	38.4	1050	60.8	308	144	120	230	105
9.1	300	40.8	1060	63	300	150	120	235	210
9.2	350	41.0	1180	64	310	158	110	240	105

(continued)

Table 2.51 (continued)

T (K)	λ_T	T (K)	λ_T	T (K)	λ_T	T (K)	λ_T	T (K)	λ_T
9.2	360	42.4	1090	66.3	260	162	120	248	102
9.8	425	43.8	1130	70.0	250	167	120	251	105
10.4	258	44.8	1130	75	200	171	115	255	110
12.0	443	47.4	1140	76	190	173	110	261	110
14.4	625	47.6	1110	77	190	180	110	268	100
14.5	520	48.4	1060	80	170	184	110	274	120
14.9	540	48.5	1100	82.0	163	187	120	278	100
16.6	625	49.7	845	83	154	190	105	255	100
18.9	750	51.6	700	85	160	192	100	290	100
19.0	780	50	1100	86	130	195	110	295	95
25.0	880	50	1050	90	150	197	120	300	85
25.0	940	50	900	95	140	201	111		

Table 2.52 Temperature dependence of specific heat c_p (J/kg/K)

T (K)	c_p	T (K)	c_p	T (K)	c_p
5.0	0.2	38.8	7.67	40.1	16.5
9.0	0.5	30.0	8.31	43.4	16.5
18.3	2.37	33.0	9.00	46.6	20.87
21.0	3.18	36.8	9.84	49.3	24.66
26.7	4.96	38.8	11.79	51.5	29.20
53.9	32.3	133.0	290.0	216.0	725.0
56.6	39.41	136.0	310.0	223.0	740.0
58.7	49.95	140.0	325.0	226.0	745.0
59.8	42.31	145.0	350.0	231.0	750.0
61.0	45.81	147.0	375.0	233.0	760.0
65.1	50.85	152.0	413.0	237.0	770.0
70.2	60.74	156.0	425.0	243.0	780.0
73.8	65.82	165.0	475.0	247.0	778.0
80.0	98.0	167.0	500.0	253.0	790.0
88.0	100.0	170.0	510.0	260.0	800.0
90.5	110.7	175.0	540.0	263.0	810.0
97.5	130.0	177.0	560.0	267.0	810.0
98.0	134.0	180.0	575.0	269.0	810.0
100.0	145.0	185.0	600.0	272.0	815.0
105.0	164.0	187.0	610.0	276.0	820.0
108.0	170.0	192.0	640.0	278.0	825.0
115.0	200.0	195.0	684.0	284.0	830.0
120.0	225.0	200.0	675.0	286.0	840.0
123,0	240.0	203.0	680.0	290.0	845.0
128.0	255.0	205.0	690.0	295.0	850.0
130.0	275.0	212.0	710.0	298.0	842.0

Table 2.53 Temperature dependence of sound absorption in ruby k_λ at 9,400 MHz frequency (db/cm)

T (K)	k_λ	T (K)	k_λ	T (K)	k_λ
4.2	0	90	0.3	152	10.5
10	0.1	95	3.6	163	11.4
15	0.1	98	3.8	166	11.4
17	0.15	100	4.8	167	11.8
22	0.16	101	4.5	170	12.0
30	0.3	105	4.5	177	12.5
34	0.25	107	5.2	180	12.07
37	0.45	110	5.5	185	13.0
40	0.5	112	6.0	190	13.4
50	0.7	115	6.2	192	13.5
55	0.75	117	6.4	200	13.8
58	0.8	118	6,7	205	14.0
60	0.9	120	7.0	210	14.2
63	1.0	124	7.4	220	14.4
65	1.0	125	7.7	222	14.5
70	1.5	126	7.6	225	14.7
71	1.3	127	8.0	230	14.5
73	1.5	130	8.2	335	14.7
78	1.8	130	7.8	240	14.7
80	2.2	132	8.2	250	14.8
81	2.2	140	9.5	252	14.8
85	2.5	144	9.5	253	15.0
87	2.7	150	10.5	270	15.0
				300	15.3

Table 2.54 Thermodynamic properties

	$H_T - H_{298}$		S_T		ΔH_T		ΔF_T	
T (K)	kJ/mol	kcal/mol	kJ/mol K	kcal/mol K	kJ/mol	kcal/mol	kJ/mol	kcal/mol
298	–	–	50.91	12.16	−1,676,000	−400,300	−1,583,000	−378,000
400	9,211	2,200	77.38	18.48	−1,676,500	−400,400	−1,550,000	−370,000
500	19,260	4,600	99.78	23.83	−1,676,000	−400,300	−1,519,000	−362,800
600	30,230	7,220	119.7	28.60	−1,675,600	−400,200	−1,488,000	−355,300
700	41,830	9,990	137.6	32.86	−1,674,400	−399,900	−1,456,000	−347,800
800	53,760	12,840	153.5	36.67	−1,673,500	−399,700	−1,425,000	−340,400
900	65,940	15,750	167.8	40.08	−1,672,700	−399,500	−1,394,000	−332,900
1000	78,340	18,710	181.0	43.22	−1,693,200	−404,400	−1,362,000	−325,209
1100	90,900	21,710	192.9	46.07	−1,691,500	−404,000	−1,328,000	−317,200
1200	103,600	24,740	203.9	48.71	−1,690,000	−403,600	−1,298,000	−309,400
1300	116,400	27,790	214.2	51.15	−1,688,000	−403,200	−1,262,000	−301,500
1400	129,200	30,850	223.7	53.42	−168,600	−402,800	−1,230,000	−293,500
1500	142,000	33,920	232.5	55.54	−168,500	−402,500	−1,198,000	−286,100
1600	154,900	37,000	240.8	57.52	−163,600	−402,000	−1,164,000	−278,100
1700	167,900	40,090	248.7	59.39	−162,600	−401,700	−1,133,000	−270,600
1800	180,800	43,190	256.1	61.17	−1,680,000	−401,300	−1,102,000	−263,100

Table 2.55 Thermodynamic properties at low temperatures

	c_p		S_T		$H - H_0$		$(H - H_0)/T$		$-(F - H_0)/T$	
T (K)	kJ/ kmol K	kcal/ mol K	kJ/ kmol K	kcal/ mol K	kJ/ kmol K	kcal/ mol	kJ/ kmol K	kcal/ mol K	kJ/ kmol K	kcal/ mol K
17	0.067	0.016	0.017	0.0041	0.218	0.052	0.013	0.0031	0.004	0.0010
25	0.163	0.039	0.056	0.0133	1.063	0.254	0.043	0.0102	0.013	0.0031
50	1.465	0.350	0.457	0.1092	1.734	4.142	0.347	0.0828	0.110	0.0264
75	5.686	1.358	1.721	0.411	98.82	23.601	1.318	0.147	0.404	0.0964
100	12.88	3.075	4.279	1.022	325.4	77.705	3.253	0.7770	1.025	0.2448
125	21.96	5.245	8.093	1.933	757.3	180.86	6.058	1.4469	2.037	0.4865
150	31.99	7.640	12.98	3.100	1431.0	341.74	9.539	2.2783	3.440	0.8217
175	42.03	10.038	18.66	4.457	2398.0	572.62	13.70	3.2721	4.961	1.1848
200	51.25	12.240	24.88	5.943	3524.0	841.55	17.22	4.2078	7.267	1.7356
225	59.45	14.198	31.40	7.500	4909.0	1172.5	21.82	5.2110	9.584	2.2891
250	66,95	15.990	38.06	9.089	6489.0	1549.9	25.96	6.1997	12.10	2.8895
275	74,04	17.683	44.78	10.694	8253.0	1971.2	30.01	7.1680	14.76	3.5260
298	79,92	19.089	50.94	12.165	1003.0	2396,4	33.65	8.0373	17.28	4.1276
300	80,35	19.190	51.50	12.299	1018.0	2432,6	33.95	8.1087	17.54	4.1900

Table 2.56 Enthalpy $H_T - H_{273}$ of aluminum oxide

Temperature (K)	$H_T - H_{273}$	Temperature (K)	$H_T - H_{273}$
400	81.98/19.58	1.400	1,251/298.80
500	178.70/42.67	1.500	1,379.00/329.80
600	284.70/68.00	1.600	1,505.00/360.00
700	396.40/94.68	1.700	1,638.00/391.10
800	512.00/122.28	1.800	1,769.00/422.40
900	631.00/150.62	1.900	1,901.00/454.10
1,000	752.00/179.56	2.000	2,035.00/486.00
1,100	875.00/208.96	2.100	2,170.00/518.20
1,200	1,000.00/238.75	2.200	2,305.00/583.20

Table 2.57 Physical properties

Mechanical properties	Density	3.97–3.99 g/cm^3		
	Hardness			
	Mohs	9		
	Knoop	‖C 1525–1,800 kg/mm^2; ⊥ C 1900–2,100 kg/mm^2		
	Vickers (9.8N)	‖C 16–17 GPa; ⊥ C 18–20 GPa		
	Young's modulus	345–494 GPa (most often used values are 345–470 GPa)		
	Tensile strength	298 K, 400 MPa 770 K, 275 MPa 1,000 K, 335 MPa		
	Compressive strength	2–2.9 GPa		
	(Balk modulus)			
	Flexural strength	‖C 1.03 GPa; ⊥ C 1.54 GPa		
	Compression modulus	250 GPa		
	Rigidity modulus	145–175 GPa		
	Rupture modulus (MOR)	350–690 MPa		
	Elasticity modulus	3.6 · 105 N/mm^2		
	Poisson's ratio	0.27–0.30		
	Jet abrasion resistance (acc. to Mackensen)	0.12 mm		
Thermal characteristics	Melting point	2,323 K (2050 °C)		
	Boiling point	3,253 K (2980 °C)		
	Thermal expansion coefficient ($\times 10^{-6}$ K^{-1})	293–323 K	‖C 6.6	⊥ C 5.0
		310–670 K	‖C 7.0	⊥ C 7.7
		1,270 K	‖C 9.03	⊥ C 8.31
		60° to C-axis:		
			293–323 K	5.8
			1,270 K	8.4
			1,770 K	9.0
	Thermal conductivity	30 K	10,000 W/m K	
		298 K	‖C 32.5 W/m K	⊥ C 30.3 W/m K
		670 K	12.56 W/m K	
		1,500 K	4 W/m K	
	Specific heat	19 mK	159 pJ/K	
		12 mK	41.7 pJ/K	
		9 mK	16.1 pJ/K	
		293 K	181–187 cal/g K	
		1,273 K	0.300 cal/g K	
Electrical characteristics	Resistivity	293 K ‖C (1.2–2.9) · 10^{19} Ω cm, ⊥ C 5 · 10^{18} Ω cm		
		573 K	10^{11} Ω cm	
		1,270 K	10^9 Ω cm	
		1,770 K	10^5 Ω cm	
		2,270 K	10^3 Ω cm	
	Dielectric constant	293 K (20°C)	⊥ C 9.35	‖C 11.53
		370 K	⊥ C 9.43	‖C 11.66
		570 K	⊥ C 9.66	‖C 12.07
		770 K	⊥ C 9.92	‖C 12.54
		970 K	⊥ C 10.26	‖C 13.18
		293 K, 10^3–10^9 Hz	‖C 11.5	⊥ C 9.3
	Dielectric tangent loss	293 K, 10^{10} Hz	‖C 8.6 · 10^{-5}	⊥ C 3.0 · 10^{-5}
	Dielectric strength	48 × 10^6 V/m		

Chapter 3
Radiation Effects in Sapphire

In modern technical devices, constructional elements made of sapphire work under extreme conditions (high temperatures, mechanical loads, and irradiation doses). For instance, in spacecraft, nuclear reactors, and thermonuclear units the crystal structure of sapphire undergoes changes on the atomic level that result in the appearance of various defects in the crystal. If the quantity of induced radiation defects (RDs) is large, the physical properties of the material, such as its electrical conductance, strength, and dimensions, may undergo noticeable changes. Such changes are somewhat unusual in character. RDs differ from defects that arise in the crystal growth process and subsequent mechanical and thermal treatments. Some radiation formations (e.g., superlattices of vacancy pores) have not been obtained by any other method.

High-energy particles that participate in elastic and inelastic interactions with target nuclei cause shifts of the crystal lattice atoms. At low energies of bombarding particles, such shifts lead to the formation of individual vacancies and individual interstitial atoms. Such pairs are formed when the energy imparted to an atom of the crystal lattice by a bombarding particle is higher than some threshold value. At energies greatly exceeding the threshold shift energy, a cascade of shifts is formed. Along the path of this cascade movement, individual vacancies and interstitial atoms arise, as well as their complexes. Then, during the process of establishing a thermal equilibrium between the heated cathode-adjacent region and the rest of the crystal, these defects undergo structural rebuilding by a diffusion mechanism. Some of these disappear due to annihilation of the Frenkel pairs, when a vacancy meets with an interstitial atom. Other defects undergo changes in their size, shape, and location. Thereby, interstitial clusters are transformed into interstitial-type dislocation loops. Vacancy clusters may develop in two ways depending on the temperature, the type of crystal structure, and other factors. One of these is the formation of vacancy-type dislocation loops, which are a kind of hole in the crystallographic planes. The other way is the formation of vacancy pores of different size. Small vacancy pores have a faceted shape corresponding to the type of mother crystal; large vacancy pores have the form of rounded hollows.

Besides the defects caused by nuclear reactions between the bombarding particles and the atoms of the crystal, different kinds of transmutants appear which are distributed in the matrix of the material as impurities. These are inert gases such as

E.R. Dobrovinskaya et al., *Sapphire: Material, Manufacturing, Applications*,
DOI: 10.1007/978-0-387-85695-7_3, © Springer Science+Business Media, LLC 2009

helium, krypton, and xenon. Moreover, other foreign elements also are formed in the irradiated material. Such impurity-induced violations may be retained in the crystal lattice site (substitutional impurity) or reach the interstitial space (interstitial impurity). While migrating within the crystal in the process of diffusion, the impurity-induced violations (and especially, high-mobility inert gases) interact with the crystal's own defects, thus forming mixed defect clusters. The impurity defects deposit on the grain boundaries, dislocations, or other large defects and form aggregates that may gradually transform into second-phase formations. While interacting with vacancy pores, gaseous impurities may accumulate in the pores. The induced radiation defects bring about changes in the strength characteristics of the material (such as shear stresses, yield limit, ultimate strength, and hardness).

Concerning the nature of radiation strengthening, currently there are two explanations to understand this phenomenon. In one, of these such strengthening is attributed to the fact that RDs are additional centers of vacancy fixation that reduce the efficiency of the action of dislocation sources. The second approach explains this phenomenon by the formation of crystal lattice defects that serve as barriers preventing the motion of dislocations in the slip planes.

It is clear that when RDs retard the motion of dislocations, plastic deformation is hindered. This leads to an increase in the yield limit and strengthening of the crystals. When reaching these barriers the dislocations cling to them, but their side wings continue sliding. As the angle between the dislocation wings decreases, the pressure exerted on the barriers rises. With the increase in the stress experienced by the dislocations, the latter break away from the barriers when the previously mentioned angle reaches some critical value, after which they realign and continue moving. The stronger the barrier is, the smaller the critical angle corresponding to it. Radiation defect clusters are located in the slip planes chaotically and have different dimensions, so dislocations often find the path of easiest slip over weak barriers. As the applied stress increases, the dislocations keep their motion until they cover the entire slip plane and all the barriers. The additional stress required for such a motion creates the additional initial yield limit for nonirradiated crystals responsible for radiation strengthening. As a rule, radiation strengthening of materials is accompanied by a decrease in their plasticity due to irradiation embrittlement. Clarification of the nature of such an effect allows establishment of the possible factors that give rise to it, as well as the methods for its suppression.

Irradiation embrittlement is often observed in crystals containing blocks. The appearance of transmutants, such as inert gases, inside these blocks during irradiation leads to migration of these impurities at elevated temperatures to the boundaries of individual grains. In particular, helium (insoluble in metals) penetrates the grain boundaries in the form of bubbles, thus weakening these boundaries. So, the irradiated material's plasticity diminishes due to the decrease in the strength of the grain boundaries caused by the appearance and growth of helium bubbles and other transmutant formations. The degree of embrittleness becomes more intense also owing to the above-mentioned irradiation strengthening of the material inside the grains. While the grains are becoming stronger, the strength of the boundaries between the grains diminishes. Everything seems to indicate that the considered facts are the main causes of irradiation embrittlement.

RDs, in combination with impurities, vacancies, their complexes, and biographic defects that form local energy bands in the forbidden zone, have a different thermal stability and influence on the properties of the crystal.

The set of physical effects related to the presence of high concentrations of RDs in sapphire can be extended. Such effects include radiation-stimulated diffusion, radiation-induced phase transformations, blistering, and formation of the defects' superlattice. Knowledge of the response of sapphire to such factors will allow enhancement of its resistance and adaptation to more complicated working conditions.

3.1 Changes on the Surface

Within the surface-adjacent layer of the crystal point and cluster defects of the Al^{3+}:$[O^{2-}]_6$ type arise. The density of dislocations increases and dislocation loops are generated. The unique characteristics of the defect formation in the electronic structure of the surface-adjacent layer depend on the type and energy of radiation. Radiation damage to the crystal surface influences the kinetics of atomic diffusion and other processes.

Surface relief. Bombardment of sapphire crystals at room temperature with 10 keV Ar ions at 10^{20} cm^{-2} doses and an incidence angle of 70° lead to removal of the surface-adjacent layer to a depth of ~40 μm. Irradiation of the ground crystal surfaces with normal-incidence ions diminishes the surface roughness value R_z from 0.85 to 0.65 μm. For polished crystal surfaces with a 6-μm-deep layer removed, the roughness decreases from $R_z = 0.03–0.04$ μm to $R_z = 0.01–0.02$ μm [1]. At inclined incidence of the ions the surface roughness increases owing to the formation of a microstructure.

The increase of the energy of the bombarding ions (Kr, 40 keV, with doses exceeding 10^{15} cm^{-2}) gives rise to radiation damage of the crystal lattice. The ratio for the shifted Al and O atoms is approximately 1:3 [2].

Amorphous sapphire surfaces are formed under irradiation at an accumulated energy of approximately $3 \cdot 10^{23}$ keV/cm^3 (~44 kJ/mm^2), for instance, at irradiation with Y ions (300 keV, $5.8 \cdot 10^{17}$ cm^{-2} dose) [3]. The threshold of amorphous layer formation at irradiation by Zr and Zn ions is of the same order. The depth of this layer reaches 40–100 nm. In accordance with the data [4], the saturation dose for sapphire corresponding to the onset of amorphous layer formation is 10^{14} ions/cm^2 (^{85}Kr, 10 keV) (for zirconium and diamond such doses are $3 \cdot 10^{13}$ ions/cm^2 and $3 \cdot 10^{14}$ ions/cm^2, respectively). The depth of the amorphous layer defines the value of microhardness at loads < 50 g. Despite the fact that the depth of ion incorporation is essential enough (on the order of 1,000 Å), the crystal lattice distortions can be removed partially by annealing in oxygen at 870–1,870 K.

The composition of surface-adjacent layers may vary at electron irradiation. The loss of some portion of oxygen ions results in an excess of aluminum ions.

Thinning of sapphire samples with 6 keV Ar^+ ions also is accompanied by the formation of amorphous surface layer. Subsequent irradiation in an electron microscope leads to the appearance of nonequilibrium γ-Al_2O_3 phase [5].

The absorption bands in the IR-region at 3,450 cm^{-1} ($2.1 \cdot 10^{19}$ protons/cm^2) and at 2,570 cm^{-1} ($1.7 \cdot 10^{19}$ deuterons/cm^2) are attributed to the formation of hydroxyls (deuteroxyls) within the surface-adjacent layer, the depth of which is dependent on the ion energy and mass.

Hygroscopicity of the surface was established after irradiation with 15 keV H^+, D^+, He^+ ions at doses exceeding $2 \cdot 10^{17}$ ions/cm^2 [6].

Blisters (gas-containing pores) are formed within the surface-adjacent layer irradiated with hydrogen ions due to the bonds between vacancies and hydrogen. When the concentrations of hydrogen reach higher than 10^{18} cm^{-2}, the blister density reaches its saturation. At certain critical values of hydrogen pressure of hydrogen in the pores and upon heating, thermally activated desorption of hydrogen (blistering) is observed in sapphire [7]. The resistance of the surface to blistering is the highest for the plane $(10\bar{1}0)$, medium for the plane (0001), and the lowest for the plane $(10\bar{1}2)$. This resistance depends on the interplanar distance but not on the density of the atoms.

Oxygen-ion bombardment of the surface also may lead to the formation of oxygen bubbles within the surface-adjacent layer [8] and cause the appearance of excessive aluminum on the surface.

Changes in the bulk of the surface-adjacent layer caused by incorporation of ^{1}H, ^{12}C, ^{14}N, ^{24}Mg, and ^{40}Ar ions (100–500 keV, 10^{16} cm^{-2} doses) are defined by the lattice anisotropy. Larger expansion is observed along the *C*-axis.

Changes in the structure. After O^+-irradiation (200 keV energy, $5 \cdot 10^5$ cm^{-2} doses) and annealing in a vacuum for 30 min at 1,170 K, small (~30 nm) dislocation loops are observed with a density of $\sim 5 \cdot 10^{15}$ cm^{-3}. With an increase in the annealing time, the density of the loops diminishes by two orders, and complexes of point defects arise (with an electron microscope these are seen as ~15 nm black spots) [9].

Desorption of the atoms and matrix ions adsorbed on the surface is stimulated by electron and neutron irradiation.

Sufficient conditions for stimulated desorption occur if the difference of electronegativity values between matrix components $\zeta \geq 1.7$, the valence of each matrix component is maximized, and the ions have enough space for desorption [10].

Auger-spectroscopy shows violations of the stoichiometry of the surface-adjacent layers. These are established from the decrease of the Auger-peaks associated with oxygen (~505 eV) and the appearance of a new peak belonging to reduced Al (~64 eV). This indicates stimulated oxygen desorption.

Neutron irradiation. Reactor γ-neutron irradiation of extrapure sapphire samples with relatively low doses ($<10^{17}$ cm^{-2}) results in the appearance of induced absorption bands with maxima at ~260 and 350 nm in the spectra. With an increase in the radiation flux the inherent absorption edge shifts sharply up to the visible region of the spectrum. This is connected with the formation of point defects and the appearance of color centers.

The IR reflection spectrum of an initial standard sapphire (Fig. 3.1a) is characterized by the presence of bands with maxima at 735–760, 600–620, 505, and 465

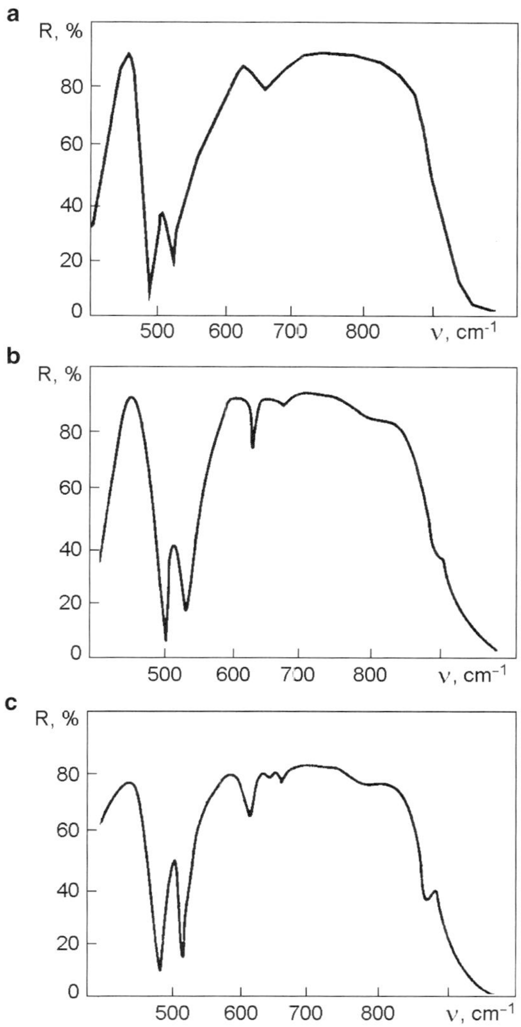

Fig. 3.1 IR-spectra of sapphire at neutron irradiation with different doses: (**a**) initial sample (sapphire grown by the Verneuil method); (**b**) $\Phi = 5 \cdot 10^{19}$ cm^{-2}; and (**c**) $8 \cdot 10^{19}$ cm^{-2}

cm^{-1} [11]. The 735–760 and 600–620 cm^{-1} bands correspond to the valent oscillations of the bonds Al–O, in particular, to asymmetric v_{as} and symmetric v_s valent oscillations in the octahedrons near frequencies of 736 and 614 cm^{-1} [12]. The 505 and 465 cm^{-1} bands correspond to the deformation oscillations v_d of these bonds. D_1 is the optical density at 257 nm; D_2 is the optical density at 358 nm; *a* and c are the lattice parameters; *d* is the interplanar distance; and 2θ is the reflection angle.

At irradiation in the BBP-CM reactor channel with doses of $\Phi = 10^{16}$ to $5 \cdot 10^{19}$ cm^{-2}, the spectral characteristics of the samples remain identical to the initial values

Table 3.1 Changes in the optical and structural characteristics of sapphire depending on irradiation dose (D_1 is the optical density at 257 nm, D_2 is the optical density at 358 nm; d is the interplaner distance; 2θ is the reflection angle)

Φ (cm^{-2})	$5 \cdot 10^{16}$	$1 \cdot 10^{17}$	$5 \cdot 10^{17}$	$1 \cdot 10^{18}$	$5 \cdot 10^{18}$	$8 \cdot 10^{18}$
D_1 (a.u.)	0.18	0.27	0.34	0.62	1.29	1.29
D_2 (a.u.)	0.10	0.12	0.15	0.35	0.51	0.64
Φ (cm^{-2})	$8 \cdot 10^{18}$	$1 \cdot 10^{19}$	$5 \cdot 10^{19}$	$8 \cdot 10^{19}$	$1 \cdot 10^{20}$	$5 \cdot 10^{20}$
D_1 (a.u.)	1.29	1.27	1.28	1.70	1.95	2.15
D_2 (a.u.)	0.64	0.93	1.05	1.05	1.21	1.99
a (nm)	0.4752	0.4752	0.4752	0.4755	0.4758	0.4764
c (nm)	1.2980	1.2980	1.2980	1.2990	1.3000	1.3010
d (nm)	0.2379	0.2379	0.2380	0.2381	0.2383	0.2388
2θ (°)	89.04	89.03	88.99	88.98	88.95	88.86

(Fig. 3.1b). The increase of the dose up to $8 \cdot 10^{19}$ cm^{-2} leads to a diminution of the intensity of the bands and a shift in frequencies of some modes (Fig. 3.1c).

As shown, reflection coefficient and the frequencies of the 735–760 cm^{-1} and 600–620 cm^{-1} bands have decreased and the latter band has a shoulder in the vicinity of 550 cm^{-1} frequency. The 465 cm^{-1} band has decreased and become broader. The changes in the functions $R(\Phi)$ and $v(\Phi)$ for these three bands bound up with the modes v_{as}, v_s, and vd point to the breakdown and extension of some part of the bonds Al–O and their deformation in the process of irradiation. A transformation of the reflection spectrum in the region of the band of asymmetric valent oscillations of Al–O–Al bonds (v_{as} = 735–760 cm^{-1}) also is observed. High-dose irradiation is accompanied by manifestation of the structure of the degenerate oscillations and the appearance of a shoulder at a frequency of 800 cm^{-1} (Fig. 3.1c). The tendency to the change of the valent angles and breakdown of some part of the bonds Al–O–Al– [11] is observed. Such a tendency also is confirmed by the kinetics of the generation of F^+-, F-, and aggregate F-centers as well as interstitial Al_i ions in the given dose range (Table 3.1).

At high doses of neutron irradiation, for the high-frequency wing of valent oscillations, a new band near 900 cm^{-1} arises. It is assumed that it is caused by the deformation oscillations of hydroxyl groups as a result of the formation of the bonds Al–OH in the process of irradiation [11]. As found while studying the IR-spectra in the region of the main oscillations of hydroxyl groups in the vicinity of 3 μm, the high-dose irradiation is followed by the appearance of a broad band at 3,400–3,500 cm^{-1} frequencies. In the spectrum of the initial crystal this band is absent. Therefore, it may be considered to be caused either by the breakdown of the bonds Al–O and Al–O–Al or by the shift of the anions and radiation-induced diffusion of hydrogen from the environment to the structural channels of the crystal followed by the formation of the bonds Al–OH. This band seems to be bound up with the valent oscillations of the groups OH located within the disordered regions of the crystal near the defective octahedrons. At the ultimate doses the following changes of the crystal lattice parameters c, a, and d are observed: Δc = 0.038 Å, Δa = 0.014 Å, Δd = 0.011 Å; $\Delta 2\theta$ = 0.20° for the

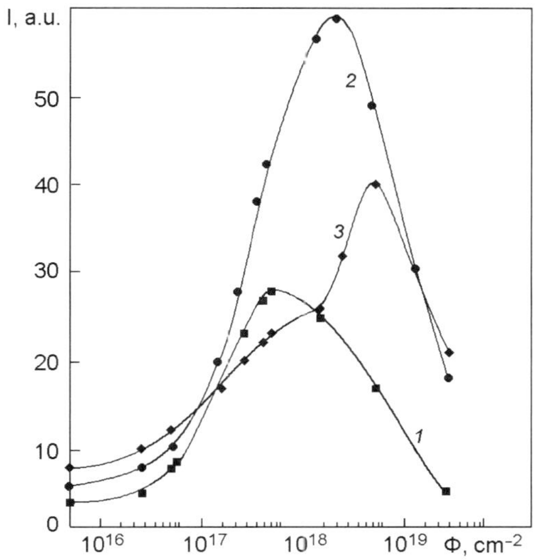

Fig. 3.2 Dose dependence of the intensity of 510 nm band at 570 K (*1*), 870 K (*2*), and 970 K (*3*)

Table 3.2 Kinetics of changes in the intensity of 510 nm band I/I_0 for $\Phi = 10^{18}$ cm^{-2}

	T (K)				
Time (s)	320	670	870	1,070	1,170
120	0.98	0.84	0.65	0.98	0.99
360	0.93	0.70	0.42	0.96	0.94
600	0.93	0.60	0.15	0.95	0.94

reflex 02.10.With the growth of the temperature, the dose dependence of the intensity of the photoluminescence band at 510 nm tends to increase. The maximum near the dose 10^{18} cm^{-2} is shifted toward higher doses (Fig. 3.2). The concentration of Ali grows too.

The kinetic curves of photodecoloration for the 510 nm depending on reactor irradiation doses are obtained (Table 3.2) [11]. The temperature dependence of photodecoloration reaches its maximum at 870 K.

IR reflection spectra show the changes in the surface-adjacent layer bound up with the deformation and breakdown of interatomic bonds. For the irradiated crystals grown by the Czochralski and HDSM the value of reflection coefficient decreases too (Fig. 3.3) [9].

Radiation stability of the crystal structure has been established by X-ray phase analysis [13]. X-ray patterns show considerable changes in the intensity and location of a number of reflexes depending on the irradiation dose. The behavior of these characteristics for one of the main reflexes is presented in Fig. 3.4. With the growth of Φ, the intensity and location of the reflex varies and shifts toward smaller reflection angles. Figure 3.5 illustrates the radiation kinetics of the function $2\theta(\Phi)$ of

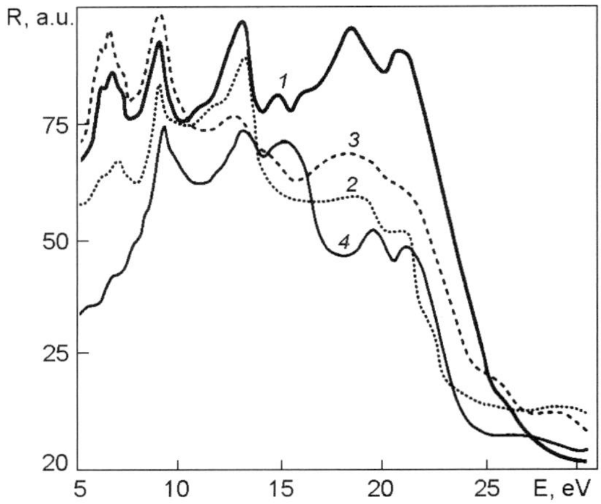

Fig. 3.3 Reflection spectra for sapphire in the region of 5–30 eV at 300 K: (*1*) initial (nonirradiated) crystal (HDSM); (*2*) sapphire irradiated with 10^{17} neutrons/cm^2 fluence; (*3*) sapphire irradiated with 10^{17} neutrons/cm^2 fluence and annealed at 770 K; and (*4*) sapphire (Czochralski) irradiated with $1.2 \cdot 10^{12}$ Pb/cm^2 fluence (the spectra were measured on the synchrotron accelerator C-60)

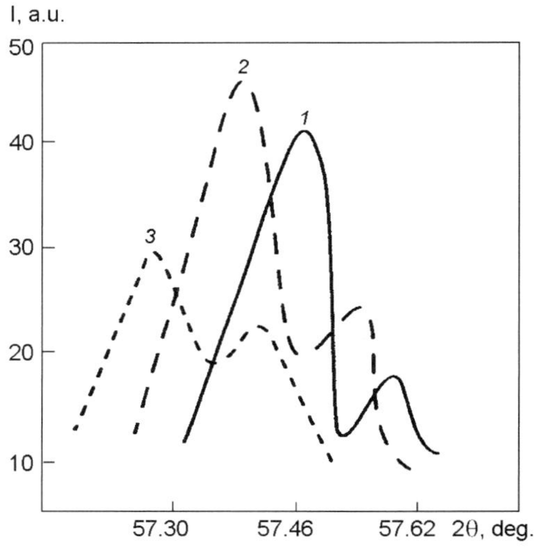

Fig. 3.4 Influence of irradiation on the parameters of the reflex 220: (*1*) $\Phi = 4 \cdot 10^{19}$, (*2*) $\Phi = 10^{21}$, and (*3*) $\Phi = 7 \cdot 10^{21}$ cm^{-1}

Kα_1 and Kα_2 doublets for another reflex. A certain regularity in the decrease of the intensity of the reflexes and the shift of their peaks toward smaller angles exist. Such results confirm the supposition of activation of the bonds Al–OH that are activated in disordered places which arise in the irradiated crystal.

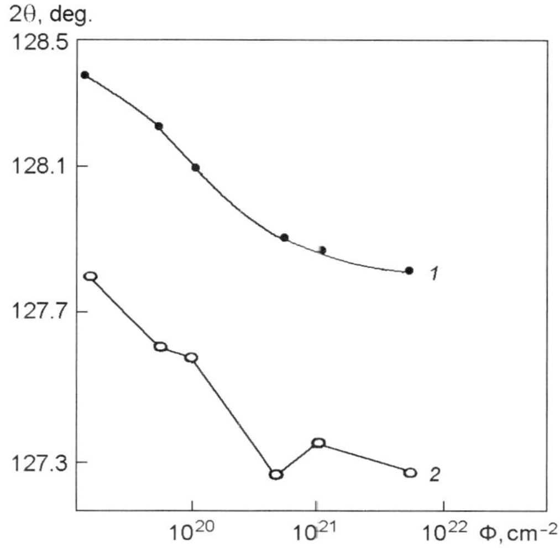

Fig. 3.5 Dose dependence of the reflection angle for the doublets *1* and *2* of the reflex $0\bar{3}.12$

3.2 Changes in the Bulk

Sapphire is found to "swell" by 1–3% upon neutron irradiation in a reactor ($E > 0.1$ MeV, 2...6) · 10^{21} cm^{-2} doses). Such an effect is most pronounced along the *C* axis and increases with increasing irradiation temperature.

Hardness of sapphire irradiated with fast neutrons increases to some extent. This may be explained by changes in the dislocation structure [14]. The effect vanishes in the crystal annealed at 870 K. The value of hardness also increases with irradiation of the crystal surface by Cr, Fe, and Cu ions (by 50%, 20–27%, and 20%, respectively).

Proton irradiation. At a temperature of 4 K in sapphire irradiated with protons (3 · 10^{18} cm^{-2} doses) and fast neutrons (10^{18} cm^{-2} doses) the centers appear with an absorption band at 358 nm and a luminescence band at 379 nm [15]. These bands are related to the center formed by two oxygen vacancies in the second coordination, sphere which have captured one electron or more.

The position of Pb ions (100 keV at 570 K, 10^{14} cm^{-2} doses) incorporated into the lattice is established. The ions are located along the Al rows ⟨0001⟩ and occupy the octahedron positions shifted along the *C*-axis [16].

Irradiation with Cr ions favors the formation of *F*- and F^+-color centers. Cr ions substitute Al ions. Disordering in the oxygen sublattice is higher in comparison with that in the aluminum sublattice.

Electron irradiation. Before electrons undergo elastic collision with the matrix atoms, they lose an essential part of their energy by ionization. Therefore, the number of the vacancies and interstitial oxygen and aluminum ions formed by

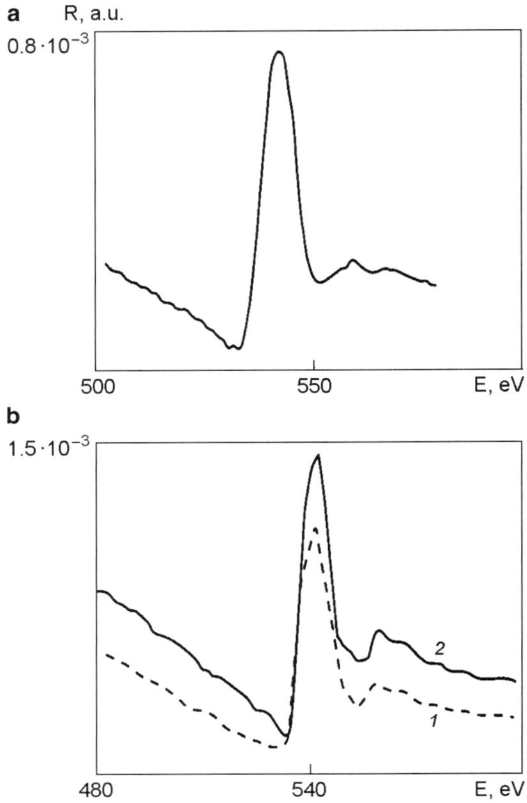

Fig. 3.6 Reflection spectra of sapphire (grown by the HDSM) at 570 K: (**a**) nonirradiated crystal; (**b**) sapphire irradiated with (*1*) $6 \cdot 10^{17}$ electrons/cm^2 and (*2*) 10^{17} electrons/cm^2 fluences

electron irradiation is considerably less compared with neutron irradiation. The reflection spectra show the tendency analogous to the one observed at neutron irradiation (Fig. 3.6) [9].

Transparency of irradiated sapphire may either diminish or increase, depending on the crystal's history. For instance, as reported in ref. [17], after irradiation of sapphire by electron flux with an energy of $2 \cdot 10^6 ... 5 \cdot 10^7$ eV and a dose of 10^{10} to 10^{15} electrons/cm^2, its transparency in the UV-region of the spectrum rises [17].

Thermal conductivity of sapphire irradiated with 10^{17} electrons/cm^2 dose diminishes by 20–25% at $T < 50$ K. The peak of thermal conductivity equal to 68 W/cm K at 34 K decreases down to 52–68 W/cm K at 39 K. Annealing the crystals (870 K, 24 h) recovers the initial thermal conductivity value almost completely.

Dislocation loops and complexes of vacancies with the formation of excess of aluminum on the crystal surface are observed after irradiation with a dose of $1.7 \cdot 10^{18}$ electrons/cm^2 at 970 K. The basal planes are more resistant to such a factor.

Electrorestriction. Internal electric fields may be induced in ruby by electron bombardment followed by the formation of a superstructure along one of the axes.

X-ray irradiation may change the color of the crystal. The lines of induced absorption are observed at 230, 400, 650 nm, and so on, depending on the content of controllable and noncontrollable impurities.

Annealing of radiation defects and their transformation. Electron exchange between RD and the medium in the process of annealing leads to certain changes in the optical and dielectric properties of the crystals. Radiation-induced impurity–vacancy complexes have different thermal stability. Depending on the temperature, pressure, and medium, the complexes possessing higher stability may be formed, and unstable RD may be annihilated. The complexes thermally stable at temperatures up to 1,600 K have been found only in polycrystalline corundum.

Changes in the absorption parameters at postimplantation thermal treatment (carbon ions, $2 \cdot 10^{17}$ cm^{-2} fluence) are caused by the generation of additional color centers. The formation of clusters on the base of individual vacancies manifests itself in the decrease of color centers density and the growth in the population of the states (Fig. 3.7). The oxygen-containing complexes diminish the population of the donor subzone levels and the degree of overlap in the localized states, reducing the efficiency of interzonal absorption.

Radiation resistance of ruby. The edge of the inherent absorption of ruby irradiated with 10 MeV electron flux on a linear accelerator is shifted by 40–50 nm toward longer wavelengths, the absorption increases, especially in the UV region. Additional bands associated with the appearance of F centers and the change of chromium valency arise in the absorption spectrum.

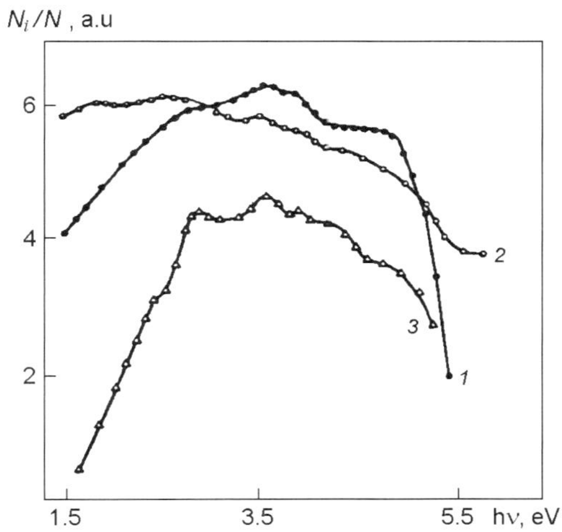

Fig. 3.7 (*1*) Influence of ion implantation and (*2*, *3*) the conditions of postimplantation annealing on the change of relative concentration of localized states in the forbidden zone of sapphire. Annealing in (*2*) vacuum at $P = 10^{-2}$ Pa, $T = 1{,}100$ K and in (*3*) oxygen at $P = 10^5$ Pa, $T = 700$ K. N, N_i are the concentrations of absorption centers before and after irradiation [18]

Even nonirradiated ruby contains a certain equilibrium number of color centers. The radiation resistance of ruby elements (RE) is characterized by the change of the pulse energy E_p and the threshold energy E_{th} (Tables 2.16–2.18).

A correlation is observed between the generation parameters and the total distortion of the spectra of electron paramagnetic resonance (the parameter β in Tables 2.17 and 2.18). Some spread in the data presented in these tables is connected with the impurity content of the RE and other individual peculiarities of the crystals.

Annealing of the irradiated RE leads to recombination of radiation defects. Even short-term light treatment (at pumping) results in recovery of E_p.

The induced absorption intensity may be suppressed with the introduction of vanadium or titanium into ruby. Titanium acts as a protector even in crystals annealed in vacuum or in a reduction medium.

References

1. Chaikovskiy E.F., Baturicheva Z.B., Taran A.A. Physics and Chemistry of Single Crystals and Scintillators (collected volume), Kharkov [in Russian], No.7, – pp. 207–210, 1981.
2. Nagulb H.M., Singleton J.F., Grant W.A. J. Mater. Sci. 8(11), 1633–1640, 1973.
3. Burnett P.J., Pege T.F. J. Mater. Sci. 19(11), 3524–3545, 1984.
4. Cesimir J., Roger K. J. Phys. Chem. Solids 31(1), 41–48, 1970.
5. Lee W.E., Lagerlof K.P.D., Mitchell T.E. Philos. Mag. A 51(4), 123–127, 1985.
6. Gruen D.M., Siskind B., Wright R.B. J. Chem. Phys. 65(1), 363–378, 1976.
7. Katrich N.P., Budnikov A.T. Physics and Chemistry of Crystals (Collected Volume), Kharkov [in Russian] pp. 81–83, 1980.
8. Mussket R.G., Brown D.W., Pinizzotto R.F. Appl. Phys. Lett. 49(7), 379–381, 1986.
9. Arutyunyan V.V. Surface. X-ray, Synchrotron and Neutron Investigations (Collected Volume), [in Russian], No.7. pp. 69–73, 2001.
10. Knotek M.L., Feibelmann P.I. Surf. Sci. 90(1), 78, 1979.
11. Abdukadyrova I. Surface. X-ray, Synchrotron and Neutron Investigations, (Collected Volume), [in Russian] No.12. pp. 68–72, 2005.
12. Vlasov A.G., Florinskaya V.A. Infrared Spectra of Inorganic Glasses and Crystals [in Russian]. Leningrad: Khimiya, p. 304, 1972.
13. Abdukadyrova I., Mukhtarova N.N. Proceedings of III International Conference "Modern Problems of Nuclear Physics", Bukhara, Uzbekistan, 1999, p.318, 1999.
14. Borber D., Tighe N. J. Am. Ceram. Soc. 51(11), 611–617, 1968.
15. Welch L.S., Hughes A.E., Pells G.P. J. Phys. C 13(9), 1805–1816, 1980.
16. Carnera A., Drigo A., Mazzoldi P. Proceeding of International Conference on Ion Beam Modification in Materials, Budapest, 1978, pp. 4–8, 1978.
17. Atabekyan R.R., Vinetskiy V.L., Gevorkyan V.A. // USSR Inventor's Certificate 1111515 [in Russian], 1985.
18. Kabyshev A.V., Konusov F.V. Surface: X-ray, Synchrotron and Neutron Investigations (Collected Volume) [in Russian], No. 9, pp. 51–60, 2003.

Chapter 4
Crystal Growth Methods

Classification of crystal growth methods [1–4] is based on the phase state, the composition of the initial components, and the conditions of the growth process. Based on these factors, one can distinguish the following groups of growth methods:

- Growth of crystals from the gaseous phase using a pressure gradient
- Growth of crystals from a solution using a concentration gradient at the crystal–solution interface
- Growth of crystals from a melt using a temperature gradient
- Growth of crystals in the solid phase

Identical single crystals can be grown by different methods. Classification by the type of phase transition (from the vapor phase, solution, melt, or solid phase) is somewhat arbitrary, as the process of crystal formation often proceeds through several phases. More detailed classification by attributes, authors, and so forth seems to be disorienting. For example, in the literature one may encounter mention of both the "active solvent" and "zone melting with a temperature gradient" methods. Both terms concern the same process: in some papers it is the Tamman or Obreimov–Shubnikov method, whereas in others the Bridgman–Stockbarger technique. This is explained by the fact that these methods were developed at different times for growing certain materials, and then extended to other substances. Although the classification of crystal growth methods by the form of phase transition is formal, we shall use it nevertheless.

Sapphire can be grown from the gaseous, liquid, and solid phases. An ample literature devoted to the methods of obtaining sapphire from different media contains data on the theory and practice of growing the crystal. Our task is to arm the readers with the logic to choose the most suitable method. This logic is based on the physical–chemical essence of the growth processes and on the peculiarities of a particular method.

4.1 Crystallization from the Gaseous (Vapor) Phase

From the gaseous phase massive, platelike crystals, whiskers, and expitaxial films can be grown. Some methods are particularly specialized. For instance, cathode sputtering and deposition from molecular beams in vacuum are used to obtain

E.R. Dobrovinskaya et al., *Sapphire: Material, Manufacturing, Applications*,
DOI: 10.1007/978-0-387-85695-7_4, © Springer Science+Business Media, LLC 2009

films, whereas larger crystals are mainly grown by crystallization in a sealed ampoule. Other methods have a wider application scope. In particular, the methods of chemical deposition in a gas flow are suitable for obtaining films, larger crystals, and whiskers.

Crystallization from the gaseous phase has a number of advantages, such as low process temperatures, inessential supersaturation, easy control of the composition, weak interaction between the container and the crystallized material, and high perfection of the crystals. However, the size of the crystals grown is not large.

Methods of crystallization from the gaseous (vapor) phase fall into physical and chemical categories. The former methods, based on the effect of condensation, include crystallization of substances from their own vapor, also called the "sublimation method," and the "molecular beam method." Crystallization from vapor is especially suitable for those substances in which solid–gas transition is not realized via the liquid phase. The sublimation method is used for crystal growth in closed and open systems in vacuum or in a gas medium, with crystallization often carried out in sealed ampoules. In such cases, the initial material is placed in one part of an ampoule. Both parts of the ampoule are heated to the same temperature, and the pressure of the saturated vapor corresponding to such a temperature is set. Then one part of the ampoule is cooled until crystal nuclei arise. Afterward, the temperature is raised again to prevent the formation of new nuclei, and the process of crystal growth proceeds under constant conditions.

During the growth of crystals by chemical reaction methods (reduction, thermal decomposition, oxidation, etc.), the gaseous phase composition differs from that of the growing crystal. The essence of the chemical transfer method is as follows. While interacting with a gaseous transporter by a reversible reaction, a solid or liquid substance forms gaseous products. The latter decompose with subsequent precipitation of a crystalline substance after being transported to another part of the system when equilibrium conditions change. For estimation of the transport properties of the reaction, the most significant factor is the difference between the partial pressures (ΔP) above the initial and secondary phases of the crystallized substance. Chemical transfer proceeds either in closed systems, via convection and diffusion, or in the flow. An open system with a forced hydrodynamic regimen is more effective. In this case, the transfer is no longer a rate-limiting stage of the growth, if flow velocity is optimized. The common feature of these methods is the transport of material from the source to the place of crystallization, since the concentration of the crystallized material in the medium is low.

Growth from the gaseous phase is distinguished by the presence of an adsorption layer on the surface of the crystal–medium interface. The properties of this layer differ from those of the crystal and of the surrounding medium. Correspondingly, there are two stages of surface processes: the transition of the substance from the vapor to the adsorption layer and its incorporation into the lattice. The physical phenomena that occur within the adsorption layer have not been completely clarified so far. However, it is accepted that they include the migration and collision of atoms and molecules as well as the formation of their associations (complexes) and nuclei: two-dimensional (with a thickness of one lattice parameter) and three-dimensional (with a larger thickness). The character of the surface processes is defined by the properties of the adsorption layer, in particular, by its composition and density. The density of this layer, or the quantity of substance per unit area, depends on the surface activity, the density of the incident flow, and the concentration of the substance in the medium.

The methods of massive crystal growth from the gaseous phase are attractive owing to their universality. These methods can be used for obtaining crystals of practically any substance. It should be noted that the term "massive" has a relative meaning, since, as a rule, large crystals are not obtained by such methods. Only once has the growth from the gaseous phase of ruby and sapphire crystals with a mass up to 150 g been reported [5].

For many crystals, the growth morphology from the gaseous phase depends on the degree of supersaturation [6, 7]. One can establish quantitative relationships between the relative supersaturation and the resultant main form of crystals (whiskers, bulk crystals, powders). For the growth of whiskers, a low supersaturation is required, while at medium degrees of supersaturation bulk crystals are predominantly obtained, and at intermediate supersaturation dendrites and platelike crystals are formed. Powders are obtained at high supersaturation degrees due to homogeneous nucleation in the gaseous phase.

High-perfection sapphire grown from the gaseous phase usually is used for scientific investigations only. These methods also are used for obtaining sapphire whiskers and felt. Due to their low production rate, methods of crystal growth from the gaseous phase generally are not used for commercial growth of sapphire.

Whiskers are grown by the method described by Frank [8] on a free surface under compression or due to the appearance of internal stresses, if plastic flow is absent and the temperature sufficiently high to provide fast diffusion. According to Sears [9], the growth of whiskers is feasible in the presence of nuclei with dislocations located along one of the crystallographic directions when the degree of supersaturation is lower than that required for the formation of polyhedrons. Sears showed the validity of the latter condition for a number of substances, but the correctness of the former condition was not completely proved [10]. Whiskers may contain either crystal defects or may be defect-free. According to the Sears model, atoms impacting on the surface of a whisker are absorbed by it and diffuse over the surface to a runoff at the crystal vertex; the atoms that do not reach the vertex evaporate again [11, 12].

Sears and Gomer established approximate laws of stable growth for whiskers, describing the experimentally observed asymptotic behavior of the growth rate. Ruth and Hirth [13] solved the problem of whisker growth in a stationary regimen by showing the necessity of maintaining immobility in most cases both for the initial and the final stages of the growth. They attributed nonimmobility to violation of the boundary conditions at the vertex rather than an inability of the atoms to reach this vertex.

In most cases whiskers are straight, but samples from the same batch may have different shapes. As a rule, bending of whiskers is explained by a change in the direction of dislocation propagation and by the presence of impurities; the bend angles often correspond to twin orientations. Webb and Forgeng studied twisting of sapphire whiskers and made a conclusion about the existence of dislocations with the Burgers vector equal to the unit lattice vector or divisible by it [14, 15]. Polytypism of whiskers has been reported [16]. However, its reliable correlation with the step height or with the elementary cell period was not established, and the degree of influence from impurities remained unclear.

Now, let us consider the chemical methods of obtaining sapphire whiskers and the methods of their growth from the vapor phase. With chemical methods, the source material and the surrounding gaseous phase interact with subsequent formation of volatile components and their transfer to the deposition zone. Webb and Forgeng [17] described the growth of sapphire whiskers by passing humid hydrogen over a boat containing melted aluminum or $TiAl_3$ (Fig. 4.1). Whiskers of different sizes and shapes, as well as tapes and powders, were formed between the metal particles. The duration of the process varied from 30 min to 10 h in a furnace at constant pressure of 30 mmHg. At temperatures ranging between 1,570 and 1,720 K, relatively many crystals were formed around the feeds. The length of the whiskers reached 4 cm, while their minimum diameter was 0.5 μm.

In such a system, other reduction reactions also may occur [18]. A porcelain boat containing metallic aluminum was heated to 1,520 K in a hydrogen flow (the dew point is 328 K). Some part of the metal oxidized, and a whisker "cocoon" was formed over the boat. At such a temperature, whiskers intensely grew only inside the boat. An analysis of Al_2O_3 crystals with a diameter from several microns to 100 μm and a length up to 1.5 cm showed that in the content of silicon varied from 2 to 3%.

The crystals were formed by the reaction [19]:

$$2Al_{(g)} + 3SiO_{(g)} \rightarrow Al_2O_{3(s)} + 3Si_{(s)}. \tag{4.1}$$

Later, to crystallize sapphire by means of reduction reactions, dry hydrogen and enriched silica were used. Under these conditions [20]:

$$\begin{gathered} H_{2(g)} + SiO_{2(g)} \rightarrow H_2O_{(g)} + SiO_{(g)} \\ 4Al_{(l)} + 3H_2O_{(g)} + 3SiO_{(g)} \rightarrow 2Al_2O_{3(s)} + 3H_{2(g)} + 3Si_{(l)} \end{gathered} \tag{4.2}$$

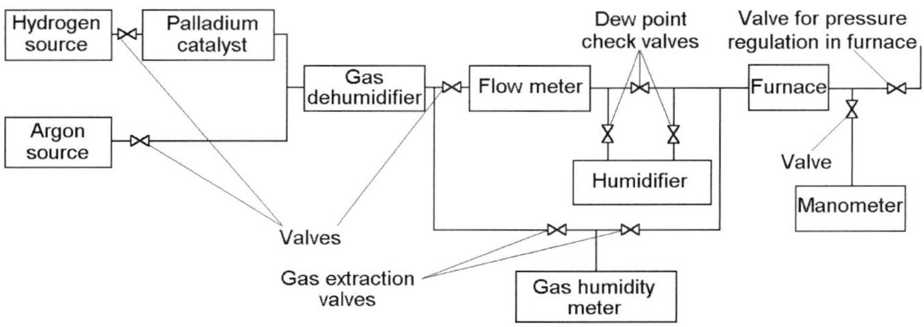

Fig. 4.1 Schematic view of a unit for growing threadlike sapphire crystals

The production rate of such a process is high enough. Barber [21] described the growth of sapphire whiskers in a chamber made of pure recrystallized Al_2O_3 due to the interaction between melted aluminum and oxygen of unknown origin. In the opinion of Barber, under such conditions water vapor could be formed from a deposit of H_2S which appeared on the chamber walls as a product of the interaction of hydrogen and sulfur residues in aluminum. In this case, the metal seems to reduce the oxide as well.

Papkov and Berezhkova [22] grew whiskers by heating massive sapphire in a graphite furnace up to 2,270 K in an argon atmosphere or in a vacuum of about 10 mmHg. At the temperature of ~2,120 K, a large quantity of whiskers with a length from 1 to 30 mm and a diameter of 0.3–50 μm appeared on the inner furnace surface and on the sapphire.

In the methods growing sapphire from the vapor phase, the gaseous reagents are mixed in a heated crystallization chamber to deposit the material. In some cases the partial pressures of both the reagents and the reaction products are established in order to provide the growth of the crystals. The substrates used are either stationary (and the gases are passed over them) or have the form of small particles (nuclei) transferred by the gas flow (Fig. 4.2).

For the growth of sapphire, aluminum chloride monomer is prepared by chlorinating metallic aluminum. Carbon monoxide (99.5%), unpurified hydrogen, and carbon dioxide (99.8%) are passed through drying columns, and then the mixture is conveyed to an injector system. To control the gas flows at standard temperature and pressure, the regulating and flowmeter valves in all the channels are adjusted in such a way that the flow rate correlates with a difference in pressure between the values equal to 1 atm. To introduce the nuclei into the injector, the required gases are passed through the chamber. The reaction chamber is made of aluminum oxide and typically has an inner diameter of 2.2–5.0 cm. The reacting gases and reaction products are continuously passed through the chamber. They then are conveyed to the collector for deposition of solid particles [23].

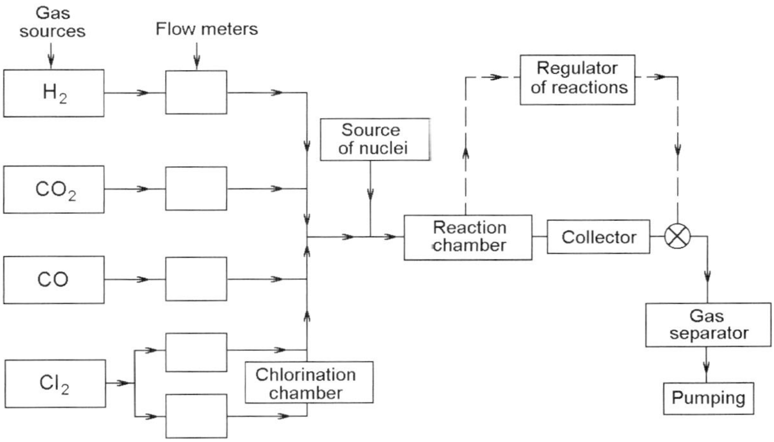

Fig. 4.2 Schematic view of a unit for sapphire crystallization by vapor-phase method

Another method of sapphire whisker growth is based on evaporation and oxidation of Al at either 1,520 K or at 1,270–1,800 K in humid hydrogen.

Achievement of large (mass exceeding 80 g) and sufficiently perfect crystals on a seed from the vapor phase by the reaction

$$2AlCl_{3\,vapor} + 3H_{2\,gas} + 3CO_{2\,gas} \leftrightarrow Al_2O_{3\,solid} + 3CO_{gas} + 6HCl_{gas} \qquad (4.3)$$

was reported in ref. [24]. Dehydrated $AlCl_3$ was synthesized directly from the elements in a generator made of inconel alloy. To avoid untimely reaction, the reacting components were introduced into the crystallization zone via a heated concentric tubular injector. Mixing the components was provided by the flow turbulence.

According to free energy calculations, the formation of Al_2O_3 in the reaction (4.3) was favored by a temperature of 500–2,000 K: $\Delta G_{500} = -43.83$ kcal/mol; $\Delta G_{2000} = -43.83$ kcal/mol. However, at temperatures up to 1,820 K the crystal growth rate turned out to be much lower than calculated (Fig. 4.3). The orientation of the substrate and the pressure also were found to essentially influence the process. For the rhombohedral face, the growth rate at 2,020 K was 76.5 mg/cm^2 h ($5 \cdot 10^{-6}$ cm/s), whereas for the prism face this value was less by half. It should be assumed that the process is characterized by certain chemical and crystallization limitations.

Concluding the description of the processes that occur during crystal growth (in particular, whiskers) from the gaseous phase, it is necessary to dwell on the absorption layer located at the crystal–medium interface. In 1964, the growth mechanism vapor–liquid–crystal (VLC) was proposed [25]. The physical essence of this mechanism is connected with the structure and properties of the vapor–liquid and liquid–crystal phase boundaries. The absolute values of the condensation coefficients for crystallization from the vapor and from chemical compounds may be

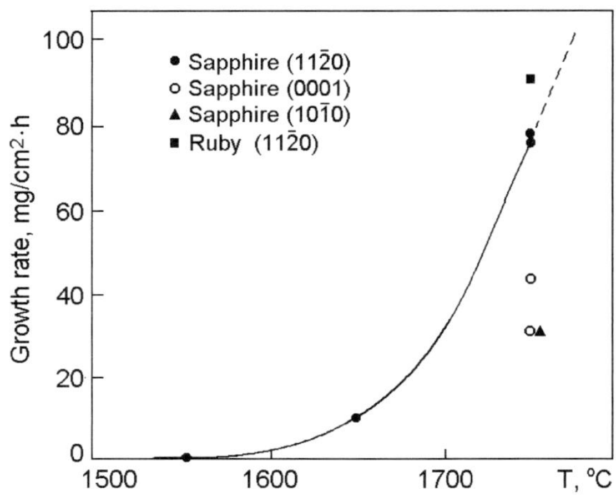

Fig. 4.3 Growth rate of sapphire and ruby as a function of the deposition temperature

essentially different, but for the VLC mechanism it is significant that for the liquid surface they are considerably higher than for the crystal facet.

Crystallization from the vapor via the liquid phase layer has the following stages:

- Transfer of the substance in the gaseous phase
- Condensation and chemical reactions at the vapor–liquid interface
- Diffusion in the liquid phase
- Incorporation of particles into the lattice of the growing crystal

The final stage (which is the slowest) defines the total rate of the process. The energy of the crystal–melt interface is lower than the corresponding value of the crystal–vapor interface by a factor of 5–10. Therefore, the rate of formation of two-dimensional nuclei at the liquid–crystal interface is by several times higher than at the vapor–crystal interface. Such a circumstance, as well as an elevated coefficient of condensation on the surface of the liquid, low diffusion resistance in micron drops, and the anti-impurity action of the liquid phase raise the growth rate of the VLC mechanism by 5–10 times in comparison with that of the vapor–crystal mechanism. The difference in growth rate becomes more significant as the temperature diminishes. Therefore, whiskers often are formed at temperatures that are 100–200° K lower than the temperature of the massive crystal and film growth in the gaseous phase.

The earlier proposed mechanisms do not clarify the role of the impurities, which stimulate whisker growth, their branching and bending, and periodic changes of the growth direction and cross section, as well as cessation of growth. In the VLC mechanism, cessation of the growth is explained by the loss of the liquid "cap" due to instability of the process. The stimulating role of the impurities is also explained by the liquid "cap," which is the most active growth sector. Branching is attributed to splitting of drops of the melt; bending and periodic changes of the growth direction are caused by the drops creeping from one facet to another.

The observed facts point to the dominating role of the VLC mechanism. The liquid phase itself can provide only one-dimensional crystal growth, whereas other factors (axial dislocation, twinning planes, strongly adsorbing impurities, etc.) may favor such a growth only with simultaneous action of a liquid phase. Since the diameter of a crystal growing by the VLC mechanism is defined by the drop's diameter, this mechanism may serve as a basis for controlling the growth of whiskers. By creating a system of drops of solution-melt with the required composition on a substrate, one can obtain a system of equally oriented whiskers.

Kaddis [25] reported the growth of sapphire whiskers by condensation and oxidation of aluminum on aluminum oxide substrates. The whiskers constantly contained a small spherical aluminum particle on their tops. Experimental data show that sapphire grows by the VLC mechanism, and aluminum serves as a solvent agent [23]. The substance is absorbed from the gaseous phase by the melt on the crystal face, and then it crystallizes by a scheme resembling the Verneuil method (Fig. 4.4). Crystallization proceeds from the solution in the melt, since the melt is highly enriched with the impurity that provides nucleation of whiskers on a substrate.

Fig. 4.4 Growth scheme of a threadlike crystal according to the VLC mechanism: (*1*) vapor; (*2*) solution in melt; (*3*) crystallization front; (*4*) crystal

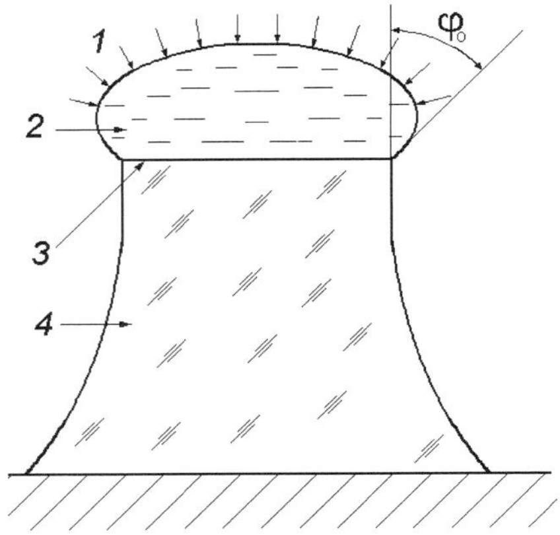

The authors have no intention of formulating detailed practical recommendations or conception for the growth of sapphire whiskers from the vapor via the liquid phase layer. We would like to show the significance of the correct choice of a transport process and solvent agent. For instance, such transport processes as evaporation and cathode sputtering are unproducative and uneconomical. The choice of the solvent agent is even more complicated in view of scarce data on the solubility of the impurities that stimulate growth of sapphire by the VLC mechanism. Knowledge of the wettability of sapphire by the melt during growth by such a mechanism is also insufficient.

The growth of crystals from a vapor via a liquid phase layer is not a rare occurrence. It obviously reflects the Ostvald's "law of steps," which states that at the transition from one state to another a substance sequentially acquires all the intermediate states.

4.2 Crystallization from Solution

While growing crystals from a solution onto a seed, supersaturation must be achieved without spontaneous formation of nuclei in the solution. Supersaturation permits establishment of growth rates that provide controlled crystallization onto a seed. Methods of growth from the solution have a number of significant advantages such as relatively low crystallization temperatures, the possibility of obtaining crystals unstable at T_m or those existing in different temperature modifications [26], and so forth.

The choice of crystallization procedure depends, first of all, on the solubility of the given compound. For instance, if the initial substance possesses relatively high solubility and a high temperature solubility coefficient (i.e., its solubility is substantially temperature-dependent), it is expedient to realize crystallization by lowering the temperature (Fig. 4.5).

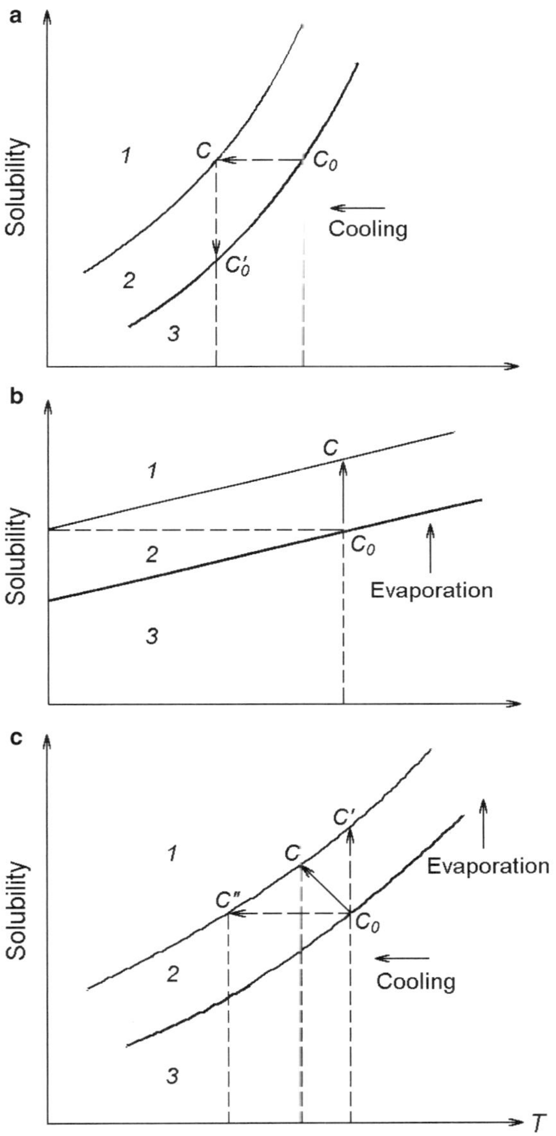

Fig. 4.5 Temperature dependence of solubility: (**a**) strong, (**b**) weak, and (**c**) moderate: (*1*) supersaturation region; (*2*) metastable region; (*3*) undersaturation region

In the process of cooling, the solubility behavior is shown by arrows (Fig. 4.5a). The difference C–C_0 characterizes the value of supersaturation necessary for the formation of a nucleus and its subsequent growth. By controlling the process of crystallization, one can grow perfect crystals. When the temperature dependence of supersaturation is weak (Fig. 4.5b), the upper boundary of the metastable region can be reached by isothermal evaporation. The variant of simultaneous use of evaporation and cooling is shown in Fig. 4.5c.

Thus, based on the above considerations, we can formulate the regulations for the choice of a method of crystallization from solution:

1. If the temperature solubility coefficient noticeably differs from zero, methods of crystallization based on changing the solution temperature can be used. Such methods have several variants

   - At relatively low-temperature solubility coefficients (0.01–0.1 g/L deg.), irrespective of the absolute solubility value, the growth by temperature gradient is preferable. These methods provide durable, continuous growth of the crystals in one part of the crystallizer due to constant dissolution of the substance in another part of the crystallizer. The methods of cooling are not applicable in this case, since isolation of noticeable quantities of the substance from the solution requires considerable cooling.
   - At high-temperature solubility coefficients (exceeding 1 g/L deg.) and low absolute solubility value (a few weight percent), growth by temperature gradient methods also is preferable, as cooling the solution even through a wide temperature range will lead to isolation of only small quantities of the substance.
   - At high-temperature solubility coefficients and high solubility, it is expedient to use the solution cooling method (or heating, if the solubility diminishes as the temperature grows). In this case, the efficiency of temperature gradient methods decreases with increasing temperature solubility coefficient, since the process of spontaneous crystal nucleation is fast and difficult to control.

2. If the temperature solubility coefficient is very low, crystallization can be performed by the methods of solvent evaporation or chemical reaction. With these methods, the absolute value of solubility is not of particular importance, but when utilizing solvent evaporation solubility must not be too low.
3. For slightly soluble substances it is expedient to realize crystallization by chemical reaction.
4. When the temperature and supersaturation in the growth zone are constant, crystallization is most readily provided using solution replenishment by the crystallizing components.

The temperature gradient method also can be considered to be a variant of replenishment. Other variants include the addition of a supersaturated solution to the crystallizer, crystallization via forced convection of the solution, and so forth. An advantage of isothermal methods exists in the fact that they provide a better control of the temperature-dependent properties of the crystals. If a phase is stable within a narrow temperature region, it should be obtained by an isothermal process. The latter also will provide more homogeneous crystals when the coefficient of impurity distribution is temperature-dependent. The main advantage of slow cooling is connected with a shorter time of diffusive transfer which does not require complicated control systems.

Brief historical information. Obtaining crystals from water solutions at approximately room temperature seems to be the oldest growth method. Close to it is the method of crystal growth under hydrothermal conditions. After the experiments in crystal growth at room temperature had failed due to low solubility or irreversible reaction of the given substance and the solvent, some attempts were made to grow crystals from solution using a solvent with a higher dissolving capacity. One of the approaches was to use nonaqueous solvents at room temperature. The introduction of a mineralizer – a substance promoting crystal growth – turned out to be effective if the formed complexes were not a stable solid phase. Another approach was to increase the solubility of the given compound at temperatures higher than room temperature. During growth of crystals under hydrothermal conditions, the temperature and the pressure essentially exceed the normal values. As a rule, in this case mineralizers are utilized and water or water alkali solutions serve as solvents.

The first attempts to obtain crystals under hydrothermal conditions were undertaken with the purpose of studying the conditions of natural mineral formation, and the dimensions of the grown crystals did not exceed several thousandth or hundredth parts of a millimeter [27]. The first developments in the hydrothermal method were connected with the growth of α-quartz crystals.

In parallel with hydrothermal methods, techniques for growing crystals from solution in a melt (the flux method) are widely used for obtaining synthetic analogs of natural gems or their substitutes. Such methods allowed growth of sapphires and rubies of high structural perfection, although in some regions of the optical spectrum their absorption may be higher than that of crystals grown by other methods (Fig. 4.6). During spontaneous crystallization, the growth shape depends on the composition of the solution and the temperature gradient in the growth zone (Fig. 4.7). During growth onto a seed, the growth rate is also dependent on the composition and overcooling of the solution (Fig. 4.8).

In the case of growth from solution of an alkali or alkali-earth tungstate melt, selective absorption of the solvent is observed on the facets $(22\bar{4}3)$. This explains the bipyramidal shape of the obtained crystals, rarely met in other cases [28].

The hydrothermal method, like other techniques of solution growth, is based on utilizing the dependence of the crystallized substance equilibrium concentration in

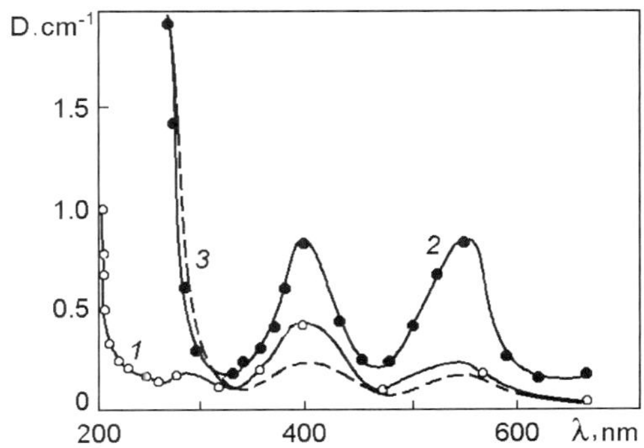

Fig. 4.6 Absorption spectra of ruby grown by the Verneuil technique (*1*) and in hydrothermal solutions (*2, 3*)

Fig. 4.7 Effect of solvent and temperature gradient on the corundum crystal faceting

Stationary growth chapes	Medium	Temperature in dissolution zone	ΔT
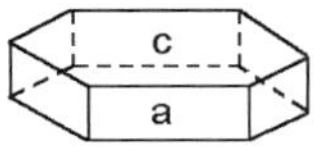	Na_2CO_3	<520 °C	30°
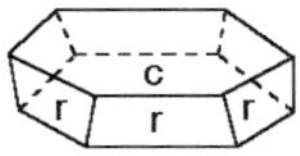	Na_2CO_3	>500 °C	40°
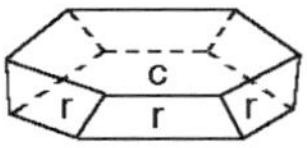	K_2CO_3	>480 °C	30°
	K_2CO_3	>440 °C	50°

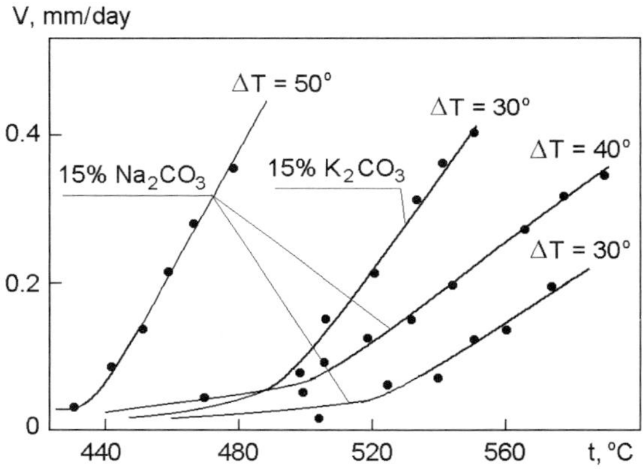

Fig. 4.8 Dependence of growth rate at the $\{11\bar{2}0\}$ face of corundum crystal on the solution composition and overcooling

solution, C_A, on the thermodynamic parameters that define the state of the system: the pressure P, the temperature T, and the solvent concentration C_B. This method is characterized by the use of a mineralizer B introduced into the system A–H_2O to raise the solubility of a weakly soluble component A. The mineralizer often is called a solvent, although strictly speaking the water solution of the mineralizer (i.e., B + H_2O) should be considered to be a solvent. As a rule, these growth systems consist of at least three components: A–B–H_2O, where A is the compound from which crystals are to be obtained and B is a readily soluble mineralizer.

Thus, the essence of the hydrothermal method is the creation of conditions that make it possible to dissolve the substance to be crystallized, to provide the required supersaturation of the solution, and then its crystallization. The value of supersaturation is controlled by changing the parameters of the system that define the solubility C_A: the temperature, pressure, type, and concentration of the mineralizer and the temperature gradient between the zones of dissolution and growth. This method is applicable for the growth of crystals with high T_m at much lower temperatures and for obtaining those crystals that cannot be grown by other methods. Large crystals are grown onto seeds by recrystallization of raw material with the same composition.

For the growth rate of crystals under hydrothermal conditions that are well studied, the rates of growth for individual facets (v) were found to rise linearly with the supersaturation ΔC_A:

$$v = \beta \Delta C_A, \tag{4.4}$$

where β is a kinetic coefficient. In practice, the dependence of the growth rate on ΔT — the difference in temperature between the dissolution and growth zones — is most often measured. As a rule, the supersaturation ΔC_A changes linearly with ΔT:

$$\Delta C_A \sim \Delta T. \tag{4.5}$$

Therefore, the relation between the growth rate and ΔT also is linear. The value of critical supersaturation is mainly defined by the composition of the solution. For instance, the growth rate of the sapphire facet (0001) in potassium and sodium carbonate solutions is practically equal to zero, whereas in bicarbonate solutions it reaches a noticeable value even at relatively low ΔT (of about 10° v).

At constant values of supersaturation, the growth rates of the facets rise with temperature. This fact manifests itself in the temperature dependence of the coefficient β in (4.1). In many cases (particularly for rapid growing facets), the dependence $\beta(T)$ satisfies the Arrenius equation:

$$\partial \ln(\beta) / \partial T = E / RT^2, \tag{4.6}$$

where E is the activation energy of the growth process, R is the gas constant.

The temperature dependence of the growth rate in the coordinates ln(v)–1/T is linear. In some cases, for slow-growing facets a deviation from linearity is observed. The slope of the function ln(v) to the axis 1/T indicates the value of activation energy. For sapphire, quartz, zincite, and so forth it has been found that this value varies from 10–15 kcal/mol to 40 kcal/mol for different facets.

The activation energy of sapphire and quartz has been established not to undergo essential changes in different solutions (Na_2CO_3, K_2CO_3, $NaHCO_3$). High-activation energies of facet growth serve as a weighty argument in favor of the fact that the growth process is not limited by diffusion of the substance in solution (with an activation energy not exceeding 4–5 kcal/mol). Rather, the growth process is defined mainly by the phenomena that occur on the growing crystal surface. Pressure influences the growth rate via mass exchange, solubility, and so forth and usually impedes dissociation of composite complexes in the solution.

The mineralizer can influence the growth rate via the solubility of the crystallized substance. If the temperature solubility coefficient, $\partial C_A/\partial T$, is dependent on mineralizer concentration [29], then an increase in the latter at a preset ΔT will alter the growth rate due to supersaturation shifts in the growth zone. Practical experience in crystallization of sapphire, zincite, and sfalerite shows that when the mineralizer concentration provides relatively high solubility C_A (2.5 wt% for Al_2O_3), the growth rate is practically independent of concentration [30].

The mineralizer can be absorbed on the crystal facets, thus retarding their growth. In this case, it is considered an impurity. In addition, it can interact with an impurity earlier absorbed on the surface, which may increase the growth rate. For instance, the growth rate for sapphire and zincite in potassium-containing solutions (KOH, K_2CO_3) is higher in comparison to that in sodium-containing solutions (NaOH, $NaCO_3$). This fact cannot be explained by changes in the supersaturation of the growth zone, since in potassium solutions such changes must be lower than in sodium solutions. One probable explanation comes from the interaction between potassium and an impurity absorbed on the surface, e.g., a chemisorbed water layer.

Now consider different crystal growth variants of hydrothermal solution methods used for obtaining rather large sapphires and rubies. *The temperature gradient method* is most widespread. The required supersaturation is created by lowering the temperature in the crystal growth zone. The raw material is placed into the lower part of an autoclave containing the solvent. The autoclave is heated in such a way that the temperature in the lower part (the dissolution zone) T_1 is higher than that in the upper part (the growth zone) T_2. Let C_1 and C_2 denote the concentrations of the crystallized substance A in solution in the lower and upper autoclave parts, respectively. The solution density ρ_c is defined by the solution temperature and concentration and changes along the autoclave axis. As a rule, the increase in the concentration raises the density, whereas rising temperature diminishes it. If

$$\rho_c(C_1, T_1) < \rho_c(C_2, T_2), \tag{4.7}$$

then the solution in the upper part of the autoclave (which has a lower temperature and higher density) moves downward, whereas the opposing flow of the solution with lower density moves upward. Thus, on attaining a certain temperature gradient $\Delta T = T_1 - T_2$ (specific to each compound), the thermal expansion of the liquid in the dissolution zone will cause a decrease of the solution density more essential than the increase resulting from the dissolution of the raw material, and convection of the solution will occur. The water solution saturated with the component A at T_1 is transferred upward. In the upper part of the autoclave the solution becomes supersaturated, since the temperature lowers to T_2 and crystallization is initiated. The value of the supersaturation ΔC_A is set by ΔT.

The method provides continuous transfer of the substance from the lower zone of the autoclave to the upper zone until the complete dissolution of the raw material. Therefore, the quantity of the material crystallized per unit of time, d*m*/d*t*, where *m* is the mass of the formed crystals and *t* is time, is to a considerable degree defined by the rate of convection. The relation between these two values may be either linear or more complex. This method is the only common industrial technique for the growth of crystals from hydrothermal solutions. The temperature gradient method has been used to grow large rubies, as well as single crystals of many compounds [31–34].

The temperature gradient method ($T_1 > T_2$) is applied for those substances that have positive temperature solubility coefficients. If the solubility of a compound decreases with the increase in the temperature and $r_c(C_1, T_1) > \rho_c(C_2, T_2)$, then the reverse temperature gradient method is used. In this case, the raw material is placed in the lower temperature zone. It is expedient to position the autoclave horizontally to easily provide repeated recrystallization of the raw material by changing the temperature and the position of the growth and dissolution zones. This cyclic recrystallization is important for purification of the initial substance and enlargement of the raw material particles. This allows preliminary synthesis from individual components while growing crystals of complex compounds.

Any variant of the temperature gradient method is applicable only in the cases where the solubility of the substance noticeably changes with temperature. The higher the absolute value of the temperature solubility coefficient $\partial C_A / \partial T$, the higher supersaturation $\Delta C_A = (\partial C_A / \partial T) \Delta T$ can be achieved with the same temperature gradient. Each substance has a characteristic value of the lowest supersaturation required to provide a noticeable growth rate. The relative values of supersaturation $\Delta C_A / C_A$ range between 0.01 and 0.1.

The general temperature reduction method implies crystallization in the absence of a temperature gradient between the growth and dissolution zones. The supersaturation necessary for growth is created by a smooth decrease in the solution temperature. The solute is transferred to the growing crystal primarily though diffusion; convection does not occur. As the solution temperature decreases, a large number of crystals spontaneously nucleate and grow. The drawbacks of this method are in the difficulties arising in the introduction of seed crystals (the seeds must be isolated from the solution until the latter becomes completely saturated and the temperature starts dropping) and in the control of the growth process.

The metastable phase method usually is applied to the growth of compounds with a very low solubility. It is based on the difference between the solubility of the phase to be grown and the phase of the raw material used. The raw material consists of compounds (or polymorphic modifications of the crystallized substance) that are thermodynamically stable under the conditions of the process. In the case when such polymorphic modifications exist, the solubility of the metastable phase is always higher than that of the stable phase, and the latter will be crystallized owing to dissolution of the metastable phase. This technique usually is used in combination with the temperature gradient or temperature reduction methods. The metastable phase method combined with the temperature gradient method has been applied to the growth of sapphire onto sapphire seeds with $Al(OH)_3$ as a raw material [35]. A solution saturated with Al in respect to $Al(OH)_3$ turns out to be highly supersaturated in comparison with Al_2O_3, and the excessive Al provides the growth of sapphire.

The equipment for hydrothermal crystal growth must withstand high temperatures and pressures over a long period of time. The autoclave material must be inert

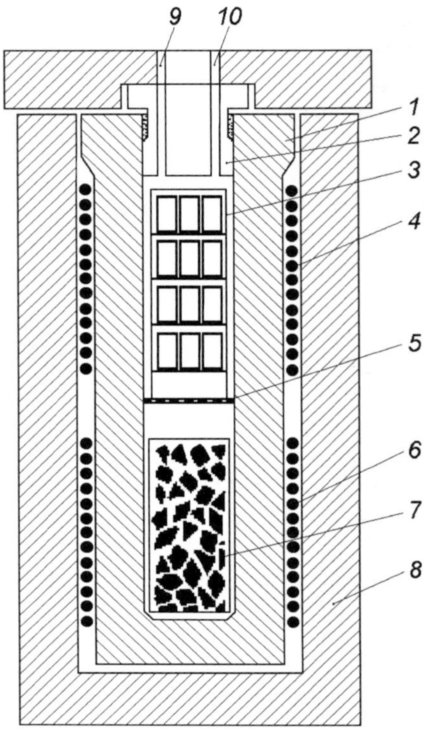

Fig. 4.9 Scheme of an autoclave for crystal growing in hydrothermal solutions: (*1*) body; (*2*) cover; (*3*) frame with seed plates; (*4*) upper heater; (*5*) diaphragm; (*6*) lower heater; (*7*) container with the raw blend; (*8*) heat insulation; (*9*) thermocouple bushing; (*10*) manometer fitting

to solvents. The interior space of the autoclave is separated by a perforated partition or diaphragm into two approximately equal-sized zones: the lower zone is meant for dissolution of the raw material and the upper zone contains the seed plates for crystallization (Fig. 4.9).

The solution fills 70–80% of the autoclave volume, and the liquid expands on heating. Prior to the complete filling of the autoclave, the pressure in the system is equal to that of the saturated vapors of the liquid. After the autoclave is completely filled, the pressure rises, and at a given temperature it is defined by the composition and density of the solution.

The autoclaves used for research purposes have a volume of 20–1,000 cm^3. Industrial vessels are much larger. This type of equipment must withstand working pressures up to 3,000 atm. and temperatures up to 870° K.

In most cases, the hydrothermal method implies the use of steel-corroding solutions. To avoid contamination of the crystallization medium, protective inserts are applied whose shape may replicate the inner space of the autoclave. In this case, they are pressed into the autoclave (contact-type inserts). Floating-type inserts take up a part of the autoclave space, and the rest is filled by water or solution. The degree of filling is to be chosen in such a way that the pressure inside the insert should be equal to that in the space between the insert and the autoclave walls. The floating-type inserts sometimes have special "corrugated" areas, which make it possible to change their volume somewhat, and thus equalize possible pressure differences. Depending on the temperature and type of solution, inserts are used made of carbon-free iron, copper, silver, titanium, platinum, melted quarts, and so forth.

The mass transfer is controlled by means of a weight gauge, with the autoclave axis positioned horizontally. In some cases melted quartz or sapphire windows are made to observe the process.

Thus, for the hydrothermal crystal growth the following is necessary:

- To choose such solvent, pressure, and temperature at which the crystal will be thermodynamically stable and soluble enough to obtain the supersaturation required for considerable growth rates. The process of nucleation on the walls or inside the vessel must not be very intense. For most substances, a solubility of 2–5% is sufficient.
- To provide a rather large ratio between the surface area of the dissolved raw material and that of the seeds, in order that the dissolution should not limit the growth rate (usually this ratio must be approximately equal to 5).
- To provide a sufficiently high temperature coefficient of the solution at its constant average density to give rise to an intense convection not limiting to the growth rate at a given ΔT.
- To choose the temperature solubility coefficient in such a way that it should provide the required supersaturation at a given ΔT.

In 1958, taking into account the above conditions, Laudise and Ballman were the first to obtain rubies and sapphires measuring ~19 × 10 mm [35]. The crystals were grown in a 1–2 molar solution of KOH and K_2CO_3 in small autoclaves with hermetic silver inserts. The seed sapphire plates with the basal or prism orientations were positioned in the upper zone. The raw material – a fine powder of $Al(OH)_3$ or crystalline Al_2O_3 powder – was placed into the lower zone separated by a perforated diaphragm. In K_2CO_3 solutions, sapphire was grown at 760 K, and the coefficient of autoclave filling was equal to 0.70–0.83 with a $\Delta T \sim 30$ K. The introduction of potassium bichromate with a concentration of 0.1 g/l made the crystals bright-red. The percentage of chromium in such ruby was up to 1%.

While using seeds with the orientation *R*, the most favorable growth conditions were found to be created in sodium carbonate solutions with the additions of sodium bicarbonate, ammonium, or ammonium bicarbonate [36, 37]. The crystals were grown in autoclaves with a capacity up to 1.5 L lined with silver by the contact method, with a temperature of crystallization of 750–770 K and a pressure of 60–190 MPa. The temperature in the lower zone, containing the raw material, was 15–20 K higher than in the crystallization zone. In all the cases, the growth rate varied from 0.1 to 0.3 mm/day depending on the orientation of the growing crystal. The highest growth rate was achieved for the rhombohedron plane $(10\bar{1}1)$, followed by, in order of decreasing rate, the hexagonal dipyramid $(22\bar{4}3)$, the prism $(11\bar{2}0)$, and the basal plane (0001). The growth rate of the rapidly and slowly growing facets differed by orders of magnitude.

To avoid the formation of diaspore, sapphire must be grown at $T > 670$ K:

$$Al_2O_3 \text{ (corundum)} + H_2O \rightleftarrows 2AlOOH\text{(diaspore)}$$

Study of the Al_2O_3–H_2–NaOH and Al_2O_3–H_2O–Na_2CO_3 systems makes it possible to establish the stability region of the phase α–Al_2O_3. At temperatures below 670 K the diaspore is stable. As the temperature and the pressure rise, the solubility increases, too. However, the increase in supersaturation is much more noticeable in a carbonate solution, and at the same ΔT the crystal grows more rapidly in it. The growth of sapphire on seed plates is most successful at 678 K, when the temperature of dissolution is 708 K, the solvent is Na_2CO_3 and the autoclave filling is approximately 80%.

Etching of the surface of the crystals grown hydrothermally (Fig. 4.10) reveals the characteristic relief connected with passivity of the facet (0001) [38]. The singular surface $(10\bar{1}0)$ is unstable, and while in contact with the solution it breaks down into steps with passive (0001) facets and active $(10\bar{1}1)$ facets.

The method of crystallization from the melt in solution (the flux method) has much in common with the hydrothermal method. Usually it is applied to high-melting or incongruently melting compounds where crystallization from monocomponent melts is either impossible or highly laborious. In this method, the melts of a number of fusible oxides (PbO, MoO_3, B_2O_3, BaO, V_2O_5, etc.) and salts (KF, Na_2CO_3, PbF_2, $CaCl_2$, NaCl, BF_3, etc.) are used as solvents (fluxes). In these melts, the solubility of high-melting compounds must not be lower than 10% at a temperature solubility coefficient on the order of 1% per 10 K. It is expedient to choose a solvent with common components compared to the crystallized substance and other components sharply different in ionic radius (to decrease the probability of isomorphism).

Fig. 4.10 Lamellar structure of ruby grown by the hydrothermal technique (×100)

Using this method, crystals are grown at normal pressure in platinum, iridium or molybdenum crucibles. The process of crystallization is realized via gradual cooling of a melt saturated with the components of the growing crystals, under isothermal conditions upon evaporation of the melt (it must have a sufficiently high vapor pressure), or by the temperature gradient method. Small crystals (measuring up to 1–1.5 cm) are obtained by spontaneous crystallization, larger crystals are grown onto a seed. The material is transported to the crystallization front by diffusion and convection. The role of the latter becomes more significant when the viscosity of the melt decreases. Rotation of the seed in the melt or stirring of the melt raises the rate of crystal growth. This is connected with reduction in the thickness of the diffusive layer, and increase in the rate of the material transport to the growing surface. These methods are applied in cases when the growing crystal is phase-stable across a narrow temperature range and the impurity distribution coefficient is essentially dependent on the temperature and growth rate.

The flux method was used to grow large rubies with a weight up to 100 g from solution in a lithium molybdate melt [39]. In this case, mixtures of lead oxides and fluorides with boron oxide serve as solvents. The solubility of sapphire in the melts of these compounds at 1,570–1,670 K may reach 30–40%. Crystallization is realized at 1,470–1,670 K [36, 37] in platinum crucibles with a capacity of several liters. As a rule, rubies are grown from the solution of aluminum oxide in a melt of $PbO–PbF_2–B_2O_3$ with a small addition of chromium oxide The flux method allows growth of star-shaped rubies with diameters up to 20 mm and thicknesses up to 4 mm [40].

After growth onto a seed at $T \ll T_m$ the crystals do not contain thermal stresses, traces of plastic deformation, or block boundaries unless they are caused by defects of the seed crystal. For instance, in sapphire grown onto perfect seeds the highest possible block disorientation does not exceed 3′. Crystals grown by spontaneous crystallization have a low density of dislocations; with crystals grown onto a seed one can obtain samples of a rather large area containing no dislocations. At the same time, internal stresses that may arise in the crystal cannot relax due to plastic deformation, as occurs with other growth methods. Then, if the elastic limit is surpassed, the crystal breaks down. Because of this, cracks are one of most widely spread defects in crystals grown by the flux method.

Discrepancies between the crystallochemical characteristics of the seed and the growing crystals (an essential distinction in the values of lattice parameters, different chemical composition) also gives rise to the formation of defects. In particular, quartz, sapphire, and ruby crack when the difference in the lattice parameters is ~0.0005 Å, ~0.001 Å, and ~0.0005 Å, respectively [41]. Cracking also is promoted by the presence and nonuniform distribution of impurities. The distribution of impurities is bound up to the growth conditions, different ability of crystal facets to adsorb impurities, and different coefficients of impurity capture for the growth pyramids. The distribution nonuniformity manifests itself in the zonal and sectorial crystal structure, which becomes essential during the growth of doped crystals.

According to the terminology used for quartz, impurities fall into nonstructural and structural categories. The former are liquid-phase inclusions. They do not cause high stresses in the crystals, but essentially influence their physical characteristics, e.g., the optical properties. Structural impurities are the ions either substituting those of the main substance in the crystal lattice or incorporated into the interstices. These are the components of the solvent, the activator, and the impurities in the raw material and solvent. Structural impurities change the lattice parameters and favor the appearance of stresses. For instance, in pure sapphire grown by the hydrothermal method stresses do not exceed 1.4 kg/mm^2, whereas in ruby they are five times as high.

The typical impurities present in crystals grown hydrothermally include H_2O, H^+, and OH^-. They influence the thermal conductivity, microhardness, electrical properties, and optical characteristics. In ruby, the captured ions (molecules) of water negatively affect the values of strength, hardness, and internal friction. Obtaining high-quality crystals requires purification of the initial reagents from impurities which raise the coefficient of water capture, the choice of crystallization conditions at which the capture of water is the least, and annealing of the grown crystals under special conditions.

Using low-quality seeds, crystals of higher perfection can be grown from the solution by carrying out sequential growth cycles. Upon completing each cycle, the grown layer is separated to be used as a seed for the next step. Such a succession may consist of ten or more individual processes. Selection of the seed material is especially significant for the heteroepitaxial crystal growth method.

In conclusion it should be noted that, despite rather high structural perfection of the grown crystals, methods of crystallization from the solution are not widespread. First of all, this is due to comparatively low growth rates. Moreover, the process of crystallization itself is complex and insufficiently investigated. In particular, this deficiency of information is connected with the fact that it is practically impossible to directly observe the process of crystallization.

4.3 Growth of Sapphire from the Melt

The methods of sapphire growing from the melt are most popular, since they allow the attainment of large-size crystals at high growth rates. The phase transition melt $\rightarrow$ crystal is the simplest and best-controlled process, which has been made the basis for modern methods of commercial sapphire production. At first sight, it may be assumed that these methods are able to satisfy all the existing demands for the crystals that possess the required properties. However, in fact, many compounds cannot be grown from their own pure melts owing to the following causes:

- The material decomposes prior to melting or melts incongruently.
- The substance sublimates prior to melting or has a too high vapor pressure at the melting temperature.
- The required polymorph modification is structurally unstable in contact with the melt, and high-quality crystals of this modification cannot be obtained by solid-phase transitions.
- T_m is so high that it is practically impossible to grow the crystals from the melt.

In the absence of the mentioned causes, the methods of crystal growing from the melt are used mainly to obtain crystals from one- or two-component systems. The subdivision of the systems into one- and two-component ones is arbitrary to a great extent. Strictly speaking, the growth in a one-component system implies the absence of other systems that contain other components, occasional admixtures/impurities, or activators. Since, however, any material contains impurities, the growth in a one-component system can be excluded essentially out of real processes. In practice, the concept of "the growth in one-component system" assumes the presence of impurities and activators, with the proviso, however, that the concentration of these is sufficiently low to exclude any significant importance of diffusion processes. The growing in multicomponent systems implies the growth when the second component is introduced intentionally, e.g., to reduce the T_m of the material being crystallized.

It should be noted that crystallization from the melt is a complex multifactor process. The phase transition from liquid to solid state is accompanied by structural changes, the melt properties – its viscosity, density, etc. – also undergoing essential changes. Distinctions in the solubility of impurities in the melt and in the crystal result in redistribution of the matrix elements between the melt and the crystal. Although these processes are complicated, they have been investigated thoroughly; technical approaches that make it possible to vary the method potentialities within wide limits have been developed; numerous crystallization equipment types have been designed. The scientific and patent literature concerning the crystal growing methods from melts and realization of those methods is so numerous, diversified, and interesting, that a special book should be devoted to those issues. We shall limit ourselves to a small fraction of information concerning the sapphire growing by the most advanced and highly industrialized methods.

There are two groups of melt growth methods:

- The methods of crystal growth from large melt amounts: horizontally oriented crystallization, Kyropoulos, Czochralski, HEM, Stockbarger–Bridgman methods
- The methods of crystal growth from small melt amounts: Verneuil, Stepanov, and zone melting methods

The amount of the melt influences the character and intensity of the physicochemical processes. For instance, the melt may undergo thermal dissociation, and the dissociation products evaporated into atmosphere. Such substances should be in the melted state for a limited time; thus, these are grown from a small melt volume. The same condition is to be fulfilled for substances interacting in the melt with the container material and atmosphere. The smaller the melt volume, the lower will be the crystal contamination level with the melt/atmosphere interaction products. A difference between the convection conditions for the two groups of methods is to be noted. In a large melt volume, the convective flows are developed easily and the convective substance transfer plays an important part. Within a small volume, the convection cannot be of the same importance, so the mass transfer occurs mainly due to diffusion. The influence of the melt volume on the growing crystal quality can be exemplified by the difference in the impurity distribution in cylindrical crystals

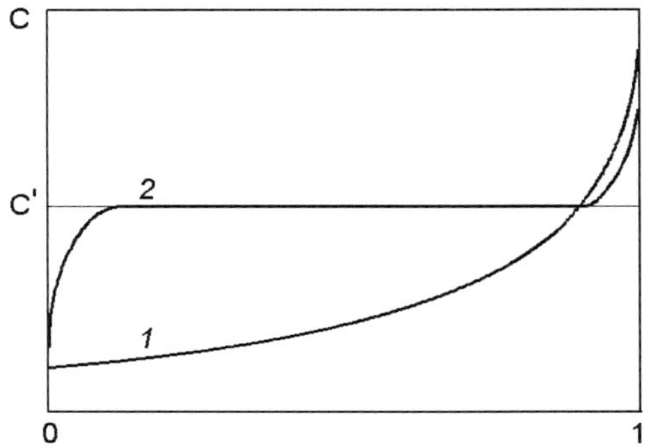

Fig. 4.11 Distribution of an impurity having the distribution ratio $K < 1$ over the crystal length for crystals grown by (*1*) zone melting and (*2*) directional crystallization. The impurity concentration in the initial material is *C*

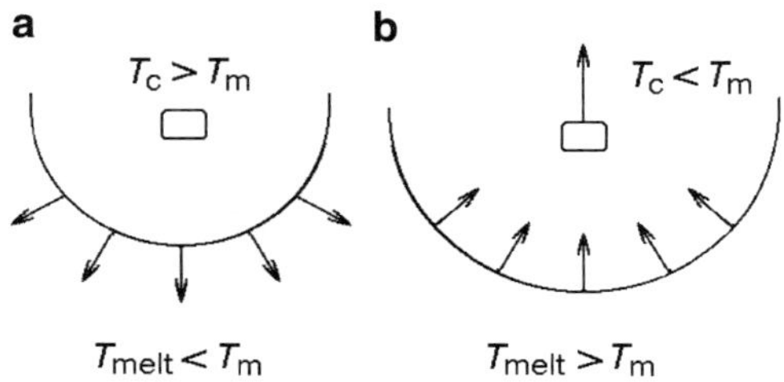

Fig. 4.12 The temperature gradient direction. Heat removal through (**a**) the melt and (**b**) the growing crystal

grown using the directional crystallization and zone melting methods (Fig. 4.11). The directional crystallization belongs to the growing methods from a large melt volume, while the zone melting belongs to those from a small one. In the case of directional crystallization, the impurity content in the middle crystal part remains constant; thus, this method is preferable for activated crystals. In the case of zone melting, the initial crystal part contains a lower impurity level than the raw material; therefore, this method is used successfully to purify the crystals.

Two limiting cases are considered, depending on the temperature gradient direction. In the first case (Fig. 4.12a), the heat is removed outward as the crystal is cooled, that is, the furnace temperature $T < T_m$ [41]. At the crystallization of an overcooled melt, heat is released, so the growing crystals take a temperature higher than that of the melt. The heat flow is directed from the crystal through the melt toward the environment. A case also is possible, however, when the heat flow is directed from the melt to the crystal and then through a local cooler (to which the crystal is connected) to the environment (Fig. 4.12b). Here, the melt temperature exceeds T_m. Therefore, the crystal surface character is defined completely by the position of isotherms in the plane of solid–liquid interphase boundary and especially by the position of that plane with respect to the isotherm corresponding T_m. Such a growth will continue until the crystal of increasing size approaches closely

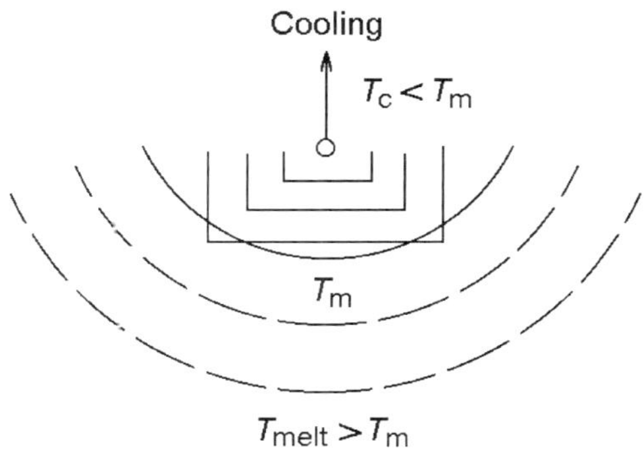

Fig. 4.13 Displacement of a polyhedral growth front toward the isotherm corresponding to T_m

the isotherm. In this case, the crystal vertices and edges going beyond the isotherm will melt, and the crystal takes on a smoothed shape (Fig. 4.13) [41]. By controlling the heat removal through the crystal/cooler, the growth process can be provided when the isotherm will move toward the melt together with the CF.

Depending on whether the crystallization is realized within or without a crucible, the crucible-free (Verneuil, flowing zone) and crucible (Kyropoulos, Czochralski, HEM, horizontal directional crystallization, Bridgman-Stockbarger, zone melting, Stepanov) methods are distinguished.

Almost all the melt-growing methods existed for several decades, but they were revived during the last few years. At present, single crystals of high structure perfection with several tens of kilogram mass are obtainable by crystallization from melt. These advances are due not only to upgraded apparatus, but also to the development of concepts of the crystallization processes.

4.3.1 Physicochemical Aspects of Crystal Growth from the Melt and Properties of the Melt

Crystallization from the melt is accompanied by a number of physical and chemical processes among which the following ones should be singled out:

- Processes influencing the melt composition: thermal dissociation of the raw material, its chemical interaction with the environment, evaporation of the dissociation products and impurities
- Processes occurring at the crystallization front that define the phase transition kinetics (aggregation processes)
- Heat transfer influencing the temperature distribution in the crystal and in the melt
- Mass transfer caused by convection, diffusion and the aggregation degree in the melt

Properties of the melt are well-known and explain the phenomena that occur near the crystallization front. The main properties are presented in Tables 4.11–4.16.

Melting and crystallization temperature. The T_m of aluminum oxide is influenced by its initial structure, chemical composition, interaction with the container material, and the environment. The spread of the values of T_m does not exceed 1 K when the impurity concentration changes by a factor of 30. The melting temperature of high-purity aluminum oxide does not depend on its initial structure. Any gaseous medium reduces T_m. This seems to be connected with the bombardment of the crystal surface with atoms, molecules, or ions of gases that have rather high energy at high temperatures. For aluminum oxide, the recommended value of T_m is (2,327 ± 6) K. The data on T_m under different conditions and the temperature dependence of the melt enthalpy are presented in Tables 4.11 and 4.12.

The degree of melt overcooling in the crystallization process is of great interest for the sapphire-growing technology. While T_m for sapphire is well-studied and even recommended for the use as a reference point, the data on the crystallization temperature T_c and the overcooling degree are contradictory [42]. It is reported that T_m and T_c do not coincide at low rates of temperature decrease. As established by the investigations carried out under conditions close to the equilibrium ones, the discrepancy between T_m and T_c makes 1.3–1.5 K. This testifies to a pronounced tendency of the melt to overcooling and points to the fact that the crystals can be grown from overcooled melt. At an abrupt temperature change, the overcooling value may reach 100 K. The measurements performed under crystal heating show that their T_m is well reproduced at the rates of temperature increase, ranging between 19.2 K/min and 7.5 K/min. The measurements of T_c under cooling conditions have shown that it depends on the temperature drop rate. When the latter is 29.5 K/min, the value of melt overcooling runs into 64 K, while at the cooling rate of 80 K/min, it is 65–67 K (Fig. 4.14).

As follows from this figure, at a rate of 1.5–12 K/min, the dependence of overcooling on the temperature drop rate is almost linear; at a rate exceeding 25 K/min the curve reaches saturation, and the overcooling amounts to 63–67 K. The presented curve can be described as

$$T = -1.226 + 3.96v + 3.4 \cdot 10^{-2} v^{-2} - 3.47 \cdot 10^{3} v^{3} \tag{4.8}$$

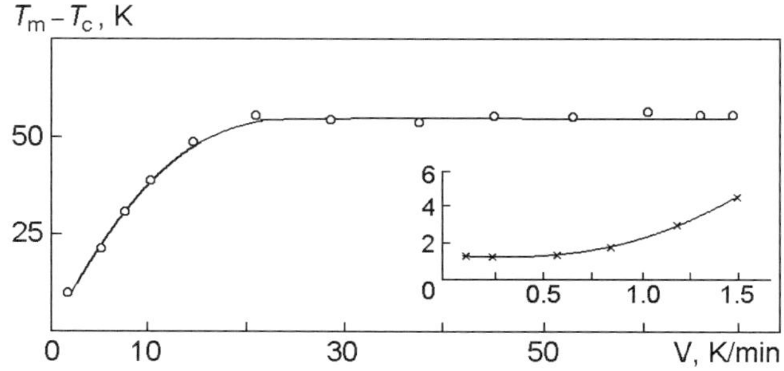

Fig. 4.14 Overcooling of Al_2O_3 melt depending on the temperature decrease rate

with an error of max. 1%. In other words, within the whole range of technological cooling rates, the crystal grows from the overcooled area before the crystallization front.

Optical properties of the melt. The data on the optical properties of the melt – the absorption and refraction coefficients at $T \leq 2{,}950$ K – are generalized most completely and authentically in ref. [43]. Sets of experiments have been performed with the melted particles and stationary melts. The latter experiments show that, at least in the visible and the near-IR spectral regions, the absorption coefficient rises sharply at the melting point, whereas the experiments with the melted particles do not reveal such a change. As the wavelength increases from 0.63 to 8 μm, the absorption coefficient of the melt grows from 0.003 to 0.02 cm^{-1}. There is an essential (up to 50%) spread in the refraction index values reported by different authors. At the thickness values used conventionally in practice, the melt is opaque, and for thermal calculations, the values of its radiant emittance can be used.

The melt density runs into 3.03 ± 0.07 g/cm^3. The temperature dependence of the melt density is expressed as:

$$\gamma = 3.03 - 1.08 \cdot 10^{-3} (T - 2327) \tag{4.9}$$

Surface tension. The surface tension value depends on the atmosphere composition. In CO medium, this value is the lowest: $\sigma_s = 0.36$ N/m^{-1}, in vacuum, it is the highest: $\sigma_s = 0.69$ N/m^{-1}. The surface tension also is defined by the pressure. As the latter rises from 0 to 0.1 MPa at T_m, the surface tension value decreases from 0.69 to 0.20 N/m in carbon medium and to 0.30 N/m^{-1} in nitrogen medium. The decrease of the surface tension in the atmosphere of the reactive gases is caused by their adsorption at the melt surface.

The recommended value of σ_s in vacuum is 0.67 N/m^{-1}; the temperature dependence has the form:

$$\sigma_s = 670 - 0.30(T - 2327). \tag{4.10}$$

The melt viscosity is equal to $h = 0.57P$ in the vicinity of T_m and drops sharply as the temperature rises (Fig. 4.15a, Table 4.15).

The nonlinear dependence $\lg \eta = f(1/T)$ speaks for structural changes in the melt that occur in the precrystallization zone. This also is confirmed by the calculated free energy of the viscous flow of the melt at temperatures ranging from 2,327 to 2,523 K, which is characterized by nonlinearity. The activation energy of viscous flow decreases from 184 kJ/mol at $T \sim T_m$ to 71 kJ/mol at 2,523 K and then remains constant. Thus, the melt in the precrystallization zone is an associated liquid and the degree of association diminishes sharply as the temperature rises.

Assuming that the whole activation energy of the viscous flow is spent for the formation of vacancies, the radius of these vacancies can be calculated [44]. The dimensions of the vacancies (the vacancy radius at T_m is $2.28 \cdot 10^{-8}$ cm) necessary to provide the viscous flow of the melt are larger than those of the holes (the hole radius is $1.20 \cdot 10^{-8}$ cm). Hence, the particles in the viscous flow, at least at $T \sim T_m$, are ion associations that arise due to an essential share of the covalent binding

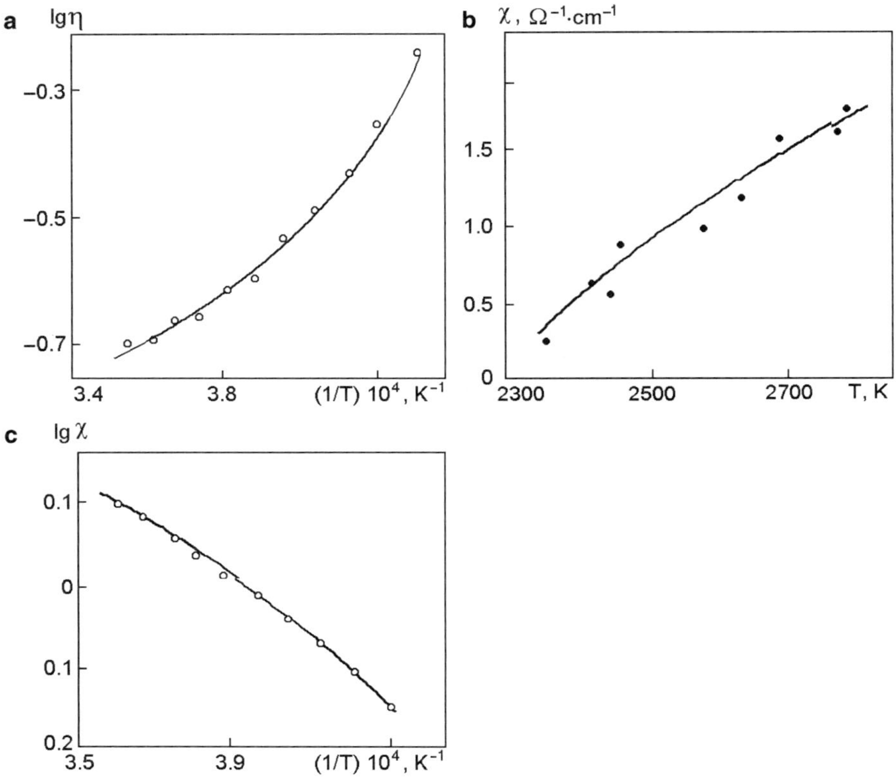

Fig. 4.15 Temperature dependences of viscosity (**a**) and electric conductivity (**b**, **c**) for melted aluminum oxide

component between the metal and oxygen, but not elementary ions, as in unassociated liquids.

The melt electric conductivity (Fig. 4.15b) increases from 0.71· Ω^{-1} cm^{-1} at $T \sim T_m$ to 1.5 Ω^{-1} cm^{-1} at 2,978 K. For this temperature range, the temperature coefficient of electric conductivity is positive. The transition of aluminum oxide from solid to liquid state is accompanied by an essential jump of the electrical conduction characteristic of the substances that possess ionic conduction: $\chi_l/\chi_s \approx 7 \cdot 10^3$. The temperature dependence of the melt conductivity lg $\chi = f(1/Te)$ deviates from the linear dependence (Fig. 4.15c), the activation energy increases from 85.8 kJ/mol at $T \sim T_m$, to 586 kJ/mol at 2,773 K. The activation energy of its electrical conductivity turns out to be lower than that of the viscous flow. The difference between these values is especially large at $T \sim T_c$. This means that the particles that are the "units" of the viscous flow differ from those that define the electrical conductivity. The mentioned distinction in activation energy, as well as the activation energy variation character of the melt viscous flow, testify to the presence of ion complexes in the liquid at $T \sim T_c$. The behavior of the lg $\chi = f(1/\eta)$ dependence also points to changes in the melt structure caused by the temperature rise (Fig. 4.16).

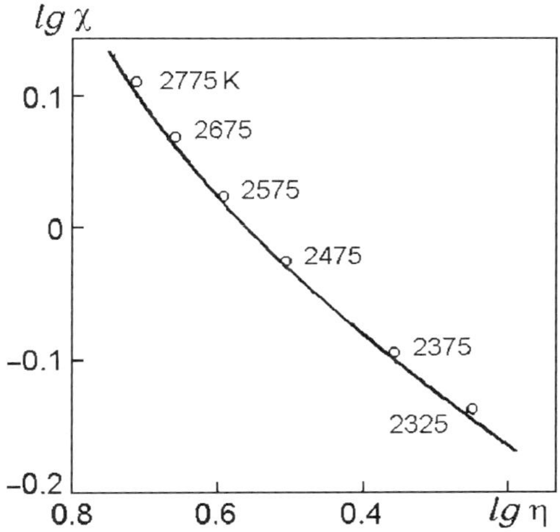

Fig. 4.16 Dependence lg $\chi = f(\lg \eta)$ for melted aluminum oxide

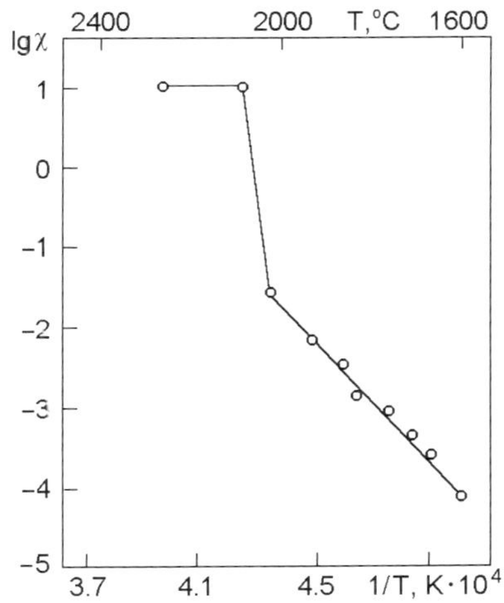

Fig. 4.17 Temperature dependence of aluminum oxide electric conductivity in the phase transition region

Due to its relatively high-heat conductivity, the high-frequency heating of the melt can be used. In the region of melt crystallization, the electrical conductivity decreases sharply (Fig. 4.17). The activation energy calculated from the slope of the curve in the temperature range between 1,970 and 2,320 K is equal to 7.8 eV. Within this range, the dependence of the electrical conductivity on $1/T$ is exponential [45].

The conductivity values for different temperatures are presented in Table 4.13.

Diffusion in the melt. The diffusion coefficient of aluminum ion at temperatures of 2,335–2,573 K is

$$D = 2.3 \cdot 10^{-3} \exp(21.2 / RT). \tag{4.11}$$

The diffusion activation energy is approximately equal to that of electron transfer. So, it should be assumed that in both cases the transfer is realized by carriers of the same type, which have the highest mobility in the melt. Presented below are the temperature dependences of the viscosity and the aluminum ion diffusion coefficient, as well as the calculated size of the particles responsible for the diffusion transfer in the melt.

T (K)	$D \cdot 10^5$ (cm^2/s)	η (poise)	$r^* \cdot 10^8$ (cm)
2,335	1.95	0.58	1.50
2,375	2.05	0.52	1.52
2,400	2.15	0.45	1.70
2,425	2.35	0.35	1.82

As is seen, the most probable dissociation reactions at melting are

$$Al_2O_3 \leftrightarrow AlO_2^- + AlO^+;\ 2Al_2O_3 \leftrightarrow Al^+ + 3AlO_2^-.$$

There, $r^*_{AlO} = 1.82 \cdot 10^{-8}$ cm, $r^*_{Al} = 0.53 \cdot 10^{-8}$ cm.

Heat conductivity. At a temperature of ~2,350 K, the temperature conductivity of the melt is $(4.8 \pm 0.4) \cdot 10^{-7}$ m^2/s, the heat conductivity makes (2.050 ± 0.15) W/(m·K). The temperature dependence of heat conductivity is expressed as:

$$\lambda_T = 2.05(1 - 5.3 \cdot 10^{-4} \Delta T). \tag{4.12}$$

Thermal dissociation [46] and chemical reactions in the melt disturb the stoichiometric composition, thus leading to the formation of defects in the crystal. The intensity of these reactions is defined by the thermal and temporal regimens which can be established from the state diagrams of the main components of the crystallized compounds, the regions of solid solution stability, phase transitions, possible changes in the composition caused by evaporation, and so forth. For instance, aluminum oxide melting at normal pressure is accompanied by dissociation followed by formation of [AlO], [Al_2O], [AlO_2], [Al^+], [O^-], and [Al_2O_2] ions. Due to high vapor pressure of the thermal dissociation products, the melt is saturated with gaseous inclusions that are accumulated at the crystallization front and influence the growth kinetics and the crystal quality.

The thermal dissociation intensity can be lowered using its dependence on external pressure, temperature, and the time period when the substance is in melted state. The use of elevated pressures is limited due to technical difficulties, so the crystallization process should be realized either under normal pressure or in vacuum. For this purpose, the melt must be overheated to such an extent where the thermal dissociation intensity remains still insignificant. The upper overheating limit is defined by the intensity of the substance dissociation and evaporation as well as by its chemical interaction with the container material and the atmosphere. The lower limit depends on the melt viscosity, which impedes its convective mixing. For example, at temperatures ranging between 2,320 and 2,220 K, the viscosity value changes by approximately two times. In practice, the average melt overheating value is ~100 K (Fig. 4.18).

The time period where the substance is in a melted state decreases at zone melting when only an insignificant part of the substance is melted. The loss due to evaporation often is compensated by introducing measured quantities of the deficit

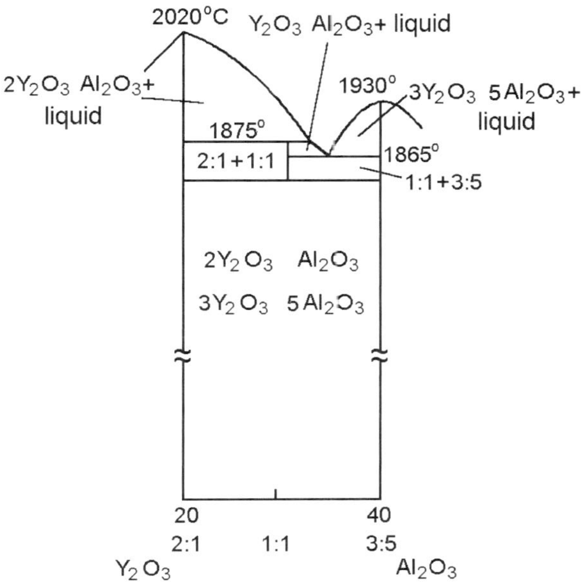

Fig. 4.18 Phase diagram for Y_2O_3–Al_2O_3 (Toropov et al. 1964)

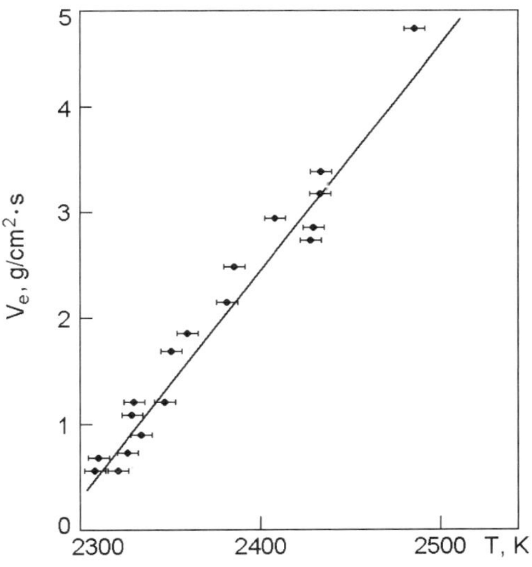

Fig. 4.19 Melt evaporation rate under $2 \cdot 10^{-5}$ Torr residual pressure from a molybdenum crucible

components into the melt. In particular, the loss of chromium at the ruby growth is compensated by increasing its concentration in the raw material.

The melt evaporation rate (V_e) is temperature-dependent (Fig. 4.19). In vacuum, $V_e = (1.95 \cdot 10^{-2}T - 44.4) \cdot 10^{-5}$ g/cm^2 s.

Interaction with high-melting materials. The most popular crucible materials are tungsten and molybdenum. Reactions involving it may proceed via gaseous oxygen:

$$Mo_s + O_g \rightarrow MoO_{s,g}; \quad MoO_{s,g} + Al_l \rightarrow AlO_l + Mo_s$$

Presented below are the calculated data on the interphase dynamic surface energy and the motive force of spreading, as well as the experimental values of the wetting angles for the case when a drop of the melt is placed on the surface of a metal under isothermal conditions:

	Mo	W	Nb	Ta
σ^*_{s-l} (GPa)	64.5	68	46.5	37
σ (GPa)	54	54	39	28
θ (deg.)	15	7	40	30

The wetting angle value depends on the atmosphere (Table 4.1).The distinction in the wetting angles seems to be connected with the measurement techniques used. The dynamic impact accompanying the contact of the melt drop with the substrate may decrease the wetting angle. The simultaneous heating of the oxide and the substrate gives rise to a more intense interphase reaction. The melt spreading over the graphite surface at ~2,400 K results in formation of aluminum carbide. The atmosphere influence of the spreading is insignificant. As follows from the thermodynamic analysis of the Al_2O_3–C system, at the first interaction stage aluminum oxycarbide Al_4O_4C is formed; its subsequent interaction with carbon gives aluminum carbide Al_4C_3.

At the graphite interaction with aluminum oxide vapors (Fig. 4.20), a coating consisting of aluminum oxycarbide, aluminum carbide, and aluminum oxide is formed on the graphite surface. Despite the growth of this coating, the mass ΔP decreases linearly. The mass of graphite decreases obviously due to the interaction with oxygen and aluminum suboxides. In the latter case, the formed aluminum oxycarbide and aluminum carbide provide the substance condensation centers from the vapor phase.

Table 4.1 Contact angle values (degrees) for high-melting material surfaces at a temperature about 2,350 K

Substrate material	Vacuum	Hydrogen	Carbon monoxide	Nitrogen	Argon
W	50	16	18	35	–
Ta	38	43	–	–	–
90%Ta + 10%W	42	40	–	–	–
TaC	–	32	36	41	–
4TaC–ZnC	–	29	28	23	–
ZrC	–	13	15	10	–
TiN	7	16	–	36	–
Zr	–	21	21	17	–
Graphite	–	–	–	–	120–150

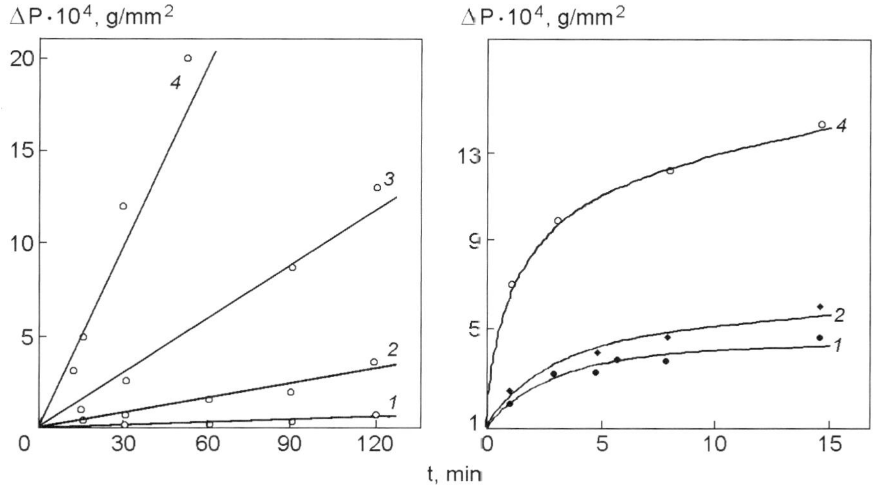

Fig. 4.20 Interaction of graphite (**a**) with aluminum oxide vapor in vacuum and (**b**) with the melt under argon atmosphere at temperatures (K): (*1*) 2323, (*2*) 2373, (*3*) 2473, (*4*) 2673

Interaction of the melt with the container material often defines the selection of the crystallization method. The container must provide the absence of mutual solubility and chemical interaction with the substance to be crystallized.

The requirements imposed on the container material are the following:

- Its chemical binding forces have to differ sharply in nature from those of the crystallizing substance (dielectric crystals are grown in metallic containers, organic crystals are crystallized in containers made of inorganic dielectrics)
- Mechanical strength
- High machinability of the material
- Close values of the expansion and compression coefficients of the container material and the crystallizing substance
- High electric conductivity (in case of high-frequency heating)

The interaction between the container and the melt may be caused by the impurities contained in the raw material or by those adsorbed at the container walls, surface of the crystallization chamber, furnace elements, and so forth. Moreover, it may be induced by some substances (such as oxygen or moisture), which may ingress the melt from the atmosphere.

In some cases, it is impossible to choose the container material that is neutral with respect to the melt. Therefore, one should use the coatings that weaken the interaction between the container and the melt. Such coatings must possess a sufficient mechanical strength and have expansion coefficients close to those of the container material. It should be noted that the use of molybdenum coated with tungsten increases the service life of the container. The selection of the container material is described in Section 4.3.3.

Interaction of the melt and the container material with crystallization atmosphere. The atmosphere composition and pressure are defined by the vapor pressure and the chemical activity of the crystallizing substance. The atmosphere may interact with the melt, thus retarding or speeding up certain processes. For instance, in a neutral medium, the valence of Cr^{3+} remains unchanged, whereas in hydrogen, this ion is reduced. One should preferably use the atmosphere, which contains volatile components of the crystallizing substance. In particular, oxides, fluorides, and sulfides are to be grown in oxygen-, fluorine-, and sulfur-containing atmosphere, respectively.

Sapphire is grown in vacuum, neutral (helium, argon, nitrogen), oxidative (air, oxygen), or reductive ($CO + CO_2$) media. Vacuum also is used to remove dissolved gases and volatile impurities from the melt. At $T > 1{,}070$ K and a vacuum worse than 10^{-4} Torr, oxidation reactions proceed actively, resulting in failure of the used heaters and crucibles. Neutral media reduce the substance evaporation intensity.

In contrast to inert media, the use of oxidative media is defined by the crystallization temperature, the container material, and the furnace components. If particular conditions do not allow the use oxidative atmosphere, the crystallization process is realized in neutral medium followed by annealing of the grown crystal in oxidative medium. Reductive atmosphere is used to prevent the oxidation reactions in the melt.

Heat transfer. Taking into account the heat transfer mechanisms, three cases are to be considered: $kl \ll 1$; $kl \sim 1$; $kl \gg 1$, where k is the absorption coefficient, l is the characteristic size of the object along the thermal flow direction. The first case corresponds to opaque media where heat is transferred only due to molecular heat conduction. The temperature distribution mets the Fourier law:

$$q = \lambda_{\mathrm{T}}\,\mathrm{grad}\,T \tag{4.13}$$

where q is the heat flow, λ_{T} is the heat conductivity coefficient and the equation of heat conduction. The second case corresponds to semitransparent media: heat radiation from the heat source which hits the crystal is damped in the latter. Heat is transferred by re-emission. Therefore, the heat flow still can be presented in the form (4.13). However, the heat conduction coefficient should be the sum $\lambda_T + \lambda_r$ [47]

$$\lambda_{\mathrm{r}} = 16n^2\sigma T^3 / 3k. \tag{4.14}$$

Here n is the refractive index, σ is the Stefan–Boltzmann constant.

In the general case, the share of the molecular and the radiative component of λ_r is defined by the position of the Planck function maximum with respect to the transmission band of the crystal. In sufficiently transparent crystals, heat transfer at high temperatures is completely defined by their optical properties. The third case concerns transparent media. At the growth of transparent crystals, heat emission therein is not damped, and heat transfer is influenced by the emission of the heater, container walls, and so forth. In this case, the relation (4.13) is not valid, and the heat conduction equation must be substituted by the integral equation of radiant energy transfer. For transparent media, the heat transfer coefficient is effective K_{eff}, as it depends on the shape and state of the surfaces that reflect and refract the radiation:

$$\lambda_{eff} = 4\pi k T^3 n \psi \quad (4.15)$$

Here ψ is the factor depending on the optical properties and configuration of the system.

Taking into account the transparency of the substance and the reflecting power of the container walls, one can establish the conditions for achieving temperature constancy over the crystal cross section and constancy of its gradient along the crystal axis. With the increase in the substance transparency, the control of the temperature gradients and their constancy becomes more complicated. Therefore, convective heat transfer in the melt plays a significant role (Fig. 4.21).

In the Stockbarger method, the isotherm and thus the CF are convex, and the flows are directed from the melt to the CF center, whereas in the Czochralski method, those have the opposite direction. The intensity of the melt convective mixing depends on the temperature conditions: it is higher, the larger the temperature gradients in the system.

Evaporation. At the first stage of aluminum oxide evaporation, molecular oxygen and Al_2O_3 reduction products are formed. At the second stage, molecular oxygen dissociates into atoms. This process is described by the reactions:

$$Al_2O_3 = 2AlO + 1/2O_2 \qquad K_{P_1} = P_{AlO}^2 + P_{O_2}^{1/2}$$
$$Al_2O_3 = 2Al_2O + O_2 \qquad K_{P_2} = P_{Al_2O} + P_{O_2}$$
$$Al_2O_3 = 2AlO + 3/2O_2 \qquad K_{P_3} = P_{AlO}^2 + P_{O_2}^{3/2}$$
$$O_2 = 2O \qquad K_{P_{O_2}} = P_O^2 / P_{O_2}$$

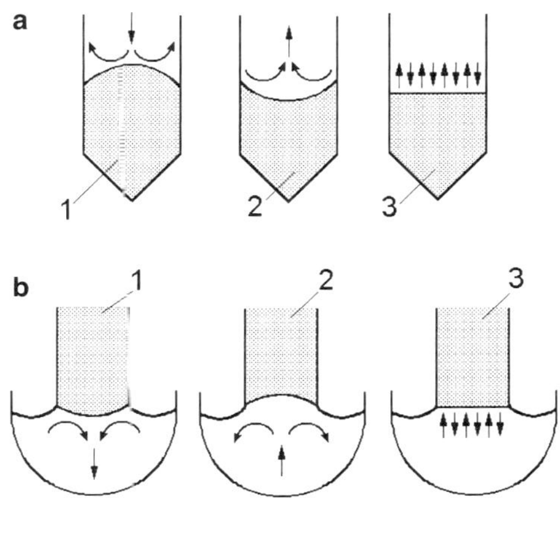

Fig. 4.21 Convective flows in the melt (indicated by *arrows*) at the crystal growing by (**a**) Stockbarger and (**b**) Czochralski techniques near the (*1*) convex; (*2*) concave, and (*3*) flat crystal growth fronts. The crystals are hatched

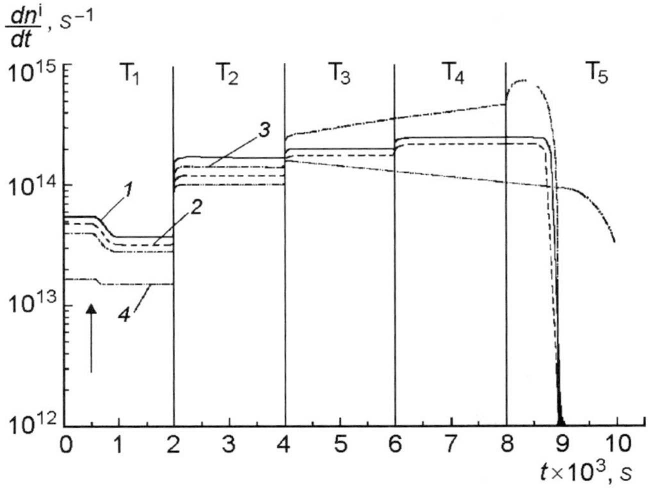

Fig. 4.22 Time dependence of evaporation rate out of Al_2O_3 melt for (*1*) CO_2; (*2*) CO; (*3*) O_2; (*4*) H_2. $T_1 = 2{,}320$ K, $T_2 = 2{,}500$ K, $T_3 = 2{,}540$ K, $T_4 = 2{,}570$ K, $T_5 = 2{,}630$ K. The onset of sapphire melting is indicated by *arrow*

The main gaseous components over the melt are one-atom oxygen and aluminum. The concentration of Al_2O and AlO suboxides rises with the temperature; at 3,000 K it reaches 10% [48].

The gases dissolved in the melt evaporate. At 2,540 K, the isothermal desorption rate of oxygen increases in time (Fig. 4.22) [49]. The melt seems to "boil" visually. Such a "boiling" is accompanied by pulsation of CO and CO_2 partial pressure. In Fig. 4.22 dn^i/dt is the gas release rate.

This pulsation can be assumed to be connected with the distribution of gas bubbles filled with the gases emerging to the melt surface. At lower temperatures, CO and CO_2 molecules seem to form clusters suspended in the melt; therefore, the melt does not "boil" and the partial pressures do not pulse. At the final evaporation stage, the isothermal desorption rates drop to zero rather fast. The hydrogen release after the melt evaporation (curve 4) is explained by its release out of the crucible where it was dissolved during the melt evaporation.

Some fraction of the remaining gases is captured by the crystallization front and enters the crystal. Mass spectrometry of sapphire shows the presence of the main gaseous impurities such as oxygen, hydrogen, OH-groups, CO, CO_2, and a number of smaller hydrocarbon peaks.

4.3.2 Crystal Growth from the Melt Not Using Crucibles

Crucible-less crystal growth technologies include the Verneuil [50] and zone melting (floating zone) methods.

4.3.2.1 Verneuil Method

The Verneuil method was the first industrial crystal growth technology. Consider the history of its creation [51]. The method was named after August Verneuil, the French professor who developed the original technique and apparatuses which allowed the growth of 20–30 carat sapphires and rubies in 2–3 h. This was an outstanding achievement that not only made available a valuable material in required quantities, but also opened up prospects for the growth of other single crystals.

Almost 50 years of research into the growth of ruby preceded the success achieved by Verneuil. The first attempts in this field were made by Mark Gaudin. After numerous experiments conducted from 1858 to 1869, he obtained microscopic bauxite crystals applicable only as an abrasive material. Transparent rubies were grown for the first time in 1877 by Edmond Fremi, who melted alumina together with lead oxide or calcium fluoride in a clay crucible. Introduced into this mixture were small quantities of potassium chromate. Afterward, the flux composition was improved by using barium fluoride instead of calcium fluoride and adding potassium hydroxide. During slow cooling of the melt, nucleation centers were spontaneously formed, and over 7–20 days small crystals grew from them. Fremi published the book *Synthesis of Rubies* in 1891, which was the precursor of the entire scientific literature devoted to crystal growth.

In 1885–1905, small quantities of so-called redesigned rubies shaped as oblate spheres were put out to the world market of gems. Investigation of such crystals still in existence led to the conclusion that these gems were man-made. Judging from the shape of the growth layers, the process of crystal growth was multistage. The crystals were heated using burners; several times they were turned over and grown further. Naturally, repeated heating and cooling resulted in the formation of a large quantity of cracks. In 1895, Michaud used a gas burner to melt together several Siamese rubies, and he added ammonium bichromate to make the color more intense. However, the quality of these crystals was not sufficiently high. The stones obtained in such a way could not be considered fully artificial. “Redesigned” rubies became completely uncompetitive after the appearance of rubies grown by the Verneuil method.

At first Verneuil grew crystals in collaboration with his teacher Fremi by the method described above. In 1888, at a meeting of the Academy of Sciences of France they reported the creation of artificial rubies and presented hundreds of sparkling rhombic crystals. But these crystals were too small to be widely used and, in view of the conditions of their growth, could not be cheaper than natural rubies. However, in 1892 Verneuil obtained the first results of growing corundum crystals by a new method from pure aluminum oxide powder. This research was completed in 1902. Owing to the simplicity and reliability of the Verneuil method, industrial production of the crystals was soon organized in France and then in other countries. Artificial rubies and other colored corundum crystals started being used widely as bearings and axes in watches and other precision instruments, cutters for finishing treatment of metals, thread carriers, and acoustic needles, as well as for making gems (Fig. 4.23).

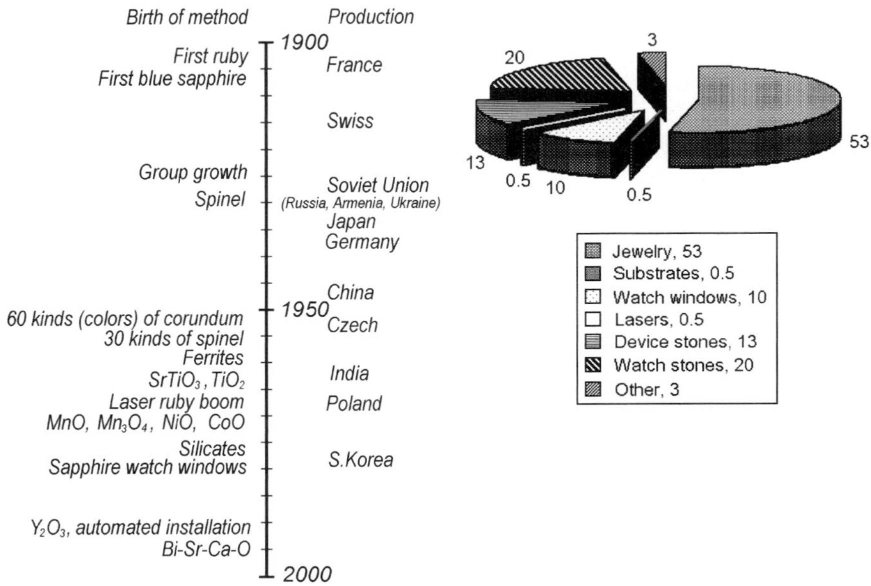

Fig. 4.23 (**a**) Milestones of the Verneuil crystals; (**b**) application of the Verneuil corundum crystals, % [52]

Prior to publication of the Verneuil growth method in 1904, some enterprising people made an attempt to initiate industrial production of ruby and sapphire by this method. In 1903–1904, in Hoquiam (USA) businessmen Polson and Ninemire hired an assistant of Verneuil to start the production of rubies for gems. But the venture turned out to be unprofitable, since the Verneuil rubies from Europe were much cheaper. Thus, the first industrial production of Verneuil crystals failed. For a long time no other attempts in this matter were realized in America until World War II, when the country lost supplies from Europe. At the beginning of the war, the government of America concluded contracts that led to the production of rubies by the Verneuil method at Linde Air Products Co. (Indiana) in 1942. The main advantage of this firm was that it possessed the production of oxygen and hydrogen used as working gases in the Verneuil method. Later, Linde became part of Union Carbide Corp. After the war, it turned out that the United States could not compete with Europe in sapphire and ruby production. To remedy this situation, Linde initiated the production of starlike sapphire for the jewelry industry. However, it failed to become a strong competitor to the European manufacturers.

The essence of the Verneuil method is the following. The crystals are grown in a shaft furnace from fine-dispersed powder that is fed through a flame of combustible gas onto a ceramic crystal holder inserted in the bottom of the shaft (Fig. 4.24). Flying through the flame, raw material particles partially melt and hit the face of the crystal holder. The top of the formed cone melts and then grows until a preset diameter is obtained. The shouldering stage can be sped up by preliminarily fixing the seed on the crystal holder.

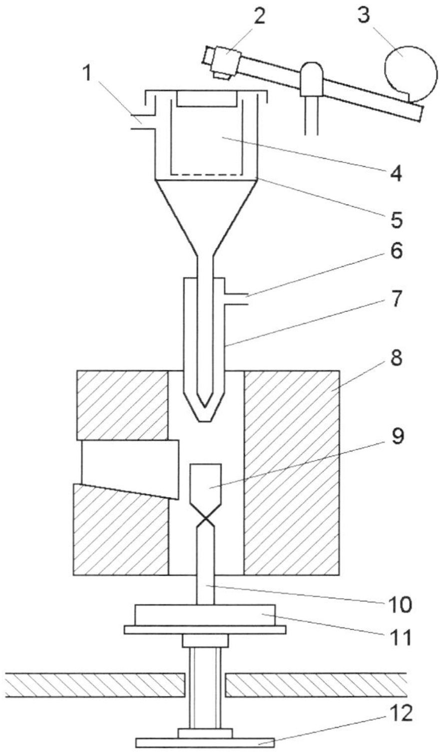

Fig. 4.24 Scheme of a Verneuil growth unit: (*1*) oxygen inlet; (*2*) hammer; (*3*) camshaft; (*4*) vessel with perforated bottom; (*5*) bin; (*6*) hydrogen inlet; (*7*) two-nozzle burner; (*8*) muffle; (*9*) the crowing crystal; (*10*) refractory crystal holder; (*11*) table with refractory coating; (*12*) device for the crystal adjustment in height

The crystal grows from the melt film, which thickness is defined by the crystal diameter and the thermal conditions at the crystallization front. The average film thickness is close to 40 μm. When particles with a size comparable to the film thickness hit the film, they give rise to the formation of pores in the crystal. Therefore, an effective means of decreasing the pore density is calcining the alum $Al(NH_4)\cdot SO_4 \cdot 12\ H_2O$ in trays at 1,270–1,370 K during raw material preparation. The sintered mass is powdered, and the resulting substance is sieved to obtain particles of a required size. Polydispersed powder must be loose and contain no fractions with a cross section (diameter) comparable to the melt film thickness. The optimum fractional composition is 5–35 μm with the main fraction of 15–20 μm. Smaller fractions disperse in flare or aggregate. Sieving does not provide the required fractional composition.

At present, automated air apparatuses are utilized for separation of powder particles in a pseudoliquid (boiling or spouting) layer (Fig. 4.25). Such separators have a number of advantages in comparison with sieve methods. The particles are sorted not only with respect to their size, but also to their mass. The pseudoliquid layer increases the powder looseness. At the same time, the raw material is purified from impurities with masses larger than those of the separated particles or from easily hydrated impurity particles (Table 4.2) [53].

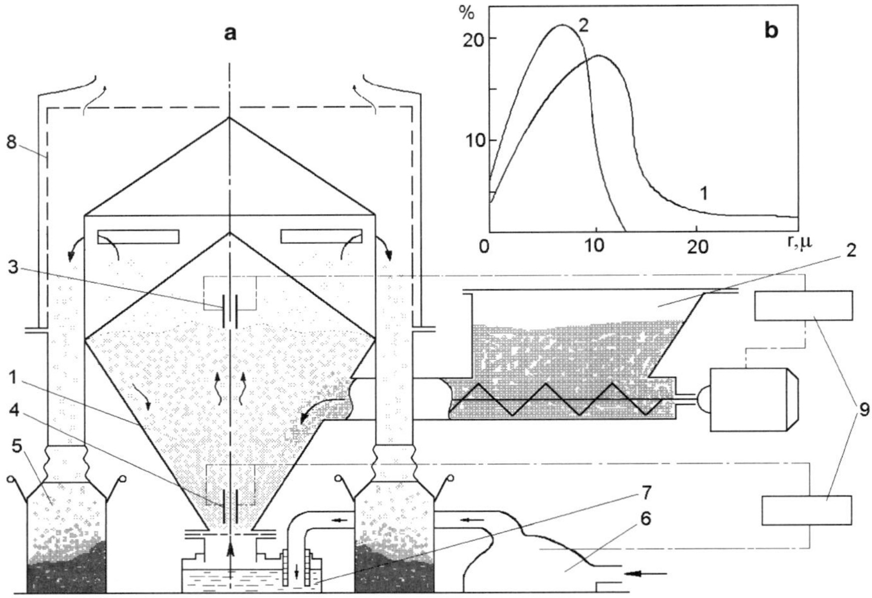

Fig. 4.25 (**a**) Industrial air separator: (*1*) casing; (*2*) charger; (*3, 4*) fluidized bed density and level gauges; (*5*) receiving vessel for fractions; (*6*) compressor; (*7*) air humidifier; (*8*) filter; (*9*) feedback device. (**b**) The raw blend disperses composition: (*1*) after passing through the mesh and (*2*) after air separation

Table 4.2 Impurity content in selected fractions and in screenings after air separation of aluminum oxide powder (mass%)

Impurities	Si	Mg	Ca	Ti	Fe	Cu	SO_4^{2-}
Selected fractions	$2.2 \cdot 10^{-3}$	$2 \cdot 10^{-4}$	$1 \cdot 10^{-3}$	$3 \cdot 10^{-4}$	$1 \cdot 10^{-3}$	$2 \cdot 10^{-4}$	0.6
Screenings	$1.2 \cdot 10^{-3}$	$9 \cdot 10^{-4}$	$4 \cdot 10^{-3}$	$4 \cdot 10^{-4}$	$1 \cdot 10^{-3}$	$5 \cdot 10^{-4}$	0.85

The grown crystals are shaped as boules thick at the top or rods. The temperature at the crystallization front is maintained constantly. This is achieved by displacing the seed crystal holder downward at a constant velocity. If the rate of raw material and combustible gas consumption conform to the velocity of seed displacement, the thickness of the film and the position of the CF remain practically unchanged. To obtain crystals of a high quality, high-perfection seeds are used. They must be oriented, as the structural perfection fundamentally depends on the spatial orientation of the seed. For sapphire and ruby, the optimum growth direction corresponds to an angle of 60° to the optical axis.

As a rule, standard crystals with a diameter of 15–25 mm and a length of 50–100 mm are grown at a rate of 10–15 mm/h. However, this method also permits growth of 1-m-long crystals with diameters exceeding 30 mm.

The Verneuil method has the following advantages:

- Absence of a container and the problems bound up with it
- Homogeneity of the doping additions over the crystal length
- Crystallization in open muffle which provides the opportunity to control the oxidation–reduction potential of the crystallization medium by varying the ratio H_2/O_2 in flame
- Possibility to introduce a large quantity of doping additions into the crystals, and growth of crystals with intentional doping gradients or zones, e.g., ruby rods with sapphire tips
- Technical simplicity of implementation and low cost of the crystals

The main drawback of this method is the high-temperature gradient in the crystallization zone (30–100 deg/mm), which leads to the appearance of high residual stresses (up to 10–15 kg/mm^2) in the crystals.

Although Verneuil crystals compared unfavorably in their structural perfection with those crystals grown by melt methods, interest in the Verneuil method has not weakened. This is explained by essentially lower costs of the crystals. At a modern plant one operator can simultaneously serve 135–150 crystal growth apparatuses assembled into blocks. The apparatuses are not meant for operation under vacuum and do not contain expensive high-melting metal components.

The essence of the method has not been changed over the period of its existence. The first Verneuil growth apparatuses and modern apparatuses have the same main components: a burner, feeder, crystallization chamber, and a mechanism of crystal holder displacement. Naturally, the modern apparatuses are supplied with controlling facilities, and all units have been upgraded.

The burner has undergone repeated modernization. At present, industrial growth apparatuses use an oxygen-hydrogen flame. Earlier, some mixtures of oxygen with acetylene or lighting gas with oxygen were utilized, but in such cases it was difficult to obtain sufficiently pure atmosphere in the crystallization zone.

The main idea behind modification of the burner design is the creation of multinozzle facilities capable of different combination of input gases [54, 55]. For the growth of color corundum crystals, two-nozzle burners consisting of two coaxial tubes often are used. Oxygen and the raw material are supplied through the tube of smaller diameter and hydrogen is introduced through the tube of larger diameter. For the growth of large-size rods, three-nozzle burners are used.

The structure of the flame and the temperature gradient in the growth zone essentially depend on the conditions of gas mixing (Figs. 4.26 and 4.27) and the number of gas mixing zones (Fig. 4.28).

To reduce residual stresses in laser ruby, gas burner facilities of a special design have been developed. They make it possible during growth and cooling to create additional heating of the upper, lower, or side parts of crystals with a diameter up to 60 mm.

Crystal growth using oxygen-hydrogen heating is based on heat release by the reaction:

$$H_2 + 1/2O_2 = H_2O + 57.8\,\text{kcal/mol}.$$

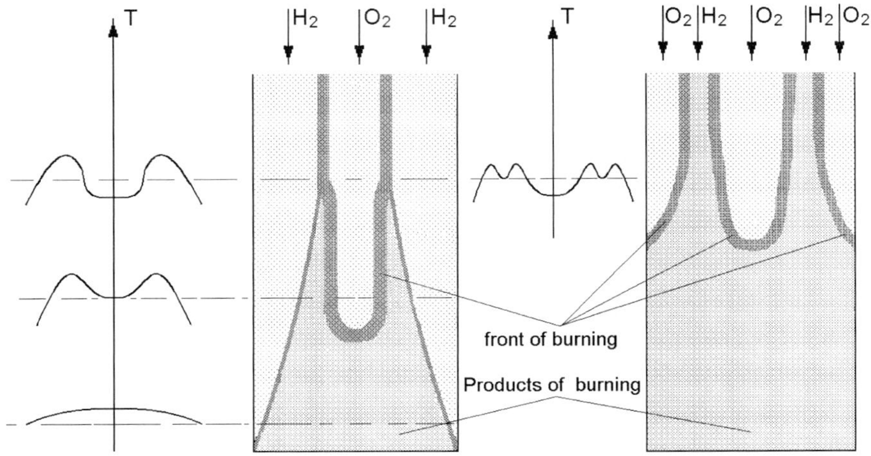

Fig. 4.26 Hydrogen flame structure and temperature distribution within the flame for two- and three-nozzle burners

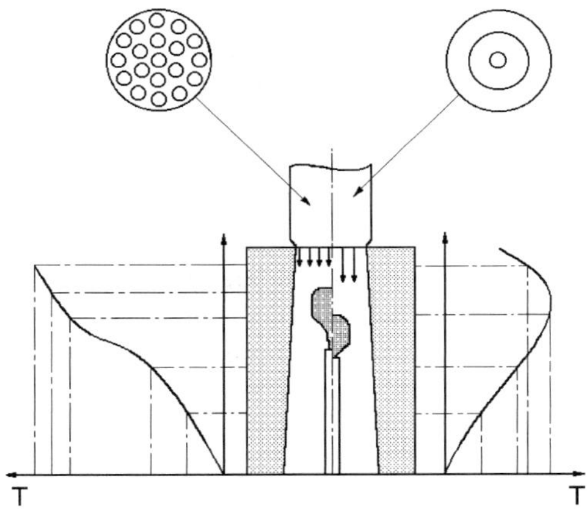

Fig. 4.27 Temperature distribution over the muffle length depending on the burning mode: three-nozzle burner (*right*) and multinozzle one (*left*)

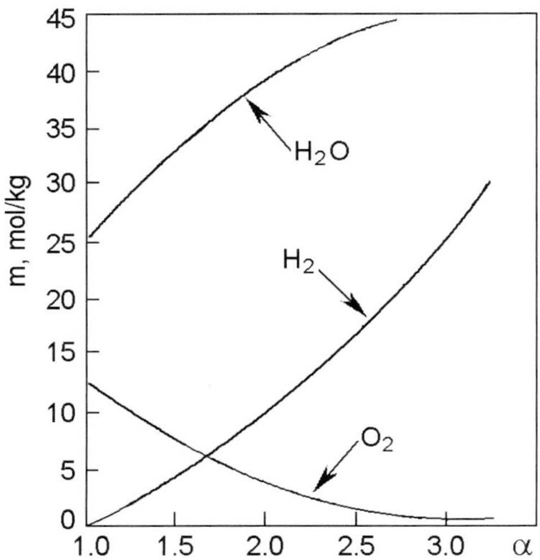

Fig. 4.28 Composition of H_2 combustion in O_2 products as a function of initial volume ratio $\alpha = V_{H_2}/V_{O_2}$

Theoretically, this method can be applied for the growth of crystals with a melting point up to 3,080 K. However, due to the loss of heat by the crystal and the powder, temperatures not higher than 2,730 K have turned out to be more realistic. This is connected with the fact that at higher temperatures dissociation of the combustion products, accompanied with heat absorption takes place. In the cases when higher temperatures are required, fluorine or other strong oxidizers may be used. But this problem has not been sufficiently investigated so far.

The feeder is the second component of the classical Verneuil scheme. Its operation defines the quality and reproducibility of the crystals. The process of replenishment of the growth zone depends on a uniform supply of small amounts of the raw material powder over a long period of time, at a rate of 20 g/h, for example. Feeders have been developed that work on the principle of impact or vibration effects [56], as well as facilities for continuous supply of the raw material. The feeders using vibrational drive and aerodynamic devices with continuous raw material feed may be considered more effective, especially if it is taken account that discrete replenishment of the melt film is connected with changes in its thickness and temperature. However, such feeders are highly sensitive to the physical parameters of the powders.

In industrial apparatuses, sieve-type feeders with hammer drives are used, although the operation of such a contrivance is discontinuous. The most widespread sieve-type feeders are presented in Fig. 4.29. One of the feeders meant for the growth crystals with varying composition, such as sapphire-ruby-sapphire, is shown in Fig. 4.30.

The dependence of feeder filling on the amount of powder fed per unit time, as well as the effect of powder consolidation on the feeder grid, are minimal in feeders with vertical grids. The quantity of the powder passed through the grid after one stroke of the hammer depends on the physical parameters of the powder (its looseness, dampness, dispersion, etc.) and on the feeder design.

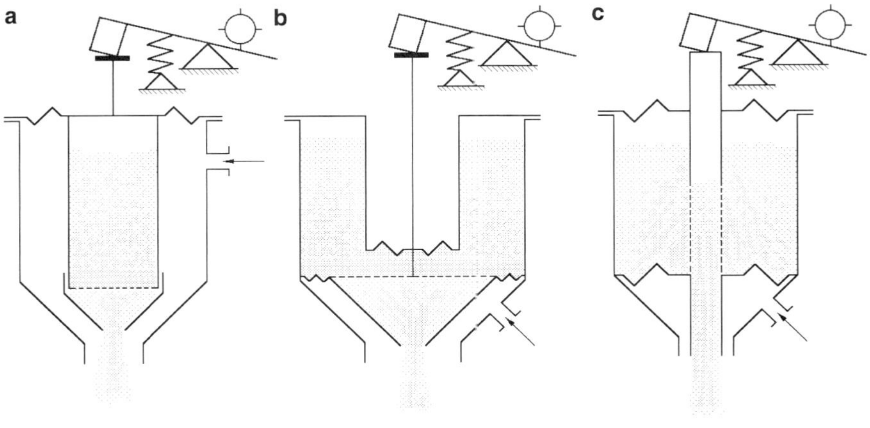

Fig. 4.29 Hammer-driven mesh feeders: (**a**) with horizontal mesh; (**b**) with horizontal unloaded mesh; (**c**) with vertical mesh

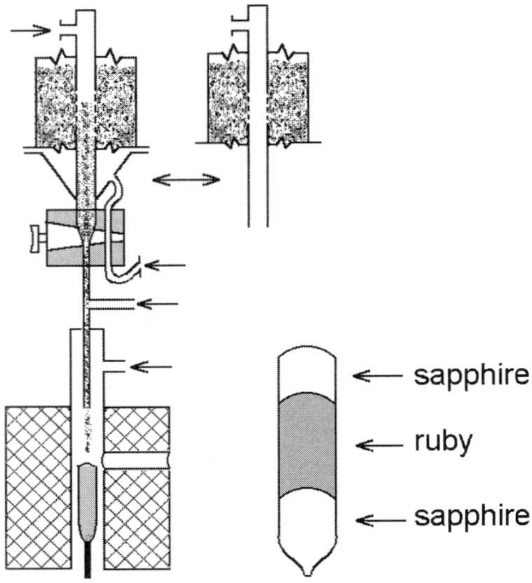

Fig. 4.30 Growing crystal with different composition of parts

In a number of papers, it has been proposed that the amount of raw material reaching the melt film should be judged by the position of the CF. For this purpose, photosensors connected with the feeder drive are mounted outside the growth chamber on a level with the CF. A signal from the photosensor arises after the position of the CF has changed. The highest sensitivity is provided by the scheme of automated replenishment control (stabilization) [52, 57]. In accordance with this scheme, a photosensor mounted inside the feeder is focused on the growing crystal face through the raw material conduit (Fig. 4.31). Illumination of the photosensor is inversely proportional to the quantity of powder in the space between the crystal and the sensor. In such a case, the sensor can register even "instantaneous" consumption of the raw material, i.e., the quantity of the powder fed in response to one stroke of the hammer (0.003–0.007 g) (Fig. 4.32).

The crystallization chamber, or muffle, is the third main component of the classic apparatus. For the growth of jewelry crystals in apparatuses assembled into blocks, small split muffles are used. Muffles meant for larger and more homogeneous crystals must have good thermal insulation, additional heating, or recuperation of the gases (Figs. 4.33 and 4.34). However, even in the case when thermal insulation is good (Fig. 4.35), the temperature gradients in the growth zone remain high, and only additional heating regimens can reduce them. While updating this unit, special attention should be paid to creating conditions for controlling the axial and radial temperature gradients, elimination of the thermal field asymmetry, and prolonging the service life of the ceramics.

Up to now more than a hundred different kinds of crystals have been grown by the Verneuil method (Fig. 4.23) [51, 58]. However, this method is most applicable to the growth of sapphire, ruby, and other colored corundum crystals.

Fig. 4.31 Feeders with automatic control of powder supply

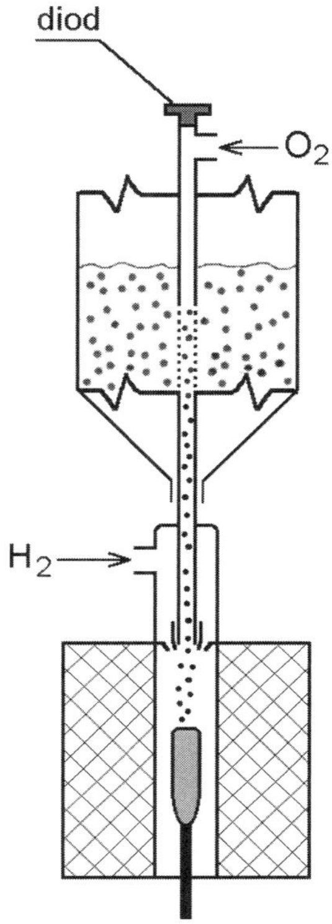

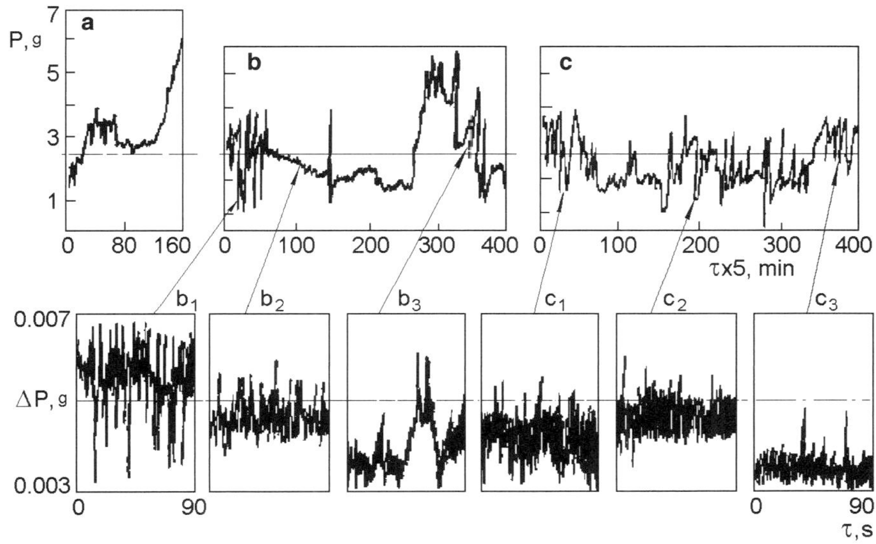

Fig. 4.32 The powder flow uniformity from mesh feeders with (**a**) horizontal, (**b**) horizontal unloaded, and (**c**) vertical mesh. ΔP is the instantaneous blend mass flow at different stages of the feeder operation ($\mathbf{b}_1$–$\mathbf{c}_3$)

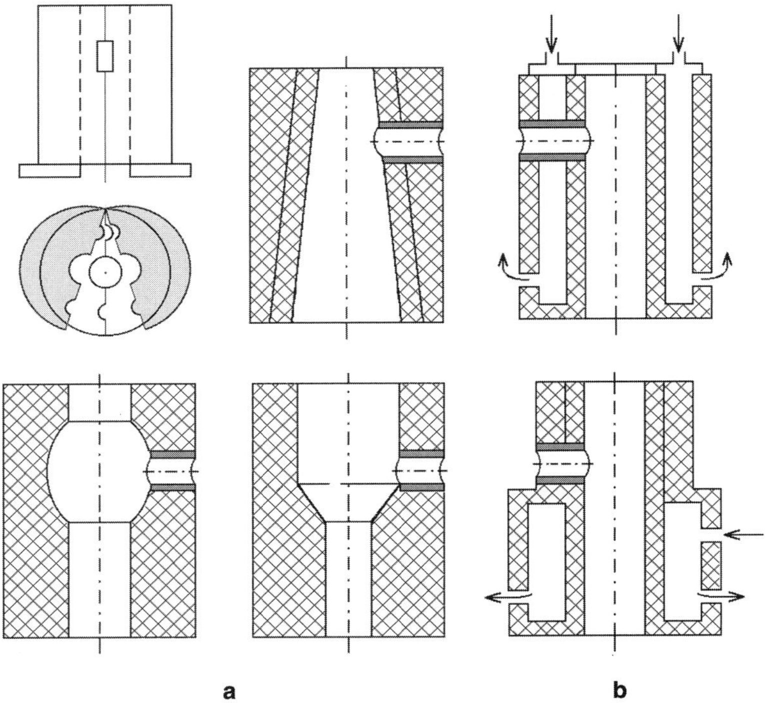

Fig. 4.33 Muffles: (**a**) types of simple muffles; (**b**) muffles with external additional heating

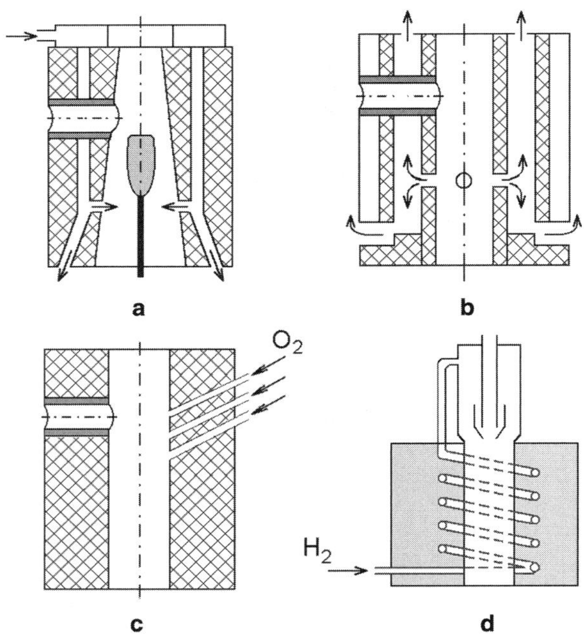

Fig. 4.34 Muffles with additional heating of growth zones (**a–c**) and of gas (**d**)

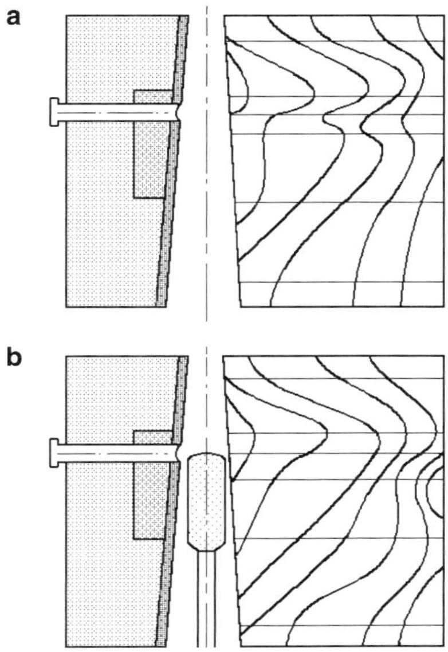

Fig. 4.35 Isotherms in the muffle cross section (**a**) without and (**b**) with the crystal

Modified Verneuil methods imply the use of other heating means, such as plasma [59], arc, optical [60], electron-beam, laser [61, 62], or radiation [63] methods. Radiation heating partly lifts the restriction of plasma heating connected with chemical stability of the crystallized material in flame. Such a heater can be isolated from the growing crystal by placing it into a shell and controlling the medium in this shell.

The use of inductively coupled plasma as a source of heating is reported in [64]. The temperature in plasma may reach many thousands of degrees, while plasma heating makes it possible to maintain oxidizing, reductive, or neutral conditions. Plasma burners work by the following principle. Between the electrodes direct current arc discharge is initiated, and plasma is removed from the electrodes through the arc by a strong gas jet. The working voltage in plasma burners is 10–100 V, the current intensity ranges between several hundreds to several thousands of amperes, and a temperature of about 15,000 K is achieved. However, under such conditions it is difficult to stabilize the gas flow. At worst, the plasma is completely removed from the space between the electrodes and the discharge is quenched. At best, the intensity of plasma fluctuates and it starts floating in the space between the electrodes. Introduction of water stabilizes the plasma, but makes it "dirty." Stabilization of plasma in magnetic fields gives satisfactory results. As plasma is a current-conducting gas flow, one can apply magnetic field to it and obtain the required form of discharge by changing the field configuration. Plasma is excited in H_2, N_2, Ar, He, and O_2 medium, but as experience shows it is rather difficult to maintain the arc for a long period of time in an oxidative atmosphere. At present, there is no information concerning the growth of high-quality crystals using direct current plasma burners. Alternating current plasma burners were investigated by Rid. Their advantage consists in the fact that they do not contain electrodes, thus the crystal is not contaminated with the electrode material. Different configurations of the burners, including those with capacitive coupling, have been described [56]. Up to now, nobody has managed to grow sapphire of high structural perfection in a high-frequency plasma.

Light heaters also have been proposed. If a substance absorbs light at the emission wavelengths of powerful lamps or lasers, it can be melted by focusing the radiation. Heating by means of high-power light sources makes it possible to raise the temperature up to 2,800–3,000 K and to use oxidizing and reducing medium or vacuum in the crystallization zone.

Oxygen–hydrogen flame is still the most widespread source of heating, despite the fact that other sources have the advantages discussed. The other sources are primarily used under laboratory conditions for the growth of crystals at higher temperatures than those used for sapphire.

For many years attempts have been made to use the Verneuil method for obtaining sapphire crystals of a preset shape: crucibles, disks, plates, and pipes. Except the growth of polycrystalline corundum pipes, such attempts have not led to industrial application of this method. At present, works in this direction are practically abandoned, since high-quality shaped crystals can be produced by other methods more easily. Future trends for updating the Verneuil method evidently will be automation aimed at elimination of subjective factors influencing the processes of seeding and shouldering and stabilization of the conditions in the growth and cooling zones.

4.3.2.2 Zone Melting Methods

Zone melting usually is considered to be a method for purification of materials. The first time it was used for this purpose was in 1952 by Pfann [65]. However, zone melting also turned out to be applicable as a crystal growth method, since in the process of purification single crystals often are formed. The role of zone melting is reduced to the formation of a temperature gradient near the CF. Zone melting

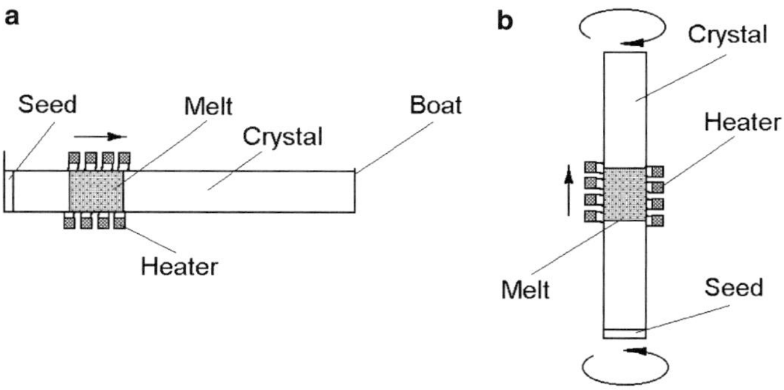

Fig. 4.36 Configurations being used at the zone melting: (**a**) horizontal and (**b**) vertical

makes it possible to control the content of impurities directly in the process of growth, this being an essential advantage of the method. Therefore, zone melting has formed the basis for a number of crystal growth methods (Fig. 4.36).

The zone (Fig. 4.36a) begins from the left side of the sample. If it is required to grow a single crystal, a seed is placed at the left end of the boat. The latter is partly melted to obtain a clean growth surface, then the zone is shifted to the right. If the left end of the boat is narrower, the crystal can nucleate without seed, too. During horizontal zone melting, a high-frequency inductor is used as a heating source. With such a heating, the connection with the field can be realized by the melt, the boat, or a special receiver of induction currents. Radiative heating from resistance elements, electron bombardment, and focused light of high-power lamps also are used. Due to wetting, the grown crystal may stick to the container, and as the linear expansion coefficients of the crystal and the boat are different, inner stresses arise during cooling. Sometimes such difficulties are avoided by using damping boats.

In the floating zone method (Fig. 4.36b), the melted zone is confined by surface tension in samples placed vertically. The initial material is a polycrystalline rod onto which the melted zone is created by means of a concentrated heat source. The connection with the induction field is provided either by the melt (on condition that its conductivity is high enough) or by a receiver of induction currents. The melted zone moves along the rod upward or downward, thus forming a single crystal (Fig. 4.37).

If the melting rod is motionless with respect to the forming crystal, the crystal diameter is approximately equal to the diameter of the initial rod. The diameter of

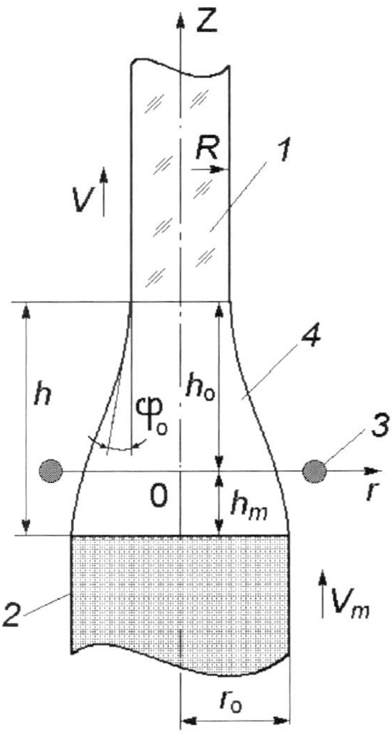

Fig. 4.37 Preparation scheme of a cylindrical crystal using the floating-zone melting: (*1*) the growing crystal of radius *R*; (*2*) feeding rod of radius r_0; (*3*) heater (inductor); (*4*) the melted zone, h_c, h_m being the positions of the crystallization and melting fronts with respect to the inductor; φ_0, the growth angle; *V*, the displacement speed of the growing crystal with respect to the inductor; V_m, the same for the melting rod

the growing crystal can be changed by constricting or extending the zone. If the sample is insufficiently dense, then the melt tends to fill the hollows (capillary effect) and it is difficult to control the zone width. Therefore, the rod must be obtained by casting, sintering, zone melting, or hot pressing.

The calculations of zone stability [66, 67] are based on the following assumptions. Surface tension and gravity are the only active forces; the melt completely wets the crystal; the crystal volume changes at melting are insignificant; and the interphase boundaries are flat and perpendicular to the axis of the sample and to gravity. The result of these calculations is presented in Fig. 4.38, where λ^* is the parameter proportional to the maximal zone length h and equal to

$$\lambda^* = h\sqrt{\gamma(l)g/\sigma_s} \tag{4.16}$$

The parameter ρ^* is a value proportional to the rod radius r_0:

$$\rho^* = r_0\sqrt{\gamma(l)g/\sigma_s} \tag{4.17}$$

In (4.16) and (4.17) γ_l is the density of the liquid, g is gravitational acceleration, and σ_s is the surface tension.

When the rod radius increases, λ^* reaches magnitudes of about 2.7. Thus, if the zone length could be controlled at low h values, then no limitations would exist for r_0. In practice, it is rather difficult to control the zone length at $h > r_0$. If it is assumed that the zone is stable at $h \approx r_0$, the only requirement to be imposed is [67]:

$$h = \frac{1}{2.7}\sqrt{\sigma_s/\gamma(l)g} \tag{4.18}$$

When the ratio σ_s/γ_l is large enough, the value h will be large, too, and the floating zone method will be realizable. Low-frequency magnetic fields and the design of high-frequency inductors make it possible to create a force constantly influencing the melted zone and directed upward. Thus, it is possible to achieve h values that are larger than those following from (4.18).

During electron-beam heating the density of the energy applied to the sample is essentially higher than that corresponding to induction heating. For the growth of nonconducting crystals, a variant of electron-beam melting – the method of hollow

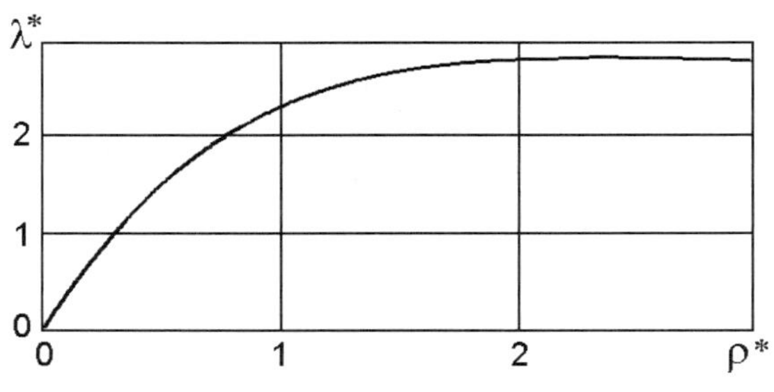

Fig. 4.38 Stability conditions for the melted zone

or "cold" cathode – is applied [68]. This method is based on self-maintained direct current gas discharge created in a circular hollow cathode surrounding the working zone. At a pressure of Ar and other gases of several millimeters of mercury and a cathode voltage of several kilovolts, electrons emitted by the cathode ionize the gas and form conducting plasma. The current in this plasma reaches hundreds of milliamperes, and any suitably grounded component of the system can be utilized as an anode. Thus, in contrast to electron-beam melting, the heated material must not be conductive so that the electric circuit should be closed. By using a cathode with a special inner surface form, it is possible to focus electrons and ions onto the sample and melt it. To sustain plasma, low pressures of readily ionized gas are needed (but not the high vacuum utilized for conventional electron-beam melting). This gas has another positive effect, as it suppresses decomposition of the melted material. The floating zone method was used to grow sapphire, yttrium–aluminum garnet, tungsten, rhenium, molybdenum, and other crystals.

4.3.3 Methods of Crystal Growth from the Melt in Crucible

Crystal structure perfection depends greatly on the constancy of the thermal conditions at the CF. Sapphire is grown at different temperature gradients. For instance, the temperature gradients characteristic of the Czochralski method exceed those used in the Kyropoulos and the heat exchange methods. The constancy of the input power, which affects the process of crystallization and the quality of the grown crystals, depends on the value of the temperature gradient. At high temperature gradients, small changes in the input power slightly influence the constancy of the thermal conditions at the CF, whereas with low temperature gradients the same changes may violate the constancy of these thermal conditions.

The choice of the container material is limited by several metals and their alloys which resist attack by an Al_2O_3 melt (Table 4.3).

Table 4.3 Some performance properties of container materials

Metal	Melting point (°C)	Maximum operating temperature (°C)	Oxidation onset in air (°C)	Thermal conductivity coefficient at 2,200 K (W/(m deg))	Operating atmosphere
Iridium	2,443	2,150	1,000	–	Vacuum, inert gas, hydrogen
Tungsten	3,410	2,800	400	101.5	Vacuum, inert gas, hydrogen (nitrogen up to 1,500°C)
Molybdenum	2,620	2,300	400	96	Vacuum, inert gas, hydrogen (nitrogen up to 1,500°C)
Rhenium	3,180	2,400	–	–	Vacuum, inert gas, hydrogen

Iridium. This metal interacts less intensely with the melt compared to other materials, but its operational temperature is not significantly higher than that of the heater. Its ultimate breaking strength diminishes from 28.4 kg/mm^2 at 290 K to 3.5 kg/mm^2 at 2,270 K. Due to plasticity of this material at sapphire growth temperatures, iridium crucibles deform in the process of melt solidification. If the crucible is not deformed, its service life is limited by recrystallization. Its grain size increases with each crystallization process and may reach 10 mm. Impurities segregate to the grain boundaries. Welded iridium crucibles have a longer service life compared to pressed crucibles. The highest recrystallization temperature (1,500–1,600 K) is characteristic of iridium obtained by electron-beam melting; the lowest corresponds to iridium obtained by galvanoplastic (1,270 K) and induction melting (1,320 K) [69].

Tungsten crucibles are made by pressing, forging, casting, and by means of gas transport reactions. The mechanical properties, such as hardness, tensile strength, and so forth, depend on the method of crucible production. In particular, for sintered tungsten the value of Brinell hardness is 200–250 kg/mm^2, while for the forged sample the corresponding value is 350–400 kg/mm^2. Tungsten also tends toward recrystallization. Crystals with a cross section of 3–5 mm were observed to merge and even to fall out of the crucible walls over the course of long-duration operation. Crucibles made by gas-transport reaction from WF_6 have a higher chemical purity, but their strength is lower.

Molybdenum crucibles are produced by pressing and mechanical treatment. The mechanical properties of molybdenum, like those of tungsten, depend on the method of its production. As a rule, this material is less inclined to recrystallization, although during the growth of sapphire by the HDS method in a molybdenum boat, large molybdenum crystals are formed due to contact with the Al_2O_3 melt. The use of molybdenum crucibles in an apparatus containing graphite heaters or elements may give rise to certain problems. These reside in the fact that molybdenum is oxidized by the produced CO_2, and with graphite heaters packed closely to a molybdenum crucible the atmospheric C–Al_2O_3 (gas) may interact with the crucible. Such an interaction leads to the appearance of a eutectic layer on the surface of the crucible, and the latter "sweats." Tungsten is more resistant to this medium. The service life of Mo crucibles can be prolonged by covering their external surface with a tungsten or tungsten–rhenium layer at a thickness of several millimeters. The melt wets molybdenum better than tungsten.While estimating the service life of the container, one must take into account the evaporation rate:

$$\lg w = A - B / T \tag{4.19}$$

The coefficients A and B are presented in Table 4.4.

In the process of growing sapphire Mo interacts with the melt evaporation products. For the mole ratios Mo/O_2 equal to 0.5:1, 1:1, 1:0.5 the gaseous phase is the mixture of O_2, O, Mo, MoO_2, MoO_3, Mo_2O_6, Mo_3O_9, Mo_4O_{12}, and Mo_5O_{15} [71]. The interaction of these products is pressure-dependent. At a pressure of 1...10^{-1} bar almost the entire gaseous phase interacts with Mo and passes into solid dioxide. At $P < 0.0705$ bar the gaseous phase mainly contains MoO_3 formed by the reaction

Table 4.4 Coefficients for evaporation rate determination of molybdenum and tungsten [70]

		Coefficients	
Material	Pressure (Torr)	*A*	*B*
Molybdenum	10^{-4}	5.259	23,760
	10^{-5}	7.450	30,470
	10^{-6}	7.390	31,540
	10^{-7}	8.230	33,549
Tungsten	10^{-4}	5.690	32,800
	10^{-5}	7.240	38,980
	10^{-7}	9.683	45,407

$$1.5MoO_2(s) = MoO_3(g) + 0.5Mo(s) \tag{4.20}$$

At a pressure on the order of 10^{-4} bar the formation of MoO(g) and MoO_2(g) is more energetically advantageous.

Rhenium has higher strength characteristics in comparison with W and Mo. Unlike W and Mo, which are brittle after recrystallization, rhenium possesses plasticity in the recrystallized state. With increases in temperature the hardness of W, Mo, and Re diminishes, but at 1,270 K the hardness of rhenium is twice as high in comparison with that of tungsten.

W–Mo alloys are less inclined to recrystallization than tungsten. These metals form a continuous series of solid solutions. By changing the ratio of the components, it is possible to obtain alloys with different mechanical properties. Crucibles also are made by sintering metal powders, e.g., those consisting of Mo(80%) + W(20%), the content of Mo and W in the inside thin crucible layer being 25% and 75%, respectively.

W–Re, Mo–Re alloys. Rhenium makes such alloys more plastic, strong, and chemically stable. An alloy containing 20% of rhenium possesses higher strength and chemical stability than tungsten. Rhodium and its alloys interact with the melt, making the crystals brown. Now consider the key methods of sapphire growth from the melt in detail.

4.3.3.1 Kyropoulos method

The essence of this method [72] was developed during 1926–1930 and consists of the growth of crystals by a smooth, slow decrease of the melt temperature and heat removal from the crystal using a cooled rod. At first, a cooler (a cooled metal tube) was slowly introduced into the melt at a temperature exceeding T_m by approximately 150°C. Then the melt was gradually cooled, and upon reaching a temperature slightly exceeding T_m the cooler was vented with air. This gave rise to crystallization at the end of the cooler followed by the formation of a hemispherolite. The spherolite was withdrawn from the melt in such a way that the part remaining in the melt was approximately equal to the cooler diameter. Thus, conditions were created favorable for geometric selection of the nucleus to be used for further single crystal growth (Fig. 4.39). This method was characterized by low (not exceeding 7–10 deg./cm) temperature gradients.

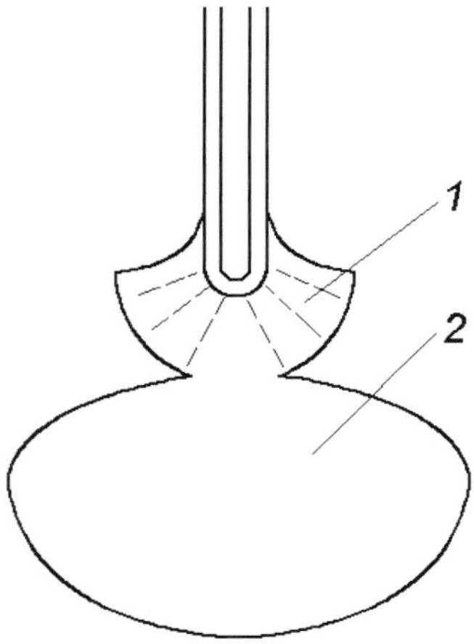

Fig. 4.39 (*1*) Polycrystalline semispherolite and (*2*) single crystal protruding therefrom

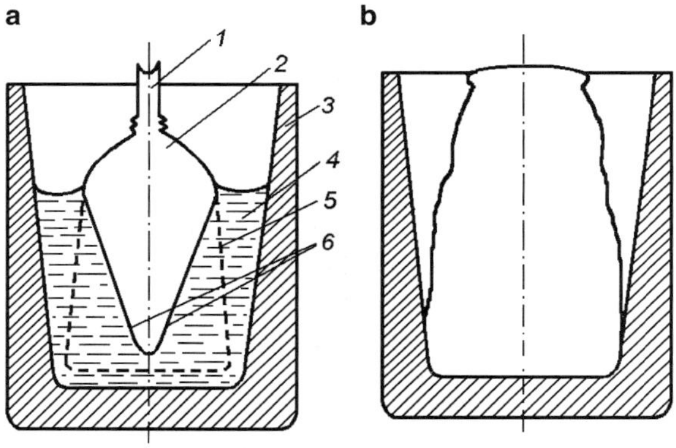

Fig. 4.40 Modification of Kyropoulos method: (**a**) real crystallization growth; (**b**) real crystal in crucible. (*1*) seed; (*2*) growing crystal; (*3*) crucible; (*4*) melt; (*5*) crystal boundary; (*6*) crystallization front

At first, the Kyropoulos method was used for the growth of alkali halide and alkali earth metal crystals. After being modified by the State Optical Institute in Russia [73, 74], this method became applicable to the growth of sapphire. In this modification, the decrease in temperature is combined with a simultaneous, insignificant pulling of the crystal, which results in crystal growth inside the crucible at low temperature gradients (Figs. 4.40 and 4.41).

The relation between the cooling and pulling rates at different stages of the growth process defines the crystal shape and quality to a considerable extent. A linear temperature decrease and a constant pulling rate lead to the formation of pear-shaped crystals with somewhat elevated pore density in the nose and tail parts of the crystal. The characteristic curve of crystal weight increment per unit time has two maxima (Fig. 4.41b). The first of these is observed at the initial growth stage and corresponds to the increase in radiant heat removal from the growing seed. The increase in radiant heat removal from the end and side surfaces of the seed crystal results in the formation of a sharp cone directed deep into the melt. Rapid growth of this cone increases the density of macro- and microdefects, so the growth rate at

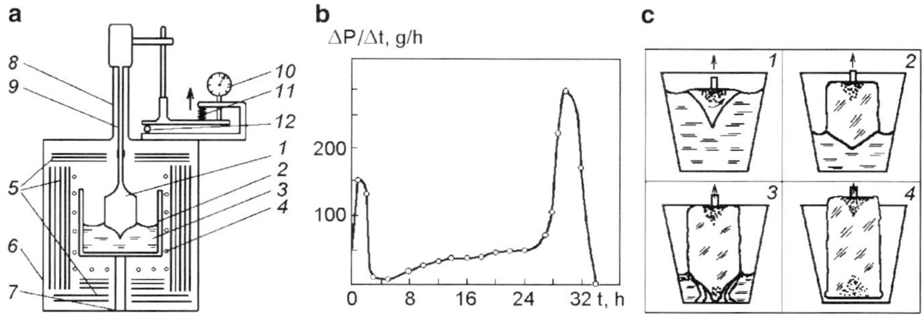

Fig. 4.41 (**a**) Scheme of installation for growing crystals by Kyropoulos method with weight system: (*1*) growing crystal; (*2*) melt; (*3*) crucible; (*4*) heater; (*5*) screen; (*6*) chamber; (*7*) support; (*8*) siphon; (*9*) rod; (*10*) indicator of crystal weight; (*11*) spring; (*12*) hinge. (**b**) Growth rate versus time of crystallization. (**c**) Stages of crystal growth

this stage must be reduced. When the crystal diameter becomes comparable to the inside diameter of the crucible, the course of the crystallization process changes and the level of melt in the crucible becomes lower. The rate of melt level decrease v_d is connected with the crystal diameter d, the crucible diameter d_c, and the rate of crystal pulling v_n by the relation:

$$v_n / v_d = 1 - (d / d_c)^2 \tag{4.21}$$

To maintain constant d, the ratio v_n/v_d must be constant also. At the final stage of growth the conditions change again. The rate of crystallization rises and its direction changes: now it proceeds from the center to the periphery; the second maximum appears on the curve (Fig. 4.41b). As a rule, sapphire is grown by Kyropoulos in vacuum without rotation.

Continuously changing conditions of heat exchange and difficulties of controlling the course of crystal growth necessitate automation of the process. One of the automation methods uses weighing facilities (Fig. 4.40a) [75]. The hinge 12 provides the deviation of the pulling gear rod under the action of the crystal weight. The spring 11 cancels the torsional moment of the pulling gear. For this device, the accuracy of crystal weight measurement is 10 g. The use of such a facility somewhat expedites the control of the crystallization course for operators. Tensometric weighing facilities provide higher measurement accuracy.

The Kyropoulos method is used to grow sapphire crystals with a diameter exceeding 350 mm and a weight larger than 80 kg. The ratio of the diameter to the height may change within the interval of 3:1 to 1:3.

4.3.3.2 Heat-Exchange Method

The heat-exchange method (HEM) [76] sometimes is called the inverted or modified Kyropoulos method. It was used for the first time to grow sapphire by Schmid and Viechnicki, the authors of this method, in 1967 [77]. The scheme of crystal growth and its stages are shown in Fig. 4.42.

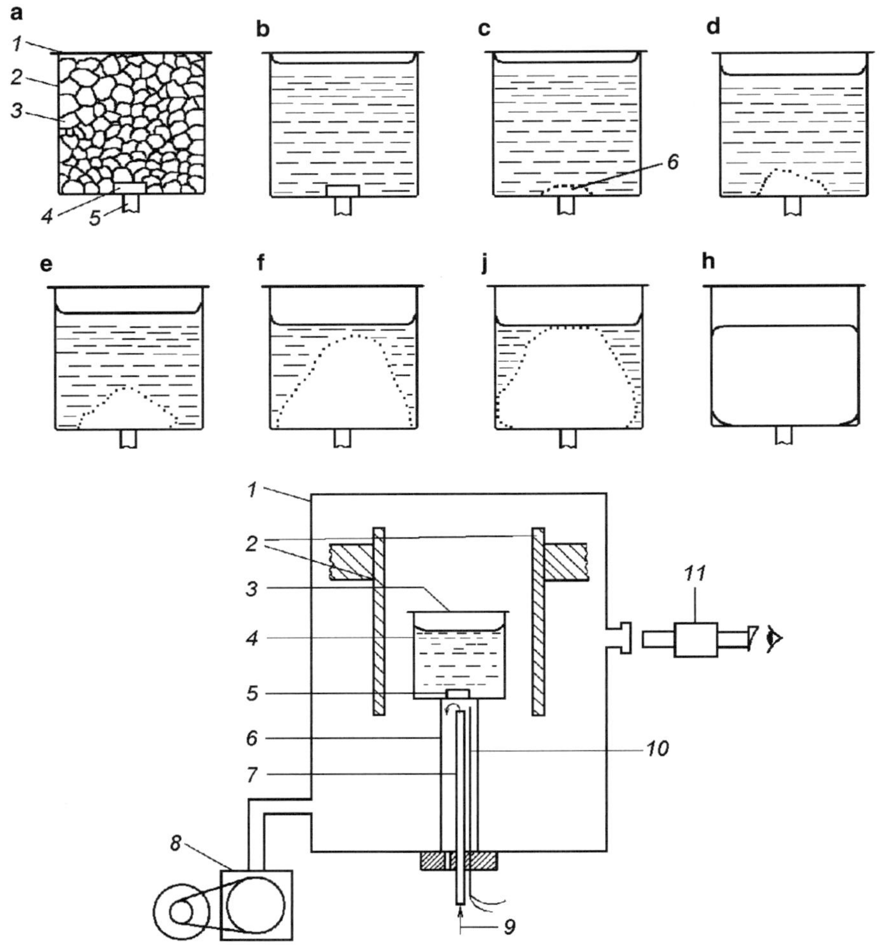

Fig. 4.42 Scheme of crystal growth by heat exchange method: (*1*) cover; (*2*), crucible; (*3*), initial material; (*4*) seed; (*5*) heat exchanger; (*6*) liquid–solid interface; (**a**) before fusion; (**b**) fusion initial material; (**c**) partial seed melting for maintenance of high-quality seeding; (**d**) crystal growth onset; (**e**) crystal gradually coats bottom crucible; (**f**) liquid–solid interface takes elliptic shape; (**g**) disturbance of liquid-phase surface by liquid–solid interface; (**h**), termination of crystal growth; (**i**), scheme of HEM furnace: (*1*) furnace shell; (*2*) heating element; (*3*) cover; (*4*) crucible; (5) seed; (*6*) heat exchanger; (*7*) tungsten tube; (*8*) vacuum pump; (*9*) helium; (*10*) thermocouple; (*11*) pyrometer

The heat-exchange method does not imply technological motion, as the motive force of crystallization is the change in the thermal field generated by the heater and the heat exchanger. Heat removal is realized by gases or their mixtures, which have high thermal capacity, or by liquid coolants. Changes in the furnace temperature and the variation of the coolant flow passing through the heat exchanger create conditions for the independent control of temperature gradients in the solid and liquid phases [78]. Small changes in the thermal flow essentially affect the shape of the growing crystal. The HEM method permits growth of sapphire crystals with a diameter up to 340 mm and a mass reaching 65 kg [79]. Attempts to obtain crystals with diameters up to 500 mm turned out to be unsuccessful, as "parasitic" nuclei of differing orientation often are aroused in certain zones during seeding, leading to cracking of the crystals during cooling [80, 81]. Currently, the largest sapphire crystal grown by the HEM method had a diameter of 380 mm and a weight of 84 kg [81]. Complexity of the CF configuration at low temperature gradients makes the growth process difficult for automation and mathematical description, which is a drawback of the HEM method.

4.3.3.3 Czochralski Method

Creation of the Czochralski method dates from 1916. While studying melts of different metals and recording his observations, Professor Ian Czochralski accidentally dipped his pen not into the inkpot, but into a crucible with a metal melt located nearby. On pulling the pen out of the crucible, he noticed a tiny, crystallized metallic ball at the end of the pen. This ball turned out to be a single crystal. At first, this method was used only to grow metallic single crystals and then it was forgotten for many years. The second birth of the Czochralski method occurred in the United States in 1950, when it was used to grow germanium single crystals. Today, not only metallic and semiconductor crystals, but also high-melting oxides such as sapphire, ruby, and other colored corundum crystal, TiO_2, garnet, tungstate, molybdate, and so forth are obtained by the Czochralski method.

In this method, crystals are grown onto a seed by gradual pulling from the melt (Fig. 4.43). The melt temperature is constant or varies in a predetermined manner. The high quality of the grown crystals, which to a considerable extent is caused by the absence of contact between the crystal and the crucible, makes this method widespread.

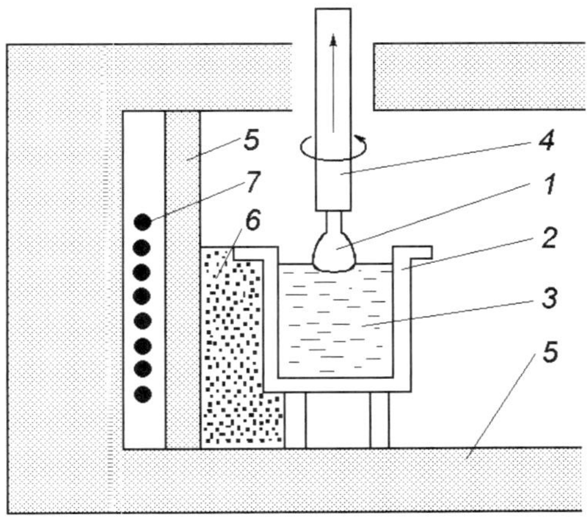

Fig. 4.43 A setup for crystal growing by Czochralski technique: (*1*) crystal; (*2*) crucible; (*3*) melt; (*4*) crystal holder; (*5*) ceramic insulation; (*6*) insulation of ZrO_2 ceramic granules; (*7*) inductor

Modern apparatus for the growth of sapphire by the Czochralski method consists of tungsten or molybdenum crucibles (in special cases, iridium) with tungsten, molybdenum, or zirconium oxide thermal insulation; facilities for creating a controlled atmosphere (vacuum or inert gas); means of providing technological motion (rotation, pulling); and mechanical, optical, or electronic feedback and controlling systems. The latest apparatuses are equipped with microprocessors for process control and regulation. Heating is realized by induction or resistive methods.

The growth of large-diameter and long crystals brings about the problems connected with melt replenishment, as the decreasing melt level leads to certain changes in the hydrodynamic processes and the temperature field. Therefore, replenishing facilities have been developed, which, in turn, require special vessels for preparation of the melt, since the introduction of additional raw material quantities into the crucible changes the thermal conditions at the CF. The melt column (Fig. 4.44), which connects the growing crystal with the melt, is supported by a surface tension force equal to $2\pi r\sigma_s$.

An approximate (somewhat overvalued) maximum length L of this column is determined from the relation [82]:

$$L = 2\sigma_s / r\gamma_l g, \tag{4.22}$$

where σ_s is the specific surface tension, γ_l is the melt density, g is the free fall acceleration, r is the crystal radius.

The highest possible growth rate is defined by the intensity of heat removal through the crystal into the surrounding medium. In the optimum case, the rates of crystal pulling and crystal growth are equal (this is confirmed by the fact that the CF is stationary). If this condition is not obeyed and the melt level decreases, then the true crystal growth rate is the sum of the crystal pulling and the melt level decrease rates. A diminution of the pulling rate with the temperature gradient remaining unchanged increases the crystal diameter or even leads to polycrystalline growth. If the pulling rate increases, the crystal diameter diminishes.

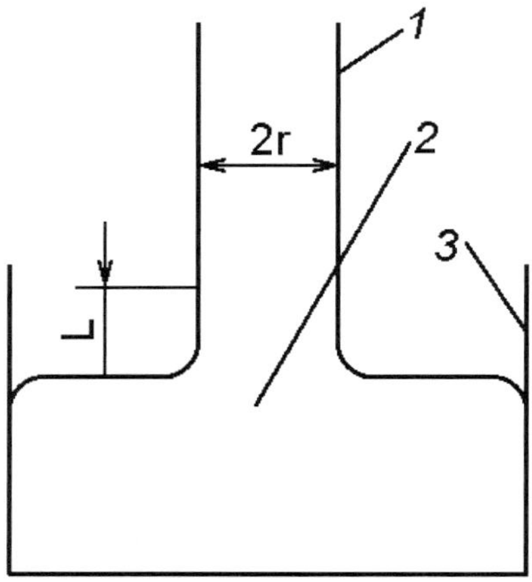

Fig. 4.44 The melt meniscus profile at the crystal growing by Czochralski technique: (*1*) crystal; (*2*) melt; (*3*) crucible

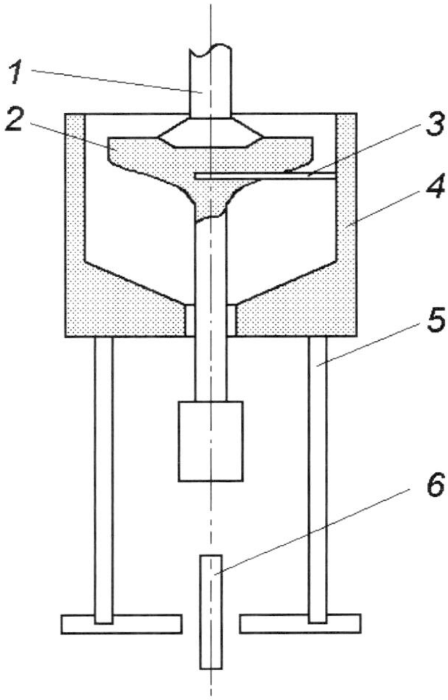

Fig. 4.45 Scheme of a unit with floating crucible: (*1*) single crystal; (*2*) floating crucible with counterweight; (*3*) channel; (*4*) outer crucible; (*5*) outer crucible support; (*6*) floating crucible elevator

The degree of crystalline perfection is defined chiefly by the CF shape. When the latter is flat, the radial temperature gradients are the lowest. Such a front arises in the case when the thermal flow is directed to the growing crystal from below and the radial thermal flow is insignificant [83]. Therefore, the isotherms are perpendicular to the growth direction and this favors the formation of crystals with low density of defects. Now consider the modifications of the Czochralski method.

Floating crucible method is interesting for obtaining crystals that have either constant impurity contents or two types of impurities with different ratios of their concentrations (Fig. 4.45). The bottom of the floating crucible has a hole for the rod that supports the crucible. To balance the mass of the crucible, a load is fixed to the lower end of the rod. The floating crucible is connected with the external crucible by a narrow channel. The crystals are grown from the inner crucible in which the quantity of the melt and its composition are stable during the entire growth process [84].

"Cold crucible" or skull method is meant for the growth of crystals from melts that interact with the crucible. In the simplest form, the idea of skull melting was implemented in 1903, when tantalum was melted by electric arc on a water-cooled copper plate. The method was then updated for the growth of crystals, including dielectrics [85]. In the modified version, the raw material is melted by induction currents. Thereafter, a skull with a crystalline rim that contains the melt is formed by a layer of crystallized substance adjacent to the cold container. Stability of the melt–skull interface is maintained by controlling the heater power.

For dielectrics, the resistivity of the melt is much lower than that of their solid phase, whereas for metals the situation is opposite. Therefore, the behavior of dielectrics and metals in HF fields is fundamentally different. In the case of dielectrics, the solid shell basically does not interact with the HF field, and the entire field energy is absorbed by the melt. In the case of metals, the solid shell absorbs a greater part of the HF-field energy, and being heated it completely melts. Thus, if a crystallized substance has low electrical conduction at room temperature or even at temperatures close to T_m, then this substance must be preliminarily heated to the temperature at which its conduction sharply rises (up to 2–10 Ω^{-1} cm^{-1}). Subsequent induction heating is realized by 2–7 MHz currents. For preliminary induction heating, pieces of metal that also are a component of the crystallized substance are introduced into the raw material. For example, pieces of aluminum introduced into the raw material for the growth of sapphire create the starting melt. The minimum (critical) volume of the starting melt V_c is expressed by the formula [86]:

$$V_c P_m = K(T_{melt} - T_0)\pi d_c^2, \tag{4.23}$$

where P_m is the power absorbed by a unit of the melt volume; K is the coefficient of heat transfer from the melt to the heat carrier; T_0 is the heat carrier temperature; d_c is the critical drop diameter.

$$d_c = 6K / P(T_{melt} - T_0) \tag{4.24}$$

By absorbing the HF-field energy, a drop of the melt with a diameter $d > d_c$ can grow due to melting of the adjacent solid mass until reaching the phase equilibrium defined as:

$$\lambda / \delta K = (T_{melt} - T_m)/(T_m - T_0), \tag{4.25}$$

where λ is the value of heat conduction; δ is the solid-layer thickness.

The value $\Delta T = T_{melt} - T_m$ characterizes the melt overheating. As ΔT goes to 0, δ goes to ∞, i.e., the melt must be overheated to maintain the phase equilibrium. If λ, K, and T_0 are constant, the equilibrium condition is expressed as

$$\delta \Delta T = \text{constant} \tag{4.26}$$

The skull method uses a water-cooled inductor which consists of several isolated segments (Fig. 4.46). The grown crystals are ~20 mm in diameter and ~100 mm long.

This method also makes it possible to grow corundum crystals in air. At the onset of crystallization in air, the starting aluminum melt rapidly oxidizes with subsequent heat release. The diameter of the growing crystal is controlled by movable screens located above the melt. During the growth of ruby in an oxidative atmosphere, considerable burnout of chromium from the melt was not observed even when the temperature of the overheated melt exceeded the melting point by 670–770 K [87].

As shown in [87], under the conditions when the temperature of water at the container input and output is 291 K and 313 K, respectively, the volume of the flowing water is 6.55 L/min, the power taken away by the heat carrier totals 10 kW.

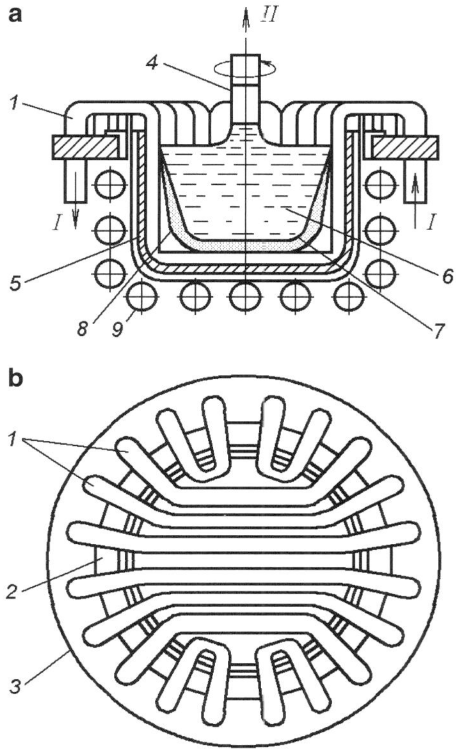

Fig. 4.46 Scheme of a device for realization of "cold crucible" method (skull melting): (**a**) section; (**b**) top view of the container; (*1*) the container tube; (*2*) quartz protective ring; (*3*) teflon ring; (*4*) growing crystal; (*5*) quartz glass; (*6*) melt; (*7*) crystallized melt layer; (*8*) unmelted starting material; (*9*) Ruhmkorff coil; the *arrows* show motion of coolant (I) and crystal (II)

The power emitted by an open surface with an area of 16.7 cm^2 is 5.46 kW. The power consumed in the anode circuit of HF generator is 31 kW, the heating efficiency is 50%.

The basic technical difficulties arising in realization of the skull method are connected with the maintenance of a stable position of the melt–skull interface and the creation of optimum and stable temperature gradients in the melt and in the crystal. The latter circumstance is caused by the fact that the temperature field of the melt is defined by its electrical conduction depending on the type and concentration of impurities in the melt, which are difficult to control.

The Czochralski method turned out to be one of the first techniques for which systems of complete automation of the growth process were created. The main element of such systems – a weighing device – possesses high sensitivity, accuracy, and reliability in the course of long-term operation. The automated systems continuously weigh the crucible containing the melt or the crystal. By setting up a program for controlling the weight of the crucible and the melt, control of the crystal

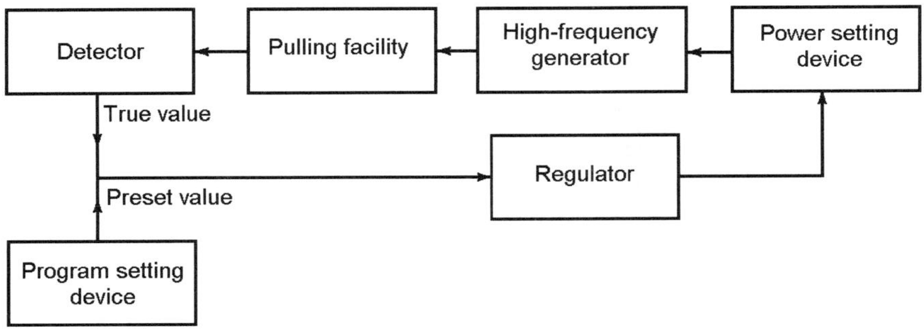

Fig. 4.47 Scheme of automated control of crystal growth process (Kyle 1973)

weight also is defined (Fig. 4.47). Rather effective schemes exist for the creation of a controlled atmosphere in the growth chamber. Using the Czochralski method, sapphire and other corundum crystals with a diameter of ~150 mm and a length up to 400 mm are grown.

4.3.3.4 Directed Crystallization Methods

Bridgman noticed [88] that during crystallization of a melt in a capillary tube, one of the tiny, nucleated crystals started intensively developing as a result of selection and grew into a single crystal. An analogous effect can be observed in large-diameter tubes with conical bottoms during the process of crystallization in a temperature gradient field. Bridgman also proposed a method to control the overcooling necessary for the formation of single crystals. After further improvements in this technique by Stockbarger [89, 90], it became known as the Bridgman–Stockbarger method. It sometimes is also referred to as the Tammann [91] or Obreimov – Shubnikov [92] method. Many versions of the directed crystallization method exist. The Bridgman–Stockbarger technique implies vertical displacement of the crucible or the temperature gradient. In the Chalmers method, the crucible or the temperature gradient are displaced horizontally. In his historical overview of directed crystallization methods, Buckley [93] described different modifications and named them after their authors. In this book, the methods of directed crystallization and their variants are considered as a single whole.

Crystal growth is realized under the conditions of spontaneous nucleation in a container with a conical bottom. The container (or furnace) is displaced through a zone with a temperature gradient. Another version does not imply such movement: the temperature is smoothly decreased, maintaining a constant temperature gradient. Crystallization can be realized both in vertical and horizontal directions (Figs. 4.48 and 4.49). Resistance heaters basically used are shown in Fig. 4.50.

When the container or furnace is displaced inside the crystallization chamber, two zones of different temperature are created. This permits combination of the crystal growth process with annealing.

The crystal diameter is set by the diameter of the container. Elastic interaction with the container walls upon cooling favors the creation of stresses in the crystal.

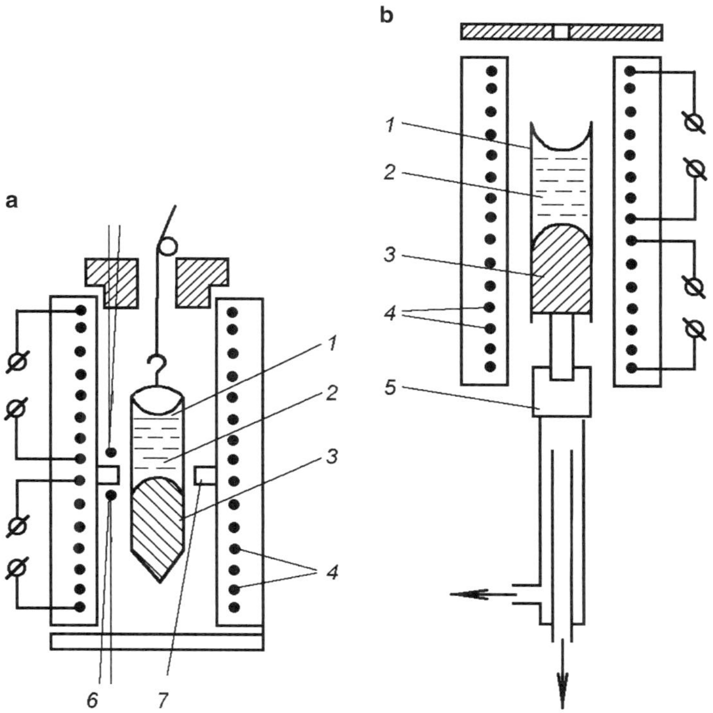

Fig. 4.48 Schemes of installations for growing crystals by the Bridgman–Stokbarger method: (**a**) two-chamber furnace; (**b**) single-chamber; (*1*) container; (*2*) melt; (*3*) growing crystal; (*4*) heater; (*5*) device for lowering the container; (*6*) thermoelectric couple; (*7*) heat shield; the *arrows* show water or cold gas flow

During the process of crystal growth, it is necessary to provide a temperature gradient along the crucible. Melts of many substances become noticeably overcooled before the onset of crystallization. If the melt is overcooled and the temperature gradient is very low, nucleation leads to accelerated growth, which inevitably is accompanied with the formation of low-quality, tiny crystals. High temperature gradients assure initiation of nucleation prior to overcooling of the entire melt. In this case the growth proceeds under controlled conditions as the isotherm corresponding to T_m moves along the sample.

For many years directed crystallization methods have been used for the growth of metallic, semiconductor, oxide, fluoride, sulfide, halide, and other crystals. The development of a number of original technical approaches [58, 94] widened the application field of this method. One of these versions permits the growth of high-temperature oxide crystals shaped as rods and plates. In particular, sapphire rods with a diameter of 50 mm and a length of 200 mm were grown in molybdenum containers.

The horizontal modification of this method, proposed by Bagdasarov in the middle 1960s [95], successfully combines the elements of directed crystallization and zone melting. In other methods of crystal growth from the melt, the whole of the raw

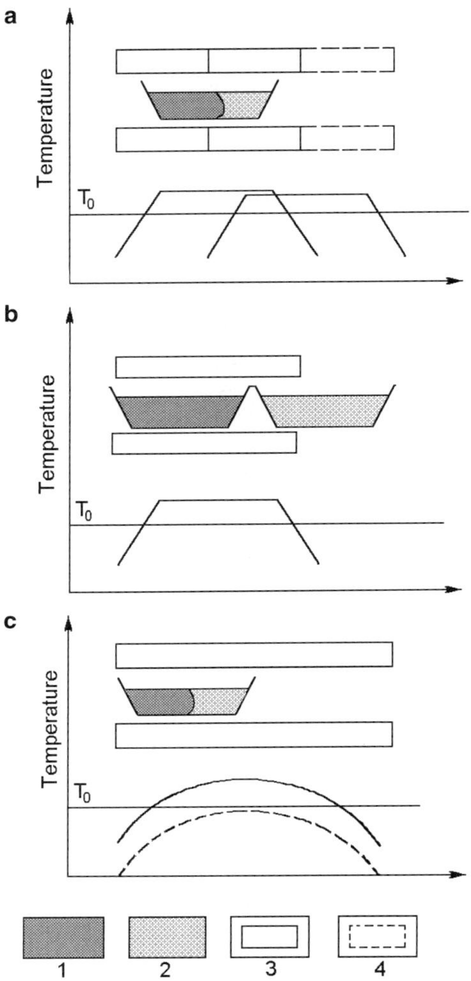

Fig. 4.49 Versions of horizontal directional crystallization: (**a**) the mobile heater; (**b**) the mobile container; (**c**) system with variable furnace temperature; (*1*) crystal; (*2*) melt; (*3*) heating device; (*4*) the heating device displacement

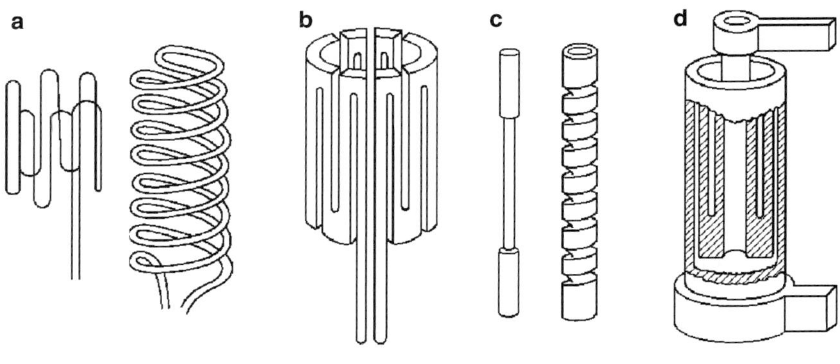

Fig. 4.50 Resistance heaters: (**a**) twisted; (**b**) cut; (**c**) rod; (**d**) coaxial

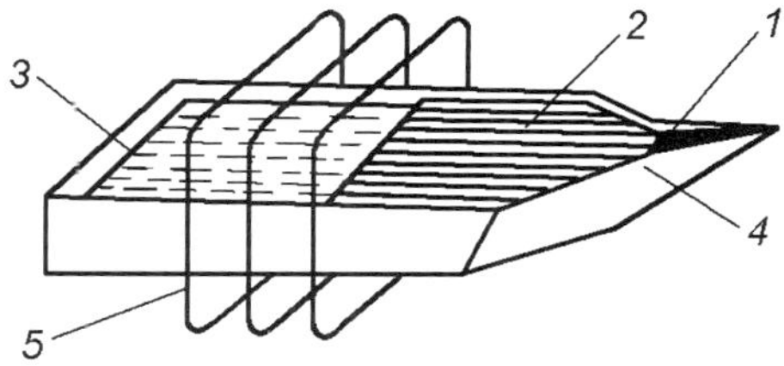

Fig. 4.51 Crystal growing scheme using horizontal directed crystallization: (*1*) seed; (*2*) crystal; (*3*) melt; (*4*) boat; (*5*) heater

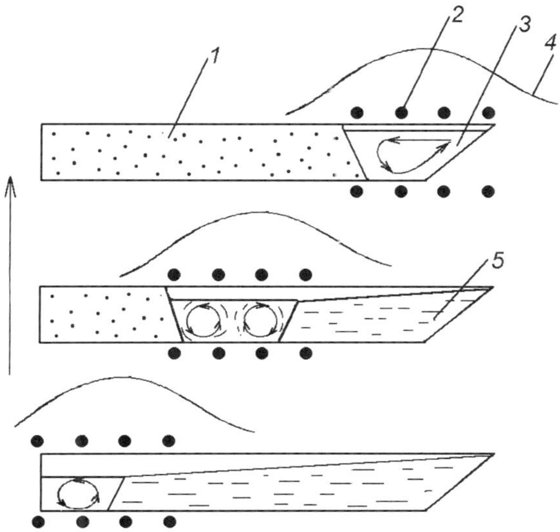

Fig. 4.52 Crystallization stages and convective flows in the melt: (*1*) raw blend; (*2*) heater; (*3*) melt; (*4*) temperature variation within the heating zone along the container; (*5*) the crystal

material is melted, whereas with HDS method a local melted zone is created between the seed crystal (nucleated or specially introduced) and the polycrystalline aggregate (the raw material). During the process of crystal growth, this zone slowly moves along the container with the raw material. The container is shaped like a boat, one end of which is very narrow (in some cases a capillary). Due to this shape, local overcooling can be quickly created in a limited space, thus favoring nucleation of a single crystal.

By the Bagdasarov method, sapphire is grown in the following way (Fig. 4.51). The raw material is placed in a molybdenum boat which moves horizontally at a velocity of 8–10 mm/h and passes through a rather narrow temperature zone. The latter provides melting of the raw material and evaporation and rejection of impurities (Fig. 4.52).

Due to the difference in solid- and liquid-phase densities ($\gamma_\sigma - \gamma_l = 0.7$–$0.8$ g/cm^3) the height of the crystal decreases in the process of its growth. The introduction of additional raw material before the CF permits growth of crystals with a constant height and increases their weight by 30–40% [96].

As the area of the melt is constant, the HDS method has a number of advantages. The process of seeding and the phase boundary are controlled visually or by optical instruments. By this method, 300 × 350 × 40 mm^3 sapphire plates have been obtained.

The HDS method has been used to grow not only sapphire, but also other types of colored corundum, such as crystals with a purplish-violet hue resembling that of natural alexandrite. Such crystals contained Cr_2O_3 (0.75–1.5 mass%), Fe_2O_3 (0.35–0.5 mass%), and TiO_2 (0.05–0.075 mass%) [97].

As a rule, the growth apparatuses used in the discussed method have tungsten or molybdenum thermal units, and the growth process is realized in $5 \cdot 10^{-5}$ Torr vacuum. To diminish the cost of the crystals, protective gaseous media containing reductive additions (H_2, CO) can be used. The growth of sapphire in a CO medium under 0.1–0.3 Torr pressure permits utilization of rather cheap carbon graphite materials, lowering the consumption of expensive metals [98]. Crystals grown in such a medium, under certain conditions, compare favorably in structural perfection with the crystals grown in vacuum. Reductive additions, based on inert gases such as Ar or He, decrease erosion of the constructional materials when introduced into the gaseous medium (Table 4.5).

The gaseous atmosphere in the growth furnaces has been investigated in detail [99]. The media formed spontaneously (under evacuation using a low vacuum pump) when growing sapphire in a furnace equipped with a graphite heating unit at a residual gas pressure of 0.05–1 Torr (Fig. 4.53) were studied.

A characteristic feature of the medium at $T < 1{,}300$ K is high water vapor content caused by the rather high absorption ability of graphite. At temperatures higher than 1,200–1,300 K, the medium becomes reductive with CO and H_2 as the main

Table 4.5 Content of gas-making impurities and optical properties of leucosapphire

Growth medium	C (at.%)	H_2 (at.%)	T (%), $\lambda = 250$ nm	Presence of second-phase microparticles
Verneuil method (raw material)	0.23	0.24		
W–Mo thermal unit				
Low vacuum (10^{-3}–10^{-4} Torr)	0.24	0.18	70–75	Absent
Low vacuum, $2 \cdot 10^{-4}$ Torr	0.43	0.45	15–40	Absent
Ar, 800 Torr	0.35	0.36	30–50	Absent
Ar + 20%H_2, 800 Torr	0.32	0.33	5–8	Present
Ar + 20%(H_2 + CO), 800 Torr	0.27	0.31	5–8	Present
Ar + 20% (H_2 + CO), 2–$3 \cdot 10^{-1}$ Torr	0.22	0.23	15–65	Absent
Graphite thermal unit				
(Ar + CO + CO_2), 800 Torr	0.3	0.31	5–8	Present
(CO + CO_2), ~$8 \cdot 10^2$ to $3 \cdot 10^{-1}$ Torr	0.3–0.4	0.24	15–65	Absent

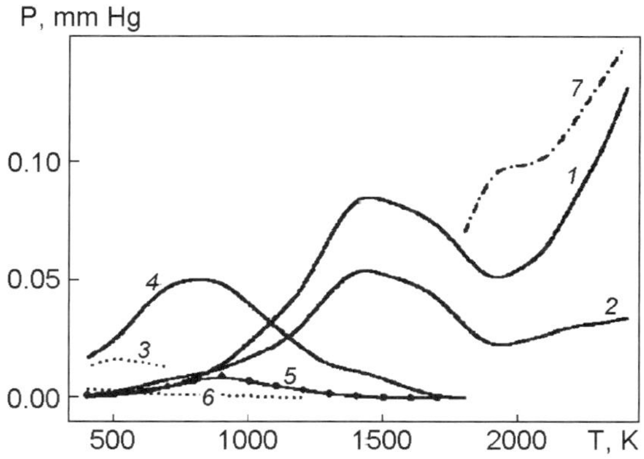

Fig. 4.53 Gas atmosphere composition in the furnace during its heating up to 2,350 K. (*1*) CO; (*2*) H_2; (*3*) N_2; (*4*) H_2O; (*5*) CO_2; (*6*) O_2 without a crucible with Al_2O_3 powder; (*7*) CO in the presence of a crucible with Al_2O_3 powder. The heating rate about 400 K/h

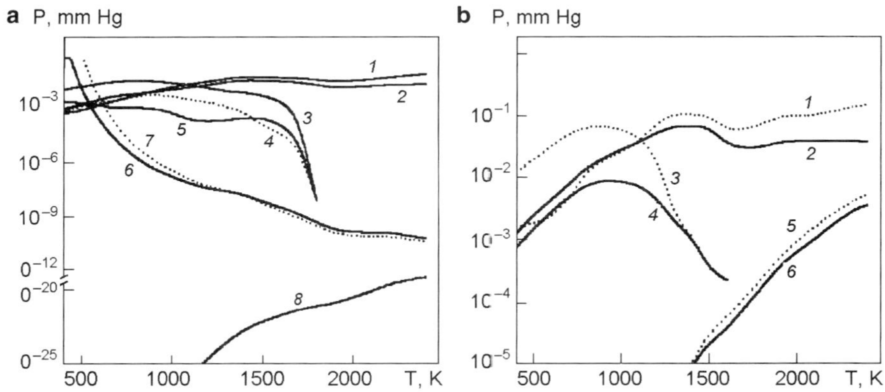

Fig. 4.54 Experimental partial gas pressures in the growth chamber as compared to the calculated equilibrium values. (**a**) Without a crucible with Al_2O_3 powder: (*1*) CO; (*2*) H_2; (*3*) H_2O; (*4*) CO_2; (*5*) O_2 (experimental data); (*6*) H_2O; (*7*) CO_2; (*8*) O_2 (calculated equilibrium data over the graphite surface at experimental P_{CO} and P_{H_2} values). (**b**) In the presence of a crucible with Al_2O_3 powder: (*1*) CO; (*2*) H_2; (*3*) H_2O; (*4*) CO_2; (experimental data); (*5*) H_2O; (*6*) CO_2 (calculated equilibrium data over the Al_2O_3 surface at experimental P_{CO} and P_{H_2} values)

atmospheric components. The relative CO component increases as the temperature rises. This is perhaps caused by the fact that CO is formed not only from water vapor, but also from chemisorbed oxygen that is desorbed mainly as CO.

A comparison of the experimental data with a thermodynamic consideration of the processes running in a C-gas atmosphere–Al_2O_3 system shows that the furnace gas medium at equilibrium is not up to premelting temperatures, and its composition is defined predominantly by gas desorbtion (Fig. 4.54).

Table 4.6 Composition protective media formed in a graphite furnace (without inert gas blowing)

Growing medium formed in the growth furnace	$CO + H_2$ partial pressure (Torr)	P_{H_2}/P_{CO}
CO, H_2 (under low-vacuum pumping-out)	0.1	0.1
Ar, He + (COH_2) (50–100 Torr), no blowing	1–3	0.3
Ar, He + (COH_2) (800 Torr), no blowing	7–10	0.5

Table 4.7 Composition of sapphire growth medium (under blowing)

	Low-pressure reducing atmosphere (about 0.1 Torr), Torr (vol%)		He blowing, 50 dm³/h (about 1.2 Torr), Torr (vol%)		He blowing, 50 dm³/h (about 50 Torr), Torr (vol%)	
Gas	1	2	1	2	1	2
H_2	$1.51 \cdot 10^{-2}$ (22 ± 4)	$3.12 \cdot 10^{-3}$ (5 ± 1)	$5.95 \cdot 10^{-3}$ (0.5 ± 0.2)	$1.7 \cdot 10^{-4}$ (0.05 ± 0.02)	$1 \cdot 10^{-1}$ (0.2 ± 0.2)	– (<0.1)
He	0	0	1.1 (91.6 ± 0.2)	1.05 (87.8 ± 0.02)	49.45 (98.2 ± 0.2)	49.4 (98.2 ± 0.2)
CO	$5.1 \cdot 10^{-2}$ (78 ± 4)	$6.75 \cdot 10^{-2}$ (95 ± 1)	$9.37 \cdot 10^{-2}$ (7.9 ± 0.2)	$1.4 \cdot 10^{-1}$ (12.15 ± 0.05)	$4.5 \cdot 10^{-1}$ (0.9 ± 0.2)	$5.5 \cdot 10^{-1}$ (1.1 ± 0.2)

Backfilling the growth chamber with helium reduces the hydrogen concentration and hinders accumulation of carbon monoxide (≤1 Torr). The backfill efficiency drops as the medium pressure rises (Table 4.6). Table 4.7 is presented for comparison, which shows the gaseous medium content in a furnace with graphite used in the thermal unit, HO inert gas flowing being absent.The interaction of the melt with reductive medium increases the concentration of anionic vacancies in sapphire, which leads to insignificant changes in the lattice parameters and to oxygen nonstoichiometry. The absorption spectrum shows the appearance of bands of anionic centers with maxima at 205- and 225-nm wavelengths [100]. This spectral region also is characterized by the presence of absorption bands from some impurities [101, 102]. These impurities are responsible for centers of induced absorption caused by UV-irradiation [103]. The anionic vacancies influence the charge of the impurity. The concentration of anionic impurities is defined by the reduction potential value of the medium. Annealing in a medium with a reduction potential lower than that of the growth medium decreases the concentration of F^- and F^+ centers. Annealing in a medium with a higher reduction potential creates an elevated concentration of anionic vacancies and lowers the concentration of Ti^{4+} impurity. Moreover, such an annealing raises the resistance of sapphire to UV-irradiation and its transparency in the region of 200- to 230-nm wavelengths (the Ti^{4+} absorption bands).

High concentrations of anionic vacancies may lead to the formation of second-phase microparticles with a diameter of 1–5 μm and a density higher than 10^5 cm^{-3} [104]. When the degree of nonstoichiometry exceeds the value "critical" for the lattice of sapphire, these microparticles may precipitate in the crystal. Their concentration is highest in the tail of the crystal, although the character of *F*-center distribution depends on the pressure of the growth medium. As the microparticle concentration in crystals grown under a pressure of 10–100 Torr increases, the concentration of *F* centers in the tail becomes lower. For instance, in the nose of a crystal grown under a pressure of 10 Torr the concentration of *F*-centers reaches $6 \cdot 10^{16}$ cm^{-3}, and this part of the crystal does not contain microparticles. Concurrent with the appearance of microparticles in the tail of the crystal, the concentration of F centers decreases to ~$1 \cdot 10^{16}$ cm^{-3}. In a crystal grown under a pressure of 100 Torr, in which a low (~$5 \cdot 10^{5}$ cm^{-3}) microparticle concentration is observed in the nose, such a decrease is less essential. In crystals grown under higher pressures (400–800 Torr) the character of the distribution of *F*-centers changes, their concentration growth is similar to that of microparticles [104].

As the concentration of anionic vacancies changes along the crystal axis, the impurity concentration varies also (Table 4.8). This is explained by the fact that, besides evaporation from the melt surface, impurities with the distribution coefficient $\kappa < 1$ are rejected by the CF. Therefore, the tail of the crystal contains an elevated impurity concentration. The nose, with an elevated anionic impurity concentration, is characterized by a higher radiation resistance.

Table 4.5 [105, 106] contains estimations of the content of H_2 and carbon impurities in crystals grown under different conditions, the transparency values at the 205-nm wavelength characteristic of the anionic vacancy concentration, and data on the light-scattering particles of the second phase. As is seen, light-scattering centers are formed in the media containing H_2 and CO with elevated pressures.

The use of gaseous media (CO_2, CO + Ar) at low pressures (0.2–0.3 Torr) [107] raises the profitability of the method and allows the growth of large-size crystals (Fig. 4.55) without noticeable reduction of the transparency in the UV-region of the spectrum.

Table 4.8 Content of Fe, Mg, Ti impurities in the crystals (mass%)

Crystal part	Fe	Mg	Ti
Nose	$5 \cdot 10^{-3}$	$9 \cdot 10^{-4}$	$6 \cdot 10^{-4}$
Tail	$3 \cdot 10^{-3}$	$1.8 \cdot 10^{-3}$	$1.5 \cdot 10^{-3}$

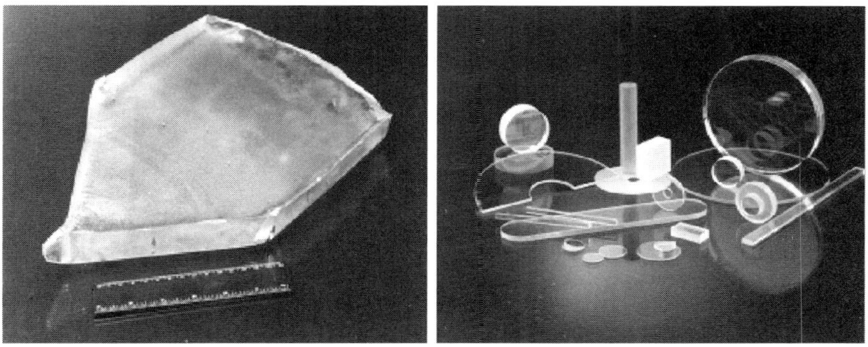

Fig. 4.55 Sapphire growth by HDS method and the articles made of it

4.3.3.5 Growth of Shaped Crystals

In the previously considered methods the opportunity for controlling the shape of the growing crystal is limited. Therefore, crystal size and shape often do not conform with the products for which they are meant. The fabrication of sapphire articles is a labor-consuming process in which 40–90% of an expensive material is lost and a large amount of abrasive material, such as diamond, is consumed.

The problem of obtaining finished products immediately from the melt arose as long ago as the middle nineteenth century. The first steps were made in 1857 by Bessemer, a British metallurgist. In 1953, Stepanov, then Corresponding Member of the Academy of Sciences of the USSR, proposed the current best solution to this problem [108, 109]. Since 1970, the Stepanov method has been used widely for the growth of shaped sapphire.

Gomperz [110] and Kapitza [111] were the first to grow shaped crystals by using plates with holes placed on the melt surface to give a desired shape to the pulled crystals. However, this method was named after Stepanov, since it stemmed from a big developmental effort and systematic investigations performed by Stepanov and his colleagues.

4.3.3.6 Stepanov Method

Stepanov developed a technique for obtaining profiled products from metals and semiconductors using surface tension and gravitational forces, electromagnetic interaction, hydrodynamic phenomena, and so forth, mechanisms that create a melt meniscus in the process of pulling the crystals. A liquid can take a certain shape not only by means of the vessel walls containing it, but also outside the vessel in a free state. Stepanov formulated the principle of shape formation as follows. The desired shape or its element is created in the liquid state through the different mechanisms that make it possible to retain this shape. The formed volume of liquid is then transformed into the solid state through certain crystallization conditions. He proposed the formation of the melt meniscus by means of special shapers. Thus, the method developed by Stepanov differs from the Czochralski method by the use of a shaping facility placed on the melt surface (Fig. 4.56).

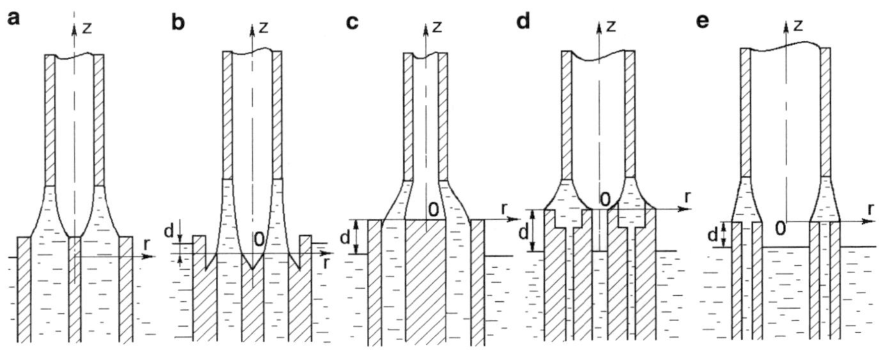

Fig. 4.56 Scheme of a tubular crystal manufacturing at different melt pressures

In some papers, the Stepanov method is contrasted with the EFG [112, 113] and CAST [114] methods. It is claimed that the former technology uses only such shapers are not wetted by the melt [115]. However, in the authors' opinion these methods have no fundamental differences. Moreover, these methods do not reach the limits of all possible schemes of shape formation. Therefore, discussion of the Stepanov method here shall further mean crystallization using a shaper, taking into account that the analysis of the crystallization processes for the Stepanov method is applicable to its other variants, including EFG and CAST.

In the Stepanov method, the region and value of possible perturbations of the free liquid surface are limited, and this favors a controlled shape formation. The meniscus is fixed near the free melt surface by the outside or inside shaper edges (Fig. 4.56).

Depending on the position of the shaper edges with respect to the level of the free liquid surface (*d*), the melt may be under certain positive or negative pressures. The dimensions and shape of the growing crystal are defined by:

- The shaper geometry
- The pressure applied to the melt being introduced into the shaper
- The position and shape of the crystallization surface
- The shape of the seed (at the initial stage)

For obtaining intricate profiles, a rod wetted with the melt may be used as a seed meant to lift the melt column above the shaper level. The crystal then is pulled in a nonstationary regimen, and the crystal cross section will be defined by the shaper configuration.

While choosing the shaper, one must take into account the physical properties of the material, such as its wettability, density, thermal conductivity, and thermal capacity. Shaping facilities make it possible to control the shape, the thermal state of the melt column and of the pulled crystal, and even the impurity distribution in the crystal.

Theoretical aspects of shape formation. The conditions for stable growth of crystals with constant cross sections are the following: the sum of the heat released upon solidification of the melt and the heat transferred to the CF from the liquid phase must be equal to the thermal flow removed from the CF through the solid phase. The angle of conjugation of the liquid phase with the surface of the growing crystal α_c (contact angle) is one of significant capillary characteristics that define the process of shape formation. So, the shape of the crystal cross section depends on the thermal and capillary conditions.

It has been assumed [116] (Fig. 4.57) that rather large negative values of the angle α_c correspond to the decrease of the crystal diameter and large positive values point to its increase. The largest negative and positive α_c values are defined by the wetting angle θ_0 at the boundary between the solid and liquid phases:

$$-\pi/2+\theta_0 \le \alpha_c \le \theta_0 \qquad (4.27)$$

In view of the condition (4.27) and assuming crystal pulling is sufficiently slow and the kinetic energy of the melt moving behind the crystal can be neglected, the shape of the meniscus corresponding to the minimum energy of the system is described by the Laplace equation

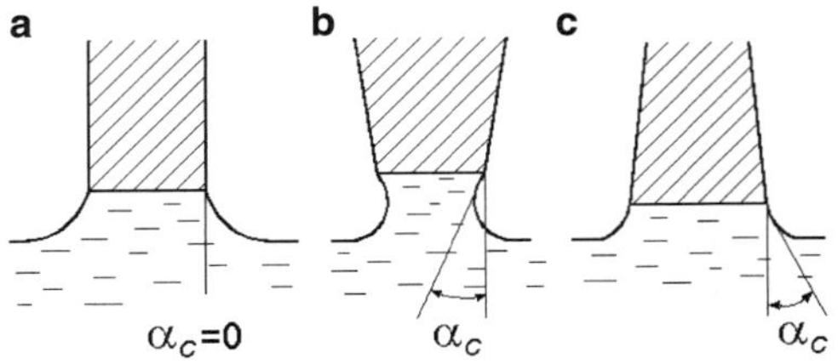

Fig. 4.57 The melt meniscus shape and the contact angle variation during the crystal pulling from the melt: (**a**) stationary growth, $\alpha_c = 0$; (**b**) the crystal constriction, $\alpha_c < 0$; (**c**) the crystal widening, $\alpha_c > 0$

$$P / \sigma_S = 1/R + 1/R_1, \tag{4.28}$$

where P is the pressure exerted on the meniscus at the given point; R and R_1 are the principal radii of the meniscus curvature.

The solution of (4.28) shows that under conditions of stationary crystal growth the meniscus height h_0 is bound to the curvature radius R_0 of the CF perimeter by the relations:

$$h_0 = 2R_0 \quad \text{at } R_0 \ll \alpha_c / 4, \tag{4.29}$$

$$h_0 = \alpha_c (1 - \alpha_c / 4R_0) \quad \text{at } R_0 \gg \alpha_c / 4, \tag{4.30}$$

where α_c is the capillary constant.

If the meniscus borders on a flat crystal facet, then $R_0 = \infty$, and based on the relationship (4.30)

$$h_0 = \alpha_c \tag{4.31}$$

Changes in the angle α_c at small deviations from h_0 are defined by the expression

$$\alpha_c = 1 - h / h_0 \tag{4.32}$$

where $h \sim h_0$.

For crystals with a circular cross section (R_0 = const) the value h_0 is the same for all the CF points. However, for the growth of plates the curvature radius differs at different points on the CF perimeter. Consequently, the following two possibilities exist:

- The meniscus height h_0 may change for parts of the perimeter with different curvature radii, i.e., in accordance with the expression (4.31), $h_0 = \alpha_c$ on the flat facets and, in view of (4.29), $h_0 = 2R_0$ at the plate edges
- The angle α_c may vary along the perimeter of the plate

Thus, for pulling profiles such as plates it is necessary to provide the required CF curvature ($h_{0(\text{facet})} = \alpha_c$; $h_{0(\text{edge})} = 2R_0$) or to deform the melt meniscus with a flat CF and a flat meniscus base that remains unchanged [116].

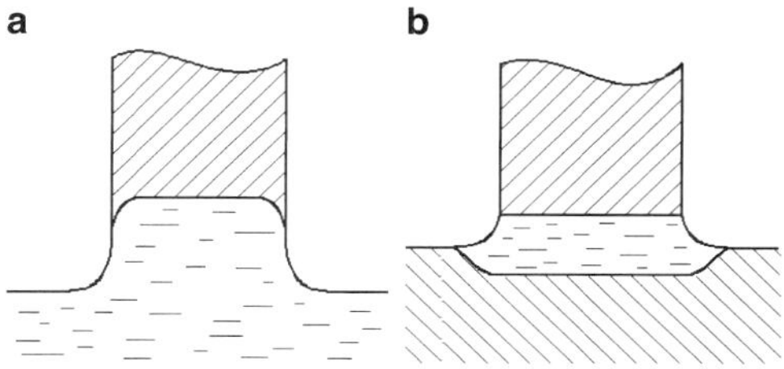

Fig. 4.58 Pulling scheme of a plateshaped crystal from the melt. (**a**) The CF level lowering at the crystal edges due to nonuniform cooling; (**b**) the meniscus base elevation resulting from the use of a crucible with elevated edges, the flat CF being maintained

The first variant can be realized through local cooling of the plate edge (e.g., by gas flows). The level of the CF at its edges diminishes to the value equal to $2R_0$ (Fig. 4.58a).

However, CF bending leads to nonuniform impurity distribution and the formation of structural defects. Under such conditions the plate width easily deviates from the preset value. If a crucible width close to that of the plate is used, with edges higher than those of the plate by the value $\alpha_c - 2R_0$ (Fig. 4.58b), then relations (4.29) and (4.31) will be satisfied for a flat CF.

The meniscus can be deformed by applying external pressure P to its parts adjacent to the flat surfaces of the plate. In this case, the meniscus curvature in the vertical plane increases, and consequently the height h_0 diminishes. Unfavorable capillary conditions at the edges of a thin ribbon are eliminated by changing the configuration of the ribbon cross section. To decrease the radius of curvature at the edges, dumbbell-like profiles are used. Ribbons of different thicknesses and widths, including broad, thin ribbons, can be obtained at the same edge thickness.

The meniscus height h and the pressure P of the melt introduced into the shaper hole are optimal when the growth angle φ is equal to its equilibrium value φ_0. This condition provides crystal size stability. The optimal values of h and P are determined from the numerical solution of the Cauchy problem for the Laplace capillary equation [117]. As an example, consider the dependence of h on P for the growth of a sapphire profile ($\varphi_0 = 17°$, crystal radius $r_1 = 19.9$ mm, and shaper radius $r_2 = 20$ mm). From the dependence of h on $(r_2 - r_1)$, one can determine the optimal height $h = h_1$ and the pressure $P = P_1$ necessary for obtaining constant-diameter crystals (Fig. 4.59). This dependence is valid for a constant level of melt in the crucible (i.e., replenishment).

The shaper is not a gauge that unambiguously establishes the crystal dimensions and shape. In order to show the problems to be solved in its creation, consider the simple example of the growth of a cylindrical rod. It can be obtained through different dimensions of the shaper and under different pressures. But if the cross section of the crystal is not circular, the problem of its growth becomes more complicated. In this case, the shaper must compensate for the difference in melt level arising from the varying crystal surface curvature, in order to provide a flat CF during the process of crystal pulling (Fig. 4.60).

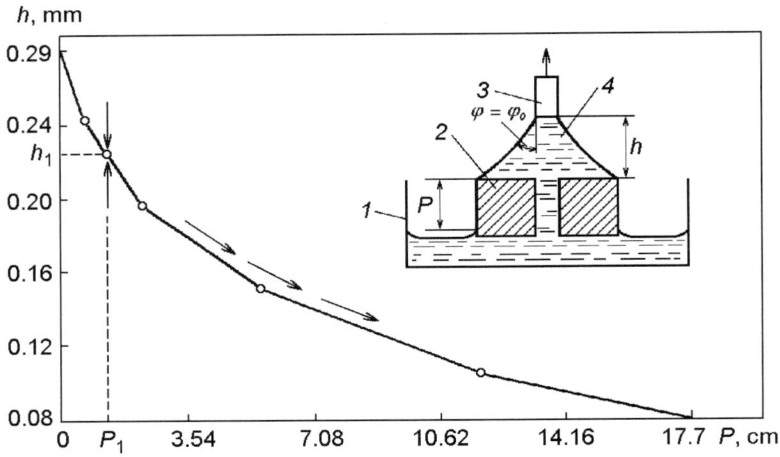

Fig. 4.59 Dependence of optimum melt column height *h* on pressure (*P*) of melt contained in a crucible defined by the equilibrium growth angle $\varphi = \varphi_0$. (*1*) Crucible; (*2*) die; (*3*) crystal; (*4*) melt

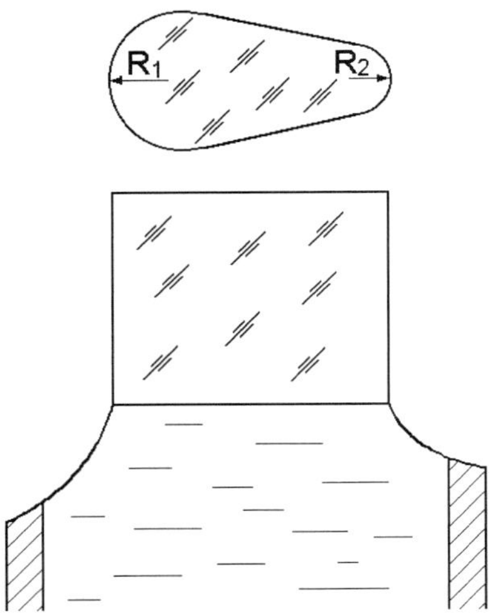

Fig. 4.60 Manufacturing of a complicated shape crystal under compensation of the difference in the elevation height in various regions of different curvature by the die

The constancy in height of the CF defines the stability of the properties of the crystal along its length. To maintain such a constancy at the initial stage (from the moment of seeding until a crystal length comparable with the height of the screened zone is obtained), the power of heating must be increased from the initial W_3 to the stationary W_c value. An increase in the profile cross section leads to a rise in Wc, while an increase in the growth rate results in the decrease of W_c to W_{c1} (Fig. 4.61) [118]. Such a diminution is caused by the increased quantity of absorbing and scattering centers. In obtaining crystals with smoother surfaces, W_c increases. As seen

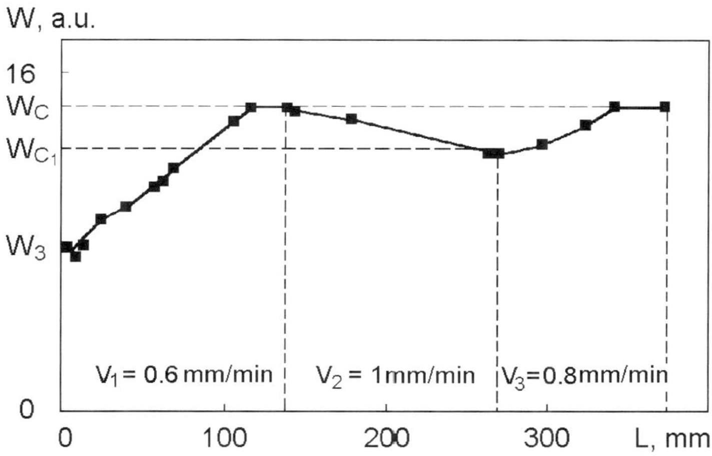

Fig. 4.61 Dependence of heating power on pulling rate and crystal length

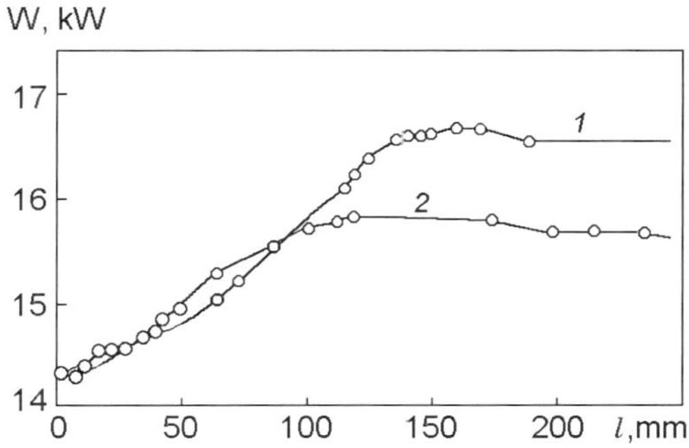

Fig. 4.62 Dependence of heating power on the d = 4.6 mm rod length. (*1*) Smooth surface; (*2*) rough surface

from Fig. 4.62, for the growth of long, thin rods with smooth surfaces, an additional power on the order of 1 kW is required [119]

In the process of sapphire growth, the main mechanism of heat transfer is thermal radiation flow (~70% of the whole of the heat transfer [120]). The crystal itself, which actually is a light guide, becomes a source of heat loss. A scheme of radiation propagation in the thermal zone has been proposed [119] (Fig. 4.63) that explains the agents influencing the heating power. When the seed length and cross section increase, W_3 also rises due to increased heat removal from the seed.

A detailed consideration of the foundations of stable shape formation for the growth of sapphire by the Stepanov method has been reported [119]. It is not sufficient to stabilize the melt temperature and the pulling rate to maintain a stable CF position, since the increase in crystal length and the decline of the melt in the crucible change the thermal conditions in the growth zone.

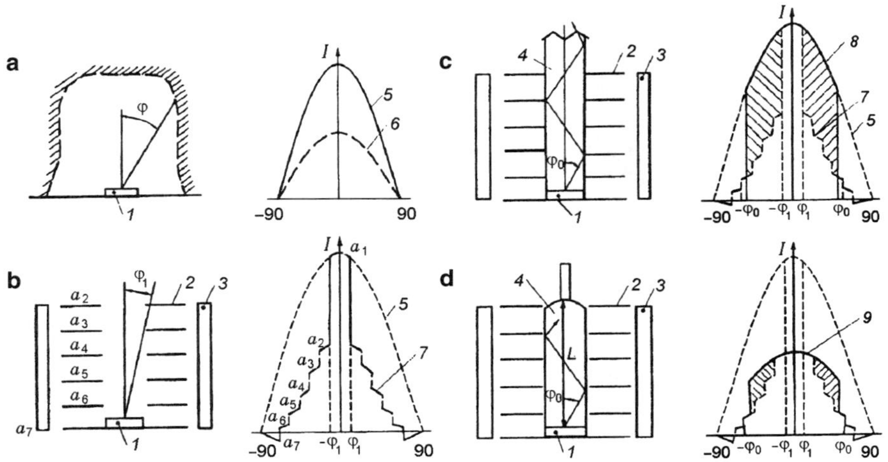

Fig. 4.63 Angular dependence of the heat emission intensity I for a point in the center of shaper at different conditions of heat exchange (scheme): (**a**) to vacuum ($T_m = 0$ K) and to medium with constant temperature $T_n > 0$ K; (**b**) in real thermal zone without crystal; (**c**) in real thermal zone with infinitely long crystal; (**d**) in real thermal zone with the crystal of the finite length L. Thermal zone elements: (*1*) shaper; (2) shields (a_2–a_6), (*3*) heater (a_7); (*4*) crystal. Heat emission intensity curves: (*5*) to free space ($T_m = 0$ K); (*6*) to medium with constant temperature $T_n > 0$ K; (*7*) in thermal zone without crystal (φ_1 characterizes holes in the shields); (*8*) the same with infinitely long crystal ($\varphi_0 = 90° - \alpha$, where α is the angle of complete internal reflection); (*9*) the same with the crystal of the finite length L. a_1–a_7 are the sections of curve *5* corresponding to emission: a_1 – to vacuum, a_2–a_6 – to shields a_2–a_7, respectively, a_7 – to heater

The apparatuses for the growth of profiled sapphire are equipped with different types of thermal units (Fig. 4.64). All the variants imply the use of a crucible with a shaper located above (Fig. 4.64a–d) or below (Fig. 4.64e–h). Prolongation of the equipment service life may be achieved when the shaper is fixed outside the crucible and withdrawn from it after completing the growth process. However, for this purpose it is necessary to increase the length of the thermal zone. Heating is to be provided by resistive or high-frequency methods. To realize more precise control of the process and to avoid overheating of the melt layer adjacent to the crucible walls, a high-frequency energy concentrator — a current-conducting element — should be placed between the crucible and the inductor (Fig. 4.64b). As heat-insulating materials, yttrium oxide (Fig. 4.64a, b) or carbon fiber fabric (mat) (Fig. 4.64d) are used.

The growth of special types of profiles has certain peculiarities. As an example, consider the methods for making capillaries.

Capillary channels in sapphire can be obtained only through profiled growth by the Stepanov method. The size and shape of a capillary channel usually is set by a special recess in the shaper (Fig. 4.65). The diminution of the channel diameter leads to a rise of the Laplace pressure, and then the melt fills the capillary channel and the shaper hole.

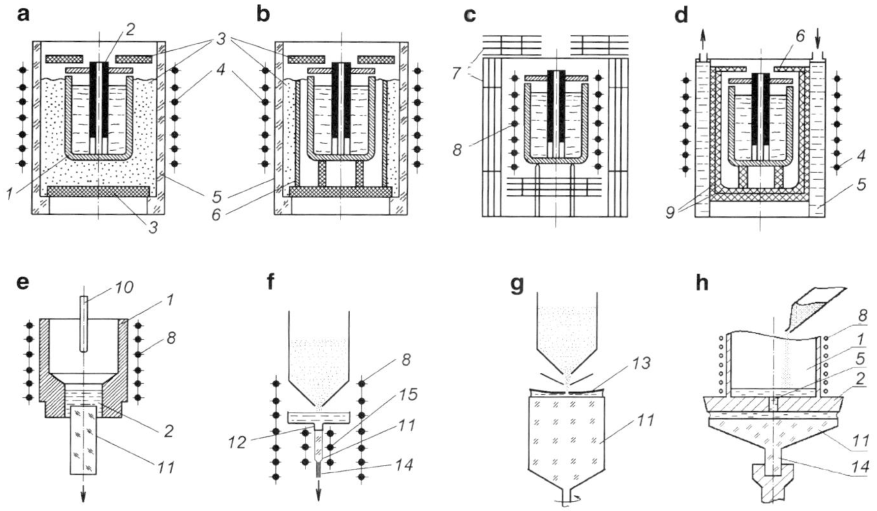

Fig. 4.64 Basic realization schemes for the Stepanov method at growing sapphire profiles: (**a**) crucible with thermally insulating ceramic filling; (**b**) inductor-crucible energy transferring device with HF-energy concentrator; (**c**) resistive heater with screen system; (**d**) graphite HF-energy concentrator and heat insulation from carbon fabric and mats; (**e**) variant with inferior die; (**f**) μ-PD method and its development for growing bulk crystals (**g**, **h** [121]); (*1*) crucible; (*2*) die; (*3*) yttrium oxide heat insulation; (*4*) Ruhmkorff coil; (*5*) quartz cover; (*6*) HF-energy concentrator; (*7*) screen; (*8*) resistive heater; (*9*) heat insulation from carbon fabric and mats; (*10*) feeding rod; (*11*) growing crystal; (*12*) micronozzle; (*13*) metal plate; (*14*) seed; (*15*) after-heater

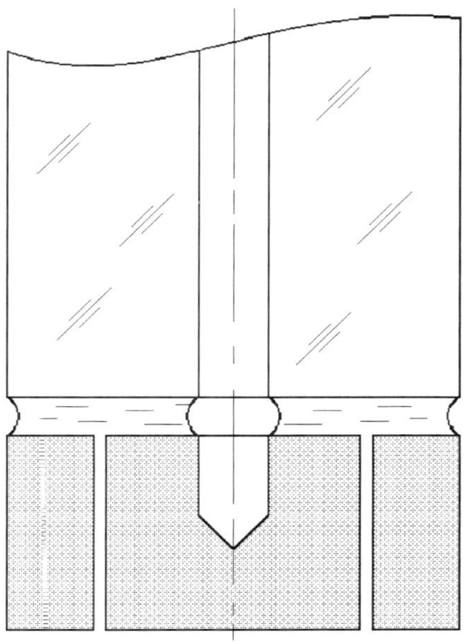

Fig. 4.65 Scheme of capillary channel forming by a hollow in the die surface

The pressure ΔP depends on the surface tension of the melt and on the conditions of the crystal and the shaper wetting with the melt

$$\Delta P = (\gamma_1 - \gamma_g)gh_0 = -2\sigma_{12} / r \tag{4.33}$$

where γ_l is the density of the liquid, γ_g is the density of the gas, σ_{12} is the interphase surface tension, h_0 is the height of the liquid column, and r is the radius of the average meniscus surface curvature:

$$\frac{1}{r} = \frac{1}{R_1} + \frac{1}{R_2}, \tag{4.34}$$

where R_1 and R_2 are the radii of the meniscus surface curvature in two mutually perpendicular section planes.

The Laplace pressure is lowered, thus decreasing the melt level in the crucible, i.e., increasing the melt column height h_0. This permits growth of capillaries with a diameter up to 0.7–0.8 mm.

Obtaining capillaries of smaller diameters entails certain difficulties. As the channel diameter decreases due to instability of the CF, the melt fills the capillary and the shape-making holes. This leads to loss of shaper function.

Alternatively, capillary channels can be formed using tungsten rods projecting over the shaper surface [122] (Fig. 4.66). During crystal growth, the melt bends around the rod, thus forming a capillary channel. In such a way, capillaries with a diameter down to 0.5 mm and a length of 50–80 mm are obtained.

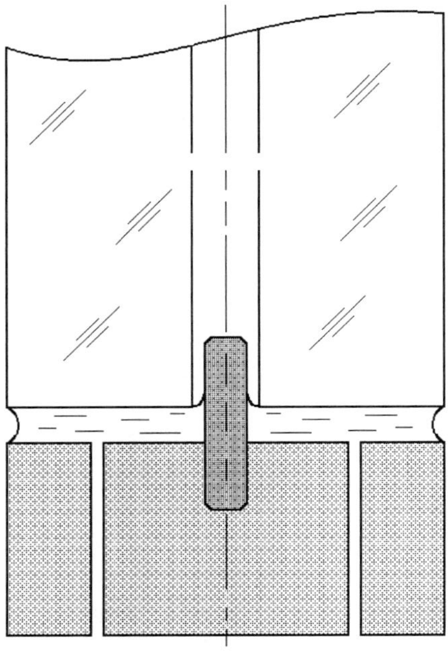

Fig. 4.66 Scheme of capillary channel forming by a protruding rod

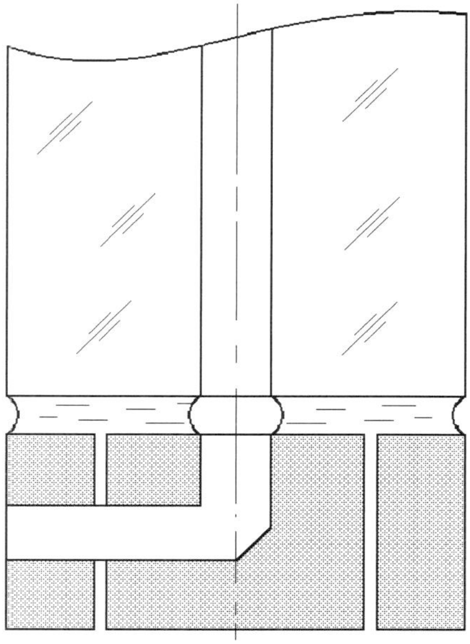

Fig. 4.67 Capillary channel forming by a hollow connected with the chamber atmosphere

Limitation of the capillary length is connected to disturbances caused by the pulling mechanism and processes at the CF leading to closing of the capillary channel (to its filling with the melt). As the rod projects above the shaper surface and some part of the melt moves toward the center, the gaseous impurities rejected by the CF accumulate at the rod point. The rejected gaseous phase finds its way to the rod and creates another capillary channel. This capillary length depends on the intensity of gaseous impurity rejection and limits growth of long capillaries.

To stabilize capillaries and to reduce the probability of their filling with the melt, classic shapers with channel-forming recesses can be upgraded (Fig. 4.65). The channel-forming recess is opened to the atmosphere of the chamber (Fig. 4.67).

Such a problem is complicated by the small outside dimensions of the shaper (~2.5–4 mm). In this space, a circular capillary for introducing the melt, a channel-forming hole, and a channel providing connection with the atmosphere of the chamber are to be situated. If the melt fills the capillary channel and the channel-forming hole, they must be reopened. For this purpose, it is necessary to create a rarefaction over the channel-forming hole and then fill it with the gas contained in the atmosphere of the chamber via the connecting channel. This action stabilizes the formation of the capillary channel and makes it possible to use the shaper many times. In such a way, capillaries with a length up to 500 mm and a diameter of 0.6 mm have been obtained.

Further reduction of the inside diameter entails the problem of withdrawing the melt from the channel-forming recess. The capillary system has inside and outside phase boundaries with different curvature radii. The melt filling the channel-forming recess can be withdrawn from the latter only in the case when the pressure in the

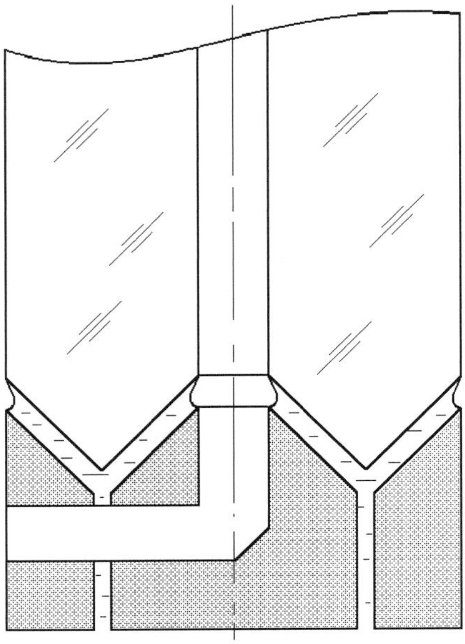

Fig. 4.68 Capillary channel forming at the concave crystallization front

outside meniscus is higher than in the recess. If the diameter of the capillary is less than 0.5 mm, the pressure difference is insufficient for withdrawing the melt from the recess.

A shaper with a concave surface [123] (Fig. 4.68) has a number of advantages. Foremost is the possibility to withdraw the melt from the channel-forming hole and to create such conditions that exclude the ingress of the melt into the capillary channel.

During the process of seeding, the melt fills the channel-forming hole in practically all cases when its diameter is less than 0.5 mm. Therefore, the melt must be withdrawn from the hole at the initial crystal growth stage. For this purpose, the thermal conditions are to be created in such a way that the CF should closely approach the shaper surface to make the ingress of the melt into the hole impossible. As a result, the amount of the introduced melt decreases, the crystal recedes from the edge of the outer diameter, and rarefaction is created in the center above the channel-forming hole. Due to rarefaction, the melt can be withdrawn from the channel-forming hole, and gas from the chamber finds its way to the capillary channel. The channel is opened, and the growth parameters can be stabilized to provide optimum conditions for the formation of the capillary channel and the full profile crystal shape. This approach makes it possible to obtain capillaries with a channel diameter of 0.4 mm. The shaper design allows multiple openings of the channel, and consequently multiple uses of the shaper. Controlled channel opening and closing also can be used to obtain gas-tight sapphire ampoules containing gas from the growth chamber.

The Stepanov method essentially differs from other methods of crystal growth from the melt by a high rate of heat removal from the crystallization zone. This is provided by a large ratio of heat-emitting surface to crystal volume, with other conditions being equal [124–126]. This high rate of heat removal makes it possible to grow crystals at rather high rates. However, the problem of obtaining crystals with low density of structural defects such as pores that appear at the side profile surface — remains unsolved. Different variants of Stepanov method are given below.

EFG (edge-defined film-fed growth) is a method of film replenishment with edge limitation of the growth [127]. The melt is introduced into the crystallization zone by capillary forces that arise while using capillaries wetted by the melt.

CAST (capillary action-shaping technique) is a method of shape-making by means of capillary action [128]. It is distinguished by a special design of the shaper and implies blowing the crystals using inert gas. This provides the possibility of creating the melt film with a larger thickness to shift the growing crystal away from the shape-forming surface, and consequently to lower the probability of contamination of the crystal with the metal impurities of the shaper.

Inverted Stepanov technique is the method of crystal pulling downward, sidelong [129], and in horizontal direction [130].

The μ-*PD (micropulling-down) method* is a variant of the technique of crystal pulling downward developed by Fukuda [131, 132]. This is a miniaturized version of the Stepanov method. While passing to microdimensions, the growth conditions essentially change. They can be stabilized by means of the temperature gradient near the phase boundary approximately equal to 103 K/cm. This method was used to grow Al_2O_3-based fibers of high-temperature eutectics with a diameter of 0.2–1 mm and a length of 500 mm meant for making fabrics to be used in aerospace devices or in other apparatuses working under extreme conditions.

Sapphire fibers of constructional quality with a diameter of 0.15 mm ± 2–3% and a length up to 1 m were grown in a iridium crucible at a rate of 2–3 mm/min. At lower growth rates optical-quality fibers were obtained. The scheme of the method and its version for the growth of large-volume crystals are shown in Fig. 4.64f–h.

Hybrid method consists of crystal pulling downward from a drop of the melt contained in a heater-shaper at continuous feed of the raw material into the melt. Although 21 versions of this method are proposed [131], the data on its use in practice are not available.

Shape-making from an element of the melt shape (local shape-making method). In different versions of the Stepanov method the growing crystal usually takes a shape close to that of the liquid melt column set by the shaper. However, for obtaining profiles it is possible to use a shape fragment [133] (Fig. 4.69). Therefore, a column of the melt is created between the surface of the seed and the feeding capillary. Such a column is a fragment of the shape of the crystal to be grown. In the process of rotation the seed touches the melt, and on its surface a film of the melt is formed. In the course of crystal growth the seed rotates and moves upward. Such a principle of shape-making was used by the authors for obtaining large-diameter sapphire pipes (Fig. 4.70). The use of a double crucible and several shapers joined with different crucible parts provides the attainment of profiles with a preset distribution of the doping addition (including a periodic one) (Fig. 4.71) [117].

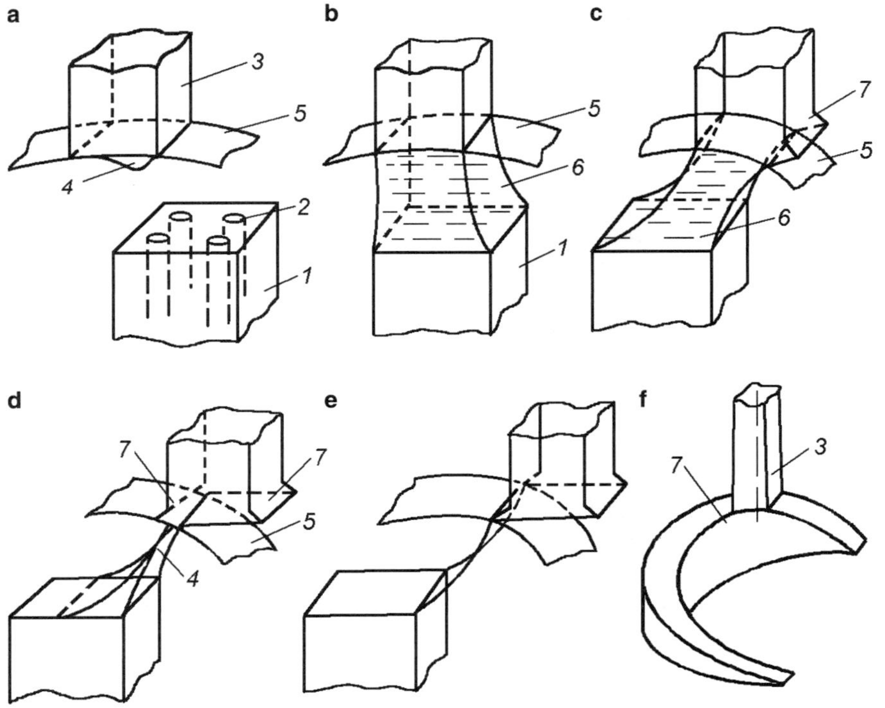

Fig. 4.69 Scheme of sapphire tube growth from a fragment of melt shape: (*1*) crystal seed; (*2*) crystal; (*3*) melt column; (*4*) feeding capillary. (**a–f**) – Stages of crystal growth: (**a**) seed fusion; (**b**) formation fo meniscus; (**c**) deformation of meniscus and crystallization of the front (in the direction of motion) side of the seed; (**d**) partial break of the meniscus from the die edges; (**e**) full breakage of the meniscus; (**f**) the grown larger tube. (*1*) Die; (*2*) capillary channels; (*3*) seed; (*4*) melt drop; (*5*) hypothetical isothermal crystallization surface; (*6*) melt meniscus; (*7*) grown crystal

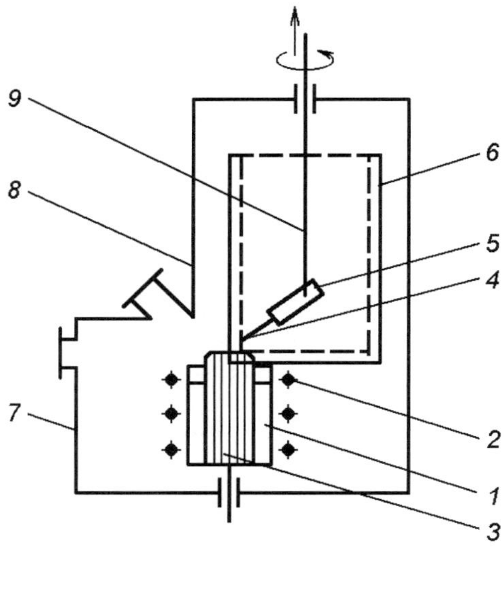

Fig. 4.70 Growth scheme of large diameter crystal pipe: (*1*) crucible; (*2*) heater; (*3*) feeding capillary; (*4*) crystal seed; (*5*) seed holder; (*6*) crystal; (*7*) crystallizing chamber; (*8*) receiver of grown crystal; (*9*) pulling mechanism rod

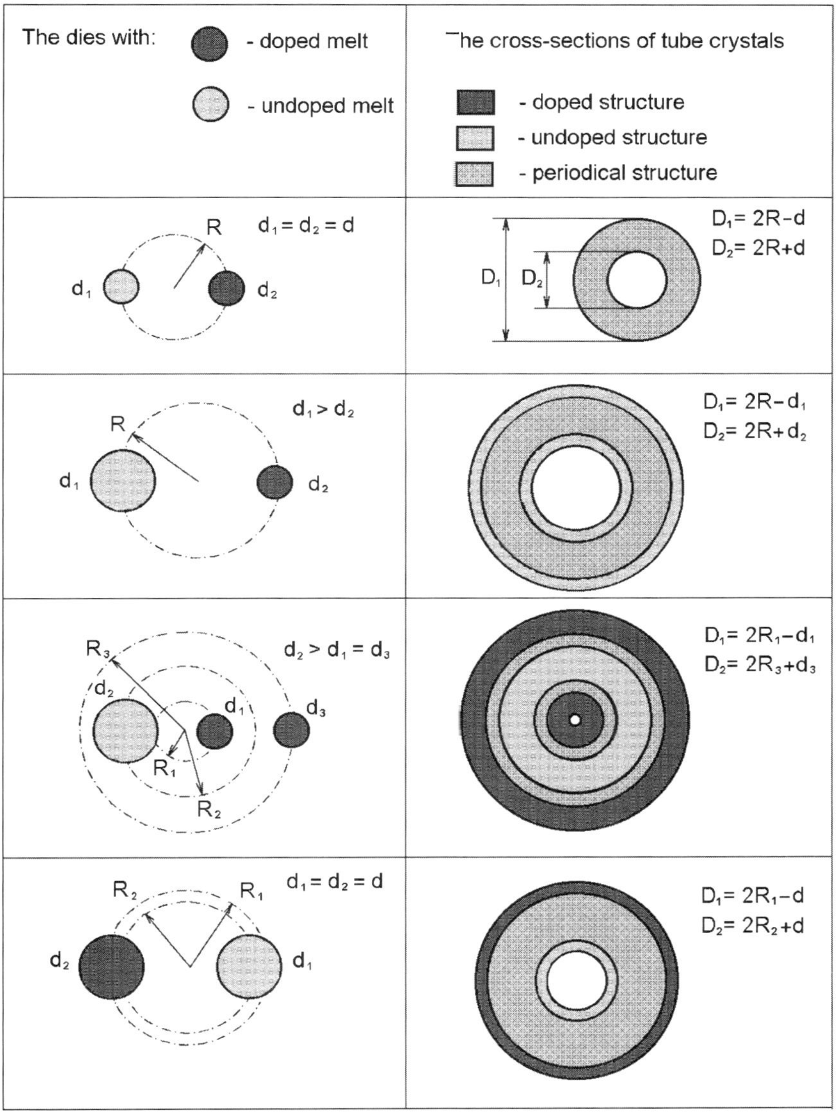

Fig. 4.71 Location of dies (*left*) and corresponding cross sections of tubular crystals (*right*) grown by local shaping technique. D_1, D_2 – internal and external tube diameters, respectively; d_i – top die diameter; R_i – distances from die top centers to crystal rotation

Variational shape-making [134–136] implies alternation of the stages of stationary growth of a preset profile with the stages of its nonstationary growth at the changeover to another shape of the product to be obtained. At the stage of nonstationary crystal growth the melt meniscus is not caught by the shaper edges. So, the base of the melt column moves over the face-shaper surface to the working edges that define the shape of the next cross section to be obtained. Therefore, the shape-making elements of the same shaper —its edges and walls —work in turns. Periodic changes of the growing pipe diameter, breakage of the side edges of the profile, or their attachment to the pipe at a required moment of time do not cause difficulties. It also is easy enough to realize the changeover from rectangular to circular cross section of the profile and vice versa. Such a method is used to obtain complex-configuration sapphire crystals and crucibles with different diameters. Moreover, it provides the growth of a group of profiles on the same seed crystal. In the process of shouldering a profile is created the cross section of which contains all the elements of the profile group to be obtained and partitions between them. Afterward the changeover to a new cross section without partitions is realized, and a group of profiles grows independently.

The growth method with displacement of the growing crystal from one shaper to another also is used for obtaining products with varying profiles, but the crucible containing the melt is not displaced [137]. In this method the profile elements are grown in turns in two shapers located close to each other (Fig. 4.72) on the end of the capillary system (2). The system shown in Fig. 4.72a, b has two shapers (3) with circular feed of the melt. At first the pipe (4), which is the crucible wall, is grown onto a flat seed crystal (Fig. 4.72a). After obtaining a required length of the pipe, the latter is placed on the second shaper to grow the bottom part of the product (Fig. 4.72b). Then the process is repeated. The view of the crystal is shown in Fig. 4.73.

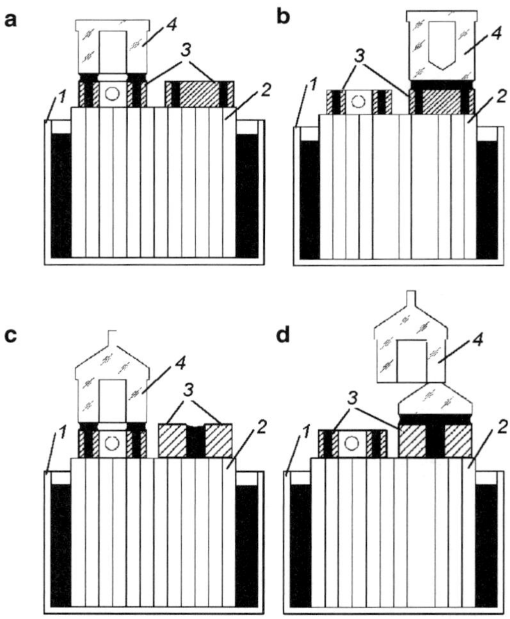

Fig. 4.72 Experimental scheme: (*1*) crucible; (*2*) capillary system (pedestal); (*3*) dies; (*4*) the crystal in growth

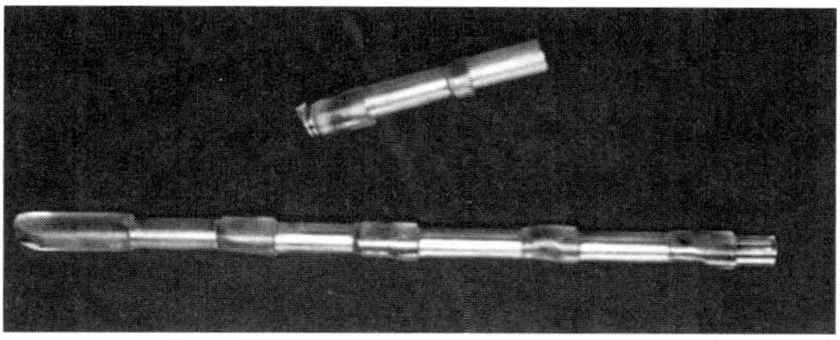

Fig. 4.73 Profiles grown using the scheme of Fig. 4.72a, b

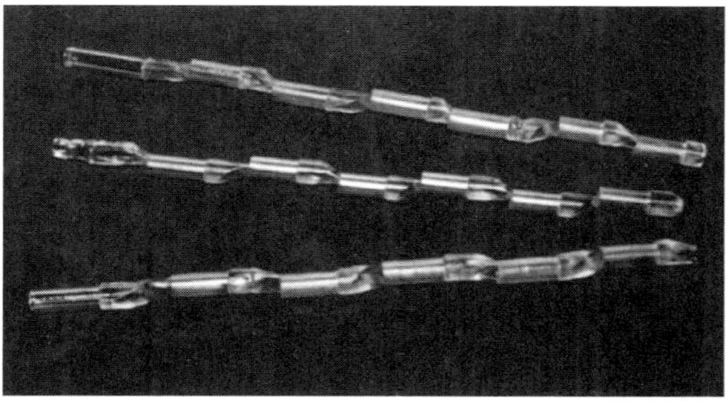

Fig. 4.74 Profiles grown using the scheme of Fig. 4.72c, d

In the method illustrated by Fig. 4.72c, d one of the shapers also has circular feed of the melt and is used for the growth of pipes. The shaper for obtaining the bottom part is equipped with the central capillary (Fig. 4.72d). At first, a rod with a length corresponding to the thickness of the crucible bottom is grown. Then this rod is displaced to the neighboring shaper and the crucible wall grown. Afterward the product 4 is returned to the first shaper to resume the growth of the rod, and so on. To provide a better symmetry of the product to be obtained, the rod is grown by turns onto the left and the right sides of the pipe. The difference in the height of the shapers (Fig. 4.72c, d) equal to 0.5–1.0 mm prevents simultaneous contact of the pipe with the two shapers at seeding. The methods were used for obtaining the bamboolike crystal bunches with a length up to 1 m (Fig. 4.74), which then were cut into separate products.

The state of the profile surface is of importance in the cases when the product to be obtained, e.g., oriented ribbons meant for epitaxy, will be used without subsequent treatment. Visually the growth surface of such a ribbon may be considered to have mirror smoothness with the presence of insignificant quantities of growth steps. Nevertheless, microscopic study reveals certain stable morphological peculiarities [138]. They are bound up with competing processes such as epitaxial growth of crystals of molybdenum (the shaper material) on the surface of the ribbon and its etching with carbon (the heater material). As established by X-ray diffraction method, fine-grained corundum with a structure independent of the ribbon orientation settles down on that part of the ribbon that is colder. It is presumed [138] that after chemical or plasma–chemical polishing the ribbons will be suitable for GaN epitaxy. However, in our opinion this is scarcely probable in the nearest future.

The economical characteristics of the considered profile growth schemes may be improved due to the following factors: the increase in the length of the crystals, the use of the methods of group crystal growth, and the application of revolving-type pulling facilities [139]. Combinations of the factors will permit the achievement of the highest productivity of profile growth.

In any variant of this method HF-heating reduces the consumption of high-melting metals (Fig. 4.75). The absence of heat-insulating screens makes it possible to use revolving-type pulling facilities (Fig. 4.76) and subsequently ensures further rise of the productivity [139]. HF-heating is provided using machine-, thyristor- or transistor-type (most compact and economical) converters.The tendency to increase the length of the crystals to be grown necessitates the increase of the height of the

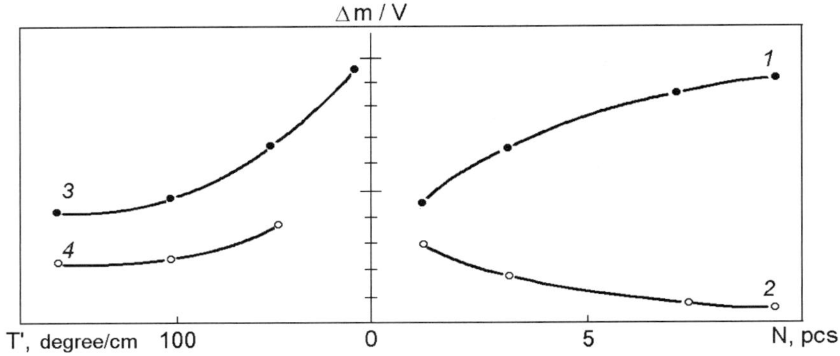

Fig. 4.75 Consumption of high-melting metals depending on heating method and value of axial temperature gradients at the crystallization front and quantities of crystals (*N*) in the group: (*1*, *3*) resistance heating; (*2*, *4*) induction heating

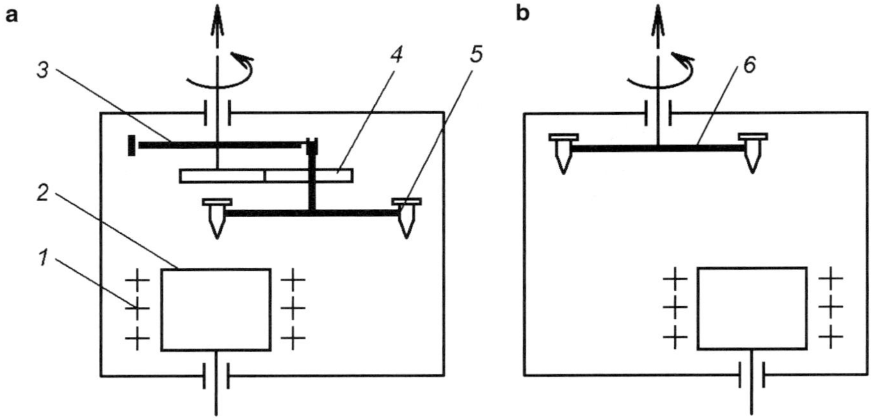

Fig. 4.76 Realization schemes of rotating group method of crystal growth: (**a**) revolving group method; (**b**) rotation method; (*1*) Ruhmkorff coil; (*2*) crucible; (*3*) coulisses; (*4*) revolving device with casing for crystal seeds; (*5*) crystal seed; (*6*) disk for fastening crystal seeds

crystallization apparatuses. While growing crystals with a length exceeding 1 m, instability of the electrical and kinetic parameters leads to serious consequences. For instance, switching out the pulling mechanism for 1 s does not cause noticeable changes in the shape of 0.2 m long crystals, but for the crystals with a length of about 1 m it leads to the appearance of a bend at a distance of 4–7 mm from the initial axis of the crystal. The method of group crystal growth requires even higher stability.

The comparison of the technical and economical characteristics of different methods of the growth of sapphire from the melt testifies that the Stepanov method has a higher productivity. Table 4.9 contains the data on the productivity of the Stepanov method used for the growth of isolated crystals. The growth of profiles by group method significantly raises the productivity. As seen from the table, the efficiency of profiled crystals manifests itself especially vividly at the stage of making articles on their base. The Stepanov method is applicable to the growth of sapphire profiles that cannot be obtained by mechanical treatment (fittings, thin-walled profiles, fibers, capillaries, figured nozzles, etc.).

The most promising directions in further development of the Stepanov method include the rise of the individual power of the growth apparatuses by increasing the quantity of the crystals to be simultaneously grown; the increase in the linear dimensions of the crystals; the technological process automation; and the rise of the homogeneity of the grown profiles and widening of their variety.

In this connection, the results reported in [140, 141] are of particular interest. Earlier sapphire profiles were mainly applied as a constructional material. But during the past few years the Stepanov method often was used to obtain sapphire optical elements. In particular, 225 × 325 × 5.6 mm windows for IR-optics meant for aerospace applications were made from 231 × 380 × 9.5 mm sapphire crystals. The plates were grown in *M* direction, with the *A* plane and lying on the larger surface and *C* plane lying on the side surfaces. A high transmission was achieved for the region of 0.7–4 μm (Table 4.10). The plates measuring 310 × 500 mm were grown for VIS-IR (500–5,000 nm) window application. The studies of these crystals by synchrotron white beam X-ray topography revealed a good and uniform quality, as determined by the value of double and triple axis rocking curve. However, lack of correlation between these measures at crystal quality and refractive index homogeneity values remains a puzzling aspect [141].

Table 4.9 Engineering and economical characteristics of growing methods from melt

Growing method	Crystal size (mm)	S/V[a]	Productivity mm/h	Productivity g/h	Crystal usage in products (%)	Average product cost, relative units
Verneuil	*D* 5–50 *L* 800–200	8–0.8	5–20	6–35	10–40	1
Czochralski	*D* 40–80 *L* 200	2–0.5	2–15	18–40	10–40	4
HDS	30 × 200 × 300	0.7	8–15	24–40	10–40	3
Stepanov	*D* 0.2–40 *L* 1,000	200–6	30–6,000	2.4–5	80–90	0.5

[a]Ratio of the profile area to the volume

Table 4.10 Transmission of 6-mm-thick sapphire plates (uncoated)

Wavelength (μm)	Transmission (%)	+/– (%)	Theoretical transmission maximum (%)
0.7[a]	84.5	0.8	85.8
0.7[a]	84.8	0.7	85.8
1.06	87.1	0.9	86.0
1.57	87.2	0.8	86.3
3.0	87.0	0.8	87.1
4.0	85.9	0.8	87.5
5.0	56.8	0.5	84.5

[a]Measured using different spectrometers

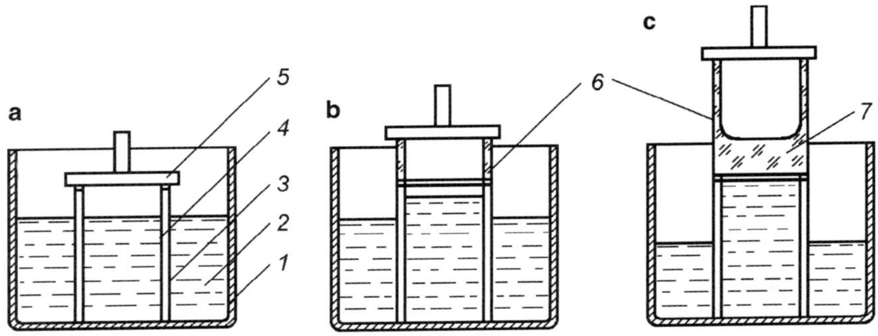

Fig. 4.77 Scheme of monolithic crystal growth by the NCS method: (**a**) seeding; (**b**) growing a hollow closed shape; (**c**) growing a monolithic crystal. (*1*) Crucible; (*2*) melt; (*3*) die; (*4*) capillary channel; (*5*) seed; (*6*) hollow closed shape of the crystal; (*7*) monolithic crystal

4.3.3.7 Noncapillary Shaping Method (NCS)

The principle of noncapillary shaping [142, 143] consists of the following. The melt moves toward the crystallization front inside the noncapillary channel due to the difference in the pressure in the growth chamber and inside the closed space under the seeding plate. The term "noncapillary" means that the channel diameter is larger than the value of the capillary constants. To create the closed space, in the process of seeding, the perimeter of the circular capillary channel is lapped over (Fig. 4.77). At the initial stage, when the melt approaches the CF only due to replenishment through the capillary, the growing crystal is hollow.

As the closed space increases, the pressure in it falls and the melt contained in the noncapillary part of the shaper rises up (Fig. 4.77b). At further pulling the melt reaches the CF through the noncapillary part of the shaper and joins the melt supplied through the circular capillary. As a result, the crystal grows as a monolithic rod (Fig. 4.77c). The distance l_{s-r} between the point of seeding and the nose of the monolithic rod depends on the pressure P in the growth chamber:

$$l_{s\text{-}r} \frac{H_{m-d}}{1-\left(H_f \gamma_m g / P\right)} H_{m-d} + H_f \tag{4.35}$$

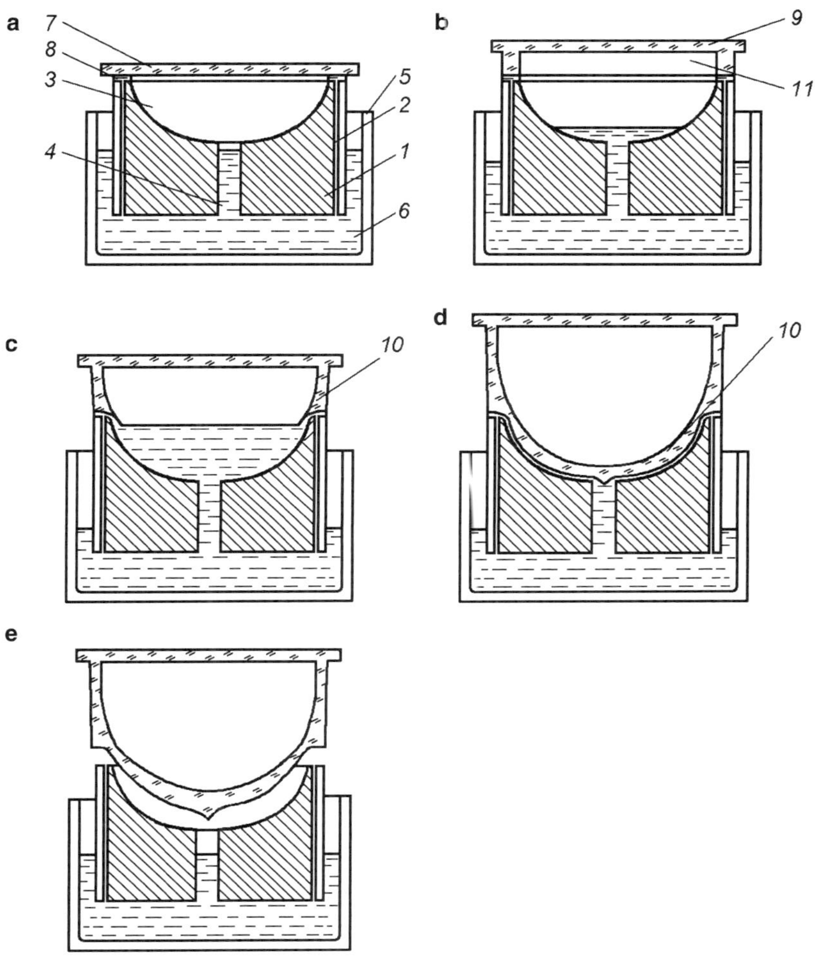

Fig. 4.78 Sequence of individual near-net-shaped dome growth stages. (*1*) Die; (*2*) ring capillary channel, (*3*) liner cavity; (*4*) noncapillary channel; (*5*) crucible; (*6*) melt; (*7*) seed-plate; (*8*) ring meniscus; (*9*) tube-shaped crystal; (*10*) near-net-shaped dome; (*11*) enclosed volume

where $H_{\mathrm{m-d}}$ is the distance between the level of the melt in the crucible and the shape-forming plane, Hf is the distance between the level of the melt in the crucible and the crystallization front, and γ_{m} is the melt density.

It also was proposed that this method should be used for obtaining sapphire meniscuses (hemispheres) [144]. In this case the shape-forming element is a hemispherical surface joined with the main volume of the melt by means of the noncapillary channel (Fig. 4.78). The shaper is supplied with the capillary channel running along the perimeter. On completing the stages of seeding and the growth of pipelike profile, the melt rises up into the hollow due to rarefaction created by the difference in the pressure. Subsequent scenario is clear from the figure. After carrying the process to completion the seed plate is cut off.

4.4 Solid-Phase Crystal Growth

This method implies the transformation of fine-grained polycrystals into a single crystal. Such a process is realized when one crystal grain absorbs the others or a new grain nucleus grows at the expense of the polycrystalline matrix. From the viewpoint of energy, this process is justified, as the polycrystalline structure possesses excess free energy. The advantages of the solid-phase method include the possibility of growing the crystals at $T < T_m$, as well as the fact that the shape of the growing crystal is preset and the amount of impurities and the character of their distribution do not vary.

The conditions of solid-phase crystal growth differ fundamentally from those of crystal growth from the solution or the vapor phase. The crystal surface captures free atoms and molecules from the adjacent layers of the matrix where all the atoms occupy certain positions. The crystal grows owing to the motion of the boundaries. The driving force of such a motion in well-annealed materials is the decrease of free surface energy due to diminution of the surface itself. That is why a bent surface of the boundary moves toward its center of curvature (Fig. 4.79).

The model is a matrix separated by the boundary into two grains (I and II) of different orientations. From the concave side of the boundary, atoms B (the grain II) are surrounded by atoms A of the grain I. It should be assumed that the atoms B tend to occupy the positions that build up the grain I. The surface diminishes

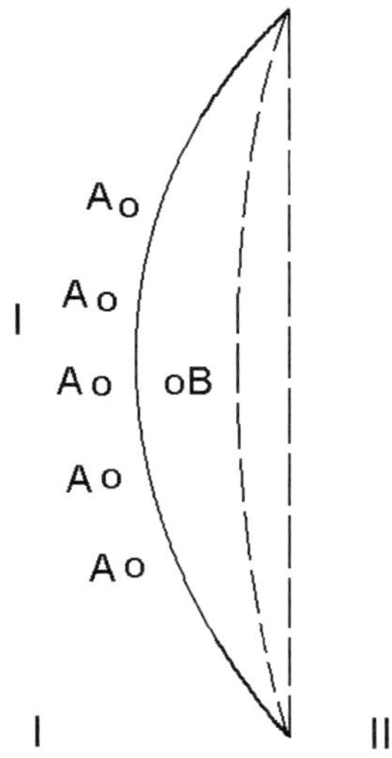

Fig. 4.79 Displacement scheme of the curved boundary between the crystals A and B

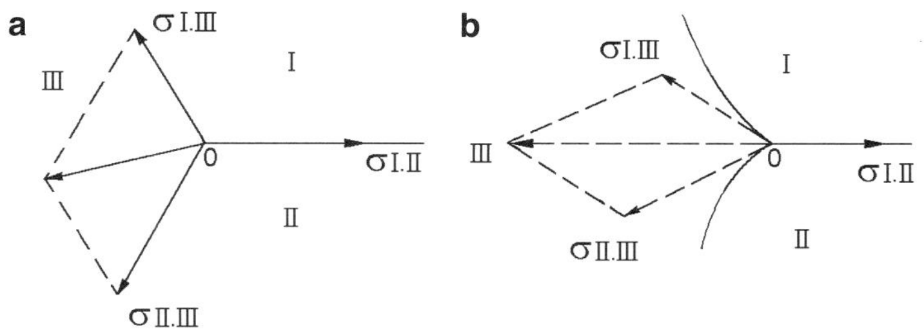

Fig. 4.80 Formation of curves boundary due to the surface tension trend to equilibrium: (**a**) the initial position of the boundaries; (**b**) after the equilibrium is attained

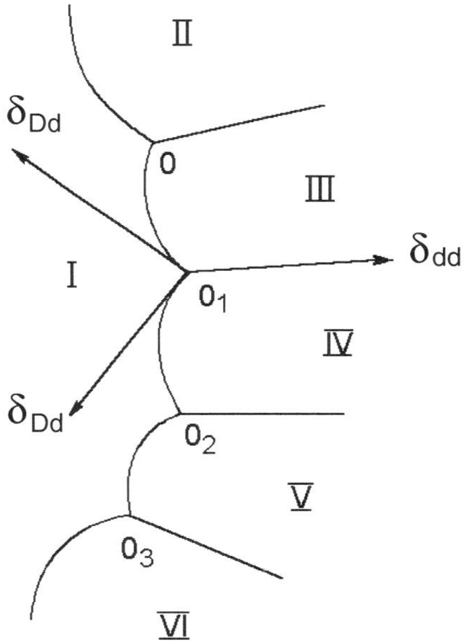

Fig. 4.81 A section of boundary between a large crystal and smaller grains

under the action of σ_s, since the boundaries are met at arbitrary angles and the balance of surface tension forces is scarcely probable, even in the case when it does not depend on the boundary type. The balance is achieved due to certain displacement of the atoms, which creates a new curvature of the boundaries, thus giving rise to their further motion (Fig. 4.80).

As follows from the energy viewpoint, a crystallite that is large enough in comparison with its neighbors tends to grow by capturing them. If such a crystallite is considered to be a polyhedron, then the angle at its vertex will increase with the number of its sides, and the surface tension between the neighbors will tend to bend the sides of the large crystallite and to make them concave (Fig. 4.81).

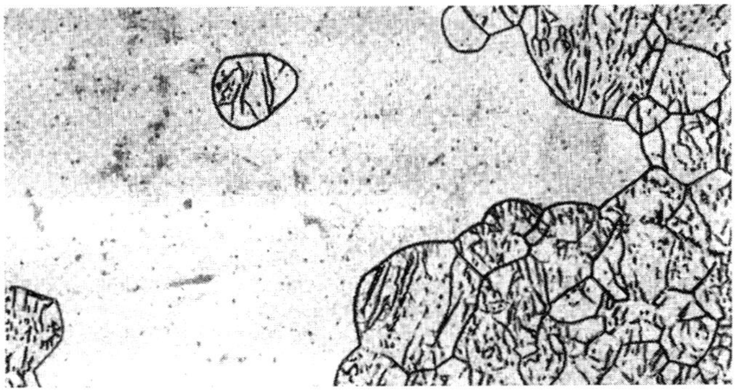

Fig. 4.82 A section of boundary between the crystal and the grains in an aluminum–zinc alloy stretched and then annealed at 450°C (×50)

The grain I grows at the expense of the crystals II, III, IV, V, and VI. The critical nucleus size corresponding to the grain growth onset depends on the ratio of the surface tension σ_{dd} between the absorbed crystallites to the surface tension between the large crystal and the neighboring fine grains σ_{Dd} (Fig. 4.81). A large value of the ratio σ_{dd}/σ_{Dd} will be favorable for the growth, as in this case the growing crystal requires less increase in size in comparison with the surrounding grains. Shown as an example in Fig. 4.82 is a large growing crystal which forms concave boundaries with the surrounding fine grains [145].

The surface energy of the boundary changes in the process of its motion. For each boundary, the driving force increases with the increase of the energy and the diminution of the curvature radius [146]. The surface energy is a function of disorientation between the neighboring crystals and at a given disorientation degree, it depends on the position of the boundary between the neighboring regions. The grain boundaries have their own structure, which provides the best possible conjugation of the crystal lattices located on either side of the boundary.

The boundaries between the grains with slightly different orientations have a dislocation structure. They are subdivided into tilt boundaries separating two crystals swung about the axis in the boundary plane and twist boundaries with a rotation axis perpendicular to the boundary plane. In the former case, the boundary model is a set of edge dislocations with identical sign [147, 148] (Fig. 4.83), while in the latter one, it is a network formed by two rows of mutually intersecting bent dislocations (Fig. 4.84).

These two figures show the structure of the boundary, which is almost parallel to the crystallographic plane. But when the boundary is not parallel to this plane, it can be presented by at least two rows of edge dislocations (Fig. 4.85) [147–151]. The defects are located on the dislocations or in close proximity to those. The boundary structure becomes more intricate as the degree of disorientation of the crystals rises. For larger angles, the boundary model that consists of single dislocations has no physical sense, since in this case, the dislocations are to be located so close to each other that their individuality is lost. Such boundaries can be considered to consist of alternating regions of good and poor conjugation. The mobility of the boundaries to

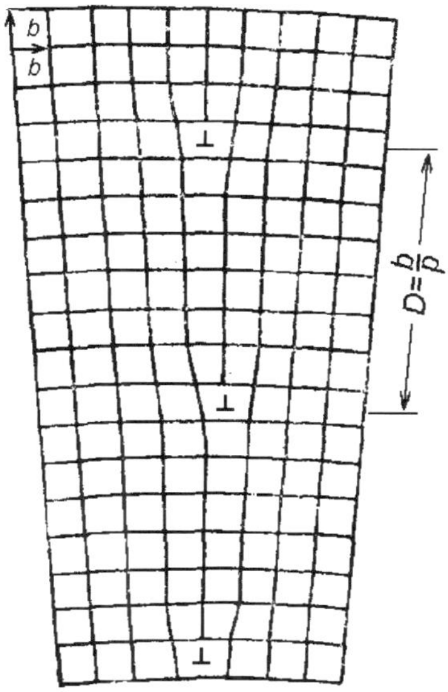

Fig. 4.83 Scheme of the slope boundary formed by a row of equidistant edge dislocations of the same sign

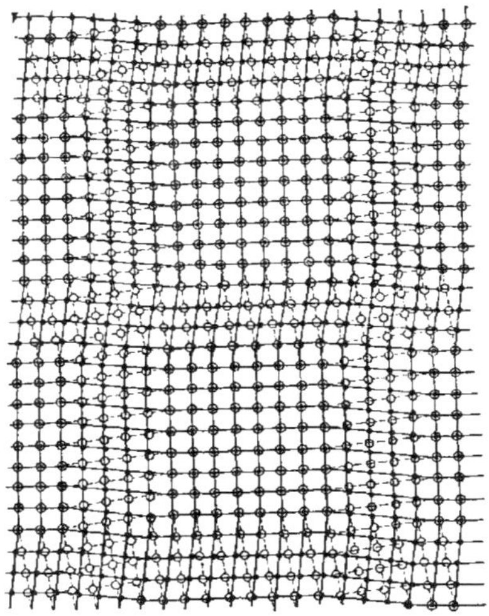

Fig. 4.84 Scheme of the twist boundary formed by two rows of intersecting screw dislocations

a considerable extent depends on their structure. The movement of a simple inclined boundary, which consists of parallel dislocations located far from each other, can be caused by simultaneous slipping of all the dislocations (Fig. 4.86).

For an arbitrary boundary, the simultaneous dislocation motion should not be expected even in the case when the lattices that form it are disoriented insignificantly. If the boundary consists of two rows of dislocations with different directions, its displacement requires simultaneous slip and creep of the dislocations. Edge dislocation creep implies shortening or elongation of the extra half-plane. It is connected

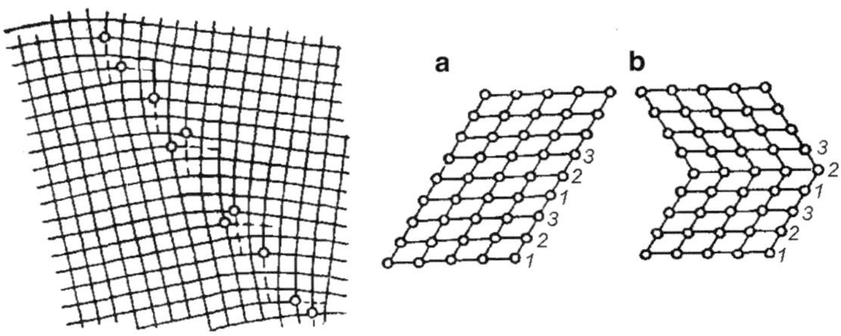

Fig. 4.85 Transition region formed by two rows of edge dislocations

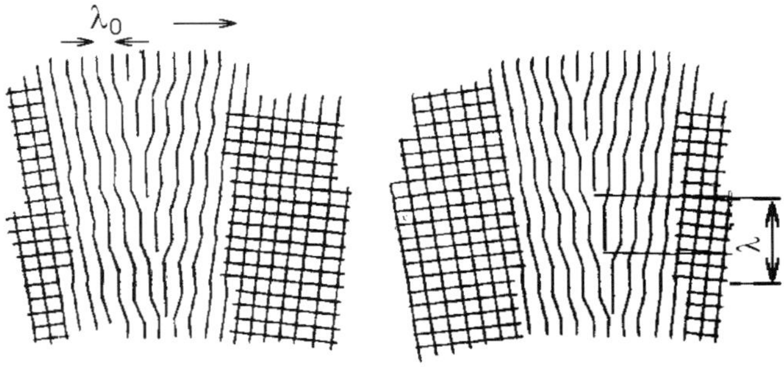

Fig. 4.86 Displacement of a small-angle tilted boundary due to applied shear stresses

with the diffusion of separate atoms and proceeds extremely slowly at low temperatures. If the conjugation is poor and the boundary energy is high, then the total atomic mobility increases and the probability of the boundary motion increases due to individual atom jumps. In this case, direct vacancy exchange at the boundary plays a significant role [151, 152]. Thus, a relatively high mobility should be observed both for low- and large-angle boundaries, whereas the boundaries with medium ($\theta \sim 1°$) disorientation angles must possess a relatively low mobility [153]. The latter fact is confirmed experimentally, as large crystals growing at the expense of the fine-grained matrix often contain small oriented domains that are obviously unabsorbed grains of the initial matrix with low-mobility boundaries.

The conjugated high-mobility boundaries are worthy of special consideration [137, 154]. Due to the presence of the conjugate points, it is difficult to predict whether their motion will be easy, since it depends on the stability of the configuration of such points [151, 152].

In strained materials, the growth of the grains and the mobility of the boundaries are influenced by the impurities contained there. It is known that in high-purity materials, the recrystallization proceeds faster and at lower temperatures. This is explained by the fact that the stress fields arising around the dislocations capture the impurity atoms. Therefore, the latter with sizes smaller than those of the main material tend to be disposed in the compressed regions, while the impurities with larger atomic size are located in the extended regions. The impurity aggregations reduce the dislocation mobility and the rate of recrystallization.

The authors used this fact when developing the methods for obtaining large-area nickel and molybdenum single crystals from the solid phase [155, 156]. The surface of the recrystallized samples was coated with a layer of activating substance with a crystallization temperature lower than the T_m of the materials being recrystallized. The melt of this substance wetted the sample surface sufficiently well. The activating substances for molybdenum and nickel single crystals were Al_2O_3 and LiF melts, respectively. A container with the samples coated with a layer of activating substance was placed into a gradient furnace and the temperature was raised to the T_m of the activating substance. Then the container was pulled through the gradient zone at a speed of 12–20 mm/h until the activating substance completely crystallized. Therefore, the metal was recrystallized under the activating substance layer and large single crystals were grown. In such a way, the $120 \times 30 \times 0.3$ mm^3 molybdenum and $120 \times 30 \times 2$ mm^3 nickel single crystals were obtained with block disorientation angles up to 20′. The recrystallization process is induced by the stresses arising at the metal/recrystallizing substance boundary. The surface energy of molybdenum wetted by Al_2O_3 melt becomes reduced. The active melt cleans the boundaries of the metal grains [155].

A defective surface-adjacent layer, which is not observed in the crystals obtained by the traditional solid-phase methods, arises on the surface of the single crystals grown by the method described above. After removing a layer of about 100 μm thickness, the microblocks disappear and the dislocation density in the crystal bulk becomes lower by a factor of 2 or 3 in comparison with that at the surface. The crystals without the surface-adjacent layer become plastic already at room temperature. Such a structure peculiarity of the surface-adjacent layers is obviously caused by their enrichment in impurities, such as oxygen and carbon, the concentrations of which exceed the limit of their dissolution in the crystal. For instance, in the initial single crystals, the average concentration of carbon is $\sim 3 \cdot 10^{-2}$ wt%; after removing the surface-adjacent layer, this value becomes lower than $5 \cdot 10^{-3}$ wt%. The presence of the mentioned layer can be explained by the fact that aluminum oxide melt works as a pump that removes the impurities from the bulk of molybdenum. At the Al_2O_3 melt/Mo interface, the impurity concentration is not equalized by diffusion flows, so the structure of molybdenum single crystals deteriorates.

While growing nonmetallic crystals, it is difficult to control the formation of nuclei and to hamper the growth of new grains due to the presence of a large number of defects. However, some materials such as corundum, fluorite, and periclase, strained by compression and heated almost to T_m, show migration of the boundaries. The growth of grains in ceramics is explained using the basic principles of the grain growth in metals. In the latter, a large grain may grow at the expense of small grains, if the concentration of the retarding particles gradually decreases. In ceramics, the grain growth is slowed down by pores that disappear at sintering. Thus, for single-phase ceramic systems, the conditions can be chosen that provide nonuniform grain growth at the sintering stage. The grains grow easily in fine-grained materials.

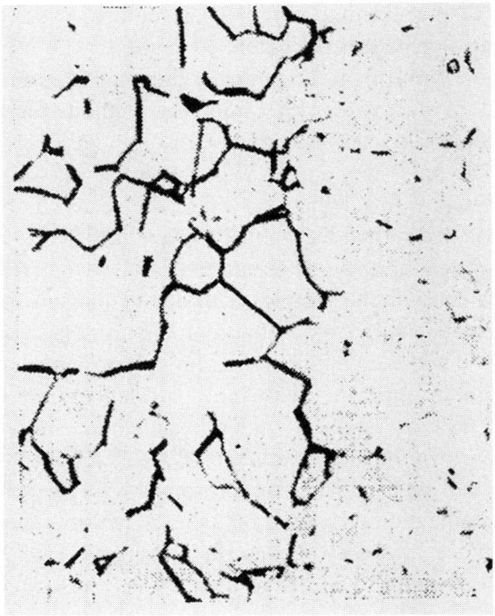

Fig. 4.87 Partition zone between a large crystal and small-grained matrix in Al_2O_3 (×470)

To grow sapphire by the solid-phase method, fine-grained Al_2O_3 powder is fixed around a sphere made of the same material. Then the ingot is sintered or the powder is melted in molybdenum or tungsten vessels. The samples are placed in a gradient furnace and moved there at a preset speed at 2,270 K. The sintering process is accompanied by the growth of a grain (Fig. 4.87).

The boundary between the large and small grains moves toward the curvature center of the latter. The initial boundary position is shown by the small pore line. The curvature of the surface between the large crystal and the fine-grained matrix is larger than that of other boundaries. This picture is similar to that observed at secondary recrystallization in metals. The absorption of small grains results in formation of a large crystal.

The structure perfection analysis of ~1-cm^3-size sapphires grown by this method shows that they differ from the crystals grown from the melt by a characteristic distribution of defects, which has the form of pore chains and impurity aggregates. As follows from the comparison with the initial polycrystalline structure, the location of the observed aggregates of the defects corresponds to that of the former grain boundaries.

Electron microscopic investigation of these samples has revealed their granular structure. In the initial grains of the polycrystalline samples used to grow the single crystals, the same granular structure was found, too. The sizes of the granular structure elements in the crystals grown from the melt and in those obtained by the solid-phase method met the same regularities and depend on the same technological parameters. In the pure undoped crystals, the grain size depends linearly on the axial temperature gradient and is inversely proportional to the squared second derivative of the temperature with respect to the coordinate.

In conclusion, it should be noted that in the present chapter devoted to the solid-phase synthesis of sapphire, we are dwelt on the growth of metal single crystals for the following two reasons. First of all, we tried to demonstrate the general laws of solid-phase growth of these crystals. Moreover, our aim was to lay emphasis on the unique properties not only of sapphire, but also of its melt.

References

1. Vil'ke K.-T. Crystal Growth, Leningrad 1977. Nedra. [in Russian].
2. Lodiz R.A., Parker R.L. Crystal Growth. Mir, Moscow. 1974 [in Russian].
3. Bakly. Crystal Growth. Inostr. Lit., Moscow 1954 [in Russian].
4. Kuznetsov V.D. Crystals and Crystallization. Goslitizdat, Moscow. 1954 [in Russian].
5. Schaffer P.S. J. Am. Ceram. Soc. 1965. **48**.
6. Langham E.J., Mason B. J. Proc. Roy. Soc. 1958. **247A**. p. 493.
7. Price P. B. J. Appl. Phys. 1961. **32**. p. 1764.
8. Frank F.C. J. Acta Met. 1958. **6**. p. 103.
9. Sears G.W. Acta Met. 1955. **3**. p 361.
10. Sears G.W. Acta Met. 1955. **3**. p. 367.
11. Sears G.W. Acta Met. 1956. **4**. p.268.
12. Gomer R. J. Chem. Phys. 1958. **8**. p. 457.
13. Ruth V., Hirth J.P. The Kinetics of Diffusion Controlled Whisker Growth. AD43497, No. 495(26), 1964.
14. Webb W.W., Dragsford R.D., Forgeng W.D. Appl. Phys. Rev. 1957. **28**. p. 498.
15. Webb W.W., Dragsford R.D., Forgeng W.D. J. Appl. Phys. 1958. **29**. p. 814.
16. Chanda G.K., Trigunayat G.C. Crystal Growth, Proc. of an Int. Conf. Pergamon, New York. 1967.
17. Webb W.W., Forgeng W.D. J. Appl. Phys. 1957. **28**. p. 12.
18. Brenner S.S. J. Appl. Phys. 1962. **33**. p. 1.
19. Tombs N.C., Welch A.J.E. J. Iron Steel Inst. 1952. **172**. p. 69.
20. Feingold E., Gatti A. TR R65SD11. General Electric Space Sci. Lab. 1965.
21. Barber D.J. Phil. Mag. 1964. **10**. p. 103.
22. Papkov V.S., Berezhkova G.V. Kristallografia. 1964. **9**. p. 442 [in Russian].
23. Cairpbell W. B. Chem. Eng. Progr. 1966. **62**. p. 3.
24. Goodman C.H.L. (editor). Crystal Growth. Theory and Techniques. Plenum Press, London, New York. 1974. p. 147.
25. Kaldys E. Crystal Growth. Theory and Techniques . Mir, Moscow. 1977. Vol.1. pp. 75–243 [in Russian].
26. Tanenbaum M. Solid State Physics. V.6. Methods of Experimental Physics. Academic Press, New York. 1961. p. 86.
27. Chirvinsky P.P. Artificial Preparation of Minerals in XXth Century. Kiev. pp. 1903–1906 [in Russian].
28. Yanovsky V.K., Voroshkova L.A., Koptsik V.A. Kristallografia. 1970. **15.2** pp. 362–366 [in Russian].
29. Lobachev A.N., Demianets L.N., Kuzmina I.P., Emelianova E.N. J. Crystal Growth. 1972. 13/14. pp. 540–544.
30. Kuznetsov V.A., Lobachov A.N. Hydrothermal Synthesis of Crystals, . 1968. p. 1972 [in Russian].
31. Butuzov V.P., Lobachev A.N. Rost Kristallov. 1972. **9**. pp. 13–24 [in Russian].

32. Demyanets L.N., Lobachev A.N. Study of Crystallization Processes in Hydrothermal Conditions, Nauka, Moscow. 1970. pp. 7–28 [in Russian].
33. Laudise R.A., Nielsen J.W. Solid State Physics, Advances in Research and Applications. Academic Press, New York. 1961. **12**. pp. 149–222.
34. Laudise R.A., Kolb E.D. Endeavour. 1969. **28**. p. 114.
35. Laudise R.A., Ballman A.A. J. Am. Chem. Soc. 1958. **80**. pp. 2655–2657.
36. Monchamp R.R., Puttbach R.C. ASD Project#8-–132. AF Contract 33(657) 10508. 1964.
37. Ganeev I.G., Kazurov B.K., Karaulnik E.N. Crystallization Kinetics and Mechanism, Minsk, 1969. pp. 399–407 [in Russian].
38. Bryzgalov A.N., Musatov V.V., Proc. of 6th Internat. Confer. Crystal Growth and Heat and Mass Transfer, Obninsk, . 2005. pp. 315–316, [in Russian].
39. Bunk H. Z.Dt. Gemmol. Ges. 1977. **Bd.26**. S.170–S.172.
40. Timofeeva B.A., Voskanyan R.A. Kristallografia. 1963. **8**. N2b. pp. 293–296 [in Russian].
41. Kuznetsov V.A., Lobachov A.N., Shternberg A.A. Kristallografia. 1962. **7**. N1. pp. 114–120 [in Russian].
42. Lingart Yu.K., Dr. Sci. Thesis, M. 1990. p. 486 pp [in Russian].
43. Lingart Yu.K., Petrov V.A., Tikhonova N.A. Thermal Physics of High Temperatures. 1982. **20**. Part 1. pp. 872–880; Part.11. pp. 1085–1092 [in Russian].
44. Maurakh M.A., Mitin B.S. Liquid High-Melting Oxides. Metallurgiya, Moscow. 1979. 287pp [in Russian].
45. Osiko V.V. Laser Materials. Nauka, Moscow. 2002. p. 393 [in Russian].
46. Kulikov I.S. Thermal Dissociation of Compounds. Nauka, Moscow. 1966 [in Russian].
47. Sperrow E.M., Sess R.D. Teploobmen Izlucheniem. Energiya, Leningrad. 1971. p. 182 [in Russian].
48. Kulikov I.S. Thermodynamic Dissociation of Compounds. Metallurgiya, Moscow. 1966 [in Russian].
49. Katrich N.P., Budnikov A.T., Krivonogov S.I. et al. Functional Materials. 2006. **13(1)**. p. 44–53.
50. Verneuil A. Ann. Chem. Phys. 1904. **8**. p. 320.
51. Litvinov L.A. All about Ruby, Kharkov, Prapor. 1991. p. 150pp [in Russian].
52. Litvinov L.A. in: Book of Lectures Notes. Second International, School on Crystal Growth Technology, Japan. 2000. p. 666.
53. Litvinov L.A. Influence of Technology Factors and Apparatuses on Quality of Ruby. Institute of Chemical Engineering. Dr. Phil. (Eng.), Moskva, 1972 [in Russian].
54. Falckenberg R. J. Cryst. Growth. 1975. 29.
55. Adamski J.A. J. Appl. Phys. 1965. **36**. pp. 1784–1786.
56. Bauer W.H., Field W.G. The Art and Science of Growing Crystals. Wiley, New York. 1963. 398p.
57. USSR Authors' Cert. 500145, 1976.
58. Klassen-Neklyudova M.V., Bagdasarova Kh. S. Ruby and Sapphire. Nauka, Moscow. 1974. p. 235 [in Russian].
59. Kinloch D.R., Birchenall C.E. Ruby and Sapphire. 1973. **Bd.19**. pp. 105–108.
60. Holden F.A., Sedlacek R. Pribory dla Nauchn. Issled. 1963. **6**. pp. 8–9 [in Russian].
61. Bagdasarov Kh.S., Dyachenko V.V., Kevorkov A.M., Khokhlov A.A., in: Crystal Growth: Sci. Commun. Coll. Rost Kristallov. M., 1980. pp. 314–320 [in Russian].
62. Takagi K., Isai H. J. Mater. Sci. 1977. **12**. C.517–C.521.
63. Keck P.H. et al. Rev. Sci. Instr. 1954. **25**. p. 298.
64. Reed T.B. J. Appl. Phys. 1961. **32**. p. 2534.
65. Pfann W.G. Zone Melting. Wiley, New York. 1966.
66. Heywang W., Ziegler G. Zs. f. Naturforsch. 1954. **9a**. p. 561.
67. Heywang W. Zs. f. Naturforsch. 1956. **11a**. p. 238.
68. Class W., Nestor H.R., Murray G.T. Crystal Growth. Pergamon, New York. 1967. p. 75.
69. Timofeev N.I. et al.: Proc. of 1st Intern. Conf. on Markets of Glass, Single Crystals and Precious Metal Equip. 2001,. Veliky Novgorod. pp. 131–134 [in Russian].

70. Golubtsov I.V., Nesmeyanov A.N. Elektrotermia. 1969. **86**. pp. 41–42 [in Russian].
71. Kostopmarov D.V., Bagdasarov Kh.S., Kobzareva S.A. et al., in: Abstr. of XI-th Nat. Conf on Cryst.al Growth, Moskva. 2006. p. 217 [in Russian].
72. Kyropoulos S. Zs. Phys. 1930. **Bd.63**. S.849–S.854.
73. Musatov M.I. Heat-Resistant Dielectrics. Atomizdat, Moscow. 1980. pp. 117–118 [in Russian].
74. Musatov M.I.: Book of Lectures of 1st International. School on Crystal Growth Technolog, Switzerland 1998. p. 624.
75. Ivanov V.I. et al. Electronnaya Tekhnika., Ser Mater. 1978. **5**. p. 55 [in Russian].
76. Viechnicki D., Schmidt F. J. Cryst. Growth. 1971. **11**. pp. 345–350.
77. Harris D.C. Pros. SPIE, vol. 5078. Window and Dome Technologies **111**. 2003. p. 1.
78. Schmid F., Khattak C.P., Felt D.M. Am. Ceramic Soc. Bull. 1994. **73(2)**. p. 39.
79. Khattak C.P., Schmid F., Smith M.B. // Proc. SPIE, vol. 3060. Window and Dome Technologies and Materials V. 1997. p. 250.
80. Schmid F., Khattak C.P., Rogers H.H. et al. Proc. SPIE, vol. 3705. Window and Dome Technologies and Materials 1. 1999. p. 70.
81. Khattak C.P., Guggenheim P.J., Schmid F. Proc. SPIE, vol. 5078. Window and Dome Technologies 111. 2003. p. 47.
82. Guzik S., Oblakovsky Ya. Artificial Single Crystals. Metallurgiya, Moscow. 1975 [in Russian].
83. Bagdasarov Kh.S., Problems of Synthesis of Large High-Melting Optical Single Crystals. Part 2. In in book: 4th All-Union Confer. on Crystal Growing and Structure Erevan, . 1972. pp. 6–25 [in Russian].
84. Laverton W.F. J. Appl. Phys. 1959. **29**. pp. 1241–1244.
85. Aleksandrov V.I., Osiko V.V., Tatarintsev V.V. Pribory i Tekhn. Eksper. 1970. **5**. pp. 222–225 [in Russian].
86. Aleksandrov V.I., Osiko V.V., Prokhorov A.M., Tatarintsev V.M. Usp. Khim. 1978. **47**. pp. 385–427 [in Russian].
87. Osiko V.V. Laser Materials. Nauka, Moscow. 2002. 384 [in Russian].
88. Bridgman P.V. Proc. Am. Acad. Arts Sci. 1925. **60**. pp. 303–383.
89. Stockbarger D.C. Rev. Sci. Instr. 1936. **7**. p. 133.
90. Stockbarger D.C. Disc. Farad. Soc. 1949. **5**. pp. 294–299.
91. Tammann G. Metallography. Elsevier, New York. 1925. p. 26.
92. Obreimov N.V., Shubnikov A.V. Zs. Phys. 1924. **25**. p. 31.
93. Buckley H.E. Crystal Growth. Wiley, New York. 1951.
94. A. A. Chernov, E. I. Givargizov, Kh. S. Bagdasarov, et al. Modern Crystallography Moscow, Nauka. 1980. Vol.3. p. 407 [in Russian].
95. USSR Author's Cert. 276921, (1968); 566170 (1972); US.Pat. 4303465, (1981); Kh.S. Bagdasarov et al. Ruby and Sapphire // M.: Nauka. 1974. pp. 20–35 [in Russian].
96. Arutyunyan V.M. et al., in: Abstr. of IX-th Nat. Conf on Crystal. Growth M., 2000. p. 475 [in Russian].
97. Dubovskaya V.M., Egorycheva O.A., Smirnova S.A., in: Abstr. of X-th Nat. Conf on Crystal. Growth, Moscow., 2002. p. 222 [in Russian].
98. Ukrainian Patent. 18923 A, (1997).
99. Dan'ko A. Ya., Nizhankovsky S.V., Kanischev V.N. et al. *Functional Materials*. 2006. **13(3)**. pp. 426–431.
100. Puzikov V.M., Dan'ko A.Ya. et al. in: Semiconductor Physics, Quantum Electronics and Optoelectronics, Kiev. 2000. Vol.3, pp. 185–190 [in Russian].
101. Bessonova T.S., Stanislavsky M.P., . Khaimov-Malkov V.Ya., in: Problems of Atomic Science and Engineering, . 1974. pp. 74–80 [in Russian].
102. Dan'ko A.Ya., Sidelnikova N.S. et al. Functional Materials. 2001. **8**. p. 462.
103. Kvapil J., Perner B., Sulovsky J., Kvapil Jos. Kristall and Technik. 1973. **Bd.8**. S.247–S.251 [in Russian].
104. Dan'ko A.Ya., Sidelnikova N.S. et al. Functional Materials. 2003. 10.

105. Dan'ko A.Ya., Sidelnikova N.S. et al. Functional Materials. 2001. 8. p. 175.
106. Dan'ko A.Ya., Sidelnikova N.S. et al. in: Functional Materials for Science and Engineering. 2001. p. 200 [in Russian].
107. Ukrainian Patent. 18923A, (1997).
108. Stepanov A.V. Izv. AN SSSR, Ser. Fiz. 1969. **33**. pp. 1946–1953 [in Russian].
109. Stepanov A.V. Zh. Tekhn. Fiz. 1959. **29**. pp. 381–393 [in Russian].
110. Gomperz E.V. Zeitchr. fuer Phys. 1922. **Bd.8**. S.184.
111. Kapitza P. Proc. Roy. Soc. London, 1928. Ser.A. **119**. No. A782. p. 358.
112. Chalmers D., LaBelle H.E., Miavsky A.J. J. Cryst. Growth. 1972. **13/14**. p. 84.
113. US Patent. 3915662, (1975).
114. Schwuftke G.H. Phys. Status Solidi (a). 1977. **43**. p. 43.
115. Surek T., Coriel S.R., Chalmers B. Crystal Growth. Moscow, Nauka. 1980. Vol13. p. 180 [in Russian].
116. Goltsmann B.M. Growing of Semiconductor Single Crystals. 1963. **5**. p. 32 [in Russian].
117. Kurlov V.N., Belenko S.V. J. Cryst. Growth. 1998. **191**. p. 779.
118. Moskalyov A.V., Bakholdin S.I., Krymov V.M., Antonov P.I., in: Abstr. of IX Nat. Confer. on Cryst.al Growth M., 2000. p. 192 [in Russian].
119. Antonov P.I., Bakholdin S.I., Krymov V.M. Physics of Crystallization, M. Fizmatizd. 2002. P 213 [in Russian].
120. Shochet O., Kassner K., Ben-Jacob E. et al. Physica A. 1992. **187**. p. 87.
121. Goriletsky V.I., Grinyov B.V., Zaslavsky B.G. et al., Crystal Growth: Alkali Halides Kharkov, Akta. 2002. p. 22 [in Russian].
122. Ukrainian Patent. 36892A, (2001).
123. Litvinov L.A., Tkachenko S.A., in: Abstr. of IX Nat. Confer. on Cryst.al Growth Moskva, . 2000. p. 482 [in Russian].
124. Tatarchenko V.A. Stable Crystal Growth. Moscow, Nauka. 1988 [in Russian].
125. Dobrovinskaya E.R., Litvinov L.A., Pischik V.V. Izv AN SSSR, Ser. Fiz. 1976. **40**. pp. 1330–1331 [in Russian].
126. Dobrovinskaya E.R., Litvinov L.A., Pishchik V.V. Krist. und Tech. 1978. **Bd.13**. pp. 289–292 [in Russian].
127. La Bell H.E. Mat. Res. Bull. 1971. **6**. pp. 581–590.
128. Sehwutike G.H. Phys. Stat. Sol. (a). 1977. **43**. pp. 43–51.
129. Bell A.E. RSA Rev. 1977. **38**. pp. 109–138.
130. Bleil C.E. J. Cryst. Growth. 1969. **5**. pp. 99–104.
131. Groisman A.F., Zagumenny A.I., Provorotov M.V., Maier A.A. in: Abstr. of VI-th All-Union Conf. Cryst. Growth. Erevan. 1985.Vol.1. p. 206 [in Russian].
132. Epelbaum B., Fukuda T. Abstr. 12th Int. Conf. on Crystal Growth. Israel, 1998. p. 8.
133. Antonov P.I., Nosov Yu.G., Nikanorov S.P., Abstr. 12th Int. Conf. Crystal Growth. Israel, 1998.ibid. pp. 63–64.
134. Borodin V.A., Steriopolo T.A., Tatarchenko V.A. Proc. Europ. Meet. on Cryst.al Growth Mater. for Electr. Prague, . 1982. p. 320.
135. Borodin V.A., Steriopolo T.A., Tatarchenko V.A., Yalovets T.N. Izv. AN SSSR, Ser. Fiz. 1983. **47**. pp. 368–374 [in Russian].
136. Borodin V.A., Steriopolo T.A., Tatarchenko V.A. in: Proc. of All-Union Conf. on Production of Profiled Crystals and Workpieces by Stepanov Technique and Use Thereof in National Economy Leningrad., 1986. p. 72 [in Russian].
137. Dobrovinskaya E.R. et al. Pat.ent USSR. 499889, 1974,. Dobrovinskaya E.R. et al.
138. Nikolaev V.I. et al., in: Abstr. of IX Nat. Confer. on Crystal. Growth, Moskva. 2000. p. 24 [in Russian].
139. Litvinov L.A., Pishchik V.V., in: Proc. of All-Union Conf. on Production of Profiled Crystals and Workpieces by Stepanov Technique and Use Thereof in National Economy, Leningrad. 1986. p. 214 [in Russian].
140. Locher J.W., Bates H.E., Jones C.D., Zanella S.A. Proc. SPIE. 2005. 5786. pp. 147–153

141. Bates H. E., Jones C.D., Locher J.W. Proc. SPIE. 2005. 5786. pp. 165–174
142. Kurlov V.N. J. Cryst. Growth. 1999. **179**. p. 158.
143. Gurzhiyants P.A., Kurlov V.N., Rossolko S.N., in: Abstr. of IX Nat. Confer. on Crystal. Growth, Moskva. 2000. p. 26 [in Russian].
144. Kurlov V.N., Epelbaum B.M. J. Cryst. Growth. 1999. **179**. p. 175.
145. Lacombe P., Berhezan A Am. Soc. of Metals. 1951. p. 211.
146. Burke J.E. Met. Tech. T. P. 1948. p. 2472.
147. Burke J.E. Trans. AIME. 1949. **180**. p. 73.
148. Bragg W.L. Proc. Phys. Soc. London. 1940. **52**. p. 54.
149. Burgers J.M. Proc. Phys. Soc. London. 1940. **52**. p. 23.
150. Burgers W.G. Physica. 1934. **1**. p. 561.
151. Shockley W., Read W.T. Phys. Rev. 1950. **78**. p. 275.
152. Lucke K. Z. Metallk. 1961. **Bd.52**. S.1.
153. Stuwe H.P. Z. Metallk. 1961. **Bd.52**. S.34.
154. Dunn C.G., Daniels F.W. Trans. AIME. 1951. *191*. p. 147.
155. Dobrovinskaya E.R. et al. Patent USSR. 413710, 1972.
156. Dobrovinskaya E.R. et al. Patent USSR 499889, 1972.

Appendix

Table 4.11 Melting temperature of Al_2O_3 [44]

Sample	Impurity concentration (%)	Gaseous medium, pressure	Container material	T_m (K)
Single crystal	0.0224	Vacuum, $8 \cdot 10^{-5}$ mmHg	Tungsten	2,324
		Helium, 760 mmHg	Tungsten	2,314
		Argon, 760 mmHg	Tungsten	2,318
Single crystal	0.1561	Helium, 760 mmHg	Tungsten	2312.5
		Argon, 760 mmHg	Tungsten	2,318
		Air, 760 mmHg	Iridium	Liquidus: 2,311, solidus: 2,288
Single crystal	0.325	Vacuum, $7.5 \cdot 10^{-5}$ mmHg	Tungsten	2,324
		Vacuum, $3 \cdot 10^{-5}$ mmHg	Iridium	2318.5
		Helium, 760 mmHg	Tungsten	2,314
		Helium, 760 mmHg	Iridium	2,309
		Argon, 760 mmHg	Tungsten	2,318
Powder	0.0222	Vacuum, $7 \cdot 10^{-5}$ mmHg	Tungsten	2323.5
		Air, 760 mmHg	Iridium	Liquidus: 2,308, solidus: 2,299

Table 4.12 Temperature dependence of Al_2O_3 melt enthalpy (referred to 1 kg of the substance)

T (K)	H (MJ kg^{-1})	T (K)	H (MJ kg^{-1})
2,350	15.1	2,600	17.2
2,400	15.6	2,650	17.6
2,450	16.0	2,700	18.0
2,500	16.4	2,750	18.5
2,550	16.8	2,800	18.9

Table 4.13 Specific conductivity of Al_2O_3 melt at different temperatures

T (K)	χ (Ω cm^{-1})	T (K)	χ (Ω cm^{-1})	T (K)	χ (Ω cm^{-1})
2,342	2.69	2,554	5.60	2,747	8.63
2,383	3.12	2,559	5.45	2,835	10.13
2,408	3.36	2,579	6.03	2,891	10.75
2,465	4.06	2,598	6.20	2,935	1.52
2,512	4.77	2,681	7.42	2,991	12.51
2,537	5.02	2,697	7.69	3,008	12.66

Table 4.14 Main physical properties of Al_2O_3 melt

Parameter	Symbol	Dimensions	Value
Density	γ_m	g/cm^{-3}	3.03 ± 0.7
Surface tension	σ_s	N/m^{-1}	0.67
Marangoni number	M_a	–	11.5
Change of volume at melting		%	+19
Thermal expansion coefficient		K^{-1}	$3.56 \cdot 10^{-4}$
Kinematic viscosity coefficient		m^{2}s^{-1}	$1.8 \cdot 10^{-5}$
Dissociation energy of Al_2O_3 molecule		kcal/mol	740
Melting entropy	S_m	J/K^{-1}	~2.2
Electrical conductance at $T \sim T_m$	χ	Ω^{-1} cm^{-1}	0.71
Thermal conductivity	λ_T	Wm^{-1} K^{-1}	2.05 ± 0.15
Temperature conductivity	q	m^2/s	$(4.8 \pm 0.4) \cdot 10^{-7}$
Melting heat	Q	J/cm^{-3}	$3.2 \cdot 10^3$
Latent melting heat	L	J/cm^{-3}	4,264
Specific heat	c_p	J/g/K	1.26
Capillary constant	α_c	mm	6
Boiling temperature	T_{boil}	K	3253
Evaporation heat		kcal/mol	115.7

Table 4.15 Melt viscosity

T (K)	2,323	2,373	2,423	2,473	2,523	2,573	2,623
Viscosity (poise)	0.584	0.457	0.388	0.335	0.295	0.265	0.264

Table 4.16 Vapor pressure

T (K)	2,633	2,683	2,763	2,820	2,853
Vapor pressure (Torr)	6	18	22	50	53

Chapter 5
The Regularities of Structure Defect Formation at the Crystal Growing

The defects are formed during the crystal growth, the postgrowth treatment, or as a result of external factors. For example, the dendrites may arise only in the course of crystallization, while a shear occurs under straining. The criteria of the crystal structure perfection are the density of point defects, single dislocations and their accumulations, the presence of macroscale and microscale blocks, the impurity distribution homogeneity, and the presence of inclusions.

This chapter will present the main mechanisms of the structure defect formation in sapphire and will look at the possibilities of growing the crystals with a prespecified distribution of the structure defects.

5.1 Point Defects

Sapphire is a mainly ionic compound where the point defects typical of those in ionic crystals are formed, namely, anionic and cationic vacancies in various charge states, interstitial ions, ions in cationic and anionic sites, vacancy pairs and aggregates, and vacancy–impurity complexes. The ionic crystals of stoichiometric composition contain equivalent numbers of anionic and cationic vacancies (Schottky defects), i.e., there are three oxygen vacancies per two aluminum ones, or there is one interstitial ion of the same sign per each ionic vacancy (Frenkel defects). The Frenkel defects are formed within the entire crystal volume, while the Schottky ones arise mainly at the surface, block boundaries, and at dislocations with the subsequent diffusion into the bulk. The point defects are formed at the crystal growth, straining, and heat treatment. One of possible mechanisms of the Schottky defect formation can involve the trapping of a melt dissociation product into the growing crystal under absence of the complementary component at the crystallization front at the instant under consideration.

Under thermodynamic equilibrium, an equilibrium concentration of point defects is set in the crystal:

$$n / N \sim e^{E/kT} \qquad (5.1)$$

E.R. Dobrovinskaya et al., *Sapphire: Material, Manufacturing, Applications*,
DOI: 10.1007/978-0-387-85695-7_5,

where N is the total number of atoms per the crystal unit volume; n, the number of point defects in the same volume; E, the defect activation energy that is equal to its formation energy; and k the Boltzmann constant. The equilibrium concentration increases exponentially as the temperature rises.

In real conditions, the point defect concentration is not the equilibrium one, which is due to the reducing conditions during the growth and annealing, the partial "chilling" of vacancies during the cooling, and the action of ionizing radiation. An electromagnetic irradiation cannot give rise to new crystal lattice defects but is only able to change the charge state of the preirradiation defects. The growth and annealing atmosphere is a factor of primary importance defining the point defect appearance/disappearance. The oxygen partial pressure defines the point defect stoichiometry and the atmosphere-oxidizing potential–the charges thereof.

The appearance and disappearance of point defects is associated with the atomic migration. Such defects have a high activation energy as a rule. The cooling slows down the migration processes, although it does not chill completely the high-temperature state. This is because the cooling rate is specified, thus the chilling efficiency depends not only on the diffusion coefficient of the atoms but also on the free path length thereof. When the equilibrium conditions require a lowered concentration of atomic defects, then the defect associates are formed that are precipitated at dislocations and partly annihilated. In such a case, it is more difficult to chill the defects because the diffusion path is much shorter. In conditions far from equilibrium, the crystal lattice may become supersaturated in vacancies; the latter becomes combined and forms pores and even "negative crystals" under some conditions.

Anionic and cationic vacancies. It can be considered to be established [1–3] that the F^+ centers (one electron in the field of an anionic vacancy) are characterized by the main absorption bands at 4.8 and 4 eV that correspond to the IA→IB and IA→2A transitions. The reverse transition IB→IA corresponds to the emission band at 3.8 eV with the decay time $< 7 \cdot 10^{-9}$ s. The intense absorption band at 6.1 eV (206 nm) in irradiated crystals is believed to be related with F centers (two electrons in the field of an anionic vacancy). The lifetime of the F center triplet excited state is longer (about $4 \cdot 10^{-2}$ s).

The F^+/F-center ratio depends on the impurity composition and the crystallization (annealing) conditions. The energy states of those centers are schematized in Figs. 5.1 and 5.2.

Sapphire is grown in oxidizing conditions (Verneuil technique), reducing conditions (98% Ar + H_2 or a carbon-containing medium), as well as in vacuum. The presence of hydrogen favors the formation of oxygen vacancies $[V_O]$ and F centers:

$$2Al^{3+} + 3O^{2-} + H_2 = Al_2^{3+} + 2O^{2-} + (V_O + e) + H_2O \tag{5.2}$$

The easily ionizable argon atoms favor the electron transfer to oxygen vacancies and the ionized hydrogen is combined with oxygen according to the scheme [4]

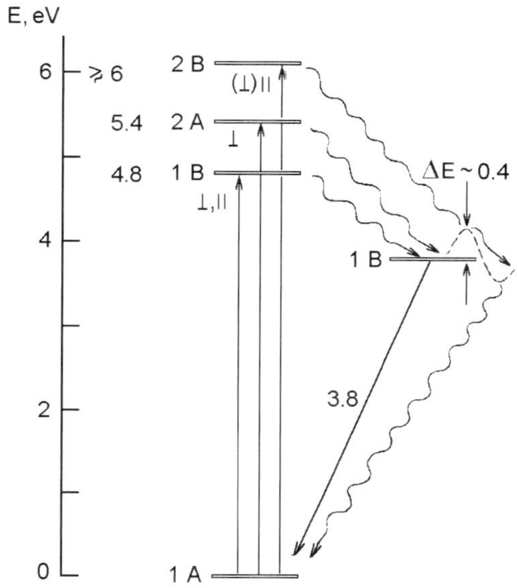

Fig. 5.1 Scheme of energy states for F^+-center [3]. (‖,⊥) The direction of the light absorption electrical vector with respect to the C_3-axis

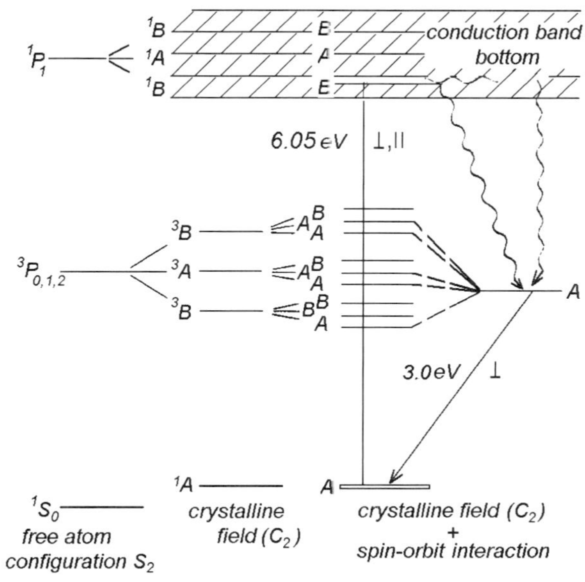

Fig. 5.2 Scheme of energy states for F-center

$$Ar_2^{2e^-} + V_O^{2+} = F\text{-center} + 2Ar^+ \quad (5.3)$$

$$O^{2-} + 2Ar^+ + H_2 = 2Ar + H_2O \quad (5.4)$$

In a graphite furnace, hydrogen also is formed at a pressure about 100 N/m^2 due to carbon interaction with water vapor:

$$C + H_2O = CO + H_2 \quad (5.5)$$

The concentration of anionic vacancies in the crystals grown in carbon-containing atmosphere increases from the crystal nose to its tail due to the increasing extent of anionic-type perturbation of the melt stoichiometry in the course of crystallization. Their number in the crystal was determined from the Smacula formula and the optical absorption intensity at 206, 225, and 255 nm as a total concentration of F and F^+ centers [5]:

$$[V_O] = [F] + [F^+] = (1.86\alpha_{206} + 0.292\alpha_{225} + 0.167\alpha_{255}) \cdot 10^{16} = 2 \cdot 10^{17}\ \text{cm}^{-3}, \quad (5.6)$$

where α_{206}, α_{225}, and α_{255} are the sapphire optical absorption coefficients at the 206, 225, and 255 nm wavelength, respectively. Vacuum is also a weak reducing atmosphere promoting the oxygen deficiency in the crystal.

Under thermodynamic equilibrium conditions, the Schottky defect formation is more favorable in energy than that of the Frenkel one both in anionic and cationic sublattices [6]. It is just the single anionic vacancy that requires the lowest energy (3.5 eV). In ref. [7], isotherms of the defect concentration are presented as functions of the oxygen partial pressure for the cases of the Schottky and Frenkel defect formation.

The penetration of transition elements (Ni, Fe) in trivalent state into the oxide lattice is related to V_O formation when the sapphire is grown by Verneuil technique at the oxygen partial pressure in the flame exceeding that in the Al_2O_3 thermal dissociation products. In the crystals annealed in reducing atmosphere, V^{2+} and V^- centers (one or two holes, respectively, localized at O^{2-} ions adjacent to a cationic vacancy) were revealed after irradiation using the EPR method [8, 9]. The optical absorption of those centers is at about 3.0 eV. The possibility of their formation under low-energy ionizing irradiation is evidence of the existence of single cationic vacancies in the nonirradiated crystals; those vacancies show no absorption bands and are not paramagnetic. In the crystals, especially in those grown by hydrothermal and Verneuil techniques, hydroxyl groups have been found that can be adjacent to cationic vacancies. When the holes become trapped by the cationic vacancies, V_{OH}-centers are formed that are very close to the V^{2+} and V^- centers in optical properties but are decayed at lower temperatures and release the holes. The holes are localized at O^{2-} ions adjacent to impurity ions and have a charge lower than that of Al^{3+}, which is substituted by the impurity. Such centers absorb in the region of 2.5 eV [8].

The cationic vacancies are formed at the annealing in oxidizing atmosphere at $T > 1,600$ K or at doping with Ti^{3+} ions followed by the oxidizing annealing. In this case, the cationic vacancy concentration increases, because it is necessary to compensate the arising positive charge of Ti^{4+}. At low Ti^{4+} concentrations, anionic vacancies contributing to the charge compensation arise along with the cationic ones; the anionic vacancies form the Ti^{4+}–V_{Al}^{3-}–V_O^{2+} complexes. The migration activation energy for cationic vacancies is 80 kcal/mol [10]. The oxygen vacancy diffusion coefficient is $D = 6.4 \cdot 10^{-6}$ cm^2/s at 2,250 K [11].

Interstitial ions. The point defects include impurity atoms located in the sites and interstices of the crystal lattice. The unoccupied octahedral voids in the lattice favor the formation of interstitial centers. The impurity atoms tend to occupy the sites that are more similarto the chemical nature of the matrix atoms. The position of the impurity atoms is defined by their penetration energy. It is just the small atoms that are located in the interstices as a rule, the introduction of larger atoms being possible only at defects of dislocation and block boundary types.

The impurity defects, as well as the intrinsic ones, may become ionized due to the charge exchange processes. When a matrix ion is substituted with an isomorphic impurity one, the electric neutrality conditions remain almost unchanged. At a nonisovalent substitution, the electric neutrality principle is conserved due to formation of intrinsic point defects, the valence change of the impurity atom or of the appropriate matrix atoms. The charge can be compensated also in the case when the total charge of the substituted ions is equal to that of substituting ones.

Comparison of experimental data on the density of sapphire containing Ca, Mg, Si, V, Cr, and Ti with theoretical density for all the possible compensation mechanisms has shown that introduction of Si, Ti, and V ions, i.e., tetravalent ones or those tending to the tetravalent state, is accompanied by formation of cationic vacancies. Chromium that enters isomorphically does not cause the formation of point defects. The stabilization of Cr^{4+} valent state is possible in the presence of Mg^{2+} as well as at a partial replacement of O^{2-} with nitrogen ions.

The introduction of unstructured oxygen into an interstice at annealing in an oxidizing atmosphere [12] seems to be hardly probable, since the oxygen ion radius exceeds considerably the interstice size. The intrinsic ions, however, can be displaced into interstices under irradiation. In the crystals irradiated with neutrons and then with γ-rays at 77 K, centers with 3/2 spin may be formed having the third-order axial symmetry along the C_3 axis. Those centers have been identified as O^+ ions in octahedral interstices [13, 14]. The heating up to about 180 K causes the center transition into another charge state, presumably into O^0.

The cationic interstitial centers in neutron-irradiated crystals were observed as Al_i^{5+} quasimolecules formed by an interstitial Al_i^{2+} in combination with a regular Al^{3+} being its neighbor along the C_3 axis [15, 16]. Based on EPR spectra and thermal stability, such centers can be subdivided into three types, namely:

- Axial (along the C_3 axis), the lifetime at 300 K is few seconds
- Axial, the lifetime at 300 K is several days
- Inclined 9° to C axis and stable at 300 K

At doses of the order of 10^{17} neutrons/cm^2, the concentration of those centers attains saturation and does not exceed 10^{17} cm^{-3}. That is why the observed centers are assumed to be related with some impurities. According to ref. [17], the interstitial cations are mobile at room temperatures while interstitial O+ ions are immobile at least up to 340 K.

The isovalent impurities (Cr^{3+}, Ti^{3+}, Co^{3+}, Ni^{3+}, Fe^{3+}, Mn^{3+}, V^{3+}, etc.) are the trapping centers for electrons and holes causing the appearance of new absorption bands under ionizing irradiation. The heterovalent impurities (Cr^{4+}, Ti^{4+}, Si^{4+}, Mg^{2+}, Li^{+}, N^{3-}) also become the trapping centers. During the growth, those form cationic vacancies that compensate their charge; the holes can be localized near the vacancies. Like the case of isovalent impurities, such centers cause new absorption bands under irradiation. Mg^{2+} does not change its charge state when entering the lattice; therefore, the negative charge compensation is necessary for its introduction. The charge can be compensated by impurities having the effective positive charge (e.g., Si^{4+}, Ti^{4+}, Cr^{4+}) or anionic vacancies. The formation of intrinsic defects requires a higher energy than the change of the compensating impurity charge state. Hydrogen that is absorbed in the melt from the growth atmosphere and introduced into the lattice acts as a charge compensator, too. Under low-temperature X-ray irradiation, F_{Mg} centers can be formed when Mg^{2+} substitutes for Al^{3+} in A or B position (Fig. 5.3) or in C or D one. The center is responsible for luminescence at 4.0 eV [18]. Along with electron F_{Mg} centers, the hole $[Mg]^0$ centers are formed under X-ray irradiation. Those are decayed at 225 K releasing holes. In the crystals annealed in oxidizing atmosphere, the $[Mg]^0$ centers are stable at least up to 300 K.

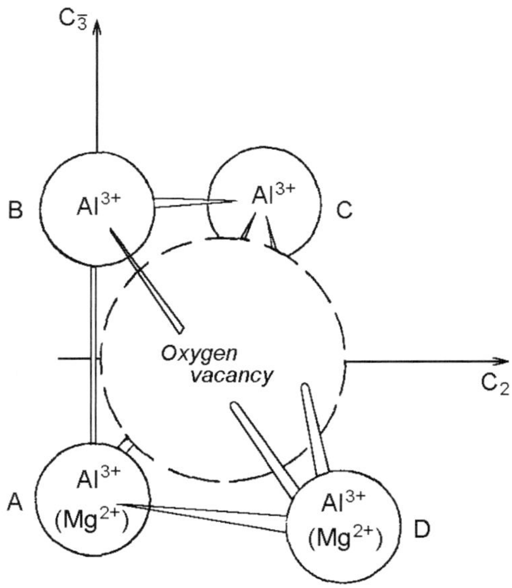

Fig. 5.3 Schematic model of F_{Mg} center

Ions in foreign sites. EPR spectra demonstrate the existence of $[AlO]^{3+}$ quasi-molecules located at two adjacent anionic sites [13]. The quasimolecule can be considered to be the result of interaction between two donor- type defects, namely, an anionic vacancy and an interstitial aluminum. The center can be imagined as Al^{3+} stabilized in the anionic site due to the trapping of four additional electrons. Such centers are expected to absorb near 2.2 eV. (Similar centers have been found in alkali halide crystals and MgO.)

5.2 Dislocations

The following mechanisms of dislocation formation at growing crystals from melt have been proposed [19, 20]: intergrowth of dislocations from the seed crystal to the growing one, i.e., inheritance of the seed defects as well as dislocations originating in the growth process; plastic deformation under the effect of thermoelastic stresses; trapping of impurity by the growing crystal, i.e., an impurity mechanism; disklike aggregation of vacancies and their consequent bang with the formation of dislocation loops, i.e., a vacancy mechanism; and incoherent coalescence under conditions of not full compliance of the nuclei, growth layers, and branches of dendrites.

Let us next consider the action of each mechanism of origination of dislocations during crystal growth in view of the problem of growing crystals with a preset law of distribution of structure defects.

Inheritance of dislocations. If the seed crystal contains dislocation lines that cross the interface, these dislocations during growth will intergrow in the newly formed layers of the solid phase [21–23]. It has been shown that provided crystals are grown under conditions under which new dislocations do not appear, they include those defects that were inherited from the seed crystals.

Considering the possibility of controlling the process of inheritance of the seed defects and those emerging in the growth process, it is expedient to introduce the concept of the coefficient of inheritance. It is known that sapphire is plastic in a narrow temperature interval and the process of growth is conducted at rather high temperature gradients. Owing to this the plasticity zone behind the CF h has a small length. Provided a crystal grows under conditions at which new dislocations are not formed and only the available dislocations are inherited, the ratio of the number of the inherited dislocations after the CF h has advanced the width of the plastic deformation zone (ρ_1) to the number of dislocations at the crystallization zone before this advancing ρ_0 we shall call the coefficient of inheritance of dislocations K. At a constant area of the cross section this ratio is equal to the ratio of the dislocations' densities:

$$K_1 = \rho_1 / \rho_0 = \rho_2 / \rho_1 \tag{5.6}$$

Provided the value K remains constant during the growth process the density of dislocations in the cross section normal to the growth direction at a random distance l from the seed will be

$$\rho_0 = \rho_0 K^{l/h} \tag{5.7}$$

It is assumed here that dislocations in this cross section are distributed uniformly.

For the determination of K that are conventionally grown to have a cylindrical or rectangular form, it is presumed that the plane in which the dislocations lie is situated at an angle θ to the plane of the cross section.

In growing crystals of a round cross section to the length equal to the width of the plastic deformation zone the dislocation lines crossing the surface S will go out to the side surface of the crystal (Fig. 5.4). The dislocation lines passing through the rest area of the cross section $S_0 - S$ will be distributed uniformly over all cross section of the crystal as a result of possible dislocation creep. Consequently,

$$K = 1 - S / S_0 \tag{5.8}$$

For the rectangular shape crystal $K = 1 - \delta/d$, where $\delta = h/\mathrm{tg}\theta$. For the cylindrical crystal

$$K = 1 - 2\left[(\arcsin \delta / d) / \pi - 2\delta(\sqrt{d^2 - \delta}) / \pi d^2\right] \tag{5.9}$$

When $\theta = \pi/2$, $\delta = 0$, and $K = 1$, the dislocation lines are parallel to the growth direction and all dislocations are inherited in the growth process. If $\theta = 0$, $\delta = d$, and $K = 0$, the dislocation lines are normal to the growth direction and fray out in the zone of plastic deformation. In the definition of K, such conditions were assumed under which dislocations are not formed. Under ordinary crystal growth conditions formation of new dislocations takes place in the zone of plastic deformation.

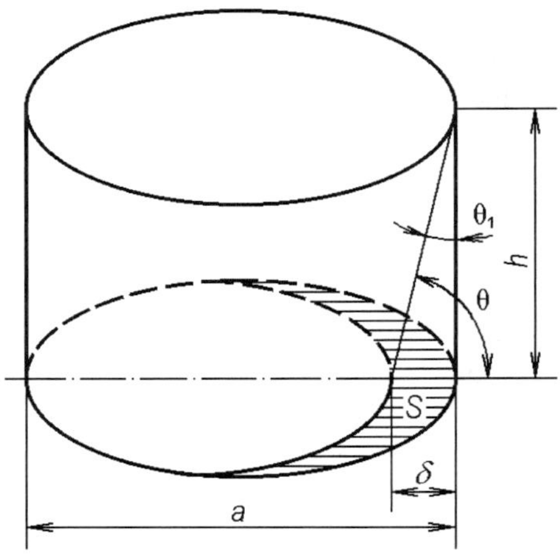

Fig. 5.4 A diagram for the determination of the coefficients of inheritance in cylindrical crystals

Let density of dislocations formed in the zone of plastic deformation be ρ_2. As the crystal grows, r in each next zone is equal to the ρ that originated just in this zone and the dislocations that were inherited from the previous zones. Provided the zone of plastic deformation is assumed to be narrow and l/h is the integer, the density of dislocations at an arbitrary distance l from the seed will be

$$\rho = \rho_0 K^{l/h} + \rho_2[(1-K^{l/h})/1-K] \tag{5.10}$$

The first term of the equation allows for the inheritance of defects from the seed and the second for the inheritance of dislocations that originated during the growth process. Figure 5.5 shows the value $(1 - K^{l/h})/1 - K$ as a function of the ratio l/h for different values of K. At $K = 0$, density of the dislocations is constant over the length and is equal to ρ_2.Unlike the considered case in which it is assumed that dislocations creep and are uniformly distributed over the cross section of the crystals, the variant at which creep of the dislocations in the growth process is insignificant is also possible. In this case such dislocations are inherited and frayed out depending on the angle θ_1 between the dislocation line and growth direction. If $\theta_1 = \pi/2$, no inheritance occurs ($K = 0$); at $0 < \theta_1 < \pi/2$, dislocations fray out. A dislocation will go out to the side surface when growing the crystal for a length $l = \delta \operatorname{ctg} \theta_1$, where δ is the distance from the dislocation at the CF to the side surface to which the dislocation goes out.

There are three basic methods of controlling the dislocation structure of crystals. The first method is to decrease the surface area through which the dislocations intergrow to the crystal. This can be achieved, for instance, by constrictions used. However, unlike Si-crystals, application of constrictions when growing sapphire by

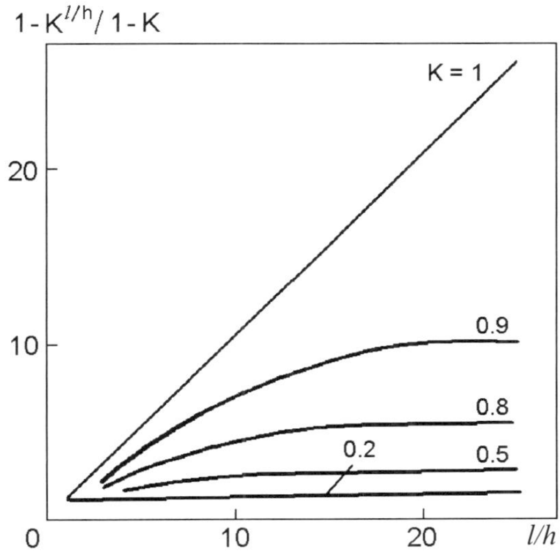

Fig. 5.5 The value $(1-K^{l/h})/(1-K)$ as a function of the dimensionless length of the crystal

the Verneuil method is not efficient, since it is difficult to make small cross-section constrictions. In the places of constriction, density of defects decreases, but in the broadening part of the crystal the density rises again, reaching the mean value. Even less efficient is variation of the cross section of sapphire when other methods are used. A significant reduction of the cross section can be achieved by using needle-like seeds when growing by the Verneuil method [24]. Structure quality of crystals grown on the needlelike seeds in a near-seed region at the length 50–70 mm is higher than those grown on the usual seeds.

The second method consists of the variation of density of defects in the seed [25]. Growth of crystals on the seeds essentially with different initial density of dislocations shows that the difference in the value ρ specified by the seed crystal is preserved on both planes (0001) and $(11\bar{2}0)$, irrespective of the growth method (Fig. 5.6).

The third, and most efficient, method is the selection of a spatial orientation that allows variation of the value K [26]. Proceeding from the known gliding systems and types of dislocations in crystals, the following four orientations have been proposed: (1) growth direction [0001], side surfaces $(11\bar{2}0)$, and $(10\bar{1}0)$; (2) $[11\bar{2}0]$; (3) $[10\bar{1}0]$ – the plane (0001) is parallel to the growth direction; and (4) $[10\bar{1}0]$. Let us consider distribution of dislocations in crystals of these orientations and a possibility of their inheritance in the growth process.

Orientation [0001] is of the most interest, since in such crystals dislocations of the lightest gliding systems (0001) $1/3\langle 11\bar{2}0\rangle$, $\langle 10\bar{1}0\rangle$ are normal to the growth direction, and for them, $K = 0$. But for sapphire this orientation is the most difficult to grow by the Verneuil, HDS, Kyropoules, and Czochralski methods.

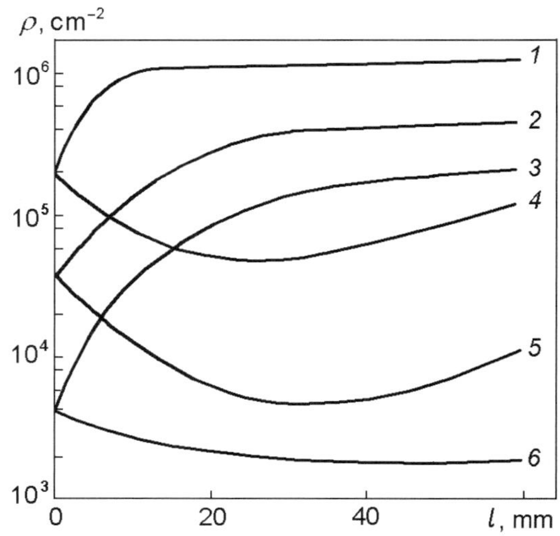

Fig. 5.6 Distribution of density of dislocations over the crystal length on the plane (0001) depending on the quality of the seed crystal: (*1–3*) the Verneuil method; (*4–6*) the HDS method

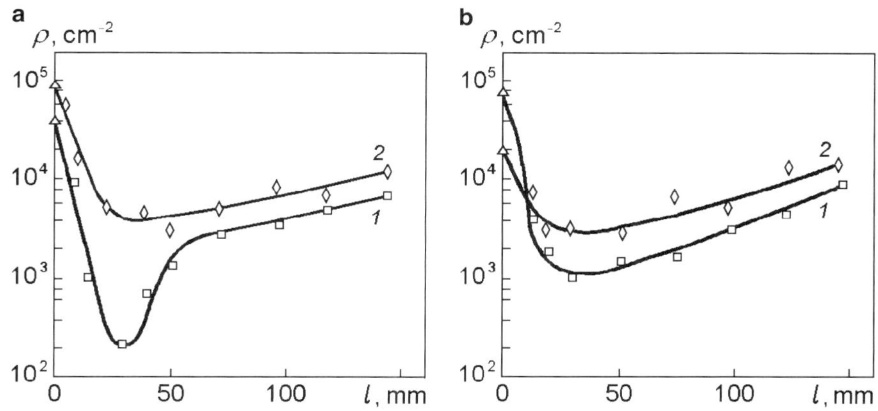

Fig. 5.7 Distribution of density of dislocations over the length of the crystal grown by the HDS method with the orientations: (**a**) 2; (**b**) 3; (*1*) on the plane (0001); (*2*) on the plane $(11\bar{2}0)$

Figure 5.7 shows the distribution of densities of dislocations over the length of the crystals that have the orientations $[11\bar{2}0]$ and $[10\bar{1}0]$. It has been shown that ρ on the plane (0001) dramatically drops compared with that of the seed crystal, since the dislocation lines that go out to this plane lie in the planes of the prism (the lightest gliding systems right after the base); they are normal to the growth direction ($K = 0$) and are not inherited by the growing crystal.

If the plane $(11\bar{2}0)$ is parallel to the growth direction (orientation $[10\bar{1}0]$), the dislocation lines normal to it are not inherited, while the lines going out at a 60° angle are inherited partly. Crystals with orientation $[10\bar{1}1]$ are the most faulty, since it is impossible to avoid inheritance of both seed defects and those originating in the growth process. An evident illustration of the role of spatial orientation in controlling the coefficient of inheritance of dislocations can be investigations of dislocation structures of crystals with orientation $[10\bar{1}0]$ using X-ray topography [27]. The results of the study of a dislocation structure of a block-free few-dislocation crystal grown by HDSM testify to the fact that the dislocations are settled mainly at the angle 90° to the growth direction and the coefficient of their inheritance tends to zero. Only some dislocations oriented along the growth direction can be observed.

Plastic deformation under the effect of thermoelastic stresses. Crystals are grown from melt in the field of temperature gradients, which leads to the appearance of thermoelastic stresses. These stresses, equal to the critical stress of the dislocations formation or exceeding it, evoke action of one of the main mechanisms of dislocation origination – plastic deformation behind the CF. Owing to the fact that thermoelastic stresses are defined by a nonequilibrium distribution of temperature in the growing crystal the problem of temperature fields in crystals is the first to be solved. The investigations usually were carried out experimentally by growing thermopairs [24] and theoretically by analytical or numerical solutions of the heat-conduction equation under different conditions of heat exchange at the boundaries.

However, ingrowth of thermopairs in the crystal does not give a full picture of the temperature field. Solution of the heat-conduction equation using PC is rather promising but labor consuming and feasible for a limited number of materials.

Thus, of the three above-mentioned methods of controlling the inheritance of defects the third one is the most efficient.

All complicated manipulations with the seed crystal turn out to be inexpedient. Selection of spatial orientation can reduce attention to the deterioration of the seed crystal structure quality at seeding. For example, when growing crystals by the HDS and Verneuil methods, density of defects in the seed crystal can increase sharply due to the thermal shock. Nevertheless, high structure quality corundum crystals can be grown on such faulty seeds (Fig. 5.8, curves 3′ and 4′). This is provided by the selection of the corresponding spatial orientation, which essentially decreases inheritance of dislocations and brings the minimum action of the basic gliding systems, i.e., plastic deformation behind the CF, which will be discussed below.

Let us now consider the formation of stresses in the crystals grown by various techniques in more detail. Two kinds of residual stresses are distinguished. The residual stresses of the first kind are caused by the temperature field nonlinearity. Those of the second kind are due to defects of the real structure, namely, the point defect accumulations, dislocations, block boundaries, impurities, and foreign inclusions.

The central parts of the crystals grown by the Verneuil technique are subjected to compressing stresses, while the peripheral ones are subjected to tension stresses

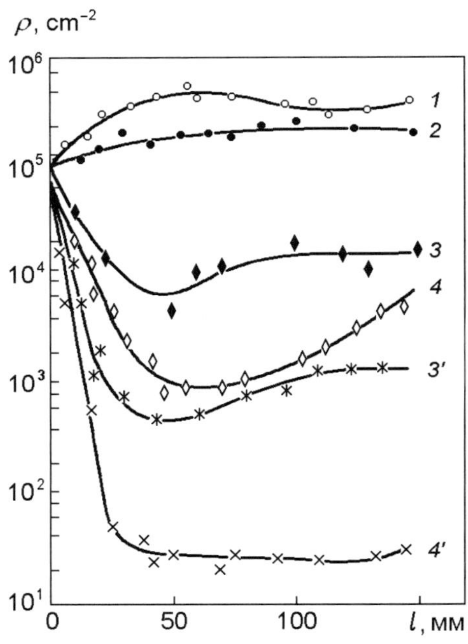

Fig. 5.8 Distribution of dislocation density over the length of crystals grown by the methods of: (*1*) Verneuil; (*2*) Stepanov; (*3, 3′*) Czochralski; (*4, 4′*) HDS; plane (0001)

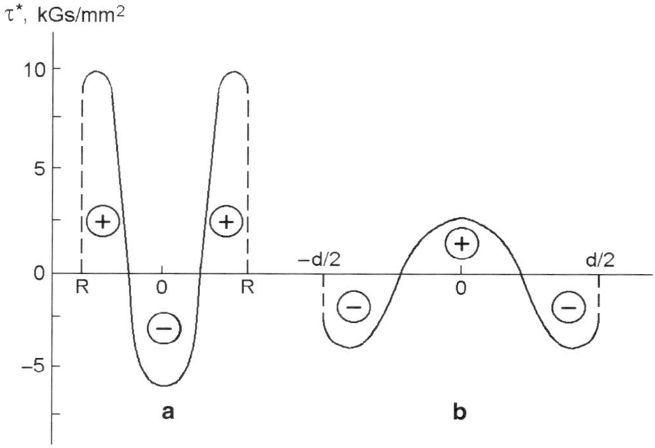

Fig. 5.9 A typical τ* stress distribution in a cylindrical crystal of radius R grown by Verneuil technique (**a**) and in a crystal plate of thickness d grown by HDS method (**b**). The compression areas are indicated by "+," the tension areas by "−"

(Fig. 5.9). In contrast, in the crystals obtained by HDC method, there are compressing stresses along the plate sides and the tension ones in the middle part, i.e., the stresses in such crystals are similar to the cooling ones. As the cooling rate increases, the residual stresses rise, and the crystal may fail at a certain critical value of the temperature gradient.

According to works by Indenbom, a complex temperature distribution along the crystal causes inhomogeneous thermal expansion of its parts and their elastic interactions, and as a result the stresses appear in the crystal even if its outer surface is free. Let us consider at first the action of such temperature-induced stresses in an already-formed dislocation-free crystal, i.e., let its quenching be considered in essence. If there are no points in the whole crystal volume where the shear stress does not exceed the critical one required for dislocation nucleation or the growth of dislocation microloops, then the crystal remains dislocation-free at the macroscale. The stress values near inclusions may exceed the average ones by 10%. Therefore, such stress concentrators are the most probable regions where dislocations may arise. If the critical stress is attained somewhere, the nucleated dislocation loops will start to extend, the dislocation multiplication will set in, and the plastic strain will extend over the whole crystal. Only the regions where the stresses are insufficient to move the dislocation cluster front or even individual dislocations (i.e., where those are lower than the crystal elasticity limit) remain dislocation free.

The plastic flow is associated not only with the dislocation sliding but includes also the viscous flow processes with the dislocation creep and transfer of vacancies and interstitial atoms. At submelting temperatures, the plastic straining and diffusion of point defects proceed under much lower stresses. Therefore, the shape and size of the crystal region where the dislocations are nucleated and extended and the point defects are redistributed depend on the thermally induced stresses and temperature in that region. The plastic flow tends to redistribute the crystal material in such a way that the stresses causing the flow will be relaxed. It will be over when the total energy of thermoelastic stresses and dislocations will attain a minimum. An equilibrium between the elastic energy and dislocation energy means that the stresses are not relaxed completely.

Now let the crystal be cooled down to a temperature that is constant over the entire volume and sufficiently low for plastic strain. Then the regions that were compressed prior to the plastic strain become stretched, and vice versa. The residual internal stresses arising in the crystal in this case are similar in values to the initial thermal ones that caused the plastic strain, but have the opposite sign. The sources of residual internal stresses are the residual strains, i.e., distribution of dislocations and point defects that has been settled as a result of flow.

The formation of dislocations at the growing crystal surface is due to the trend to minimization of the elastic energy. In this case, the material does not eliminate the stresses but it accumulates in the unstressed state. In the course of the process, the crystal moves with respect to the furnace temperature field. Each element of the surface layer becomes three dimensional and a plastic strain proceeds, thus resulting in relaxation of stresses that might exist in the corresponding point when it is within the crystal bulk. Finally, the element under consideration attains the temperature zone where there is no plasticity and any straining is impossible. Thus, the grown crystal in each volume element will show a defect structure formed at the moment it leaves the plasticity zone. After cooling down to a constant temperature, the defects chilled in such a manner result in formation of stresses equal in value and opposite in sign to those that were relaxed by the defects during the growth. If the material is plastic only within a narrow temperature range around T_m, as is the sapphire, the plasticity boundary will go in parallel to the CF at a small distance therefrom. As a result, the crystal will contain the chilled-in stresses that were at the CF (with opposite sign). The stresses at the front, in turn, are defined by the temperature field within the crystal bulk. That field is very complex. For example, when a cylindrical crystal is grown by the Czochralski technique, a radial and axial thermal flow are moved within the crystal. The first flow is associated with cooling through the side surface while the second is associated with the heat flow from the melt to the cooling holder through the crystal. The radial flow is associated with the cylinder cooling, while the axial flow is associated with the growth process itself. It is just the CF curvature that is a measure of the relative intensity of those heat flows. When the CF is convex or concave appreciably, the radial gradients have opposite signs and are comparable to the axial gradient, although considerably less so than the latter. At the flat CF, the axial gradient is the only one that exists at the front (of course, if the latter coincides with the isotherm).

Thus, the determination of the residual stresses in the crystal is equivalent to determination of thermoelastic stresses in the course of growth and cooling. Despite numerous experimental and theoretical works dedicated to the study of such fields in the growing crystal, the information on the correlation between thermal conditions and dislocation structure is insufficient [28]. Therefore, it is impossible at present to derive a relation that would precisely describe the relation of the crystal structure with crystallization conditions. In some cases it is assumed that density of dislocations is defined by radial temperature gradients T'_r:

$$\rho = (\alpha / b)T'_r \tag{5.11}$$

where α is a coefficient. In other cases the main role is played by axial temperature gradients, since usually they significantly exceed the radial ones [28]:

$$\rho = (\alpha / b)T'_n \tag{5.12}$$

Conclusions of different authors about the relative role of T'_r and T'_n. are not in conformity. But in reality the dislocation structure of the growing crystal is defined by thermal stress and temperature field over all the plasticity zone [29]. The role of T'_r and T'_n can change depending on a disposition of the region of dislocation generation with respect to the CF. If the crystal is characterized by a wide plasticity zone and is plastic at a distance longer than the diameter, the basic role can be played by a radial heat flow that is formed at the side surface of the crystal cooling. If crystals have a narrow plasticity zone (as, for example, sapphire) the dislocations and residual stresses are formed in a narrow zone adjacent to the CF and are mainly defined by the axial temperature drop. Evaluations made for sapphire show that the value ρ determined by the formula (5.6) turns out to be understated [24].

Since the T'_n is connected with thermal expansion of the lattice and the distance between the atomic planes becomes longer as it approaches the CF, it is assumed that the stress and bend of the lattice are completely removed by dislocations. Given below is an expression combining the density of single dislocations with the temperature gradient at the CF with allowance for the temperature bent of the lattice [30]:

$$\rho \approx \alpha T' + r_{\mathrm{curv}} \Delta\beta, \tag{5.13}$$

where r_{curv} is the lattice curvature; β is the elastic bulk expansion of the crystal.

An attempt to define more exactly such relations was made in refs. [31, 32], but here it was assumed that the free path of dislocations is commensurate with the diameter of the crystal (d) or cross section size of macroblocks:

$$\rho = (\alpha / b)\, T'_n - [2\tau^*_{cr} / Ebd] \tag{5.14}$$

where τ^*_{cr} is the critical stress in the plane of the CF in the direction the Burgers vector.

The length of the free path of dislocations is not large. For instance, during growth of sapphire from melt, it is about 500 μm [24], and for the case of large deformations the free path length determined experimentally does not exceed 50 μm.

Of interest is a macroscopic theory of stress and dislocation generation during crystal growth described in refs. [28–30]. It presumes that the residual stresses are equal to the scattered thermoelastic stresses taken with the opposite sign. Provided that the T'_r can be neglected, and here a linear axial distribution is observed at a distance equal to the crystal diameter, no stresses appear in the growing crystal. A number of relations have been suggested to determine the values of thermoelastic stresses. Thus, for a plate having the width *d*, pulled along the z axis in the temperature field $T = T(z)$, the stresses can be expressed as [33, 34]

$$\tau^*_{xx} = \alpha E \int_0^\infty A(z/d) T'(z) \mathrm{d}z, \tag{5.15}$$

where *A* is the influence function that defines the contribution of an individual temperature jump that occurs at a distance *z* from the CF.

The function $A = 1$ at $z = 0$ drops to zero at $z \approx 0.02$, changes the sign, passes the minimum, and quickly tends to zero, so that the stresses on the plate end depend on temperature distribution at distances close to the cross section size *d* of the sample.

The function of influence also can be introduced for the second derivative

$$\tau^*_{xx} = \alpha E \int_0^\infty \Phi(z/d) T''(z) \mathrm{d}z, \tag{5.16}$$

where Φ increases from zero to the maximum value $0.018d$ at $z = 2.2d$ and drops to zero not changing the sign, $\Phi = \int_0^z A \,\mathrm{d}z$.

The area of the plot $\Phi(z/d)$ is about $0.04\mathrm{d}^2$ and at a parabolic distribution of temperatures τ^* can be found by the second derivative from temperature by the coordinate [35]:

$$\tau^* = 0.04 \alpha E d^2 T'' \tag{5.17}$$

However, the estimation of the value τ^* made by this formula for crystals turns out to be insufficient, since T'' is not preserved constantly under the usual conditions and the relation (5.12) gives a better conformity with the experiment; this will be discussed below.

It is expected that at deformation $\varepsilon_\mathrm{d} = \tau^*/E \approx 10^{-3}$ under growth conditions the stresses can destroy the crystal, at $\varepsilon_\mathrm{d} \approx 10^{-4}$ plastic flow appears, and at $\varepsilon_\mathrm{d} \approx 10^{-5}$ growth of dislocation free crystals is provided [34].

Let us consider how to control the dislocation structure of sapphire by taking into account action of the mechanism of plastic deformation behind the CF. The first way of control is variation of axial temperature gradients.

Irrespective of the growth methods the value of ρ increases as the T'_n becomes higher (Fig. 5.10). While growth by the HDS method a direct dependence between ρ and T'_n is observed, for the crystals grown by the Verneuil method such dependence is less pronounced. The difference in the behavior of the curves is explained not by the gradient but by different degrees of its stability, i.e., by the value of the second derivative from temperature by the coordinate.

The role of the temperature gradient constancy is confirmed by the experimental fact [24]. If in the case of the HDS method a temperature field was formed in which the temperature gradient at the CF varied by 3.5 times during the growth

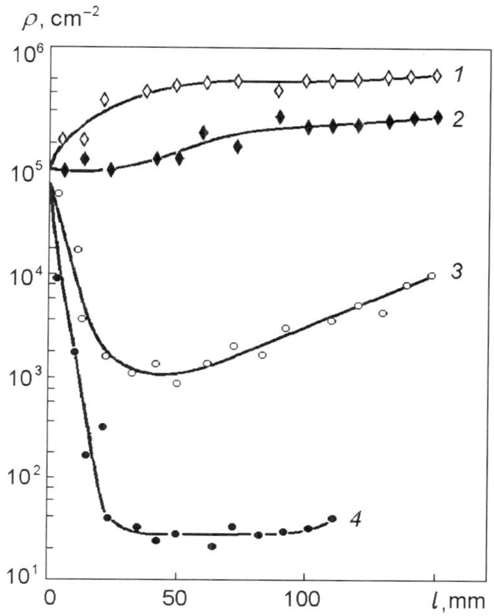

Fig. 5.10 Distribution of density of dislocations over the length of crystals grown by different methods with different temperature gradients: (*1, 2*) Verneuil method (*1* – T_1', *2* – T_2'); (*3, 4*) HDSM (*3* – T_3', *4* – T_4', $T_3'/T_4' \approx 2$); plane (0001)

process at the length of the crystal 40 mm, the structure quality of such crystals was essentially lower than that of crystals grown under conditions at which the temperature gradient varied no more than two times. For these cases, the temperature gradients and values of the residual stresses as well as the calculated values (5.16) of thermoelastic stresses are presented in Table 5.1. The values τ^*_{T} and τ^* are in rather good compliance. If crystals on block-free seeds with ρ below the critical one are grown, ρ increases as the distance from the seed becomes longer, and when it reaches the critical value the first block boundaries appear (Fig. 5.11). In the first case the length of the area free of block boundaries does not exceed 50 mm (Fig. 5.11a), while in the second case at a smaller variation of the temperature gradient the length of this area increases two times (Fig. 5.11b).The described experiment demonstrates that for the structure control it is not that the initial temperature gradient is so important (since in the considered cases temperature gradient is the same at the moment of seeding), but rather the degree of its constancy maintained during the growth process. The presented data also show that by varying the temperature gradient change cannot be the only value; the sign of thermoelastic stresses is also important.

Until now, when considering the possibility of controlling the dislocation structure of sapphire, the effect of the T_{n}' playing a decisive role and its variation in the growth process have been discussed. However, the structure quality of crystals can

Table 5.1 Thermal gradients and thermoelastic and residual stresses

No. experiments	T'_n (K cm^{-1})	τ^*_T (MPa)	τ^*_0 (MPa)
1	25	−3.5	35
	85	41	
2	25	−18	10–15
	50	−7	

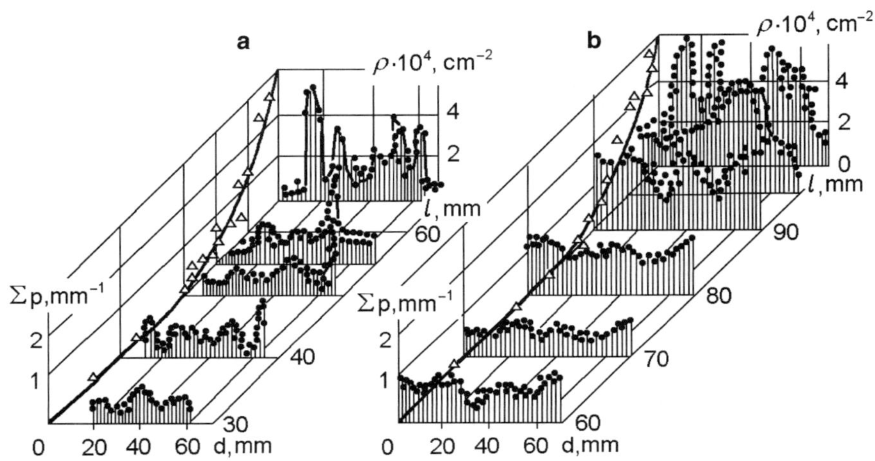

Fig. 5.11 Topograms of distribution of density of dislocations and extension of the block boundaries in the area of block boundary formation for the growth directions (**a**) [0001] and (**b**) [11$\bar{2}$0]

be regulated by varying temperature gradients normal to the growth direction as well. Naturally, for crystals with a narrow plasticity zone the control of structure perfection by varying radial temperature gradients is significantly less probable compared to crystals with a wide plasticity zone. But in this case, growing crystals by the Kyropoulos, HDS, and Czochralski methods under conditions of lower axial temperature gradients such control is possible to a higher degree than growing crystals by the Stepanov or Verneuil methods. Thus crystals can be grown with essentially different distributions of ρ in the planes parallel to the growth direction and normal to it [26]. For example, plates of sapphire grown by the HDS method in the presence of the temperature gradient between the upper and lower surfaces are characterized by the minimum density of dislocations in their central part (Fig. 5.12), as approaching the lower and upper surfaces ρ increases.

The same distribution of dislocations typical for the planes parallel to the growth direction and normal to it and the distribution of dislocations formed in the growth process also can be obtained in the model experiment on a four-point bend [36]. This

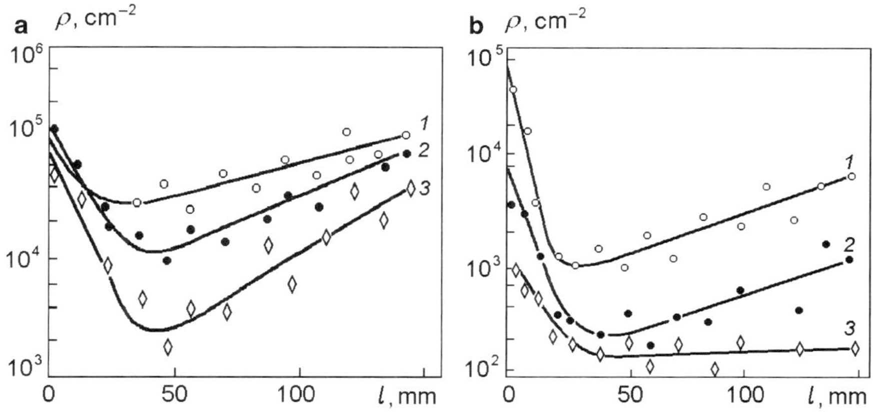

Fig. 5.12 Distribution of density of dislocations along the crystal length in different cross sections of the crystal: (**a**) on the plane $(11\bar{2}0)$; (**b**) on the plane (0001); (*1*) crystal surface; (*2*) at a depth of 1 m from the surface; (*3*) central part of the crystal

experiment, which presented an interest for modeling a stressed state that appeared in crystals at the effect of T'_r, allowed the study of the distribution of dislocation in different cross sections of the crystal depending on its spatial orientation.

If crystals have the orientation $[11\bar{2}0]$ and $[10\bar{1}0]$ at a four-point bend and load applied normally to the plane (0001), the base glide is forbidden and it occurs in the nonbasal gliding systems, perhaps in the planes $\{11\bar{2}0\}$. That is why at such deformation ρ on the base plane increases while on the plane of the prism (exit of basal dislocations) it varies insignificantly, and mainly redistribution of dislocations takes place. It should be noted that the base glide contribution rises as the deviation from the sample orientation from the specified one becomes larger. The results of investigations showed that in crystals with the orientation $[11\bar{2}0]$ and $[10\bar{1}0]$ at the stress $\tau^* = 3 \cdot 10^7$ Pa the velocity of deformation at 2200 K does not exceed $\nu = 10^{-8}$ s^{-1}, while as in crystals with the orientation [1011] (base glide is allowed) at stresses by an order lower the deformation velocity rises to the values $\nu = 10^{-5}$ s^{-1}, because $\nu \approx (\tau^*)^{4.5}$ at equal stresses the difference in the deformation velocities is 10^7 times. Although it is obvious that the system of forces acting at crystal growth and their four-point bend are different, the qualitative comparison of the results shows that high resistance appearing at bending crystals with the orientation $[11\bar{2}0]$ and $[10\bar{1}0]$ is extremely important when solving a problem of growing sapphire with low density of dislocations.

Making constrictions turns out to be inefficient for growing perfect sapphires the by Verneuil method, but in cases where crystals with varying density of dislocations along the length, according to a certain law distribution, are needed, this technique is expedient (Fig. 5.13). In the places of constriction, density of dislocation and boundary extension Σp decrease; as the cross section becomes larger, the values r and Σp rise.

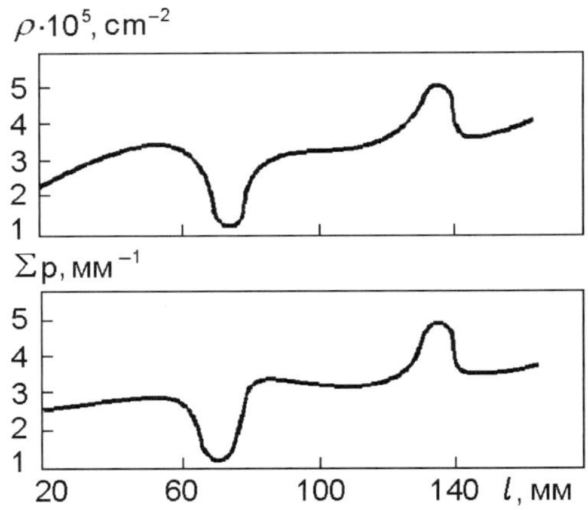

Fig. 5.13 Distribution of density of dislocations and block boundary extensions over the length of the crystals having varying cross sections

Thus, control of structure quality of crystals primarily consists of control of plastic deformation of crystals behind the CF. Variation of the T'_n in the growth process plays the main role in this case. While varying the T'_n can change a little distribution of density of dislocations over the length and cross-section area of the crystals, the main role is played by the value of the second derivative of temperature by the coordinate T'_n. The degree of T'_n influence on structure perfection of the crystal is defined by its spatial orientation, i.e., a possibility for the main easiest gliding systems to act. Thus, perfect low-dislocation crystals with the orientation $[11\bar{2}0]$ or $[10\bar{1}0]$ have been grown by the HDS method. At the minimum value of T'_n in these crystals $\rho < 1 \cdot 10^2$ cm^{-2}, large areas (about 15m long) did not contain dislocations at all [26]; note at a higher initial T'_n it is possible to obtain more prefect crystals than at a lower initial gradient but a high value of T'_n.

Not only density of dislocations depends on creation of conditions for relaxation of thermoelastic stresses. The possibility for thermoelastic stresses to relax defines whether the crystal will be damaged during its growth or when cooled down to room temperature [37]. In this is case if by some reason relaxation of thermoelastic stresses by means of plastic deformation is impeded (there is a certain configuration of dislocations or point defects that restrict mobility of dislocations), irrespective of the growth method, crystals are damaged; ρ in the damaged crystals is always lower than that in the intact ones (Table 5.2).

The observed difference in densities of dislocations two to four times remains unchanged along the length of the crystal. Thus, the mechanism of plastic deformation behind the CF plays a prevailing role both in "survival" of crystals and in the formation of their dislocation structure.

Vacancy mechanism of dislocation formation. When concentration of vacancies in the solid phase exceeds the equilibrium phase, vacancies can pile up into flat

Table 5.2 Density of dislocations $\rho \cdot 10^4$ (cm^{-2}) in the intact and damaged crystals

Growth method	Angle btween geometric and optical axis	Orientation of crystals			
		Intact		Cracked	
		$(11\bar{2}0)$	(0001)	$(11\bar{2}0)$	(0001)
Verneuil	90	60	15	25	5.0
	0	35	15	10	4.5
HDS	90	–	3	–	1
Stepanov	0	75	–	35	–

clusters (disks) [38]. At a certain size if free energy of the clusters is higher than that of the dislocation loops, these disks shut, forming closed dislocation loops. Thus formed loops can expand, attaching vacancies, and creep at high enough velocity owing to the fact that point defects migrate not only over the volume but also along dislocations; they build up an incomplete atomic plane on one side of the loop because of atoms released at the creep of the opposite side of the loop because of atoms released at the creep of the opposite side of the loop. It is assumed that vacancies can pile up into half-disks connected with the solid–liquid interface. The ends of half-loops formed of the shut disks propagate in the crystal as the CF moves forward. However, as the calculations show [39], cooling necessary for the formation of dislocation loops should be very high: 60–700 K. That is why this mechanism, which requires a significant supersaturation of the crystal lattice with vacancies, can be useful only at hardening.

At the same time, the presence of vacancy clouds–clusters was noticed in the semiconductor crystals [40]. During the selective etching these clusters are usually revealed on the surface in the form of flat-bottomed pits with a density reaching 10^9 cm^{-2}. These vacancy clusters may settle in a close vicinity to dislocations or at a rather long distance from them and favor creeping of already available dislocations and their building up into polygonal walls. A series of investigations [41] confirm the dominance of the vacancy mechanism of dislocation formation during growing plastic single crystals (zinc and copper).

During growing sapphire by the Kyropoulos, Czochralski, and HDS methods under conditions of small T'_n and low growth rates the action of vacancy mechanism of dislocation formation is not very probable, since fast cooling of the grown crystal in these methods is impossible and can be provided only at dramatic disturbance of the technological regimen. During growing massive crystals by the Verneuil method the role of this mechanism is insignificant; it probably becomes a little more important when growing small crystals ($d < 5$ mm) at high rates. The role of the vacancy mechanism when growing crystals by the Stepanov method at $v > 100$ mm/h may become more essential, since in this case the crystals are grown under conditions of high axial temperature gradients ($T'_n \approx 300$ deg cm^{-1}). This leads to the fact that during some minutes the newly grown parts of the crystal shift to the zones with the temperature by 300–500 K lower than the T_m. Fast cooling favors the lattice supersaturation with vacancies. However, since vacancies are sufficiently mobile only in a narrow temperature interval even if the lattice is supersaturated with them, it does not mean that all excess vacancies can pile up into disks during growth. Clouds of vacancies are most clearly seen in low dislocation block-free crystals (Fig. 5.14).

The action of the vacancy mechanism may appear to be the most significant at growing crystal filaments by the Stepanov method at $v \approx 6{,}000$ mm h^{-1} and $T'_n \sim 1000$ Kcm^{-1}. It is obvious that under such conditions the excess concentration of vacancies is high, but in the filaments 0.1–0.5 mm in diameter the outer surface can serve as a good vacancy sink.

Thus, when growing sapphire a full exclusion of the vacancy mechanism seems to be unjustified; in a number of cases it can make its contribution to the formation of dislocations. However, at a general high-enough density of dislocations stipulated basically by plastic deformation behind the CF (in crystals grown by the Stepanov method $\rho \approx 3 \cdot 10^4 - 1.10^6$ cm^{-2}), this contribution is so small that perhaps there is no need to take into account the vacancy mechanism when solving the problem of dislocation structure control.

Effect of impurities on the dislocation structure. The effect of impurity is defined by the concentration and difference in the atomic radii of the solvent and impurity, coefficient of its entry, and so forth [19]. For some crystals the nonuniform distribution of impurity does not play an independent role in the formation of dislocations. It leads only to local internal stresses, facilitating in these places generation of dislocations due to other reasons, e.g., due to thermoelastic stresses. Thus, in semiconductor single crystals rows of dislocation pits parallel to impurity lines were observed [40]. It should be assumed that these rows are stipulated by an indirect action of the impurity mechanism. Perhaps changes in conditions at the CF in certain moments leading to the appearance of impurity lines also stipulate changes

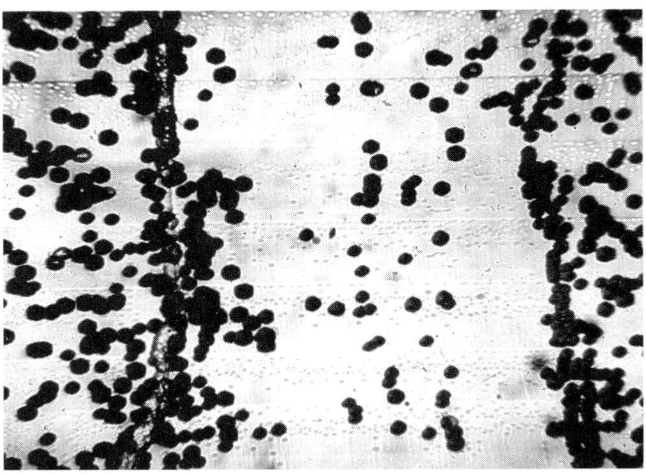

Fig. 5.14 Clusters in low-dislocation sapphire (×270)

of thermoelastic stresses, relaxation of which right behind the CF is responsible for the local increase of dislocation density. For semiconductor crystals it has been shown as well [21] that if the impurity concentration is lower than its solubility, it has no effect on the formation of dislocations. When exceeding the solubility limit, there appear to be dislocations in crystals, the density of which increases with the increasing solubility limit.

It is known that with the introduction of chromium to sapphire and the formation of a solid solution of substitution each percent of Cr_2O_3 changes the lattice parameter along the axis a by 0.04% and along the optical axis c – by 0.03%. The ρ distribution does not correlate with chromium distribution. Thus $\rho_{(0001)}$ does not depend on chromium concentration. As to prismatic dislocations a small rise of ρ is observed as the concentration increases, but this rise is probably connected first with variation of thermal conditions in the growth zone.

The analogous results were received with crystals in which iron, titanium, manganese, and cobalt were introduced separately at a concentration up to 0.5% as well as iron and cobalt, cobalt and chromium, and titanium and chromium introduced together at a total concentration of impurities to 0.8%. Density of dislocations in these crystals does not depend on the content and distribution of the impurity. At the same time in ruby grown by the hydrothermal method one could see the dependence of ρ on chromium concentration.

The presented data give evidence to the inconsistency of the results received by different authors. In particular a series of investigations concerning the effect of impurities on the content and distribution of dislocations in sapphire grown by the Verneuil, Czochralski, and HDS methods showed that up to a mass concentration of chromium 1% in the crystal (ruby-10) ρ does not depend on the chromium content. The same can be related to the vanadium and titanium that are introduced in crystals.

A nonuniform entry of impurities leads to the emergence of impurity striation. Its detailed analysis in ruby showed that distribution of dislocations in crystals even with high concentrations of chromium does not correlate with impurity distribution. Impurity striation does not define distribution of the ρ value. In the region of impurity aggregation in the center of the line density of dislocations practically does not differ from that in the interval between the lines.

This suggested for a number of crystals the assumption that an increase of dislocation density over the length is conditioned by the rejection of impurity; this assumption, however, turned out to be wrong. In fact, moving away from the seed crystals ρ can gradually increase.

If in some parts of the crystal impurity concentration exceeds the solubility limit in the given compound, ρ may sharply go up. So, if the local concentration of chromium reaches about 3% in this region it exceeds ρ in the middle density in the volume. Such aggregations of impurity and the clusters of dislocations stipulated by them in sapphire, however, rarely occur.

The impurities that essentially change the crystal lattice parameter create concentration stresses, the relaxation of which can result in the formation of dislocations. Indeed, for the majority of metals and some plastic materials the rise of density of dislocations even at an insignificant increase of the impurity content is reliably determined. Density of dislocations caused by action of the impurity mechanism in the assumption that the macroscopic elastic deformation, which appears at a sharp increase of the mass share of the impurity, relax by means of dislocation formation can be evaluated using the relation [42]

$$\rho = \Delta C(\mathrm{d}a/\mathrm{d}c)(ab)^{-1} \tag{5.18}$$

where da/dc is the variation of the lattice parameter a at variation of the mass share by a unit.

This relation differs from the conventionally used one by the fact that together with variation of the impurity mass shares ΔC it contains a mass share gradient that allows the evaluation not of the quantity of dislocations per unit of length but of the quantity of dislocations per unit of area. The value ρ, evaluated according to (5.18) turns out to be about $1 \cdot 10^5$ cm^{-2} which does coincide with the experimentally received values. While in crystals grown by the Verneuil method this difference is not large, in crystals grown by the Czochralski and Kyropoulos methods ρ_{calc} is one or two orders higher than ρ_{exp}. In crystals grown by the Verneuil method formation of impurity striation does not lead to the formation of dislocations (Fig. 5.15). Even a sharp drop of the mass shares of the impurity distribution of dislocations creates no rise.

The observed discrepancy between ρ_{exp} and ρ_{calc} is probably related to the fact that at deriving the expression (5.14) it was assumed that deformation of the lattice

Fig. 5.15 Microstriation and dislocation structure in corundum crystals (×120)

caused by the presence of impurity is completely nonelastic and the stresses that appeared decrease as dislocations are formed. This approach is not plausible for sapphire, since the estimations show that the stresses corresponding to the deformation conditioned by impurity nonuniformity make 10^{-1} MPa. Thus, the values of stresses depending on the impurity and its nonuniform distribution in crystals correspond to the region of elasticity even at a premelting temperature. It follows from here that the impurity mechanism does not contribute to the formation of dislocations.

Incoherent coalescence of the growth layers, branches of dendrites, nuclei. The formation of dendrites is due to a rapid growth under overcooling, that is, to a negative temperature gradient the face of the CF. The dendrite axle and its branches grow along a specific crystallographic direction. Each dendrite grows starting from a single crystallization center. The whole dendrite, including its branches, is a single crystal. Any impurity decelerates the growth rate. The temperature conditions being the same, either pointed needlelike dendrites or dendrites with paraboloid-shaped needles are formed.To prevent the dendrite growth, the crystallization rates are selected that provide a sufficient heat removal through the crystal in growth. Even the cubic crystals, such as silver, copper, or gold ones, which must grow equally along three perpendicular directions due to the structure symmetry, may form dendrites. The unequal growth is explained usually by two opposite trends to free energy minimization and to the fastest process completion. A crystal can attain the minimum surface energy only under equilibrium conditions, that is, at the infinitely slow growth, while the fastest formation is attained at an infinitely developed surface. In real cases, compromises always take place which result sometimes in faceted equilibrium shapes and sometimes in branched nonequilibrium ones.

An incoherent accretion of dendritic branches, nuclei, and growth layers may result in a considerable increase in ρ at the crystal growing from melt. For example, at a considerable melt overcooling when the latent heat of crystallization is removed from the CF not only through the solid phase but also through the melt, the interphase boundary becomes unstable and protrusions are formed that are converted into dendrites as the overcooled zone width increases. The dendrite tip may become misoriented with respect to adjacent dendrites due to convective flows, vibrations, and so forth. Dislocations are formed at joints of such dendrites. The mechanism under consideration requires a considerable overcooling and high growth rates. In such conditions, the crystal may grow by attaching individual atoms as well as entire complexes. In the latter case, dislocations and small-angle dislocation boundaries may arise.

There is widespread opinion that when considering the crystal growth methods from melt, it is insignificant to take into account the dislocations formed at incoherent accretion of the dendrite axles, nuclei, and so on, and that the contribution of these is small as compared to that of plastic straining and the defect inheritance. This mechanism, however, cannot be neglected, for example, when growing sapphire by the Stepanov technique at high growth rates and a considerable overcooling of the melt film above the shaper.

Growth of corundum crystals with a specified dislocation structure. The analysis of the main mechanisms of the dislocation formation in crystals shows the degree of dislocation structure controllability in different growth methods, i.e., a possibility to obtain crystals with a specified distribution of dislocations. Comparison of dislocation density in crystals grown from the melt under the different limits of each method's temperature gradients allows the division of these methods into two groups: by their thermodynamic parameters and, consequently, by structure quality (Fig. 5.16) [43]. In some approximation variation of the axial temperature gradient is found from the expression

$$T_n''(x) = 2\pi rek / \pi r^2 (T^4 - T_0^4)\lambda_T \quad (5.19)$$

where T and T_0 are the temperatures of the crystal and environment; e is the emissive ability of the surface; k is the Stefan–Boltzmann constant; and r is the crystal radius. As appears from the above, variation of the temperature gradient is defined on the one hand by the ratio of the grown crystal surface to its volume (S/V) (the first cofactor), and on the other hand by the difference in the crystal and environment temperatures (the second cofactor). As mentioned, the value T_n'' highly defines thermoelastic stresses in the crystal and correspondingly density of dislocations. Therefore, the methods of the first group characterized by low values of both factors also means that the values of $T_n(x)$ facilitate the growth of high-structure quality crystals. Variation of this value and other technological parameters of the process offers the possibility to grow crystals with a specified distribution of structure defects. Application of the seeds with orientation at which inheritance of defects and plastic deformation are impeded allow one to obtain crystals with $\rho \approx 1 \cdot 10^3$ cm^{-2} (see Fig. 5.5). Decrease of the $T_n(x)$ value while preserving the above-mentioned conditions allows one to grow crystals with $\rho \approx 50$ cm^{-2} [44].

In the methods of the second group the values of T_n are much higher. Thus, in the Verneuil method the value of the second cofactor is high ($T^4 - T_0^4$). In cases of growing shaped crystals, first of all hollow profiles, by the Stepanov method, high

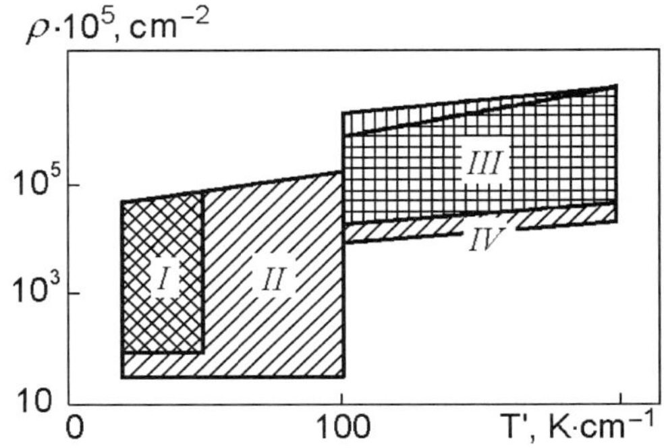

Fig. 5.16 Density of dislocations as a function of temperature gradients for sapphire grown by the (I) Czochralski, (II) HDS, (III) Verneuil, and (IV) Stepanov methods

values of T''_n (x) are explained by a large ratio of the area of the radiating surface of the crystals to its volume (V). The crystals of the second group have the fact that the crystallization process proceeds from a thin film of the melt in common. The second group methods offer much less possibility of controlling the dislocation structure than those of the first group.

Along with common regularities observed in crystals grown by the Verneuil and Stepanov methods shaped crystals possess a number of typical characteristic features, conditioned by a higher growth rate, presence of a die, and connected with this distortion of the temperature field in the growth zone. In the first place it is a nonuniform distribution of dislocations in the crystal volume: a noticeable decrease of ρ near the surface of crystal profiles up to the occurrence of dislocation free zones (sometimes the width of this zone can reach 200 mm); right after the dislocation free zone a region with the increased density of dislocations and a manifold net of low angle boundaries usually follows.

As mentioned above structure perfection of crystals is to a large extent defined by the ratio S/V; this being so, the maximum growth rate v_{max} corresponds to each value of S/V [45], allowing at $v \leq v_{max}$ the attainment of crystals of the required quality. For hollow profiles S/V and v_{max} do not depend on the cross size of the profile; for solid ones such dependence is well pronounced. The maximum rate and the ratio S/V for hollow profiles are defined by the profile wall thickness. Thus, when the cross size of the hollow profile for which the ratio S/V remains constant increases (hollow profiles with a constant wall thickness), the changes in the ρ are insignificant. When growing tubes 4–90 mm in diameter in the rate range of 20–120 mm h^{-1}, the value of r varies in the range of $(2–8) \cdot 10^5$ cm^{-2} [46]. While for solid crystals grown by the considered methods ρ strongly depends on the rate in any range of rates, in the hollow profiles at $v < v_{max}$ the dependence of ρ on v is insignificant.

Special attention should be paid to the specific character of crystallization of the aluminum oxide melt under conditions of high gas saturation of the melt. Disturbance of the stability of thermal conditions at the CF results in the trapping of gas bubbles in the melt at the interface, the bubbles being responsible for pore formation. These pores often are faceted and as the concentrators of stresses (Fig. 5.17) they play the role of additional sources of dislocations [26, 44]. The distribution of ρ in the crystal grown by the HDS method under conditions of periodic disturbance of the thermal regimen stability is shown in Fig. 5.18. As it appears from the above data, disturbances of the stability evoke an increase of dislocation density. As the normal crystallization conditions are reestablished, ρ abruptly drops, which is explained by the absence of inheritance and wedging out of dislocations depending on spatial orientation of the crystal.

Despite the fact that the dislocation structure of crystals grown by the Stepanov and Verneuil methods is less amenable to control than that of crystals grown by the Czochralski and HDS methods, the above-mentioned analysis allows the assertion that it is possible at present to grow corundum crystals from melt with any specified distribution of ρ: from low dislocation crystals to those with periodic density maxima (Fig. 5.18).

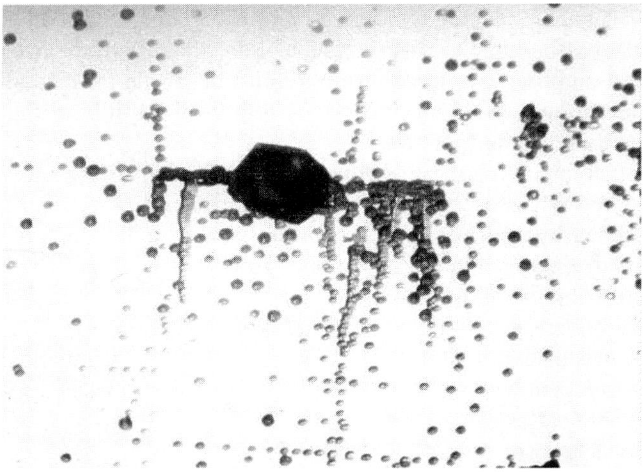

Fig. 5.17 Dislocations formed around a faceted pore (×120)

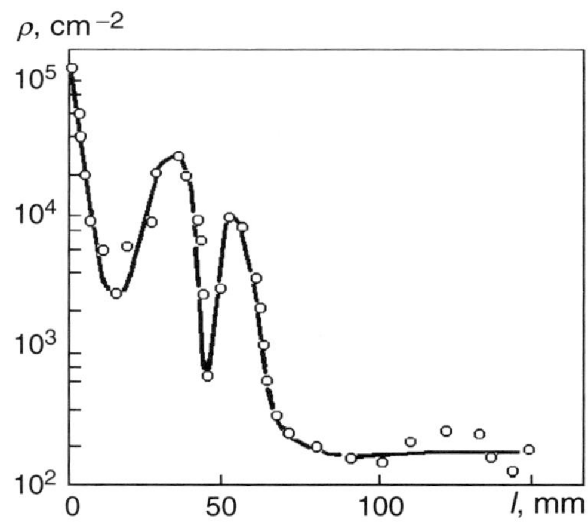

Fig. 5.18 Distribution of density of dislocations in the plane (0001) at a periodic disturbance of the thermal regimen stability

5.3 Block Structure of Crystals

When growing crystals from the melt, block-free crystals or those whose block boundaries are disoriented at low angles are the desired crystals. If the disorientation is high in crystals of many compounds including sapphire, an intercrystallite damage takes place. Different authors suggest many classifications of substructures in crystals. Still the most convenient seems to be a classification by the disorientation angles and block size [47]:

- The first-order substructure consists of the blocks stretched along the growth axis with the disorientation angles 0.5–3°. The block sizes vary from a fraction of a centimeter to several centimeters; sometimes such substructure (macroblock structure) is called "striated" or "columnar" [48].
- The second-order substructure, with block sizes from 100 μm to 1 mm with the disorientation angles 0.5–30′; usually these are microblocks splitting like a fan in the crystal.
- The third-order substructure consists of low-angle dislocation boundaries; block sizes of 10–100 μm and the disorientation angles do not exceed 30′.

The analysis of the block structure of different crystals assumed the following mechanisms of its formation [49]: vacancy, impurity, inheritance of the seed crystal boundaries, polygonization, and incoherent coalescence of layers. At the investigation of the block structure formation the main attention was focused on the *vacancy mechanism*. According to the data of Chalmers and Frank aggregation of vacancies leads to the formation of a dislocation loop. If this loop lies in the plane parallel to the growth direction it can go out to the CF and transform into a half-loop the ends of which will intergrow the crystal. If many half-loops are formed one in the other making groups, they can form a low-angle dislocation boundary. Necessary for this is high concentration and high mobility of vacancies. Taking into account that sapphires are grown at rather high rates under conditions of high temperature gradients, a simultaneous fulfillment of these conditions is hardly possible. Therefore, the role of the vacancy mechanism in the formation of block boundaries is not so important. The same concerns the *impurity mechanism* when growing crystals with a total mass content of heavy metal impurities not exceeding $5 \cdot 10^{-2}\%$, although it is supposed [50] that the impurities and dissociation products can aggregate in separate parts of the CF, locally decrease the growth rate, and be the reason for formation of the first-order substructure. Inclusions of the foreign particles, in particular molybdenum, when crystals are grown in a molybdenum container, also can be the sources of crystal lattice distortions and the reason for microblock formation. The most important role the mechanisms of inheritance play is polygonization under the effect of thermoelastic stresses and incoherent coalescence of layers.

Inheritance of block boundaries. The principles of controlling the inheritance of block boundaries are the same as those of controlling the dislocation structure because boundaries are inherited by the growing crystal the same way as dislocations going out to the CF. As an example Fig. 5.19 shows the distribution of block boundary extension Σp in the near seed part of crystals grown by the HDS method on different qualities of the seed crystals. At the initial stage of the growth on the block containing seeds, one can observe a decrease of Σp due to wedging out of a part of boundaries not parallel to the growth direction; then Σp rises again. In crystals grown on perfect seeds practically free of blocks there are no block boundaries in the near seed part of the crystal; only at a certain distance from the seed are new boundaries formed, and their extension gradually enlarges. Their appearance is not now connected with the inheritance mechanism (the reasons of such boundary formation will be discussed below). It is worth noting that the block structure of

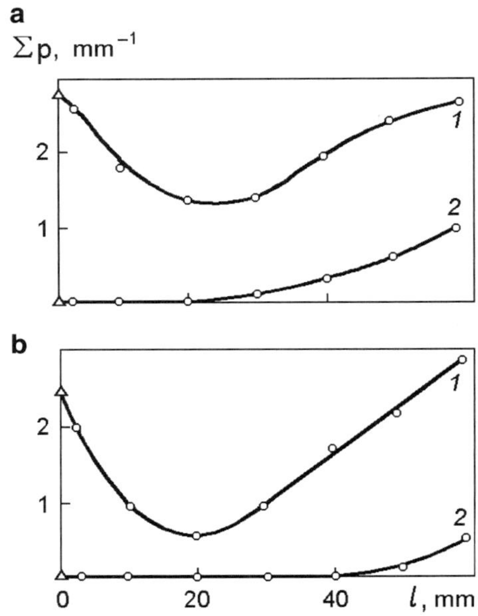

Fig. 5.19 Distribution of block boundary extensions over the crystal length depending on quality of the seed crystal: (**a**) in plane (0001); (**b**) in plane (11$\bar{2}$0); (*1, 2*) seed crystal with blocks and free of blocks, respectively

crystals grown on the block-free seeds or on crystals where the block boundaries are normal to the growth direction and do not go out to the CF (curve 2) is much less pronounced than in crystals grown on usual seeds (curve 1). The effect of the seed structure perfection is particularly strong in the near seed part (at a length about 50 mm).

Control of the inheritance of block boundaries can be provided by variation of the area of the surface through which the block boundaries intergrow the growing layers of crystals and (or) by choosing such spatial orientation at which block boundaries are not inherited by the growing crystal. This way one can obtain block-free (70 mm or longer) corundum crystals on the needlelike seeds by the Verneuil method [24] and crystals practically without block boundaries on the block containing seeds with their orientation so that the main part of the available boundaries was normal to the growth direction.

Formation of polygonization block boundaries. It is known that the low-angle boundaries can be formed due to polygonization, which is connected with creep of dislocations [51], or the interaction of the gliding under the effect of thermoelastic stress dislocations when their density reaches the critical value. Owing to the fact that in the growth process the path of individual dislocations at the build up of boundaries with minimal angles of disorientation usually is unknown, it should be supposed that under real growth conditions both phenomena take place, i.e., there occurs motion of dislocations in the gliding systems upon which the creep is imposed.

A conception of the critical density of dislocations ρ_{cr} as the density above which formation of low-angle block boundaries begins was introduced [52]:

$$\rho_{cr} = [4 / ld]\cos\varphi, \tag{5.20}$$

where d is the cross size of the crystal in which formation of low-angle boundaries is considered; φ is the angle between the direction of dislocation glide and the plane normal to the axis of the grown crystal; and l is the distance between dislocations (Fig. 5.20). At derivation of the equation for evaluation of the critical density the length of the free path of dislocations was assumed to be $l_{\text{free path}} = d/\cos\varphi$.

As seen from the figure, when crystals have a narrow zone of plastic deformation, $l_{\text{free path}} = h/\sin\varphi$. The equation for the evaluation of critical density of dislocations can be written in the form [23]:

$$\rho_{cr} = [4 / dh]\sin\varphi \tag{5.21}$$

It appears from this expression that the ρ_{cr}, when growing one and the same material, is to change going from one method to another. A satisfactory compliance between the values of ρ_{cr} found experimentally and evaluated using the expression (5.22) is observed in crystals with narrow plasticity zone. Thus, for the method of HDS $\rho_{exp} \approx (2–3) \cdot 10^4$ cm^{-2}, $\rho_{theor} \approx 1 \cdot 10^4$ cm^{-2}. In the formula for ρ_{cr} only the width of the plasticity zone is allowed for, defined for the given material by the temperature gradient. Actually, the value of h is not important, but the time the crystal stays in this zone is important, which depends on the temperature gradient, growth rate, and mobility of dislocations in the plasticity zone.

Since sapphire is anisotropic, it is natural that their plasticity is anisotropic as well. While in the plane (0001) formation of low-angle dislocation boundaries and correspondingly increases of Σp occur at $\rho_{cr} \approx (3–6)\ 10^5$ cm^{-2} (Verneuil method); for the plane the value of Σp increases up to (1.2–1.4) 10^6 cm^{-2}. Such a difference

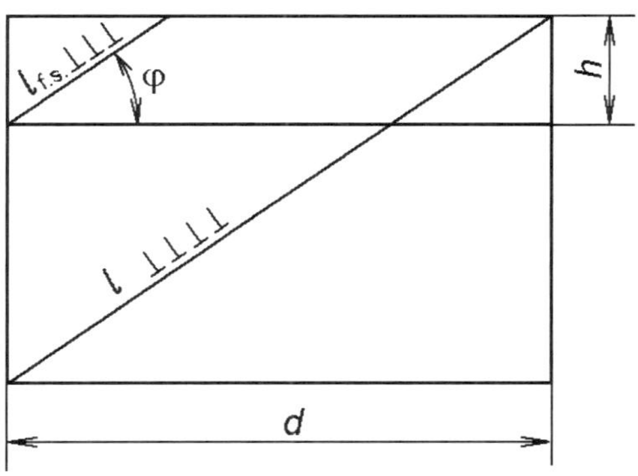

Fig. 5.20 A scheme for determination of the critical density of dislocations in crystals with a narrow zone of plastic deformation

is explained by a lower mobility of basal dislocations going out to the plane of the prism, which is connected with the interaction of dislocations. It was found that in accordance with Burgers vector for basal dislocations reactions can be affected according to the equation

$$[2\bar{1}\bar{1}0]+[\bar{1}2\bar{1}0]+[\bar{1}\bar{1}20]=0 \quad (5.22)$$

As a result of this reaction a self-fixation of dislocations and formation of dislocation joints take place; this must retard mobility of basal dislocations. Given below are the values of the critical density of dislocations determined for sapphire on the plane (0001) grown by different methods.

Growth method	ρ_{cr} (cm^{-2})
Verneuil	$(3–6) \cdot 10^5$
Stepanov	$(1–2) \cdot 10^5$
Czochralski	$(5–7) \cdot 10^4$
HDS	$(2–3) \cdot 10^4$

The difference by an order in the values of ρ_{cr} for the Verneuil and HDS methods is stipulated by the fact that at high-temperature gradients typical for the Verneuil method and narrow temperature interval of plasticity characteristic of sapphire the time for the polygonization process to proceed turns out to be insufficient. Owing to this crystals with high density of dislocations containing practically no block boundaries can be grown. A 2-hour annealing of such crystals at a premelting temperature proves to be enough for a branched net of low-angle boundaries to form. The role of the growth rate in the formation of block boundaries during growth of corundum crystals [43] can be seen from Fig. 5.21 in which the dependence $\Sigma p = f(v)$ for sapphire tubes grown by the Stepanov method is plotted. For this method this dependence is particularly pronounced since the growth rate is widely varied. At low rates the length of the block boundaries is bigger than at high ones: at $v \leq 60$ mm h^{-1} the polygonization manages to occur and low-angle dislocation boundaries are formed; at $v > 60$ mm h^{-1} no boundaries are formed although $\rho \geq \rho_{cr}$.

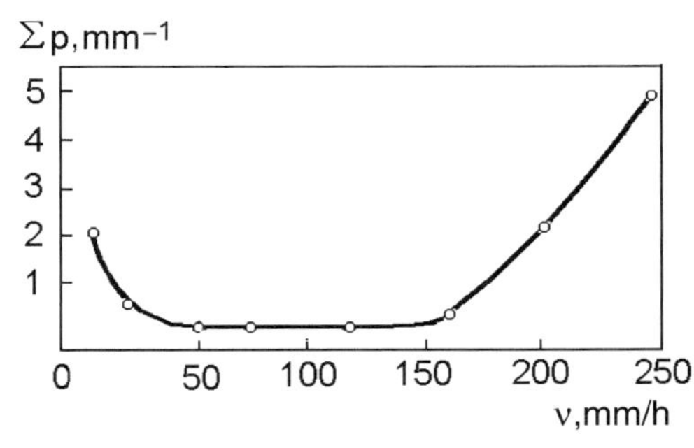

Fig. 5.21 Block boundary extensions as a function of growth rate in hollow profiles

A sharp rise of Σp at $v > 140$ mm h^{-1} is now conditioned by another mechanism, which will be discussed below.

If density of single dislocations in the seed crystal is lower than the critical one and growth proceeds on a block-free seed or is oriented so that the available boundaries are not inherited by the growing crystal just adjacent to the seed, there appears to be an area with no block boundaries (the incubation period). The size of this area depends on the growth conditions and spatial orientation of the growing crystal. Therefore, the incubation period of formation of low-angle dislocation boundaries l_1 can be controlled. At $K = 1$ [23]

$$l_1 = h[(\rho_{cr} - \rho_0)/\rho_2], \qquad (5.23)$$

where ρ_0 is the density of single dislocations formed at the CF.

The control of the block-free zone size of the crystal in fact is reduced to the control of ρ in the growing crystal. Provided the $K \sim 0$ and density of dislocations originating behind the CF under the effect of thermoelastic stresses is lower than the critical one, the incubation period (the distance from the seed crystal to the region where ρ becomes equal to ρ_{cr} and the first low-angle boundaries appear) is long enough. Thus, decrease of $T''_n(x)$ allows the incubation period to elongate and to receive a long piece of the block-free crystal. If as the crystal grows larger, ρ reaches its critical value, and low-angle dislocation boundaries which go out to the CF appear, they are inherited by the growing crystal, their disorientation angles become larger, and a gradual transformation from the third-order substructure to first-order substructure takes place.

Incoherent coalescence of nuclei. During crystallization in different points of the interface nuclei appear. The layers growing from these nuclei can approach the place of joining in a somewhat disoriented state; as they join, a low-angle boundary can form. This phenomenon is observed when growing crystals under conditions of a concave CF when two-dimensional nuclei of the solid phase are formed in the peripheral parts of the crystallization front, grow larger toward the center, and block boundaries appear. Thus, at growing crystals of C-orientation (the growth direction coincides with the direction of the optical axis) by the Verneuil method at a concave crystallization, front blocks containing crystals are grown with $\Sigma p \geq 3$ mm^{-1}.

In that way if the problem consists of growing crystals with a certain block-free area or completely without blocks, the seed crystal must not contain block boundaries; but if nevertheless it has them, they should not go out to the interface. If this condition is fulfilled, the size of the block-free area is defined by the distance from the seed crystal at which ρ becomes equal to ρ_{cr}. To enlarge this area the space orientation of the seed crystal must be such that the coefficient of inheritance of both seed dislocations and those that are formed in the growth process are reduced to the minimum in order to impede the action of the main gliding systems [44]. The value of the axial temperature gradient and its variation during the growth process should be minimal. To prevent formation of the first-order substructure, which is especially dangerous for crystals, it is necessary that the formed microblocks not go out to the CF.

5.4 Impurity Nonuniformity

The character of impurity distribution when growing crystals from melt is first of all defined by heat and mass transfer in the solid and liquid phases, kinetics, and growth mechanism. Heat transfer is the main variable. The temperature gradient availability leads to convection, impurity transfer, and specifies the growth kinetics. Heat transfer has a direct effect on the melt hydrodynamics, since it stimulates and constantly maintains heat convection. Still, melt motion is provided not only by thermal and thermal-capillary convection, but by a compulsory way as well – by rotation of the crystal, crucible, or other methods of mixing. The growth kinetics and impurity distribution are interdependent and affect each other. The growth rate and the overcooling value at the interface depend on impurity concentration. In its turn the growth kinetics due to a segregation has an effect on the impurity concentration near the CF. Just these interdependent and multifactor processes are responsible for the formation of the impurity substructures in crystals: striation, facets, cellular structures, impurity inclusions, and so forth. The traditional approach to the problems of impurity nonuniformity formation is based on the assumption that despite a variety of simultaneously running processes, a periodic variation of the growth rate lies in the core of the mechanism of such defect formation, leading to the variation of the effective coefficient of distribution. Until recently such an approach allowed a physical interpretation to the observed phenomena and practical recommendations for bringing such defects to the minimum. However, with the intensification of the growth technologies (changeover to higher rates, upsizing the grown crystals, etc.) as new rapid growth methods have appeared, the possibilities of controlling the impurity nonuniformity were reduced, the earlier established regularities were violated, and new types of defects appeared.

A detailed analysis of the impurity nonuniformity in sapphire grown by different methods demonstrates that there are two main types of impurity striation in these crystals [53]:

- *Macrostriation*, which has been observed visually and with the aid of a polarization-shadow device; the width of the striae – from tens to hundreds of millimeters and the distance between the striae can reach several (sometimes tens) millimeters (Fig. 5.22)
- *Microstriation*, which has been observed using optical microscopy, the width of the striae $5 \cdot 10^{-3}$–$3 \cdot 10^{-2}$ mm, the distance between the striae $1 \cdot 10^{-2}$–0.1 mm (Fig. 5.23)

On the planes parallel to the growth axis the impurity striae of these two types look like parallel lines that repeat the form of the CF. On the planes normal to the growth axis the striae can have the form of concentric circles. If crystals are grown with rotation, there appears to be an additional rotational striation the intensity of which increases from the central part of the crystal to the periphery. As the total concentration of impurities becomes higher, the impurity striae intensify. At a concentration higher than 1% there appears to be a new type of the impurity substructure in the form of triangular periodical inclusions (Fig. 5.24).

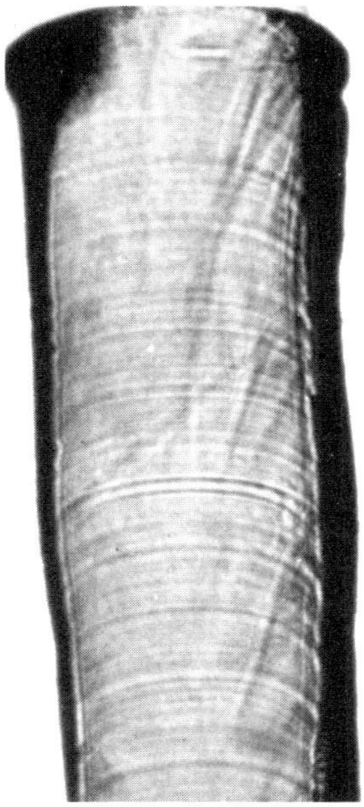

Fig. 5.22 Macrostriation

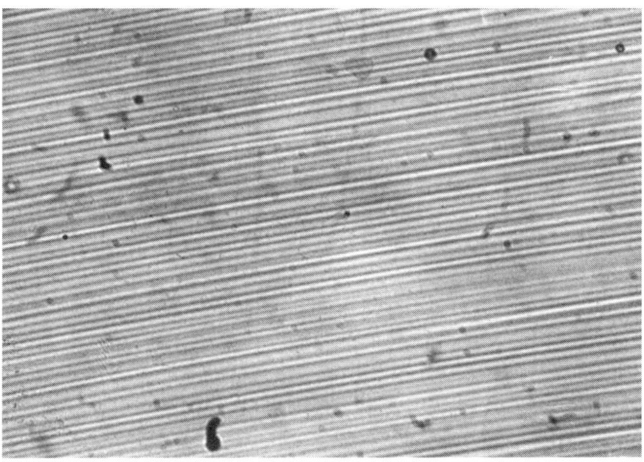

Fig. 5.23 Microstriation (×120)

A correlation between the impurity striae with the dopant is unambiguously confirmed by the data of micro-X-ray spectral analysis. The analysis of the structure of the doped sapphire has shown that formation of macrostriation bears mainly an "apparatus" character and is connected with the operation of the crystallization equipment: nonuniformity of the motion of the pulling mechanisms, a low level of

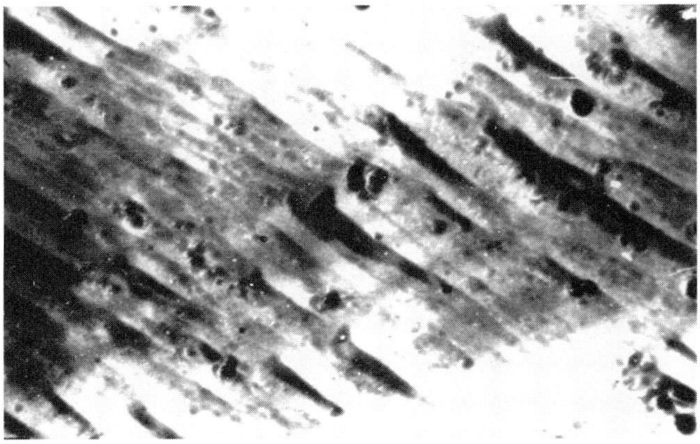

Fig. 5.24 Impurity inclusions (×10,000)

power stabilization, irregularity in feeding the raw material to the crystallization zone, and so on. Thus, the changeover from a discrete to a continuous replenishment of the melt when growing by the Veneuil method allows the change of the form of macrostriation [54]. By improving temperature stabilization, updating the pulling mechanisms and melt feeding to the crystallization zone one can practically completely avoid this type of striation.

The reasons for the microstriation emergence can be both of the "apparatus" and "fundamental" character. The main reason for the apparatus character of microstriation is connected perhaps with temperature oscillations in the melt caused by natural convection and asymmetry of the thermal field created by nonuniformity and inconstancy of temperature at the crystallization front. A decrease of the striae intensity can be achieved by a choice of optimal temperature fields, lowering of gradients, raising of their stability, application of an intense compulsory mixing of the melt, and decrease of the total concentration of impurities. For example, in crystals grown by the Verneuil method under conditions of high varying temperature gradient the microstriation is well pronounced (Table 5.3, case I). Lowering the axial gradients and provision of their constancy (case IV) can weaken this effect. For practically all growth methods the increase of the temperature gradient results in the change of the half-width of the striae. It is characteristic that in the undoped samples grown in crucibles of molybdenum, tungsten, or iridium these metals enter in the concentration of 10^{-2}–10^{-3}% and create the impurity striation, obeying the same regularities as that of the doped crystals.

The main reason for the fundamental character of microstriation formation is the emergence of an enriched ($K_d < 1$) with or depleted ($K_d > 1$) of impurity layer in the melt and connected with this process periodic trapping of impurity by the growing crystal. For the evaluation of the parameters of the impurity striation arising owing to this reason, let us consider two cases: a diffusion mixing of the melt and a limited diffusion in the melt if convection flows are present. Both variants are possible;

Table 5.3 Impurity striation

Growth method	T'_n (K cm^{-2})	T''_n (K cm^{-2})	$\lambda_{\mu s}$ (μm)	Δh (cm)
Verneuil			by Cr	
I	Over 200	250	10	$3 \cdot 10^{-4}$
II	About 200	500	18	$2 \cdot 10^{-4}$
III	About 50	150	15	$8 \cdot 10^{-4}$
IV	About 70	100	14	$7 \cdot 10^{-4}$
Czochralski	About 50	70	by Cr 20	$7 \cdot 10^{-4}$
			by Mo 30	$7 \cdot 10^{-4}$
HDS	About 70	50	by Mo 30	$6 \cdot 10^{-4}$
Stepanov	About 200	500	by Mo 20	
			by Y 20	–

note that when passing from one growth method to another the probability of realization of these variants varies.

Let us estimate some parameters of the impurity striation for the mentioned variants. In the first case, according to L. Landay [58], one can estimate the value of the microstriation period by finding the value of the overcooling gradient at the CF (Thiller criterion of the existence of concentration overcooling [55]):

$$(\text{grad}\, U)_{\text{cr}} = \varphi_1 C_{\text{L}} v(1-K_{\text{d}})T'_{\text{melt}} / K_{\text{d}} D \tag{5.24}$$

Assuming for sapphire $\varphi_1 = 1.4°$, $T'_{\text{melt}} = 10$ K cm^{-1}, $K_{\text{d}} = 2$, $C_{\text{L}} = 10^{-1}\%$, $v = 3 \cdot 10^{-4}$ cm s^{-1}, D = 10^{-6} cm^2 s^{-1} we receive (grad U)cr ≈ 10 K cm^{-1}. Knowing this value we can find the components of the microstriation period $\lambda_{\mu s}$ equal to the sum of l_1 and l_2, where l_1 is the distance at which the amplitude of the impurity striae becomes higher, l_2 is the distance at which it becomes lower. Analytically these values are found from

$$l_1 = (D(v)) \ln\left[1 + \frac{(\text{grad}\, U)_{\text{cr}}}{T'_{\text{melt}}}\right] \tag{5.25}$$

$$l_2 = \frac{1}{K_{\text{d}}}\frac{D}{v} \ln\left\{\frac{C_{\text{L}}\dfrac{(1-K_{\text{d}})}{K_{\text{d}}} - \dfrac{T'_{\text{melt}}}{\varphi_1}\dfrac{D}{v}}{C_{\text{L}}\dfrac{(1-K_{\text{d}})}{K_{\text{d}}} - \dfrac{DT'_{\text{melt}} + (\text{grad}\, U)_{\text{cr}}}{\varphi_1 v}}\right\} \tag{5.26}$$

A substitution of the corresponding value gives $l_1 \approx 20$ μm, $l_2 \approx 30$ μm, $\lambda_{\mu s} \approx 50$ μm.

In the second case when the impurity distribution is stipulated by the convective mixing, one can calculate the thickness of the diffusion layer δ [56] assuming that an impurity striae is formed in this layer, its thickness being

$$\delta = 1.6\eta^{1/6} D^{1/3} \Omega^{-1/2}, \tag{5.27}$$

where of sapphire kinematic viscosity $\eta \approx 2 \cdot 10^{-5}\ m^2\ s^{-1}$ [5]; Ω is the angular velocity. Substitution of the corresponding constants also gives the value $\delta \approx 50\ \mu m$.

There is a reason to believe that when growing sapphire by the directed crystallization in a horizontally positioned container a dominant role is played by the diffusion mixing of the melt, while in the case of the Czochralski method it is the convective mixing. The practice shows that in these methods the coincidence of the calculated periods of the impurity striation with the experimental ones is observed only for a certain interval of technological parameters. This being so, in crystals for which there is a divergence between the calculated and experimental periods as a rule a characteristic distribution of point defects in the form of the grain structure can be observed (Fig. 5.25) as can *ultramicrostriation*, which is the third kind of striation with the striae width 0.01–0.03 μm and the distance between them 0.05–0.25 μm. Unlike macro- and microstriation this type of the impurity nonuniformity has a discrete character, i.e., the striae themselves and spaces between them consist of separate elements (Fig. 5.26).

By varying the type and concentration of the impurity in the melt (chromium, boron, gallium, molybdenum, yttrium) one can change the size of these elements in a wide range (from 100 to 1,200 Å). Note that the form of *ultramicrostriation* varies slightly.

Ultramicrostriation in crystals grown by the Verneuil and Stepanov methods is observed always irrespective of the thermodynamic parameters of the crystallization process. In the case of the Kyropoulos, Czochralski, and HDS methods this type of defect appears only in the peripheral parts of the crystals.

With regard for the bulk character of crystallization a reconsideration of the problem of the impurity distribution behavior in the growing crystal is required [57].

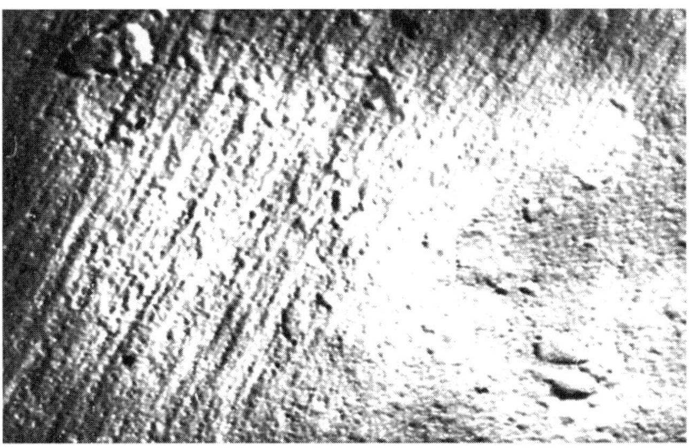

Fig. 5.25 Typical distribution of point defects – "grain" structure (electron microscopic photograph) (×10,000)

Fig. 5.26 Ultramicrostriation (×62,000)

The authors consider the impurity distribution in front of the growing crystal surface at the directed crystallization from melt, believing that the crystallization process has a bulk character and begins even in the melt by nucleation of clusters (complex constructions having a structure close to that of the crystal and a size of one or more elementary cells). The existence of clusters is possible above the liquid line as well, but in this region they are in the dynamic equilibrium with the melt and the total mass has typical properties of the liquid. In the model under consideration it is supposed that the mass growth of the already existing and nucleation of new clusters begins as the liquid line is crossed. Being located in the overcooled zone as a suspension of large particles among single molecules the clusters become almost fully ordered and prepare for embedding owing to the long-distance forces between them and the crystal surface. It also is presumed that the redistribution of the impurity between clusters and liquid phase is defined by the equilibrium coefficient of distribution. It is known that the physical sense of the distribution coefficient implies not a spasmodic but a degradation by temperature phase transition, because at melting the less strong bonds break first, while at crystallization the stronger ones are to form.

According to the accepted model in the overcooled by the value ΔT_0 region of the melt of the thickness d clusters should exist with a continuous gathering of impurity concentrations from C'_{k1} to C'_{kn}(Fig. 5.27). Such an approach gives grounds to make conclusions principally different from the classic descriptions of impurity distribution: the profile of the concentration curve in front of the surface of the growing crystal is defined by impurity concentration both in the liquid phase and in the "phase of clusters" with the account of their specific volumes; the profile of the concentration curve in not to have breakage.

The analytical solution of the problem in such interpretation showed that the impurity concentration really has no jump when passing from the melt to the crystal a jump occurs only under equilibrium conditions when $\Delta T_0 = 0$. The analysis of the obtained dependences allows one to state the following:

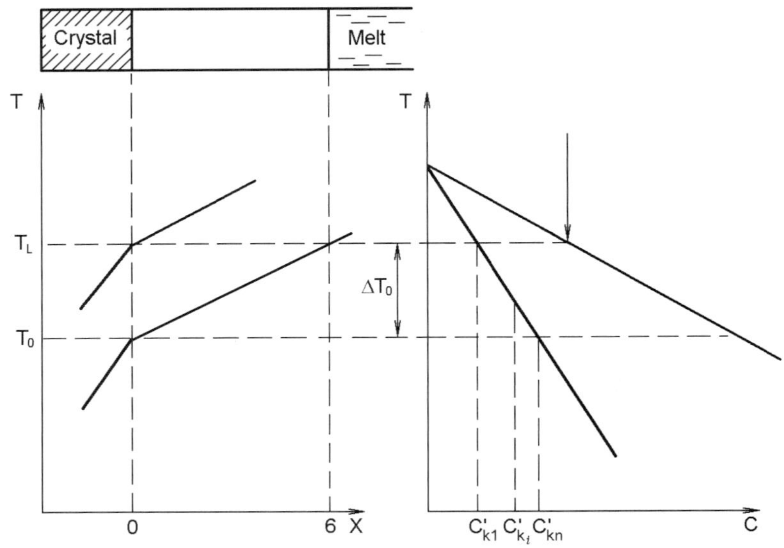

Fig. 5.27 Temperature distribution in the melt and in the crystal grown by HDSM: (*1*) crystallization front is dead; (*2*) crystallization front moves. The state diagram of the quasiequilibrium process; $K_d < 1$

- The proposed model explains with sufficient reliability the formation of such defects as cellular growth, growth layers, nonuniformity of impurity distribution in the cross section of the crystal, and confirms the experimentally observed rise of the effective coefficient of impurity distribution as the growth rate becomes higher.
- There is no necessity in the use of conception of the coefficient of impurity distribution at the interface.

The purposeful experiments on modeling the process of the impurity nonuniformity appearance demonstrated the trustworthiness of the author's assumptions as well as the fact that a necessary but not sufficient condition for the origination of ultramicrostriation is the emergence of defects of the grain structure type in the crystals. An ample condition for the ultramicrostriation origination is the presence of the impurity (in these experiments chromium) of the critical concentration. The appearance of such a characteristic distribution of point defects itself indicates the changeover of the growth mechanism from an atom to an aggregate, i.e., junction of crystallites. The critical concentration value depends on thermodynamic parameters of the crystallization process and type of impurity. The origination under such conditions ultramicrostriation is revealed on the electron microscopic photographs owing to different sizes of the grain structure elements in the striae and outside it.

5.5 "Grain" Structure

It seemed [20, 26, 63] that by controlling two mechanisms of the dislocation structure formation – mechanism of inheritance and plastic deformation behind the CF – one can grow crystals not only free of dislocations but also with the specified law of structure defect distribution. However, as the experimental data stored it became obvious that control for the crystal structure using the mentioned mechanisms has a number of limitations. First, restriction is a possibility to control the structure in the solid phase. The reliably established experimental and theoretical principles of the general character were not confirmed when passing from one crystal to another. Even if a different method was used for growing the same crystal, the optimum crystallization conditions had to be chosen again [59]. The practice shows that variation of thermodynamic parameters in the Stepanov and Verneuil methods used (Fig. 5.28) does not allow the attainment of crystals with low density of dislocations.

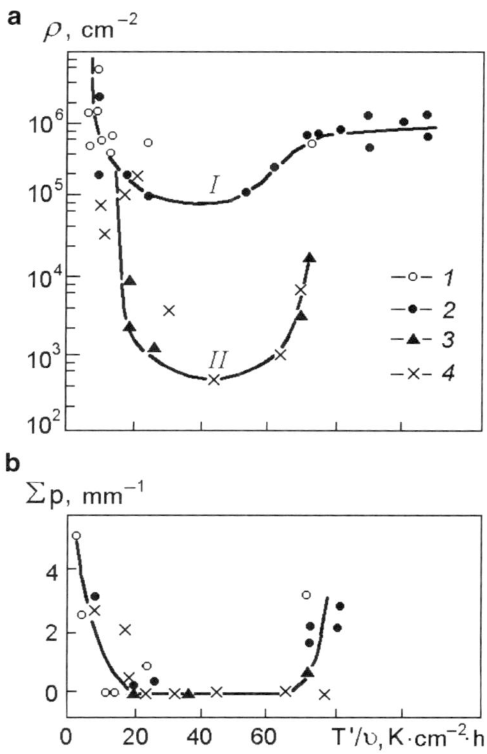

Fig. 5.28 Density of dislocations (**a**) and extension of block boundaries (**b**) versus axial temperature gradient/growth rate ratio for the growth methods: Stepanov (*1*), Verneuil (*2*), curve *I*, Czochralski (*3*), HDS (*4*), curve *II*

The value ρ varies in the range of $1 \cdot 10^5$–$5 \cdot 10^6$ cm^{-2}, here the left part of the curve is basically related to the crystals grown by the Stepanov method (high growth rates) and the right part by the Verneuil method (curve *I*). Variation of T'_n/v when growing crystals by the Czochralski and HDS methods gives a possibility of obtaining samples differing in density of dislocations by three orders. In this case the experimental dots keep within one curve II as well.

As shown by the given data for crystals grown by the Czochralski and HDS methods at $18 < T'_n/v < 70\,\mathrm{K\,cm^{-2}\,h}$ the formation of the dislocation structure is mainly realized by plastic deformation behind the crystallization front. In this case realization of the main concepts of the macroscopic theory of defect formation allows the control of the structure quality of crystals in the solid medium and to obtain block-free low-dislocation crystals or those with a specified distribution of structure defects. At $18 < T'_n/v < 70\,\mathrm{K\,cm^{-2}\,h}$ the possibilities of controlling structure perfection of crystals decrease; there is a discrepancy between theory and experiment. This being the case the values of the first and second derivatives from the temperature gradient by the coordinate define the dislocation structure of the grown crystals to a much lower extent. As to the Stepanov and Verneuil methods the majority of the experimental dots lie outside the interval $18 < T'_n/v < 70\,\mathrm{K\,cm^{-2}\,h}$; therefore, the possibility of controlling structure quality is not big in this case.

The extension of the boundaries at $18 < T'_n/v < 70\,\mathrm{K\,cm^{-2}\,h}$ is close to zero for all the methods because growth with $\rho < \rho_{cr}$ is possible, which allows reception of the block-free crystals. While for the Stepanov method elongation of the block boundaries takes place at $v \geq 140$ mm h^{-1} for the directed crystallization method the rise of Σp occurs at much lower values ($v \geq 25$ mm h^{-1}).

A number of experimentally observed trials do not keep within the frame of the existing theories of growth and structure defect formation. The high value of residual stresses (5–7 MPa) in the block-free low-dislocation crystals ($\rho = 1 \cdot 10$–10^2 cm^{-2}) annealed at the premelting temperatures; variation of the spatial orientation over the length of the perfect crystals (Fig. 5.29); and anomalously low entry of Cr^{3+} at low total concentrations of the impurity (Fig. 5.30) are examples of this.

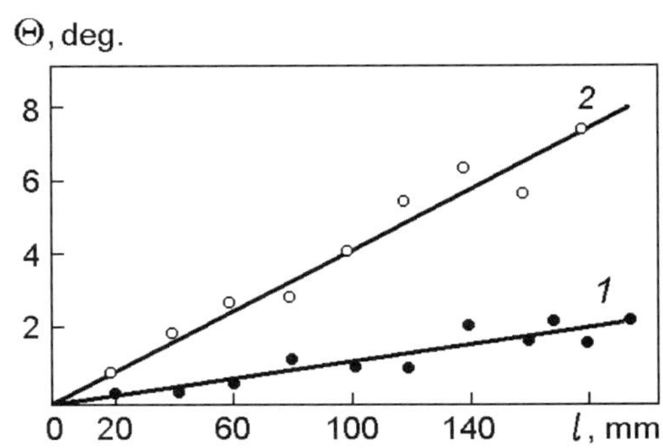

Fig. 5.29 Variation of spatial orientation over the length: (*1*) block-free low-dislocation crystals; (*2*) crystals with $\rho = 3 \cdot 10^5$ cm^{-2} and $\Sigma p = 2$ mm^{-1}

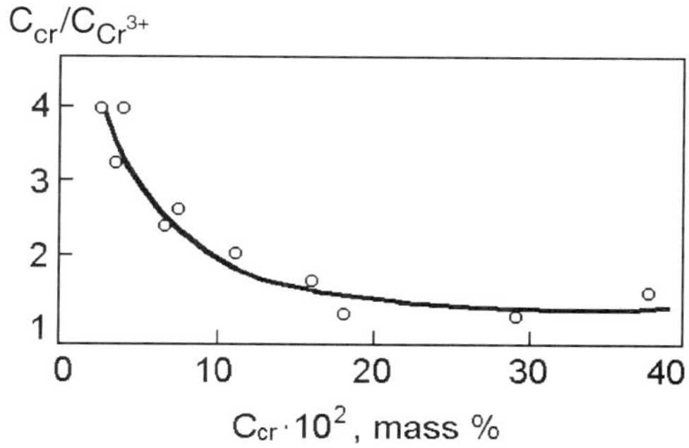

Fig. 5.30 The ratio of the total concentration of chromium to the concentration of chromium which substitutes the sites of the crystal lattice as a function of total concentration of chromium

In the region of low concentrations the ratio of the total concentration of chromium $C_{Cr^{3+}}$ to the concentration of chromium substituting the sites of the lattice $C_{Cr^{3+}}$ is high. With the rise of C_{Cr}, the concentration of the trivalent chromium goes up and the impurity mainly enters the crystal lattice sites. This being the case the studied crystals showed no specific distribution of single dislocations or block boundaries. This means that the observed anomalies cannot be explained by only the processes proceeding in the solid phase behind the CF.

The investigation of the fine structure of sapphire showed [60] that all of the crystal consists of separate "subgrains" (see Fig. 5.25). The electron microscopic photographs were made by the method of two-step replicas. For this purpose a faulty subsurface layer formed at mechanical treatment was removed by chemical polishing, plastic replicas were taken, and then using the method of self-shading platinum-carbon replicas were prepared on their basis. The resolution of the replicas was about 20 Å. Fine structure was observed on the electron microscopic photographs irrespective of the crystal growth method and preparation of the surface for the study (chemical, ionic polishing, etc.). In order to exclude completely the mechanical effect on the surface the same experiments were carried out on the chips of crystals. The result was the same.

In crystals grown at $18 < T_n'/v < 70\,\mathrm{K\,cm^{-2}\,h}$ there are aggregates the size of which varies at the changeover of the growth method and within one method but with variation of the dopant concentration. For crystals grown by the Verneuil method an increase of C_{Cr} from 0.03 to 1% stimulates growth of aggregates from 200 to 1,200 Å. For sapphire grown by the Stepanov method the size of the aggregates reaches 300 Å and for that grown by HDS method 400 Å. If crystals have impurity striation, the aggregates are observed both within the impurity striae and outside them. The presence of such grain structure cannot be explained in the scope of the existing theories of growth built on the Kosselev principle of energy stimulation, models of dislocation, banded, normal growth, and so forth.

Proceeding from the fact that the typical grain substructure has to fix the "history" of crystal growth in all the totality of the actions during growth, the factor studied was the dependence of the grain structure size upon impurity concentration within one growth method and at passing from one method to another. Within one method the characteristic size of the substructure is defined by the amount of the dopant. At changing the growth method one can vary the size of the grain structure in a wide range by varying the degree of the melt overheating. If the crystal is grown with dopants, it has an ultramicrostriation and the striae themselves and the intervals between them are discrete, i.e., consist of separate elements.

It is characteristic that some impurities (gallium, boron) added to the melt at rather high concentrations are revealed in the crystal only at trace levels (content by mass 10^{-4}–10^{-5}%). It is obvious that when changing the character of physical-chemical processes proceeding in the melt and connected with the grain structure formation these impurities do not enter the crystal.

Under the assumption that the boundaries between the elements of the grain structure is characterized by an increased concentration of vacancies a series of the diffusion experiments was carried out. Pairs of low dislocation samples oriented so that the available dislocations were basically normal to the diffusion flow and were not the favored paths of diffusion were taken. One sample of the pair was grown by the Czochralski method and the other by the gas-transport reactions. The first consisted of the subgrains 200–400 Å in size and the second was free of them. The layers of radioactive ^{51}Cr and ^{59}Fe were applied to the surface of the samples, then the pairs were put together by the treated surfaces and annealed at 1,500–1,900 K. It turned out that the diffusion coefficients of chromium and iron in the samples with the grain structure are approximately two times higher than those in the samples without such structure.

No doubt that high values of the diffusion coefficients are stipulated by the increased concentration of vacancies. While in crystals grown by the Stepanov method high concentration of vacancies can be stipulated by quenching of the grown crystal due to a high growth rate (to 240 mm h^{-1}) and high values of the axial temperature gradients (to 400 K cm^{-1}). In crystals grown by the Czochralski, Verneuil, and HDS methods there is no such source of vacancies. The fact itself of revealing boundaries between the elements of the grain structure by chemical polishing and ion bombardment is an indication of an increased energy of these boundaries, which can be explained by the excess of vacancies and presence of impurity.

Until recently there were two hypotheses on such characteristic distribution of point defects in the form of the grain structure. According to one [55], analogous to the semiconductor crystals such distribution of point defects is related to the processes proceeding in the solid phase near the CF: diffusion flows of the intrinsic defects (vacancies and interstitial atoms) directed deep into the crystal are created. The existence of such flows is connected with the fact that an equilibrium concentration of point defects is formed right behind the crystallization front and at some distance from it these defects recombine due to the temperature lowering, i.e., concentration gradients of defects are created (the higher they are, the higher is the temperature gradient in the crystal) and corresponding to them diffusion flows of defects. The total flow of each type of defect consists of the diffusion flow itself (proportional to the value of the T'_n in the crystal) and the flow connected with the crystal motion at some velocity. It is the relation of these flows, i.e., the relation of the crystal motion velocity to the temperature gradient, initial concentration of point defects, as well as their coefficient of diffusion, that defines distribution of point defects in the form of the grain structure.

By the other hypothesis [61] the processes that run in the melt are responsible for the formation of such defects. In order to determine this crystals grown by the method of gas-transport reactions were studied. They were not grown purposefully as they were waste products from growing corundum crystals by the Czochralski method (thin hexagonal plates 5–6 mm in size seldom grow on the crystal holder). No grain structure has been revealed in such crystals. It has not been found in crystals grown by the solid-phase method [62] if polycrystals received by pressing or slip casting were used as initial samples. In the solid-phase method the samples were pulled through the gradient zone at different temperatures and axial temperature gradients. The grain structure was found only in single crystals for the growth of which polycrystals received by filling the melt in molybdenum or tungsten molds.

The analysis of the structure quality of the crystals grown by this method (their size being no bigger than 1 cm^3) showed that they differ from crystals grown from melt by a characteristic distribution of defects and constituting chains of pores and aggregations of impurities. A comparison of the defect distribution with the initial structure of the polycrystal demonstrated that the observed aggregations of defects lie in the places of the former grain boundaries; that means that the crystal preserves "memory" about them. The electron microscopic investigations of the same samples revealed the grain structure. It is worth noting that in the initial grains of the polycrystalline samples of which the crystals were grown the same grain structure was disclosed. The size of the grain structure elements in crystals grown from melt obeys the same rules and depends on the same technological parameters as the maximum grain size at the solid-phase growth: in pure purposely nondoped samples the size of the elements linearly depends upon the T'_{n} and inversely proportional to the square of the second derivative of temperature T''_{n} by the coordinate.

A similarity of structure defect distribution in crystals grown from melt and by the solid-phase method of molded pieces, closeness of the dependence of the element size on the technological parameters indicates that the characteristic distribution of point defects observed on the electron microscopic photographs also is the memory of the crystal about the process of its formation.

5.6 Correlation Between Structure Quality of Crystals and Mechanisms of Their Formation

Growth of crystals with the specified structural and physical–chemical characteristics requires clear knowledge about the processes that run at the CF, in front of it, and behind it, because they define the presence of defects in the structure, impurity distribution, i.e., structure-sensitive properties and quality of crystal.

As a result of a series of interatomic collisions the initial electron structure of the isolated atom can be broken and the excited atoms will join into unstable complexes in which bonding is realized by the electrons on some intermediate metastable levels [62]. The probability of such complex formation depends on the relaxation time of the excited electron system (relaxation time is longer than the time interval between the collisions) and multistage (discrete) character of this system structure variation. Therefore, in some collisions excited in the atoms are rather long-living states in which the atoms form complexes of one or another degree of stability. The authors believe that the last stage of crystal growth can be considered a joining of complexes of the ultimate size. This stage is not accompanied by changes in their structure and proceeds at the expense of the surface energy of complexes. At low temperature and absence of impurities this would result in the formation of a perfect crystal. However, under real conditions the impurity is always a component of the condensate, at which the major part of the impurity is rejected and relegated to the boundaries of the formed complexes. This process violates the coordination of mutual orientation of complexes and the crystal acquires a mosaic structure, i.e., it consists of the blocks separated by boundary layers of atoms that are in an unstable state. Such character of the structure allows the authors to draw a conclusion about the peculiarities of crystal melting as well. Since the size of the mosaic blocks is sufficiently small (100–1,000 Å), the influence of the surface layer should be rather essential. With the temperature rise the mutual displacement of blocks increases, i.e., their perturbation action on each other becomes stronger. Perturbation from the block surface propagates to its volume and at high enough temperature excitation of atoms leads to a decomposition of the initial block structure. After some time necessary for relaxation of the electron excitation of atoms the block structure can change; it also can be accompanied by block decomposition into smaller families of atoms. Thus, at the approach to the Tm the changes take place in the crystal; these changes are opposite in their succession to those that occurred at the initial stages of crystallization. At Tm the blocks become fully isolated from each other, and some of them decompose.

In ref. [63] the motion of the interphase boundary at growing crystals from melt by the directed crystallization method is discussed. The author is basing on the facet that the crystallization rate is defined only by the conditions of heat transfer in the melt volume and removal of the latent crystallization heat from the interphase boundary and that the phase transition takes place at the temperature of phase equilibrium. On the basis of the calculation data it is concluded that at low crystallization rates the crystal grows without overcooling in the melt volume. At the crystallization rate higher than some critical one in the CF the overcooling zone in the melt appears; simultaneously, with a rise of the rate the depth of this zone increases. With such a formulation of the problem there is a solution that indicates a probability of partial crystallization of the material in the melt volume, which means that a spontaneous nucleation and growth of crystallites can occur in the melt volume. The experimental dependence of the overcooling value on the crystal growth rate was reviewed in ref. [64] for the aluminum oxide melt. The data of this work as already mentioned testify to the fact that Al_2O_3 melt is inclined to a significant overcooling and consequently the process of the material aggregation in the depth of the melt is easier. Theoretical investigation of the effect of thermal conditions on the critical velocity of the concentration overcooling and the analysis of the conditions of the three-dimensional nucleus formation in CF confirm that suggested in observations [65–67].

Hence, under certain conditions such a particular case of the growth mechanism when a precrystallization aggregation of particles happens in the melt is possible. It precedes attachment of these "bricks" to the crystal, making the growth process easier and accelerating it but at the same time lowering the "fidelity" of the crystal construction; its fine structure is a manifestation of this. Still there are few direct experimental data on the correlation between the melt structure and that of the crystal. Evidently, the limitation of such data is connected with the complexity of the experimental study and the analytical description of the processes proceeding in the melt. The point is that the processes that have an effect on the growth mechanism itself and are dependent on it are phenomena that occur in the solid phase minimally change the experimentally measured parameters of the melt, the values of which even at a significant variation of thermodynamic and chemical characteristics of the melt usually keep within a narrow range of values, which often are in the limits of the measurement error. The analytical description of such processes on the structure of the melt turns out to be a very small value. Moreover, a reliable measurement of the dependence of the crystal quality on the melt structure by a direct investigation of the processes proceeding in the melt may turn out to be principally impossible: on the one hand, due to low accuracy of reproduction of the external conditions, and on the other hand, because of the possibility of the melt structure variation by the process and measuring facilities.

To answer the question about the role of melt in the formation of the grain structure (characteristic distribution of point defects) and the experimental modeling of these processes, the Stepanov method was been carried out. This method was chosen for modeling because in crystals grown by it the grain structure is nearly always present and in the entire volume (Fig. 5.31).

A sapphire plate cut of a crystal grown by the Kyropoulos method and free of the grain structure was fused until the line shown in the Fig. 5.31; after this the remainder of the plate was taken as a seed on which the crystal was grown. Crystallization conditions, temperature gradients, broaching velocity, fusion rate, degree of a preliminary overheating of the melt, and so forth were varied. No grain structure has been revealed in the remainder of the crystal despite the fact that there were all the conditions for the origination of the diffusion flows of point defects deep into the crystal. Instead, the grain structure always appeared in the grown part of the crystal. The results of the experiment demonstrated that the processes not running in the solid phase behind the CF but proceeding in the melt in the precrystallization zone or just at the crystallization front are responsible for the formation of the grain structure.

The melt aggregation occurs in the vicinity of the interphase boundary are nucleated and grow crystallites. Their mutual interaction with the interphase surface is reduced to the fact that at a collision the growth of crystallites stops in the direction

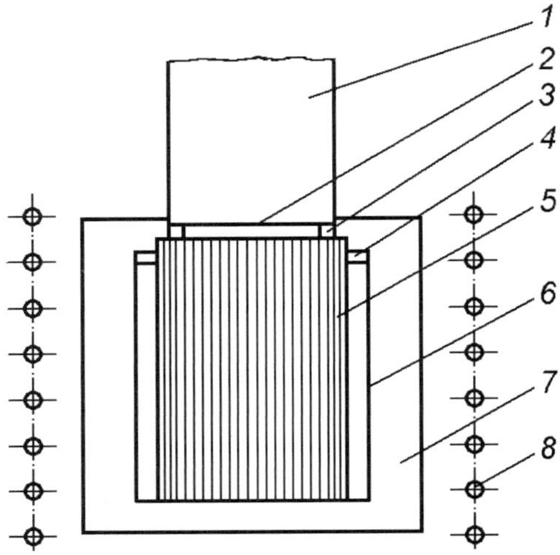

Fig. 5.31 A diagram of the experiment: (*1*) sample; (*2*) interphase boundary; (*3*) die; (*4*) cover plate; (*5*) capillary system; (*6*) crucible; (*7*) concentrator of HF power; (*8*) inductor

of the collision, while in other directions the growth goes on. Between the neighboring crystallites a grain boundary appears (the front of their collision) that contains an increased concentration of defects. In the ultimate case of the melt aggregation a mass crystallization may occur before the interphase boundary; a thin polycrystalline layer is formed, which afterward recrystallizes (forming a single crystal) in the gradient zone.

Accordingly, the considered mechanism of the displacement of the interphase surface is a particular case of the crystal growth under conditions of a strong melt aggregation and includes two stages. The first is the origination of individual crystallites near the interphase boundary (in the ultimate case a polycrystalline layer); the second – monocrystallization – consists of a migration of the grain boundaries (Fig. 5.32). The moving boundaries during the monocrystallization leave a "trace" in the single crystal in the form of a typical distribution of point defects (impurities, vacancies) in place of the original settlement of the boundaries. The thickness of the melt layer in which the crystallites are formed and attached can be qualitatively evaluated from the following considerations. In the melt both at free and compulsory convection a transition layer exists the parameters of which vary from the values characteristic of a dead melt to the values corresponding to the free melt. The thickness of this layer, δ', depends on the velocity of the crystal shifting through the gradient zone. By assuming as usual for the directed crystallization $v\delta'/D \approx 10^{-1}$ we get $10^{-4} \leq \delta' \leq 10^{-2}$ cm. Evidently, the value of the parameter δ allows the explanation of the experimentally observed dependence of the grain structure element size on the ratio T'_n/v (Fig. 5.33). When growing crystals from a thin film of the melt (Verneuil and Stepanov methods) due to a low thermal inertia of this film, small changes in the thermal conditions at the CF result in the variation of the value δ' and the corresponding change in the size of the crystallites. When growing

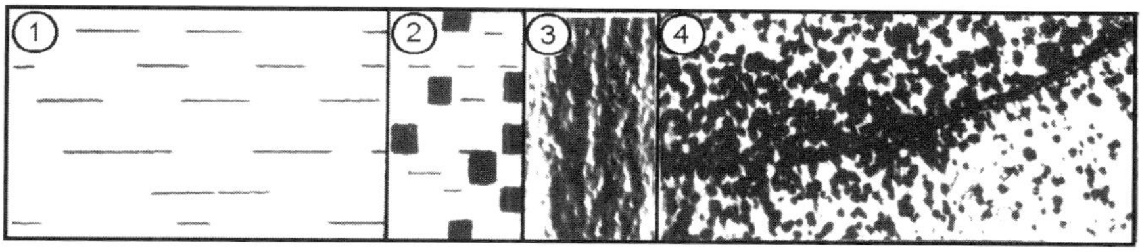

Fig. 5.32 A pattern of the growth processes under conditions of a strong aggregation of the melt: (*1*) free melt; (*2*) precrystallization zone; (*3*) zone of monocrystallization; (*4*) plasticity zone; (*1–3*) ×100,000; (*4*) ×250

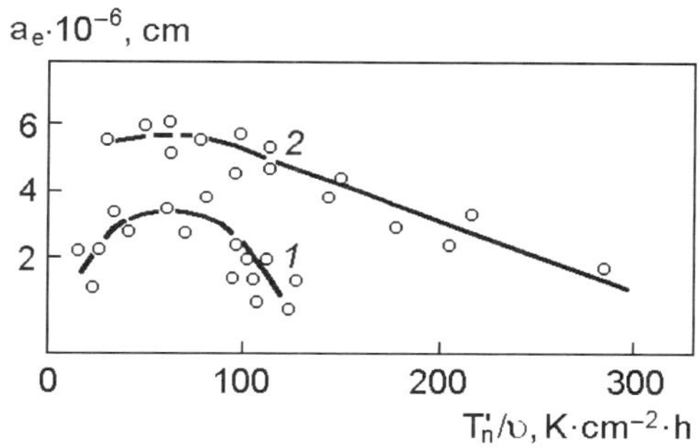

Fig. 5.33 Dependence of the size of the grain structure elements on the axial temperature gradient/growth rate ratio, methods: (*1*) Verneuil, Stepanov; (*2*) Czochralski, HDS

crystals by the Czochralski, Kyropoulos, and HDS methods a large mass of the melt smoothes out the temperature oscillations of the system and the size of the grain structure elements varies insignificantly.

The suggested mechanism of the displacement of the interphase boundary called by the authors MPS (melt–polycrystal–single crystal) explains the experimentally observed anomalies in the dependence of the structure quality of crystals on their growth conditions. For example, there appears to be a possibility to account for the existence of a high enough level of the internal stresses in the low dislocation crystals, emergence of a wide dislocation free zone near the crystal surface, and low density of dislocations in the cracked during growth crystals and to explain the mechanism of the effect of the external conditions (temperature gradients, their inconstancy, growth atmosphere, impurity composition of the melt, etc.) on the quality of crystals basically through the melt structure changes.

Whatever the mechanism of the grain structure formation is, there is no doubt that its presence, i.e., the existence of a characteristic distribution of point defects, cannot but influence the processes behind the crystallization front in the plastic region. So, density of dislocations in crystals free of the grain structure is mainly defined by the value of thermoelastic stresses. As to the ρ in crystals that have the grain structure, it also is defined by the character of distribution and concentration of point defects, which in their turn depend on the size of the elements, type of the impurity, and its concentration in the crystal [68]. When growing by the Verneuil and Stepanov methods at rather a high level of thermoelastic stresses in the impurity-free crystals (small size of the grain structure elements), such distribution of the point defects appears which does not impede relaxation by plastic flow, i.e., formation of dislocations. This defines the total high level of dislocations in crystals grown by the Verneuil and Stepanov methods ($1 \cdot 10^5$ to $1 \cdot 0^6$ cm^{-2}).

The size of the grain structure elements in "pure" crystals grown by the Czochralski, Kyropoulos, and HDS methods is much bigger; the concentration of point defects at their junctions is higher, and the total level of thermoelastic stresses is essentially lower. Namely, this offers the possibility of obtaining by the above-mentioned methods crystals with low $\rho \sim 1 \cdot 10^2$ to $1 \cdot 10^3$ cm^{-2}.

If when growing by the Verneuil and Stepanov methods the impurity concentration in crystals increases and grain structure element size enlarges, the concentration of point defects at their boundaries rises, the relaxation of thermoelastic stresses by plastic flow is impeded. The ultimate case of such a situation leads to the cracking of crystals during their growth or cooling; simultaneously density of dislocations in the cracked crystals is several times lower than in the safe ones [37]. In case of an increase of the impurity concentration in crystals grown by other methods, the situation with the enlargement of the grain structure elements and distribution of point elements is repeated; however, owing to the low total level of thermoelastic stresses the crystals remain intact. If in crystals grown by the Verneuil and Stepanov methods density of dislocations near the surface becomes appreciably lower (in some cases even a dislocation free zone 200–500 μm wide is formed [69]), the size of the grain structure elements near the surface also becomes larger (by 20–50%) than in the crystal volume. The suggested assumptions about the effect of the characteristic distribution of point defects on plastic deformation of crystals are proved by a correlation between the cracking resistance of crystals, density of the growth dislocations, and size of the grain structure elements (Table 5.4).

Surfaces of the crystals that have undergone a mechanical treatment ($R_z = 0.05$ μm) were locally loaded. The general tendency of the fracture toughness to increase with the lowering of defect density both within each method and between the methods is clearly seen. The highest fracture toughness belongs to the most perfect crystals grown by the Kyropoulos method. In the limits of this method a rise of the dislocation density at the order gives in diminution of k_c approximately in 1.3 times.

Table 5.4 Coefficient of the fracture toughness of crystals (k_c)

Growth method	ρ (cm^{-2})	k_c (MN m$^{-3/2}$) Cleaved surface	k_c (MN m$^{-3/2}$) Mechanically polished surface
Verneuil	$(6–8) \cdot 10^5$	2.3–2.8	4.0
Stepanov	$(3–5) \cdot 10^5$	3.0	3.7
Czochralski	$8 \cdot 10^4$ $1 \cdot 10^5$	3.5	4.6
Kyropoulos	$1 \cdot 10^2$ $5 \cdot 10^3$	4.0	5.4

The following dependence is traced: the processes proceeding before the CF in the melt and responsible for the distribution of point defects in the crystal are in direct dependence on the liquid phase state and have an effect on mechanical characteristics of the crystal – processibility, working capacity, and durability of the elements prepared under different reactions. This is because the size of the elements and density of point defects at their junctions affect the density of growth dislocations dependent on which is the coefficient of the fracture toughness. With the increase of density of growth dislocations the fracture toughness drops and mechanical characteristics of the crystal become worse. This implies that the problem of controlling the structure quality of crystals, i.e., getting crystals with the specified distribution of structure defects and required mechanical characteristics, is to be solved as early as the stage of crystal structure formation in the liquid phase and the determination of general regularities of the effect of the external conditions on the properties of the system crystal–melt is a necessary condition for growing crystals with the specified structure perfection. However, determination of such regularities at real growth is practically impossible due to a complicated multifactor dependence of melt aggregation on particular conditions: impurity composition, technological parameters of the crystallization process, and so forth. In this connection, the crystal formation mechanisms and creation of defects in it by local fusion of the preliminarily studied crystal with a posterior crystallization varying in a wide range the growth conditions were modeled.

The experiments were carried out on the crystallization installations for the growth of sapphire furnished with an additional heater in the form of a thin tungsten wire (Fig. 5.34). The crystal was heated until temperature T_1, then it was locally

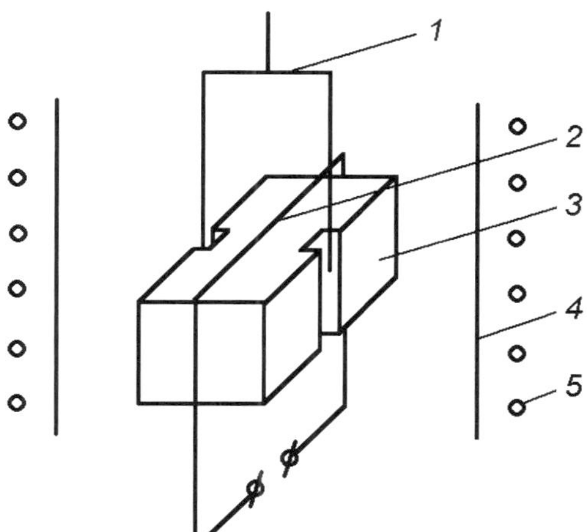

Fig. 5.34 A diagram of carrying out the model experiments: (*1*) sample holder; (*2*) wire heater; (*3*) sample; (*4*) concentrator of the HF power; (*5*) inductor

fused using the additional heater, at the temperature $T_2 > T_1$. The temperature T_1 was varied from 1,900 to 2,300 K and the temperature T_2 was varied from 2,320 to 2,700 K. The velocity of the sample displacement with respect to the additional heater varied from 10 to 20 μm h^{-1}. Samples with the impurity microstriation were taken for the study of the effect of impurity on structure quality of the fused zone. The width of the striae were 10–100 μm and the distance between them can reach several millimeters. Structure quality of the recrystallized zone was judged by the density of dislocations, block structure, density and pores, and point defects (size of the grain structure elements); its mechanical characteristics were judged by the coefficient of fracture toughness. The dependence found during model experiments are shown in Fig. 5.35. For a comparison the same dependence found for crystals that were grown by the Verneuil, Stepanov, Czochralski and HDS methods are shown in this figure. As in the real experiments in the interval $20 < T_n'/v < 80$ K cm^{-2} h density of dislocations and the coefficient of fracture toughness are maximal.

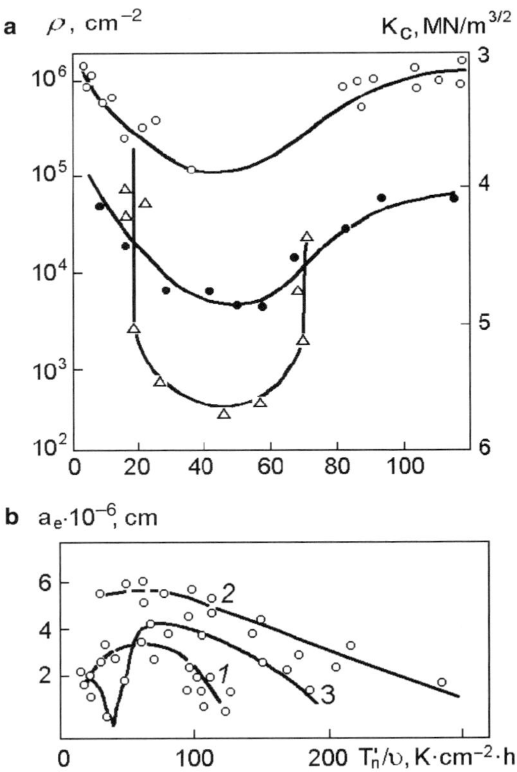

Fig. 5.35 Dependence of density of dislocations, coefficient of fracture toughness (**a**) and size of the grain structure elements (**b**) on the axial temperature gradient/growth rate ratio, methods: (*1*) Verneuil and Stepanov; (*2*) Czochralski and HDS; (*3*) model experiments

The dependence of a_e on the ratio T'_n/v has a different, more complicated, character as compared to the curves found during real growth. By varying the ratio T'_n/v one can have the size of the elements from 0 to 400 Å. At $T'_n/v = 30 - 40\,\mathrm{K\,cm^{-2}\,h}$ the grain structure is not formed.

A significant contribution to the formation of structure defects is made by the melt overheating in the recrystallized zone. If the overheating is not high (Table 5.5) irrespective of the crystallization rate and temperature gradients, the characteristic distribution of point defects in the recrystallized zone does not differ from the corresponding distribution in the remaining volume of the crystal. In the zone directly adjacent to the recrystallized zone at high velocities of broach and high temperature of the main heater (high specific load and high plasticity of the material) the blockboundaries near the fusion zone can change their direction by almost 90° and turn around toward passage of the additional heater. At $T_1 > 2{,}100$ K density of the growth dislocations can increase and redistribute. Increase of density of dislocation is accompanied by lowering of the coefficient of fracture toughness.

Something different is observed in the case of high overheating of melt (Table 5.6). The original fine structure is not inherited, the value of a_e can be varied in

Table 5.5 Parameters of the crystallization process and crystal structure at low overheating of the melt (T_2 = 2,323 K)

Original structure			Varied parameters		Structure of the recrystallized zone		
ρ (cm^{-2})	a_e (Å)	k_c (MN m$^{-3/2}$)	T (K)	v (mm h^{-1})	ρ (cm^{-2})	a_e (Å)	k_c (MN m$^{-3/2}$)
$2 \cdot 10^3$	0	5.2	1,900–2,000	10	$5 \cdot 10^3$	0	5.0
				100	$8 \cdot 10^3$	0	4.8
			2,100–2,200	10	$8 \cdot 10^3$	0	5.0
				40	$4 \cdot 10^4$	0	4.6
				100	$8 \cdot 10^4$	0	4.8
			2,300	10	$8 \cdot 10^3$	0	4.8
				40	$6 \cdot 10^4$	0	4.2
				100	$1 \cdot 10^5$	0	4.2
$2 \cdot 10^5$	200	3.8	1,900–2,000	10	$3 \cdot 10^5$	200	3.6
				100	$4 \cdot 10^5$	200	3.6
			2,100–2,200	10	$3 \cdot 10^5$	200	3.6
				40	$6 \cdot 10^5$	200	3.4
			2,300	10	$2 \cdot 10^5$	200	3.4
				40	$4 \cdot 10^5$	200	3.0
				100	$8 \cdot 10^5$	200	3.2
$5 \cdot 10^3$	600	4.8	1,900–2,000	10	$5 \cdot 10^3$	600	4.8
				100	$3 \cdot 10^3$	600	4.8
			2,100–2,200	10	$1 \cdot 10^3$	600	4.6
				40	$3 \cdot 10^3$	600	4.2
				100	$1 \cdot 10^4$	600	4.0
			2,300	10	$2 \cdot 10^3$	600	4.6
				40	$8 \cdot 10^3$	600	4.0
				100	$1 \cdot 10^4$	600	3.8

Table 5.6 Parameters of the crystallization process and crystal structure at large overheating of the melt (T_2 = 2,700 K)

Original structure				Varied parameters		Structure of the recrystallized zone		
ρ (cm^{-2})	a_e (Å)	C (%)	k_c (MN m$^{-3/2}$)	T (K)	v (mm h^{-1})	ρ (cm^{-2})	a_e (Å)	k_c (MN m$^{-3/2}$)
$2 \cdot 10^3$	0	–	5.2	1,900–2,000	10	$8 \cdot 10^3$	200	4.8
		$3 \cdot 10^{-3}$	–		100	–	100	4.6
		$1 \cdot 10^{-2}$	–	2,100–2,200	10	$8 \cdot 10^3$	600	4.6
		$5 \cdot 10^{-2}$	4.9		40	$1 \cdot 10^4$	300	4.0
		$1 \cdot 10^{-2}$	–		100	$5 \cdot 10^4$	200	3.8
		$5 \cdot 10^{-2}$	4.6	2,300	10	$1 \cdot 10^4$	400	4.2
		$5 \cdot 10^{-2}$	–		40	$5 \cdot 10^4$	300	4.2
		$5 \cdot 10^{-2}$	–		100	$8 \cdot 10^4$	200	3.6
$2 \cdot 10^5$	200	$1 \cdot 10^{-3}$	–	1,900–2,000	10	$3 \cdot 10^5$	200	3.8
		$1 \cdot 10^{-3}$	–		100	$2 \cdot 10^5$	200	3.8
		$1 \cdot 10^{-2}$	3.6	2,100–2,200	10	$3 \cdot 10^5$	200	3.8
		$1 \cdot 10^{-2}$	–		40	$5 \cdot 10^5$	300	4.2
		$1 \cdot 10^{-2}$	–		100	$6 \cdot 10^5$		400
		$5 \cdot 10^{-2}$	–	2,300	10	$4 \cdot 10^5$	400	4.0
		$5 \cdot 10^{-2}$	–		40	$8 \cdot 10^5$	400	4.2
		$5 \cdot 10^{-2}$	–		100	$8 \cdot 10^5$	500	4.6
$5 \cdot 10^3$	600	$5 \cdot 10^{-3}$	–	1,900–2,000	10	$2 \cdot 10^3$	500	4.6
		$5 \cdot 10^{-3}$	–		100	$4 \cdot 10^3$	500	4.8
		$5 \cdot 10^{-3}$	4.8	2,100–2,200	10	$2 \cdot 10^3$	400	4.6
		$5 \cdot 10^{-3}$	–		100	$1 \cdot 10^4$	600	5.0
		$5 \cdot 10^{-3}$	–	2,300	10	$2 \cdot 10^3$	400	4.8
		$5 \cdot 10^{-3}$	–		100	$5 \cdot 10^3$	500	4.6

a wide enough range by varying, first of all, type and concentration of impurity by taking the initial samples with macrostriation. The ultramicrostriation appears under such conditions. The state of the recrystallized zone can be varied from a polycrystal to a perfect crystal. In the zone closely adjacent to the recrystallized zone the block structure and ρ change as the difference $T_2 - T_1$ becomes less, i.e., the main heater temperature rises.

The same way as in the previous case the block boundaries change their direction, the pores coagulate, and large areas of the volume depurate from pores. Comparison of Tables 5.6 and 5.7 concludes that the fine structure of the crystals at their local fusion under conditions of low overheating of the melt does not decompose. A series of experiments on fusion of the zone of contact of two crystals with the same or quite different grain structure also confirm this fact. At $T_2 \approx T_m$, provided two crystals are in contact with the same grain structure, the fusion and posterior crystallization of the contact zone offers a one-piece junction with low fracture toughness. If crystals are in contact with different sizes of the grain structure elements, no one-piece junction can be obtained. Perhaps this is connected with a high level of the internal stresses arising in the contact zone at fusion. Upon

Table 5.7 Partial pressures (P_p) and density of alumina decomposition yields (γ_p) at 2,303 K

Components	*M*	$P_p \times 10^2$ (Pa)	Part (at. %)	$\gamma_p \cdot 10^{-11}$ (kg cm^{-3})·	Part (at. %)
AlO	43	1.4	6.4	3.4	12.2
Al_2O	70	0.4	1.8	1.4	5.7
Al	27	7.2	33.3	9.8	39.3
O_2	32	0.5	2.3	0.8	3.2
O	16	1.2	56.1	9.7	39.3

cooling, the crystals crack. At $T_2 \gg T_m$ irrespective of the size of the grain structure elements a one-piece junction with rather high k_c (sometimes even higher than that of the original crystals) can be obtained. At melt overheating the initial characteristic distribution of point defects in the contacting samples is not inherited, and depending on the conditions of the experiment, a specified structure quality of the recrystallization zone can be obtained.

Thus, the experiments allow not only modeling of different mechanisms of the crystal formation but distinguishing two cases as well: the first ($T_2 \approx T_m$) – control for the structure quality of the recrystallized and closely adjacent to it the zone is realized only in the solid phase behind the CF; the second ($T_2 > T_m$) – the control is realized in the liquid phase. The control for the structure quality in the solid phase behind the CF itself depends on the character of the proceeding in the liquid-phase processes.

5.7 Inclusions in Sapphire

In addition to point, linear, and surface structure, the defects described in the preceding paragraphs, a specific group of volume defects exists in sapphire that generally are referred to as inclusions, and include gaseous pores, inclusions of foreign solid phase, and (rarely in natural crystals) liquid phase inclusions. *Pores* are almost inevitable defects in any growing technique. These are formed to a greater or lesser extent in the crystals grown under hard thermodynamic conditions as well as in nearly equilibrium ones. The pores deteriorate the performance characteristics of pieces and the polish quality, as well as increase the light scattering. The *pore size* varies from submicron values to 3 to 5 mm. At a high concentration of these defects (>10^5 cm^{-3}), the material is unsuitable for optical applications.

The *pore composition* depends on the growth atmosphere. Mass spectrometry reveals the main peaks in sapphire with mass numbers of 2 (hydrogen) and 28 (CO). The peak intensities and ratios depend on the growth technique and the growth zone atmosphere. Peaks of lower intensities with mass numbers 12 (carbon), 14 (N, CH_2), 16 (O, CH_4), 32 (O_2), 44 (CO_2), and 68 (C_3O_2) have been found [70]. The estimated partial pressure of hydrogen in a pore is about 10 mmHg, the 28 peak is thrice higher [71, 72].

The *pore shapes* are evidence of the complexity and variety of processes resulting in pore formation. Spherical, pear-shaped, dumb-bell shaped, fiber-shaped, hexagonal disklike pores were observed.

The main pore formation factors in sapphire grown from melt are believed to be the melt interaction with impurities in the raw blend, container material, and atmosphere as well as the melt dissociation. The general conditions favoring the pore formation are the melt supersaturation in dissolved gases. The pore formation in crystals is preceded by the back-drivingof the gas dissolved in the melt from the crystallization front. A bubble of subcritical size is pushed back from the CF as the liquid is mixed weakly, the concentration densification and the bubble size increase, resulting in its contact with the CF and retention due to surface tension forces. Since the crystal growth above the bubble is hindered, a pore is formed in the solid phase. A reverse process, namely, the gas penetration into melt along the block boundaries, also is possible although less probable. The boundary exit onto CF is accompanied by formation of a recess that is a region of elevated concentration of dissolved gases. No pore formation, however, is observed in some cases at such regions. According to ref. [73], this is due to the fact that the boundary itself is a runoff for the dissolved gases that do diffuse along the defects included in that boundary. Thus, the gas suction out of the recess at the boundary results in a decreased excess gas concentration.

When a crystal is grown using a container, it is difficult to avoid overheating (overcooling) of some container zones, and thus of the melt. The moving CF can capture microvolumes of the cooled melt. In this case, the voids are formed during the subsequent crystallization of those microvolumes due to density differences between the crystal and melt. The melt dissociation, being a possible consequence of overheating, also increases the gas saturation. The main dissociation products are atomic oxygen and atomic aluminum (Table 5.7).

The total pressure of the dissociation gaseous products attains $2.1 \cdot 10^{-1}$ Pa; the total density of these components makes $2.5 \cdot 10^{-10}$ g cm^{-3}. Perhaps the composition of dissociation products dissolved in the melt differs from that in the gas phase. Since Al_2O_3 melt has an ionic character, it contains AlO^+ and AlO_2^- ions.

General concepts of segregation of impurities near the crystallization front are applicable to the dissolved dissociation products. During crystal growth, a concentration condensation layer is formed near the moving CF. At a distance from the CF, the melt is saturated with the dissolved dissociation products, the effective concentration C_0 thereof being no doubt higher in the vicinity of the moving front. As appears from the presented evaluations (by V. Thiller), the threshold of metastability, i.e., supersaturation, above which nucleation of gas bubbles begins, is 90% for Al_2O_3, which corresponds to the concentration $1.9C_0$. On the other hand, in the nonmixed melt, the established concentration of the gaseous components at the crystallization front is $C_f = C_0/K_d = (100–1{,}000)C_0$ at $K_d \approx 10^{-2}–10^{-3}$. This value is well above the metastability threshold, i.e., gas bubbles are always formed in the nonmixed melt.

The mixing intensity as conventionally accepted can be characterized by the effective thickness of the diffusion boundary layer δ. Having assumed that the tolerant value C_f/C_0 should be lower than the metastability threshold or at least equal to it, we can find the thickness δ at which the bubbles cannot appear using the Barton, Prim, and Stichler formula:

$$\delta = \frac{D}{v}\ln\frac{C_f / C_0(1-K_d)}{1-C_d^{C_f/C_0}} \tag{5.28}$$

Assuming for sapphire that $C_f/C_0 = 1.9$, $v = 2 \cdot 10^{-4}$ cm s^{-1}, $D = 1 \cdot 10^{-5}$ cm^2 s^{-1}, and $K_d = 10^{-2}$, one can get the thickness of the boundary layer $\delta \approx 320$ μm.

To conceive of bubble formation kinetics in the melt and that of pores in the crystal, let the critical radius of a primary nucleus r^* and its emergence frequency as a function of supersaturation be [74]:

$$\omega' = \omega'_0 \exp\left(-\frac{\frac{4}{3}\pi\sigma_s(r^*)^2\mathrm{N}}{RT}\right) \tag{5.29}$$

where N is the Avogadro number, ω'_0 is the preexponential factor, and $\omega'_0 \approx 10^{37}$ cm^{-3} s^{-1}.

The results of calculation of r^* and ω' depend on supersaturation for the melt with $\sigma_s = 0.7$ N m^{-1}, $\gamma = 2.63$ g cm^{-3}. At the metastability threshold ($C_f/C_0 = 1.9$), the nucleus-critical radius is about 10^{-7} cm and the bulk frequency of nucleation makes about 10^3 cm^{-3} s^{-1} (Table 5.8). As supersaturation increases, the latter value rises and vice versa. A nucleus r^* at $C_f/C_0 = 1.9$ contains about 250 atoms.

Table 5.8 Influence of oversaturation on radius and frequency occurrence of germs

C_f/C_0	r^* (cm)	ω' (pcs cm^{-3} s^{-1})
1	–	–
1.1	$6.2 \cdot 10^{-7}$	–
1.2	$3.3 \cdot 10^{-7}$	–
1.3	$2.25 \cdot 10^{-7}$	$1.1 \cdot 10^{-160}$
1.4	$1.75 \cdot 10^{-7}$	$2.2 \cdot 10^{-90}$
1.5	$1.45 \cdot 10^{-7}$	$2.7 \cdot 10^{-48}$
1.6	$1.25 \cdot 10^{-7}$	$1.1 \cdot 10^{-26}$
1.7	$1.11 \cdot 10^{-7}$	$3.1 \cdot 10^{-13}$
1.8	$1.01 \cdot 10^{-7}$	$4.1 \cdot 10^{-4}$
1.9	$9.2 \cdot 10^{-8}$	$1.1 \cdot 10^{3}$
2.0	$8.5 \cdot 10^{-8}$	$7.9 \cdot 10^{7}$
2.5	$6.4 \cdot 10^{-8}$	$1.7 \cdot 10^{20}$
3.0	$5.4 \cdot 10^{-8}$	$2.5 \cdot 10^{25}$
3.5	$4.7 \cdot 10^{-8}$	$1.2 \cdot 10^{28}$
4.0	$4.2 \cdot 10^{-8}$	$5.6 \cdot 10^{29}$
4.5	$3.2 \cdot 10^{-8}$	$8.3 \cdot 10^{30}$
5.0	$3.7 \cdot 10^{-8}$	$3.7 \cdot 10^{31}$

The container material influence on the melt gas saturation is beyond doubt. The intensity of the container–melt interaction increases in the sequence Ir–W–Mo. The melt gas saturation in the Al_2O_3–Mo system exceeds that in the Al_2O_3–Ir one and in Al_2O_3 by a factor of 10 and in the Al_2O_3–W system by a factor of 1.1–1.5. The maximum growth rate at which the CF still does not capture the bubbles is in inverse proportion to the melt gas saturation. Other conditions being the same, the pore content in ruby is higher than in sapphire, because the melt doped with Cr_2O_3 interacts with the container materials more intensively.

Pore formation due to vacancies manifests itself when sapphire is grown at high speeds or in reducing conditions. In this case, there also is a critical stoichiometry deviation in the Al_2O_3 lattice; when that deviation is exceeded, a second-phase microparticles may be precipitated in the crystal. In reducing conditions, the melt becomes depleted of oxygen and the composition of dissociation products becomes shifted toward compounds with excess Al content as compared to stoichiometric Al_2O_3. In crystals with high vacancy density, the pores may arise after its formation, i.e., at the cooling. That is, the vacancies (as well as some impurities) still remain in equilibrium during the crystallization and the supersaturation, coagulation, and separation of inclusions occur in the course of cooling. The high-temperature vacuum annealing of such crystals also favor the coagulation of vacancies into faceted pores (negative crystals).

Specific causes of pore formation are typical of some crystal growing methods or groups of methods. In plates being grown in boats by HDC, large near-bottom pores are formed that are retained due to surface tension. The amount thereof depends on the raw material purity. According to evaluations [71], bubbles in the melt having a diameter of about 2.1 mm are unable to tear off from the molybdenum container surface and to emerge. Therefore, these are captured by the growing crystal.

In the *Verneuil technique*, the crystals grow from a thin melt layer. When superheated, this layer "swims down"; therefore, the melt dissociation is hardly probable. Along with the pore formation caused by impurities, however, a cause specific for that method appears, namely, the size distribution of the raw blend particles. The particles commensurable in size with the melt film thickness attain the CF with no sufficient time to be melted in the torch (Fig. 5.36a) and the gas adsorbed thereon is released and captured by the front [75]. The radius of particles that do not attain the front is defined as

$$r < B\sqrt{1+\delta_{\mathrm{m}}/B-1} \qquad B = 9\mu/2\gamma_{\mathrm{m}}v_{\mathrm{x}} \tag{5.30}$$

where μ is the melt dynamic viscosity; δ_{m}, the melt layer thickness; γ_{m}, the melt density; v_{x}, the particle speed.

It follows from (5.30) and Fig. 5.36 that to decrease the pore content in a crystal being grown from a melt film fed by raw blend particles, it is necessary to decrease the particle feeding rate and to optimize the particle size distribution.

The pore size in a sapphire grown after Verneuil is as a rule < 0.1 mm. In crystals doped with chromium (0.5–1 wt%), however, pores as large as about 1–1.5 mm may appear.

In the *Stepanov technique*, the crystals grow from a thin melt layer, too. The pore formation and distribution are influenced considerably by the shaper design and the melt flow hydrodynamics at its shape-forming surface. At a multichannel feeding of the layer, the gas-saturated melt flows collide with each other at the shaper. Microturbulences arise in the collision zones and quickly attain the critical size, and then become captured by the CF. The pore formation kinetics in a long sapphire rod are shown in Fig. 5.37. As the rod length increases, the capillary pressure drops as well as the collision intensity of the flows. The turbulence can increase the shaper roughness. In this case, the width of the flow interaction zone rises. The flow interaction character is defined by interphase turbulence. The flows are braked mutually at the junction boundary due to the melt viscosity. The velocity vectors oriented in

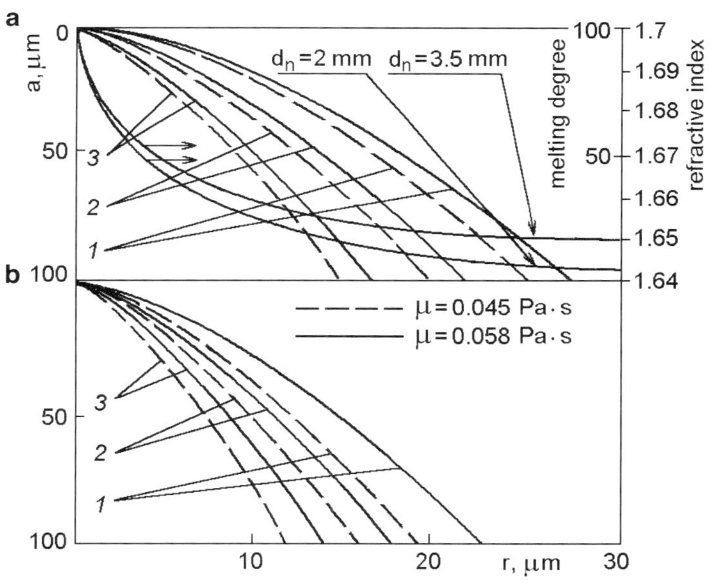

Fig. 5.36 Immersion depth *a* of a particle of radius *r* into melts of different viscosity (**a**), taking into account the melting degree (**b**). The particle speed, (m/s): 5 (*1*), 10 (*2*), 20 (*3*); d_n is the nozzle diameter [86]

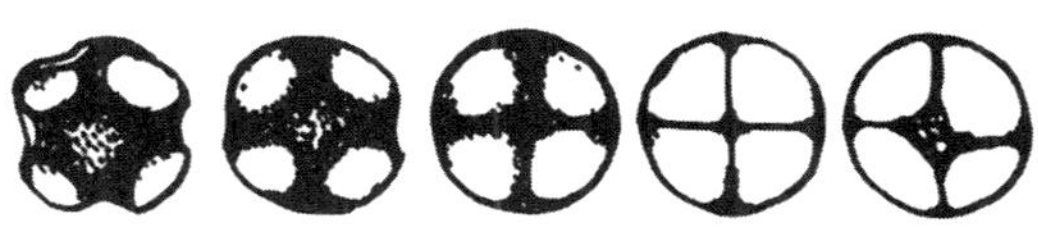

Fig. 5.37 Pore formation in a sapphire profile in four-channel feeding at the crystal length of 10, 155, 300, 450, and 600 mm

different directions give rise to force couples that rotate the flowing layers and originate structures similar to vortices. The vortices pierce the boundary layers and penetrate into the flows due to the Zhukovsky forces. Zones of the melt turbulent flow arise. In this case, the pressure drops below the saturated vapor one; therefore, the dissolved gases are released and forced by the melt flows to zones of lowered pressure where they are condensed.

As to profiled crystals, a near-surface layer with an increased pore density is typical. The melt flows attaining the meniscus surface are repulsed there from and collide with the incoming flows. Therefore, a 100–300 μm thick porous layer is formed at a distance of 0.2–0.3 mm from the profile surface. At the single-channel feeding, the pores are formed mainly within the profile near-surface layer.

The change of the crystal pulling speed from 0.2 to 0.3 mm/min does not shift appreciably the concentration maximum. There is a plateau in the dependence of the maximal concentration supersaturation on the pulling speed (Fig. 5.38). This speed range is favorable for the growing.

At a certain critical growing speed that depends on the meniscus parameters, the concentration supersaturation region becomes extended over the entire meniscus width. It is seen from Figs. 5.39 and 5.40 that the concentration supersaturation

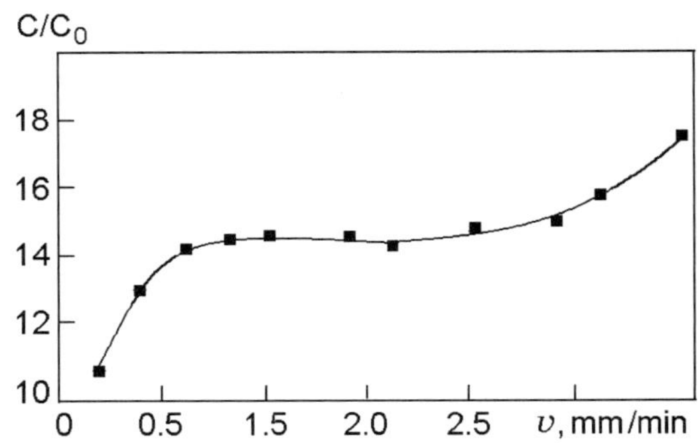

Fig. 5.38 Dependence of C/C_0 ration on the crystal growing speed [87]

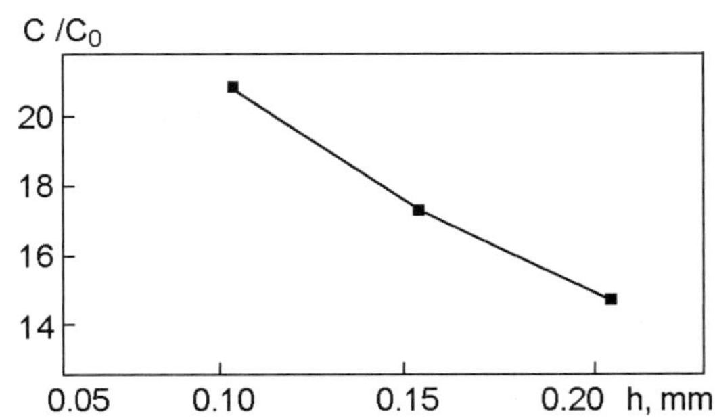

Fig. 5.39 C/C_0 ratio as a function of the meniscus height

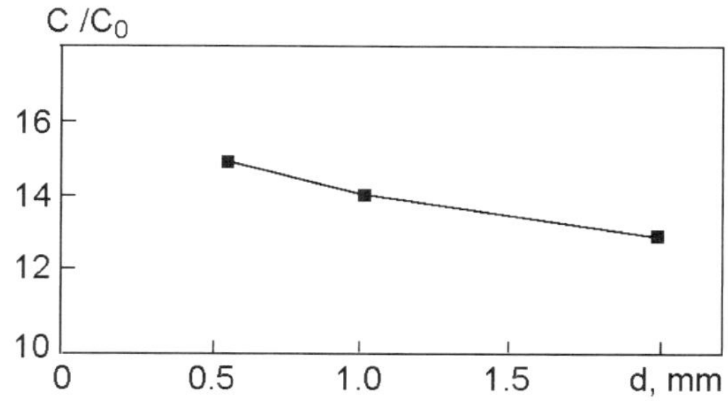

Fig. 5.40 C/C_0 ratio as a function of the capillary diameter

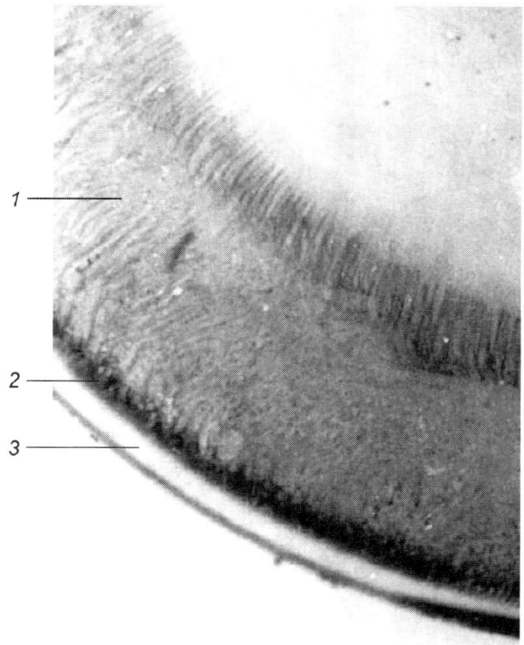

Fig. 5.41 Surface-adjacent defective layer of the sapphire rod with a diameter of 14 mm grown by the Stepanov method using a conical shaper at a growth rate of 40 mm/h. (*1*) Low defect density zone; (*2*) high defect density; (*3*) defect-free zone

can be decreased without changing the pulling speed by increasing the capillary diameter and the meniscus height.

As a rule the near-surface layer contains a layer free of foreign phase inclusions and a layer with a high inclusion density (Fig. 5.41).

There is a certain critical crystallization rate for each crystallographic direction; when its value is exceeded, a sharp increase of the defective near-surface layer depth is observed. For example, for a D = 14-mm rod grown along the directions $[11\bar{2}0]$ and $[10\bar{1}0]$, the critical growth rate is 24–26 and 18–20 m/h, respectively [77]. In the rod with *C*-orientation, the near-surface defect-rich layer has a depth of 0.2–0.4 m in the vicinity of the seed; however, the critical growth rate is limited by foreign phase inclusions observed even at 15 m/h crystallization rate.

As a rule, the defective layer depth decreases when temperature gradient between the shaper and the crystal increases. Such an effect appears with the increase of the crystal length (Fig. 5.42). As it follows from the dynamics of the heater power variation in the course of growth, it may be assumed that *T*-gradient at the crystallization front will increase by 5–6% after seeding at the crystal length of 100–130 mm. This increases the critical crystallization rate which at a crystal length of 100 mm makes 29–31 and 25–27 mm/h for the crystals grown along the direction $[11\bar{2}0]$ and $[10\bar{1}0]$, respectively.

The pore size is related to the growing speed (Fig. 5.43). As the speed grows, the pore size diminishes and the pore concentration increases [75, 76].

The formation mechanism of larger pores (150–300 μm) is due to peculiarities of the Stepanov technique. A bubble is originated at the shaper butt under certain conditions. In the liquid saturated with gas, it grows up to a size (D_{cr}), at which it emerges.

$$D_{cr} = 0.020\sqrt{\sigma_l / \theta(\gamma_l - \gamma_g)}, \tag{5.31}$$

where θ is the wetting angle; σ_l, the liquid surface energy; γ_l and γ_g, the liquid and gas density, respectively. For Al_2O_3 melt on molybdenum, $\theta = 15^\circ$, σ_l = 400–680 erg/cm² [91].

According to the calculation, D_{cr} exceeds the melt layer thickness; thus, the bubble will not emerge. According to the scheme [75, 76], the CF sags toward the melt during the bubble growth, since the bubble heat conductance is lower than that of

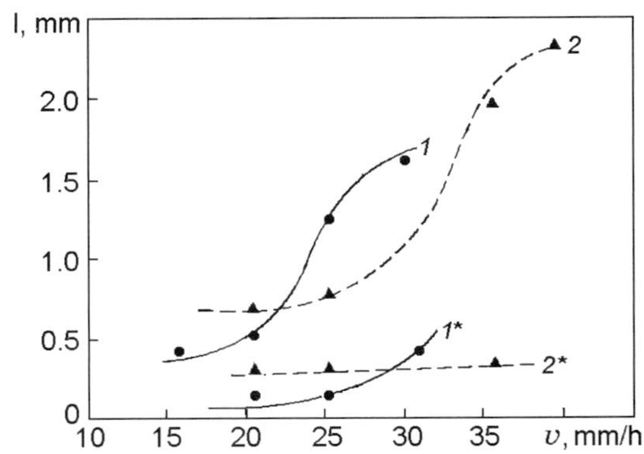

Fig. 5.42 The depth of the surface-adjacent layer of the sapphire rod grown by the Stepanov method depending on the crystallization rate: $[10\bar{1}0]$ growth direction, crystal nose (*1*), crystal tail (*1**); $[11\bar{2}0]$ growth direction, crystal nose (*2*), crystal tail (*2**)

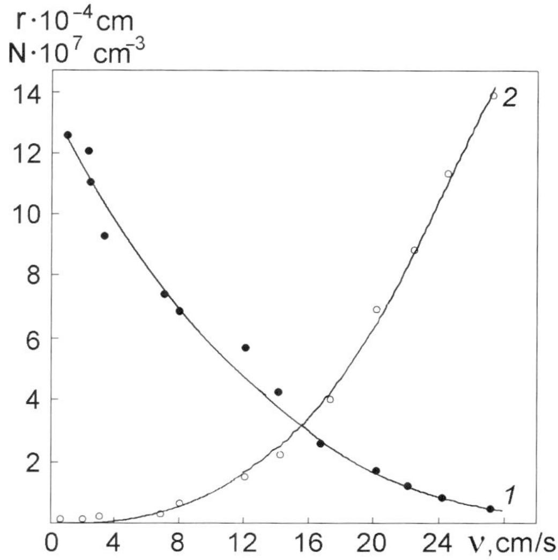

Fig. 5.43 Dependence of the average pore size (*1*) and the average total concentration *N* thereof (*2*) on the crystallization rate at the sapphire tube growing (diameter, 9 mm; wall thickness, 1 mm; circular capillary channel of 0.1 to 0.2 mm width) [89]

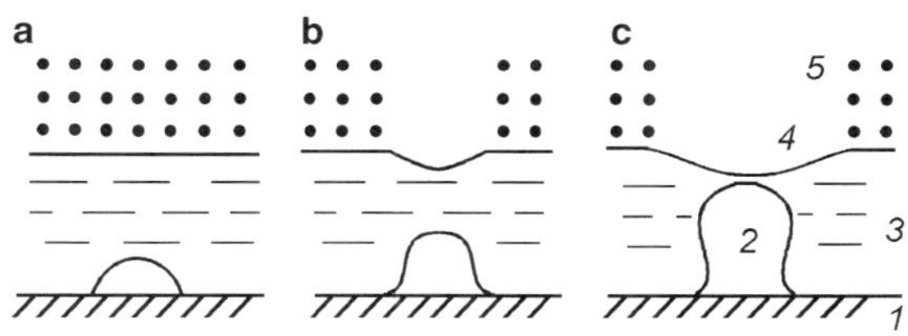

Fig. 5.44 A large bubble growth stage on the shaper; (**a**, **b**, **c**), the growth stages; (*1*) shaper; (*2*) bubble; (*3*) melt; (*4*) crystal; (*5*) pores in crystal

the melt. The CF drives the impurity away in the convex zone (Fig. 5.44). Moreover, the bubble itself is a runoff for the gas phase. After the bubble capturing, the front sags toward the crystal, since the heat removal to the solid phase becomes reduced in the capture site. The pore formation probability and pore sizes depend on the shaper material and surface state (oxidation extent, roughness, etc.).

In the HDC method, the pore formation also may be due to several causes. The oxygen vacancy and microparticle concentrations in sapphire depend on the protective reducing atmosphere pressure [81, 93]. The crystals were grown from a pure raw blend in a carbon-graphite thermal assembly in CO and Ar + CO atmospheres within pressure range of 10–800 Torr. As the pressure rises, the concentration of *F*-centers in the crystals increases, attaining about $8 \cdot 10^{16}$ cm^{-3} at about 30 Torr pressure. The further pressure rise does not result in any appreciable increase in the *F* center concentration but is accompanied by segregation of second-phase microparticles, the concentration thereof being increased as the atmosphere pressure rises (Fig. 5.45).

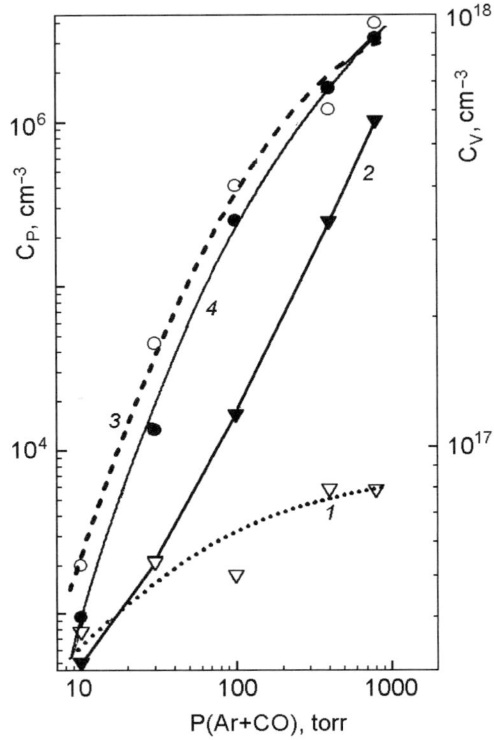

Fig. 5.45 Concentrations of *F* centers (C_V) and microparticles (C_p) in sapphire as functions of the protecting atmosphere (Ar + CO) pressure. (*1*) *F* center concentration (experimental data); (*2*) oxygen vacancy concentration in Al_2O_3 melt (calculated for $P_{CO} = 0.01P_{(Ar + CO)}$; (*3, 4*) concentration of microparticles (calculated and experimental data, respectively)

In ref. [94], the authors have supposed that the second-phase microparticles have the composition $AlAl_2O_4$ (Al–Al spinel); $1.2 \cdot 10^{10}$ to $3.25 \cdot 10^{11}$ vacancies are necessary to form one microparticle of 1–3 μm size. The distribution character of *F* centers depends heavily on the growth atmosphere pressure (Fig. 5.46).

In the crystals grown under 10–100 Torr pressure, the *F*-center concentration decreases from the crystal head toward its tail as the microparticle concentration rises. So in the head of a crystal grown under 10 Torr pressure, the *F*-center concentration attains $6 \cdot 10^{16}\, cm^{-3}$ (there are no microparticles in that crystal part). At the crystal tail, the *F*-center concentration is observed to drop down to about $1 \cdot 10^{16}\, cm^{-3}$ simultaneously with the appearance of microparticles. A similar situation is observed in the crystal grown at 30 Torr: in the crystal head, the *F*-center concentration attains $7.5 \cdot 10^{16}\, cm^{-3}$ (there are essentially no microparticles in that crystal part, their concentration being as low as $10^4\, cm^{-3}$); at the crystal tail, the *F*-center concentration drops down to about $3 \cdot 10^{16}\, cm^{-3}$ simultaneously with the segregation of microparticles. This decrease is less significant in a crystal grown

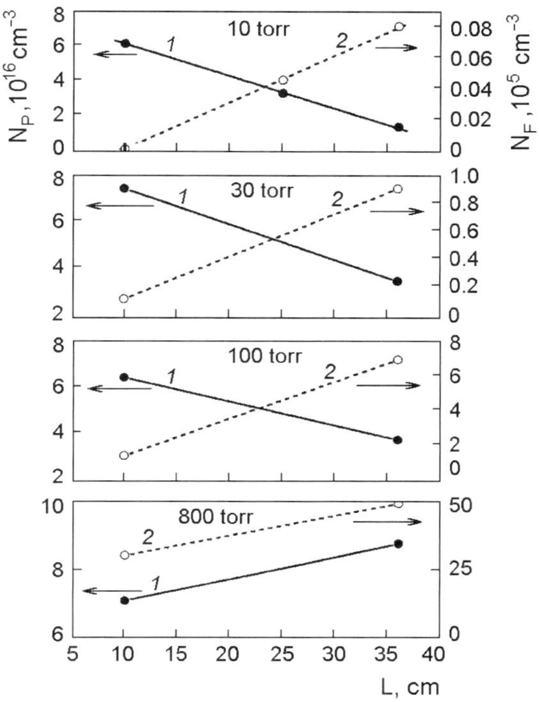

Fig. 5.46 Distribution of microparticles (*1*) and anionic vacancies (*2*) over the crystal length (*L*). The crystals have been grown in (Ar + CO) atmosphere under 10-, 30-, 100-, and 800 Torr

at 100 Torr where a low (about 10^5 cm^{-3}) microparticle concentration is observed at the crystal head. In the crystals grown at higher pressures (400–800 Torr), the *F* center distribution character is changed: the concentration thereof increases as well as that of microparticles.

These results indicate the critical vacancy concentration in Al_2O_3 lattice to be about $8 \cdot 10^{16}$ cm^{-3}. The microparticle concentration amounts to $\leq 10^4$ to $\leq 10^7$ cm^{-3}, depending on the growing conditions [94]. Based on the microparticle size, these can be subdivided into "large" (3–5 μm, more rarely up to 10–15 μm) and "small" (about 1 μm) ones. The average size has been found to tend to diminish from 3–5 to 1–3 μm as the particle concentration rises. The small particles may merge into larger formations (Fig. 5.47). The particles are cut or irregularly shaped. The minimum size of faceted particles is about 2 μm. The cutting of smaller particles has not been established for certain. The most often often occurring cut shapes are presented in Fig. 5.48. The particles cut in various manners can be observed simultaneously (Fig. 5.49). In sapphire grown using the HDC technique, the large pores (3–5 mm) occur in the near-bottom part and more rarely in the bulk. Comparison between observed and simulated cavity shapes is discussed in refs. [70, 83].

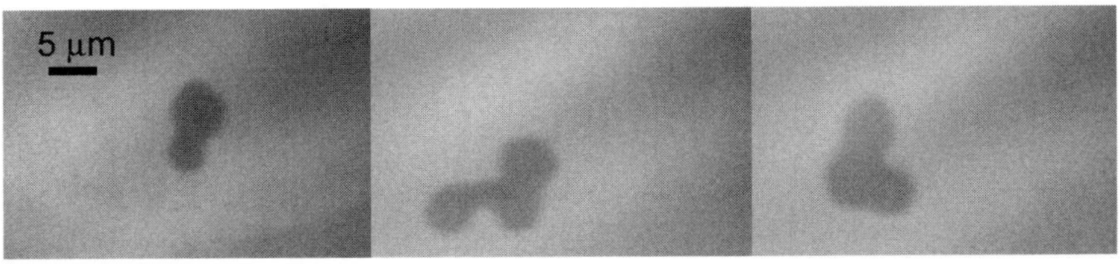

Fig. 5.47 Particles resulting from fusion of several smaller particles [82]

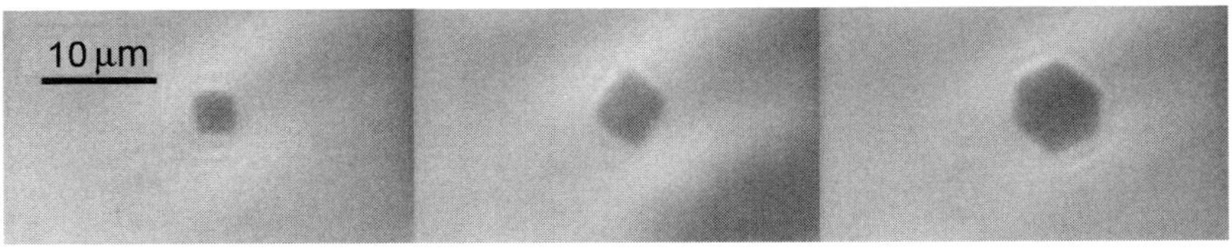

Fig. 5.48 Microparticle cutting shapes

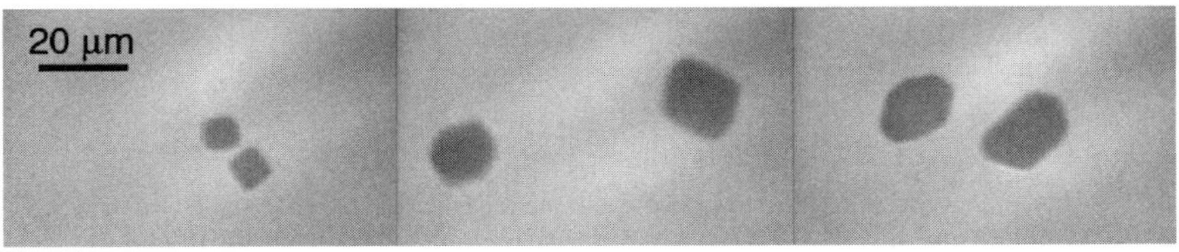

Fig. 5.49 Simultaneous existence of various microparticle shapes within one crystallographic plane $(11\bar{2}0)$

Relationships with crystallography. In the crystals grown by the Czochralski method, the pores are localized at the intersections of rhombohedron facets, pinacoid and rhombohedron, and rhombohedron and prism facets [97]. The pores may form ordered chains and planes normal to the CF. Such an ordering is typical especially of the profiled sapphire grown by the Stepanov technique. According to heat balance [85], a local heat removal decrease occurs under a bubble attached to the CF, and a decreased I_1 results in a growth deceleration at that site. The recess, being the region of increased dissolved gas concentration, favors the nucleation of a new bubble at that site

$$vL = I_1 - I_2 \tag{5.32}$$

where v is the profile growth rate; L, the crystallization heat; I_1, heat removed from the CF through the crystal; I_2, heat supplied to the CF by the shaper.

The phase interface morphology has been studied in the case of growth of sapphire straps with the large facet oriented in parallel to the $(10\bar{1}2)$ plane [78]. A specific feature of the strap profile consists of a high lug at the butt middle caused by a high heat removal from the melt to the strap. A system of steps is seen on the lug under amplification. The step generation centers are the strap edges or lugs associated with the block boundary exits to the butt. The lugs at the block edges give rise to the growth steps, which are collected together into stacks and then move toward both sides along the lug ridge. At the contact sites, a cellular structure arises at the CF that favors the capturing of gas inclusions. As the growth speed increases up to 3.6 mm/ min, cut transversal dents are formed at the lug ridge. The valleys between the dents are the formation sites of new pores being arranged wall-like along the valleys. A further speed increase up to 4.8–6 mm/h results in disappearance of the central lug. The dents arranged across the strap define completely the butt surface microrelief. In this case, the walls of pores are extended over the entire strap thickness.

The pore ordering is believed to be associated not only with the cellular CF. The discreteness is explained also by temperature pulsation. The temperature jumps may be related to the instability in the temperature control systems and to inhomogeneous scattering of the light flow on the internal imperfections and surface unevenesses of the crystal [73]. In this case, the pose spacing is $l_p = n\nu T_p$, where ν is the crystallization rate; T_p, the temperature pulsation period; $n = 1, 2, 3,\ldots$

Somewhat different morphological features of pore formation are manifested when sapphire is grown at high speed when the cellular CF morphology changes into a skeletal one [98]. The bubble-capturing probability at the developed crystallization front increases sharply at the skeletal growth (Fig. 5.50). The inclusion rows are oriented in parallel to one of the skeletal crystal planes: $(10\bar{1}2)$, $(1\bar{2}10)$, $(01\bar{1}2)$, $(2\bar{1}10)$, $(11\bar{2}0)$, $(1\bar{1}02)$. A void arises between adjacent edges (Fig. 5.51) where the shrink pores are formed and bubbles are captured. The pore size is defined by the void width.

In Ti-sapphire grown by Czochralsky technique in a carbon-containing medium, foreign phase inclusions in various shapes and sizes may exist, depending on the crystallization conditions: gas-filled inclusions (pores and voids) of 12 μm to several millimeters size; cut inclusions of 2–17 μm size; and submicrometer inclusions (about 100 nm) [87].

Based on consideration of more than 100 crystals, a regularity has been established in the inclusion distribution along the ingot length. In the seed-close area, the gas-filled inclusions are localized as a rule. As the distance from the seed increases, the cut inclusions may arise. Then there is an inclusion-free area. At the crystal tail, an area containing the submicrometer inclusions is formed. The size of each area containing the specific inclusion types depends on the crystallization conditions, but the sequence thereof along the ingot remains the same. Moreover, several defect types may coexist in one and the same crystal part.

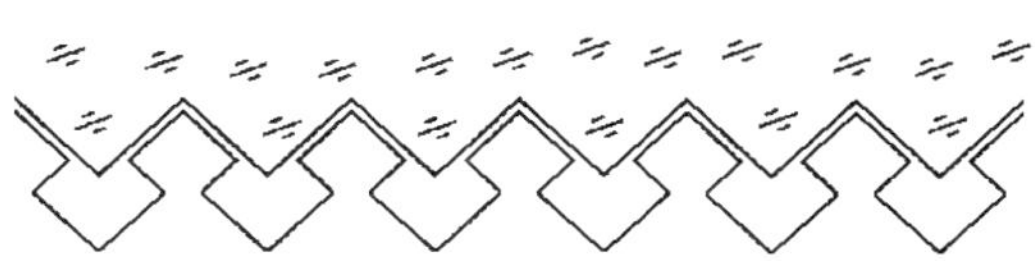

Fig. 5.50 Cellular crystallization front transition into the skeletal one

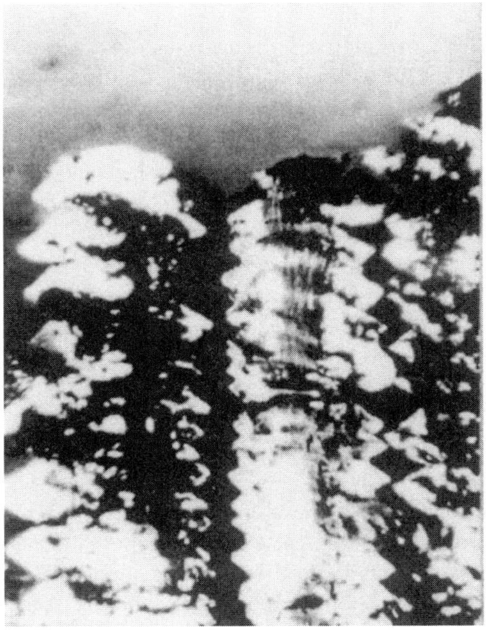

Fig. 5.51 Cavity between adjacent skeletal edges (×100)

The gas-filled inclusions often are observed as the crystal grows from a gas-saturated melt. The gas bubbles in the melt arise due to gases adsorbed at the crucible surface, volatile impurities in the raw material, as well as the thermal dissociation of the melt itself. The gas bubbles can be formed at active centers (microcracks, open micropores) of the crucible inner surface that heats the melt. In this case, no significant supersaturation of the melt in gaseous impurities is needed, because the gas is released into an existing gas bubble. At the initial crystallization stage, the melt may contain a considerable amount of gas bubbles. In this case, separate inclusions of 0.1–2 μm size are formed in Ti-sapphire as well as characteristic gas voids up to several tens of millimeters in size.

When corundum is crystallized in a carbon-containing medium, the melt becomes depleted of oxygen; that is equivalent to an excess aluminum content in the melt. This is due to interaction of carbon oxide in the gas medium with oxygen arising due to thermal dissociation of Al_2O_3. Ti-Sapphire grown in those conditions contains faceted inclusions of 2–17 μm size (Fig. 5.52). Such inclusions were observed also in sapphire grown by other methods. The faceted inclusions decorate the crystallization front at the growth speed fluctuations; this is evidence of the localization thereof in the melt at the crystal surface. If the micrometer-sized inclusions are supposed to be molybdenum particles, then the concentration thereof should increase as the reducing chemical potential of the crystallization medium drops and the free oxygen amount in the melt increases. However, no faceted micrometer-sized inclusions are observed in Ti-sapphire grown using the Kyropoulos technique

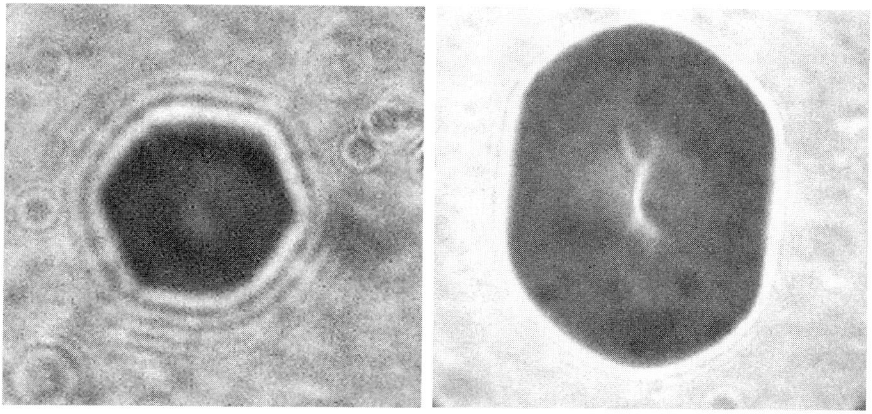

Fig. 5.52 Cut (**a**) and partially cut (**b**) inclusion of 17 and 40 μm size, respectively, in Ti-sapphire

Table 5.9 Chemical composition of raw material, crystal, and residual melt after crystallization

Sample	Element content (ppm)					
	Fe	Ca	Mg	Si	Mo	W
Raw material	10	<5	3	20	<0.5	<10
Crystal	4	5	1	10	5	<10
Residual melt	10	150	3	10	3	25

in a W–Mo thermal assembly at neutral chemical potential of the crystallization medium. Moreover, the atomic emission spectral analysis of the residual melt after Ti-sapphire grown in carbon-containing medium did not show any considerable increase of the molybdenum content compared to the initial material (Table 5.9) [73]. Thus, there is no appreciable interaction between the melt and the container material. Besides, the concentration of molybdenum inclusions must increase at the crystallization end, which does not correspond to the distribution character of the cut inclusions over the crystal length. Thus, the faceted micrometer-sized inclusions in the crystal are not the crucible material particles.The density of scattering centers depends on crystallographic directions (Table 5.10) [88, 99].

Solid-phase inclusions result from capturing of foreign solid particles by the crystallization front or from the phase segregation under the crystal cooling.

In crystals grown from melt by crucible methods, inclusions of metal of which the crucible and other elements of the technological equipment are made. As a rule, these are well-faceted inclusions of 2–50 μm in size and opaque in the transmitted light. Sometimes clusters of several inclusions are formed, those being arranged in the crystal chaotically. The speed of their trapping by the crystallization front is

Table 5.10 90° Scattering as a function of crystallographic direction

Laser beam propagation direction	Scattering record direction	90° Scattering intensity (au)
$[11\bar{2}0]$	$[1\bar{1}00]$	24.7
	[0001]	7.3
$[1\bar{1}00]$	$[11\bar{2}0]$	24.4
	[0001]	7.1
[0001]	$[11\bar{2}0]$	8.5
	$[1\bar{1}00]$	8.6

obviously low, since such inclusions are formed even at the growth rate of 1 mm h^{-1}. The capturie of inclusions by the front is facilitated by their high-heat conduction. Attraction of such particles to the CF should result in its bending, thus making the particle capture much easier. Such inclusions can appear in the crystal due to peeling of particles from the crucible, ingress of particles from outside of the heaters or screens, or dissolution of metal oxides from the surface of crucibles with subsequent reduction thereof by the products of aluminum oxide dissociation. Reduction of the evaporating metal oxides outside the melt by the components of the growth gas atmosphere or gaseous products of aluminum oxide dissociation followed by mechanical ingress of the reduced metal to the melt where metal particles can adhere by the method "from solution in melt" is also probable.

Oxides of some metals form with aluminum oxide numerous compounds of spinel structure: $Me{\cdot}Al_2O_4$ where Me is Ni, Co, Fe, Mn, Cu, Zn, Cd, Mg, Al. Spinels have a cubic structure and are optically isotropic. An interesting feature of the chemical composition of man-made spinels is their ability to contain a significant excess of aluminum oxide without any violation of the spinel structure. Heat treatment of corundum crystals that contain inclusions of cobalt spinel results in a decomposition of inclusions.

In the presence of other impurities in the melt together with α-modification of aluminum oxide, β-Al_2O_3 is formed during crystallization from melt. It represents a group of compounds of the type $MeO{\cdot}n{\cdot}Al_2O_3$, where $n = 7$–11, and Me – K, Mg, Ca, Ba, Pb, Sr, Rb, Cs, Ag, Ti, Ga, NH_4.

β-modification has a hexagonal structure, crystals are uniaxial, and optically negative. At heating to 1,300–2,000 K, β-Al_2O_3 decomposes. Usually it is not single-phase and constitutes a mixture $Me{\cdot}\alpha$-Al_2O_3 + $Me{\cdot}\beta$-Al_2O_3.

Several impurities can form phases of a more complex composition; for instance, Na and Ca give a phase of the composition $Na_2O{\cdot}4CaO{\cdot}10Al_2O_3$ which at 1,700 K coexists with the phases $CaO{\cdot}Al_2O_3$, $CaO{\cdot}2Al_2O_3$, $2Na_2O{\cdot}3CaO{\cdot}5Al_2O_3$, and $Na_2O{\cdot}11Al_2O_3$ [95] and at heating above 1,800 K decomposes under formation of $(Na_2O, CaO){\cdot}2Al_2O_3$ [95] . At Al_2O_3:Ti annealing under oxygen presence, the rutile or titanium aluminate phase is segregated as needles oriented along the crystallographic directions.

At crystal growth, this results in the formation of inclusions of eutectics on the basis of Al_2O_3–Al_4O_4C in the crystal according to the phase diagram Al_2O_3–Al_4C_3 (Fig. 5.53). Thin interlayers of the eutectic composition are formed along the block boundaries, while inclusions of Al_4O_4C are dispersed near the pores [96]. At a consequent vacuum annealing, a decarbonation occurs and aluminum tetraoxycarbide loses carbon in the course of reduction

$$Al_4O_4C + Al_2O_3 \rightarrow Al_2O_{3-x} + CO_y \tag{5.33}$$

The released gaseous carbon oxides favor the pore growth under annealing.

In the crystals with high additive content, e.g., in ruby with chromium concentration above 0.5 wt%, microinclusions of rounded or extended shape of 0.5–5 μm size arise along with the above-mentioned inclusions (Fig. 5.54). Perhaps it is the Cr_2O_3 phase that is segregated [97].

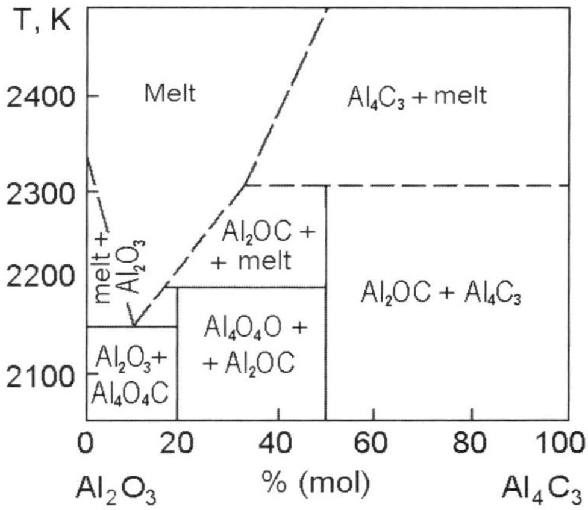

Fig. 5.53 Phase diagram of Al_2O_3–Al_4C_3

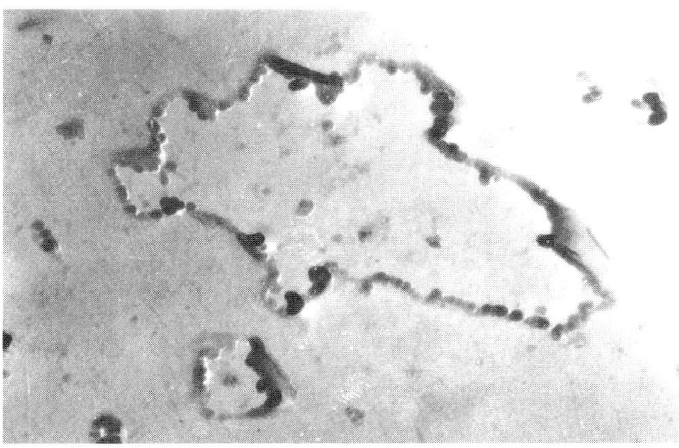

Fig. 5.54 Microinclusions in ruby (×20,000)

References

1. Turner T.J., Crawford J.H. Phys. Rev. B 1976, 12, 1735.
2. Lee K.H., Crawford J.H. Phys. Rev. B 1977, 15, 4065.
3. Evans B.D., Stapelbroek M. Phys. Rev. B 1978, 18, 7089.
4. Blistanov A.A. Kristally kvantovoi i nelineinoi optiki. Moscow: MISIS, 2002 [in Russian].
5. Andreev Y.P., Kryvonosov Y.V., Lytvynov L.A., Vyshnevskiy S.D. Funct. Mater. 2005, 12, 142–145.
6. Dienes G.J., Welch D.O. Phys. Rev. B 1975, II, 3060.
7. Brook R.J., Yee J., Kroger F.A. J. Am. Ceram. Soc. 1971, 54, 444.
8. Cox R.T. Solid State Commun. 1971, 9, 1989.
9. Lee K.H., Holmberg G.E., Crawford J.H. Phys. Stat. Solid A 1977, 39, 669.
10. Krivonosov E.V. Influence of conditions of annealing on the optical and mechanical performances of single crystals of a corundum. – Diss. kand. tekhn. nauk., Kharkov, 1989 [in Russian].
11. Jones T.P., Cobble R.L., Mogal C.I. J. Am. Ceram. Sci. 1969, 52, 331.
12. Vakhidov S.H.A. Diss. – Tashkent, 1973 [in Russian].
13. Gamble F.T. et al. Phys. Rev. A 1965, 138, 577.
14. Bartman R.H. et al. Phys. Rev. A 1965, 139, 1941.
15. Ia S.Y., Bartman R.H., Cox H.T. Phys. Rev. Chem. Solids. 1973, 34, 1079.
16. Cox R.T. Phys. Lett. 1966, 21, 503.
17. Valbis Ya.A., Strunkis M.E. Elektronnye i ionnye protsessy v ionnykh kristallakh: Sb. Latv. Unta, Riga. 1980, 10 [in Russian].
18. Springis M.E. Diss. LGU. – Riga, 1981 [in Russian].
19. Ovsienko D.E. Rost i defecty metallicheskikh kristallov: Sb. nauch. tr., Kiev 1966, 164 [in Russian].
20. Milvidskii M.G., Osvensikii V.B. Problemy sovremennoi kristallografii: Sb. nauch. tr., Moscow 1975, 79 [in Russian].
21. Belyaev A.D., Vasilevskaya V.I., Miselyuk E.G. Rost kristallov: Sb. nauch. tr., Moscow 1961, 3, C.380 [in Russian].
22. Gritsenko N.V., Mandych A.A., Podlesnaya A.D., Raikhel's E.G. Kristallografiya 1975, 20, 204 [in Russian].
23. Chernik M.M., Dobrovinskaya E.R. Izv. AN SSSR Ser. fiz. 1972, 36, 570.
24. Bagdasarov Kh.S., Dobrovinskaya E.R., Lytvynov L.A., Pishchik V.V. Izv. AN SSSR Ser. fiz. 1973, 37, 2362.
25. Dobrovinskaya E.R., Galagura A.N., Lytvynov L.A. et al. Kristallografiya 1971, 16, 666 [in Russian].
26. Bagdasarov Kh.S., Dobrovinskaya E.R., Pishchik V.V. et al. Kristallografiya 1973, 18, 390 [in Russian].
27. Dobrovinskaya E.R., Kukol' V.V., Pishchik V.V., Tsigel'nitskii G.M. Kristallografiya 1975, 20, 399 [in Russian].
28. Indenbom V.L. Izv. AN SSSR Ser. fiz. 1973, 37, 2258 [in Russian].
29. Milvidskii M.G., Osvensikii V.B. Rost kristallov: Sb. nauch. tr., Erevan 1977, T.12, 257 [in Russian].
30. Indenbom V.L. Rost kristallov: Sb. nauch. tr., Erevan 1961, T.3, 217 [in Russian].
31. Tsivinskii S.V. Fizika metallov i metallovedenie 1968, 25, 1013 [in Russian].
32. Tsivinskii S.V. Izv. AN SSSR Ser. fiz. 1976, 40(7), 1532 [in Russian].
33. Indenbom V.L., Tomilevskii G.N. Izv. AN SSSR Ser. fiz. 1958, 5, 593 [in Russian].
34. Indenbom V.L., Osvenskii V.B. Rost kristallov: Sb. nauch. tr., Moscow 1980, T.13, 240 [in Russian].
35. Indenbom V.L., Zhitomirskii I.S., Chabanova T.S. Kristallografiya 1973, 18, 39 [in Russian].
36. Aksel'rod E.I., Vishnevskii I.I., Dobrovinskaya E.R., Tal'yanskaya N.D. Dokl. AN AH SSSR 1973, 213, 331 [in Russian].

37. Dobrovinskaya E.R., Lytvynov L.A., Pishchik V.V., Tsaiger A.M. Kristallografiya 1977, 22, 879 [in Russian].
38. Seiz F. Phys. Rev. 1950, 79, 890.
39. Jackson K.A. Philos. Mag. 1962, 7, 1117.
40. Veselovskaya V.G., Sheikhet E.G., Neimark K.N., Fal'kovich E.S. Rost i legirovanie polurovodnikovykh kristallov i plenok, Novosibirsk (Ch. 2). 1977, 284 [in Russian].
41. Bakholdin S.I., Galoktionov E.V., Zasimchuk I.K., Fomin A.V. IX sovesch. po polucheniyu profilirovannykh kristallov i izdelii sposobom Stepanova, L. 1982, 101 [in Russian].
42. Belaya A.N., Dobrovinskaya E.R., Lytvynov L.A. et al. Izv. AN SSSR Neorgan. materialy 1981, 17, 918 [in Russian].
43. Grinchenko Yu.T., Dobrovinskaya E.R. et al. Izv. AN SSSR Ser. fiz. 1976, 40, 1499 [in Russian].
44. Bagdasarov G.S., Dobrovinskaya E.R., Pishchik V.V. Rost kristallov: Sb. nauch. tr., Moscow 1977, T.12, 195 [in Russian].
45. Dobrovinskaya E.R., Lytvynov L.A., Pishchik V.V. Izv. AN SSSR Ser. fiz. 1976, 40, C.1330 [in Russian].
46. Dobrovinskaya E.R., Lytvynov L.A., Pishchik V.V. Vyraschivanie kristallov i ikh struktura: Sb. nauch. tr., Tbilisi 1977, T.2, 254 [in Russian].
47. Ovsienko D.E., Zasimchuk I.K., Meleshko L.I. Rost i nesovershenstvo metallicheskikh: Sb. nauch. tr., Kiev 1966, 220 [in Russian].
48. Ovsienko D.E. Rost i defekty metallicheskikh kristallov: Sb. nauch. tr., Kiev 1972, 135 [in Russian].
49. Gliki I.V., Urusovskaya A.A. Kristallografiya 1965, 10, 650 [in Russian].
50. Sb. "Rubin i sapfir" Moscow: Nauka, 1974, 235 [in Russian].
51. Lainer L.V., Turovskii V.M. Fiz. i khimiya obrab. material. 1969, 4, 63 [in Russian].
52. Tsivinskii S.V., Maslova L.A. Izv. AN SSSR Ser. fiz. 1972, 36, 575 [in Russian].
53. Tsivinskii S.V., Maslova L.A. Izv. AN SSSR Ser. fiz. 1971, 35, 514 [in Russian].
54. Salli I.F., Fal'kovich E.S.Proizvodstvo poluprovodnikovogo kreminiya.Moscow: Metallurgiya 1970, 342 [in Russian].
55. Tiller V.A. Theory and Practice of Crystal Growth M. 1968, 294.
56. Belaya A.N., Dobrovinskaya E.R., Lytvynov L.A., Pishchik V.V. Kristallografiya 1981, 26, C.164 [in Russian].
57. Belaya A.N., Dobrovinskaya E.R., Zvyagintsev V.P. et al. Metody polucheniya i issledovaniya monokristallov i stsintillyatorov: Sb. nauch. tr. 1980, 5, 15 [in Russian].
58. Landau A.I. Physics of Metals, 1958, *T.6*, 148.
59. Toropov I.A., Barzakovskii V.P., Lapin V.V., Nurieva I.Y.A. Diagrammy sostoyaniay silikatnykh sistem. Moscow: Nauka, 1965, 376 [in Russian].
60. Voronkov V.V. J. Cryst. Growth 1982, 59, 625.
61. Brice J.C. Acta Electronica 1973, 16, 291.
62. Zharikov E.V., Zavartsev Yu.D., Laptev V.V., Privis Yu.S. Materialy Vsesoyuz. sovesch. po polucheniyu profilirovannykh kristallov i izdelii sposobom Stepanova i ikh primeneniyu v narodnom khozyaistve: Sb. nauch. tr., L 1986, 26 [in Russian].
63. Dobrovinskaya E.R. Upravlenie strukturnym sovershenstvom monokristallov korunda. Moscow: NIITEKhIM, 1979, 68 [in Russian].
64. Dobrovinskaya E.R., Lytvynov L.A., Pishchik V.V. Izv. AN SSSR Ser. fiz. 1983, 47, 332 [in Russian].
65. La Bell H.E. Mater. Res. Bull. 1972, 6–7, 581.
66. Dobrovinskaya E.R., Lytvynov L.A., Pishchik V.V. Izv. AN SSSR Ser. fiz. 1979, 43, 1940 [in Russian].
67. Dobrovinskaya E.R., Pishchik V.V. Izv. AN SSSR Ser. fiz. 1985, 49, 2386 [in Russian].
68. Dobrovinskaya E.R., Pishchik V.V. Fizika kristallizatsii: Sb. nauch. tr., Kalinin 1986, 46 [in Russian].
69. Dobrovinskaya E.R., Pishchik V.V., Zvyagintseva I.F. Vyraschivanie monokristallov i ikh svoistva: Sb. nauch. tr. 1984, T.13, 47 [in Russian].

70. Satunkin G.A. et al. Izv. AN SSSR Ser. fiz. 1976, 40, 1492 [in Russian].
71. Puchkov V.V., Kislovskii L.D. Rost kristallov: Sb. nauch. tr., Moscow 1974, T.10, 26 [in Russian].
72. Avdonin I.A. IV Vsesoyuz. sovesch. po rostu kristallov: Vyraschivanie kristallov i ikh struktura. Tsakhkadzor. 1972, 174 [in Russian].
73. Lytvynov L.A. Dokt. diss. Kharkov, NTK «Institut monokristallov» 1994, 26 [in Russian].
74. Lingart Yu.K., Bodyachevskii S.V. Izv. AN SSSR Ser. Neorgan. Materialy 1978, 14, 1834 [in Russian].
75. Skripov V.P., Kaverda V.P. Spontannaya kristallizatsiya pereokhlazhdennykh zhidkostei. Moscow: Nauka, 1984, 210 [in Russian].
76. Borodin V.A. et al. Kristallografiya 2002, 47, 950–954 [in Russian].
77. Dobrovinskaya E.R., Pishchik V.V. Kristallografiya 1988, T.33, 1000 [in Russian].
78. Papkov V.S., Perov V.F., Ivanov V.I. Elektronnaya tekhnika. Ser. Materialy 1978, 9, 50 [in Russian].
79. Zykova A.V., Lytvynov L.A. Sintez i issledovanie opticheskikh materialov: Sb. nauch. tr. 1987, 19, 62 [in Russian].
80. Timan B.G., Dobrovinskaya E.R., Babiichuk I.P. Fiz. tverdogo tela. 1974, 16, 1792 [in Russian].
81. Pishchik V.V., Dobrovinskaya E.R., Birman B.I. Sb. Monokristally i tekhnika, Kharkov 1971, 5, 26 [in Russian].
82. Katrich N.P. et al. Sb. Monokristally i tekhnika, Kharkov 1976, 14, 27 [in Russian].
83. Katrich N.P., Khizhnyak E.M., Miroshnikov Yu.P.. Monokristally i tekhnika, Kharkov 1976, 14, 33 [in Russian].
84. Musatov V.A. Monokristally i tekhnika, Kharkov 1973, 1(8), 24 [in Russian].
85. Chalmers B., Labelle H.E., Mlavsky A. J. Mater. Res. Bull. 1971, 6, 681 [in Russian].
86. Kanevsky V.S., Krivonosov E.V., Lytvynov L.A., Tkachenko S.A. Funct. Mater. 1999, 6, 636.
87. Andreev Y.P., Kryvonosov Y.V., Lytvynov L.A., Vyshnevskiy S.D. Funct. Mater. 2005, 12, 142–146 [in Russian].
88. Indenbom V.L. Kristallografiya 1964, 9, 74 [in Russian].
89. Zykova A.V., Lytvynov L.A. Fizika i khimiya opticheskikh i stsintillyatsionnykh materialov: Sb. nauch. tr. 1985, 14, 11 [in Russian].
90. Yalovets T.N. et al. Osnovnye tipy por v profilirovannom sapfire i mekhanizmy ikh obrazovaniya. Chernogolovka. Preprint IFTT AN SSSR 1988 [in Russian].
91. Nesis E.I. Kipenie zhidkosti. Moscow: Nauka, 1973, 174 [in Russian].
92. Maurakh M.A., Mitin B.S. Zhidkie tugoplavkie okisly. Moscow: Metallurgia, 1979, 237 [in Russian].
93. Dan'ko Ya.A., Sidelnikova N.S., Adonkin G.T. et al. Funct. Mater. 2003, 10, 217 [in Russian].
94. Dan'ko Ya.A., Sidelnikova N.S., Adonkin G.T. et al. Sb. Funktsional'nye materialy. IM NAN, Ukraine 2001, 200 [in Russian].
95. Zaldat G.I., Kuprianova M.S., Tubolev A.L. Neorganicheskie materialy 1985, 21, C.51–53 [in Russian].
96. Yalovets T.N. Vliyanie fiziko-khimicheskikh uslovii kristallizatsii na sovershenstvo profilirovannykh kristallov sapfira. Avtoreferat kand. dis.Chernogolovka, 1987 [in Russian].
97. Belaya A.N., Dobrovinskaya E.R., Lytvynov L.A., Chernikov E.I. Fizika i khimiya obrabotki materialov. 1979,4, 679 [in Russian].
98. Bakholdin S.I., Kuandykov L.L., Antonov P.I. Tez. dokl. X Nats. konf. po rostu kristallov, Moskva 2002, 254 [in Russian].
99. Vyshnevskiy S.D., Kryvonosov Ye V., Lytvynov L.A. Funct. Mater. 2006, 13, 238–244.

Chapter 6
Influence of Chemical–Mechanical Treatment on the Quality of Sapphire Article Working Surfaces and on the Evolution of Surfaces under the Action of Forces

The process of transforming a crystalline ingot into a crystalline product is multistage, complicated, labor-intensive, and expensive. Experience shows that the growth of high-quality crystals (containing no blocks and small dislocation densities) or of crystals with a preset distribution of structural defects does not always ensure end articles possessing such properties. The process of treatment disturbs not only the surface structure, but also that of the surface-adjacent crystal layers. A crystal surface is a surprisingly solid substance. According to Wolfgang Pauli, "God created the volume, whereas the surface was a work of the Devil." However, specialists engaged in crystal treatment cannot choose between creatures of God and tricks of the Devil.

Development of a technology for reproducible crystal treatment is a complicated problem. On the one hand, even crystals grown by the same method under identical conditions may differ in quality and properties, and consequently in the requirements of treatment (to say nothing of the crystals obtained by different methods). On the other hand, optimization of treatment technology requires knowledge of the physics and chemistry of the growth processes, as well as the mechanisms of formation and evolution of the surface-adjacent defective layer during different treatment techniques.

First, consider the general approaches and methods used for sapphire treatment and then dwell on the main empirical factors observed in the course of sapphire evolution during treatment. Currently, the generally accepted stages of sapphire treatment are:

- Grinding
- Fine grinding (lapping)
- Polishing

6.1 Preliminary Grinding and Lapping

The process of grinding imparts the required shape and dimensions to a crystalline ingot. Methods of hard and elastic grinding exist. For hard grinding, the input parameters are the grinding depth and other components of cutting; the output

E.R. Dobrovinskaya et al., *Sapphire: Material, Manufacturing, Applications*,
DOI: 10.1007/978-0-387-85695-7_6, © Springer Science+Business Media, LLC 2009

parameter is the cutting effort. For elastic grinding the primary input parameter is the specific pressure, which defines the depth of grinding and the specifics of cutting. The method of elastic grinding is most widely used for sapphire treatment, since it can achieve a better and more uniform cutting condition for the worked part and consistent working conditions for the abrasive grains. At constant normal pressure and other equilibrium conditions, the depth of cutting also is constant, while changing one of the cutting parameters also changes the depth of cutting. For example, due to wear of the abrasive grains, the depth of grit introduction diminishes; thereby, the cutting conditions are altered. The mechanism of sapphire grinding as realized by diamond tools is a brittle failure of the material accompanied by the formation of scratches, cracks, and cleaveds. The interaction of the crystalline surface with lubricating and cooling liquids (LCLs) leads to the appearance of hydrolysis products in the newly formed scratches and cracks. The volume of such products largely exceeds that of the damaged crystal and so induces wedging of the crystal along cracks, leading to a subsequent formation of cleaveds. The resultant large quantity of cleaveds produces a rough mat surface characteristic of a ground crystal. The degree of surface roughness depends on the size and quantity of the simultaneously worked abrasive grains, the rate of cutting, the relative rate of motion of the crystal work piece, and the composition of the LCL used. To summarize, grinding of sapphire disturbs surface continuity due to the appearance, development, and intersection of microcracks; this evolution of sapphire surface damage during the grinding process is as follows:

- Nucleation of submicrocracks
- Coalescence of submicrocracks into microcracks and then into macrocracks
- Growth of macrocracks until separation of the broken fragments occurs

Submicrocracks are defined by a length of ~0.1 μm; they grow by joining together or by interacting with vacancies and dislocation aggregates, until finally microcracks are formed with a length of about 1 μm. A crack will propagate until the local stresses arising at its propagation front become less than the ultimate strength of the crystal.

Thus, in the process of sapphire grinding brittle separation of material caused by periodic development of cracks occurs [1]. Proceeding from the fundamentals of damage theory [2], the depth of crack propagation depends on the temperature in the zone of contact, the rate of loading, the characterics of the material stress state, the orientation and structural perfection of the crystal, the properties of the LCL, and so on. Crack depth can range from hundreds of Angstroms to several microns. Moreover, during the grinding process thin layers adjacent to the cleaved facet surfaces undergo plastic deformation. The thickness of these layers and the energy consumption upon plastic deformation at the ultimate (breaking) stress are defined by the nature of the crystal and the conditions of damage. Consequently, the specific energy consumption upon plastic deformation per unit volume of treated material is proportional to the surface area of the damaged fragments and increases as their dimensions decrease. At the same time, the specific energy consumption by ultimate elastic deformation does not depend on this degree of dispersion. Therefore,

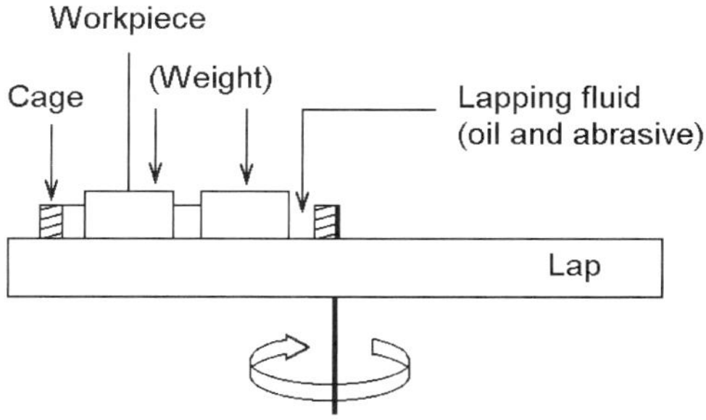

Fig. 6.1 Scheme of lapping

plastic deformation is decisive in the balance of energy consumption upon breakdown of small particles, even in a brittle material such as sapphire.

Lapping is the process of fine grinding (Fig. 6.1), intermediate between preliminary grinding and polishing. On one hand, the mechanism of its action on the treated surface is similar to that of grinding. On the other hand, lapping is more similar to polishing, if the load and the amount of material removal are considered. It is a preliminary stage of polishing which provides the required final shape and surface flatness. As a rule, lapping is realized by means of a fine, free abrasive in some form of liquid suspension. The purpose of lapping is to remove the damaged, surface-adjacent layer formed after preliminary grinding and to create a homogeneous, rough surface with the highest possible level of flatness and the lowest possible degree of damage in the surface-adjacent layer.

The crystal work piece undergoes a planetary motion that renders preset parameters to the treated surface based on the process parameters (abrasive size, specific pressure applied to the work piece, relative speeds of the piece and the lapping surface, etc.).

The main distinction between lapping and polishing is the different combination of applied abrasives and tools that these processes are based on, the choice of which is not rigorously specified. The firms engaged in sapphire treatment confront this problem, taking into account their own specific requirements.

6.2 Polishing

Polishing is the final stage of mechanical treatment, consisting of removal of microrelief created by lapping to obtain a transparent and mirrorlike smooth surface. As a rule, coarse abrasives and rigid working surfaces are used for lapping, whereas polishing is realized by fine abrasives and soft pads (Fig. 6.2).

Figure 6.3 demonstrates removal of material during the processes of lapping, mechanical, and chemical–mechanical polishing. The horizontal axis is the comparison scale for the treatment parameters (cutting depth, size of the cleaved facets and cracks, etc.).

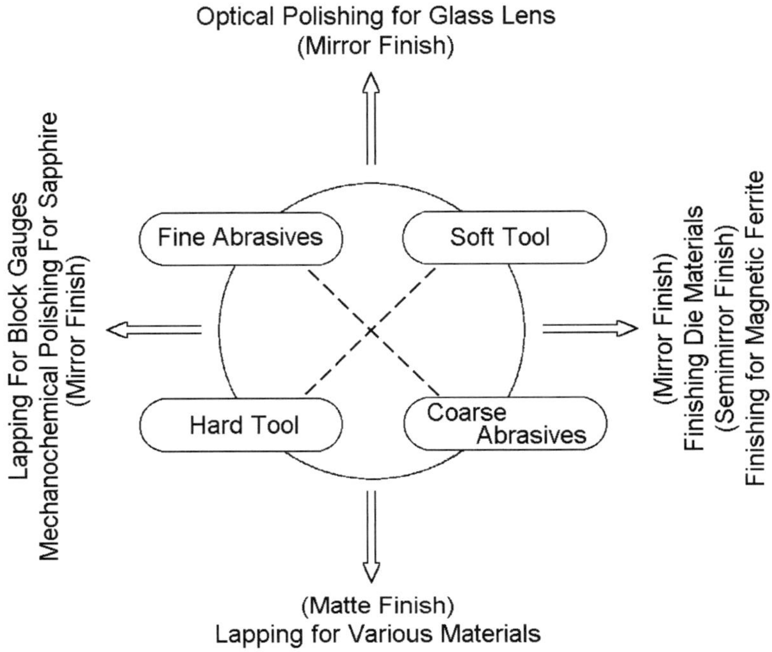

Fig. 6.2 Distinctions between lapping and polishing [3]. Classification of lapping and polishing based on combinations of tools and abrasives

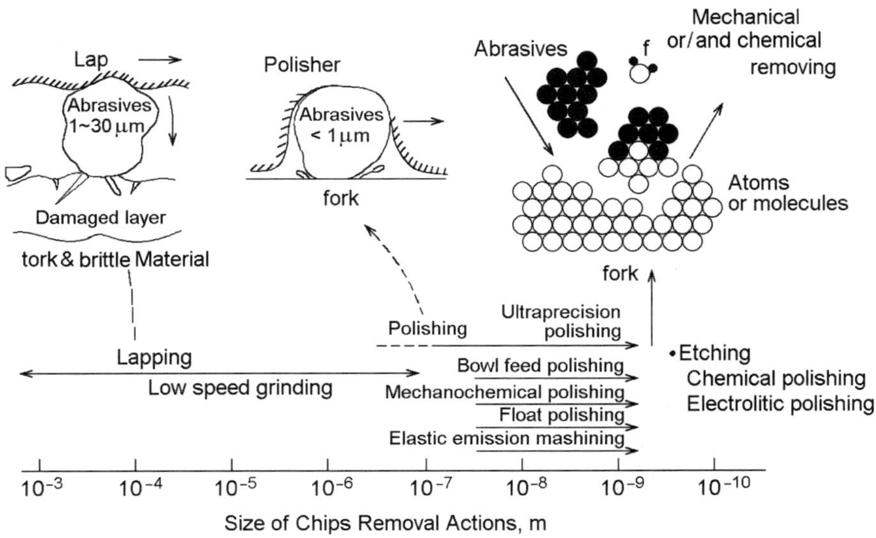

Fig. 6.3 Models of lapping, polishing, and ultraprecision polishing for hard and brittle materials [4]

As seen from this figure, while removing the previously formed damaged layers, the abrasives used for each subsequent treatment stage often give rise to new cracks whose sizes are essentially less than the original. The minimum amount of material removal, characteristic of chemical–mechanical polishing, is on the order of several atoms [3, 4]. Both lapping and polishing are highly sensitive to treatment parameters, as the rate of material removal is proportional to the relative speed of the crystal workpiece and the working surface of the tool, pressure, time, and so forth.

Present-day sapphire polishing physics is based on two approaches. The first approach is traditional: polishing is considered to be fine grinding by micron and submicron grains with subsequent chemical–mechanical treatment to remove the resultant "diamond background" (the ultrafine grain network of scratches arising on the surface during abrasive treatment). During the past 10–12 years, surface requirements for sapphire optical elements (the surface perfection, roughness, shape, and level of residual stresses and crystal lattice distortions) have sharply increased and fundamentally changed. Now the term "optical surface" denotes a surface that does not distort a reflected or transmitted wave front or contain residual stresses in the surface-adjacent layer, i.e., a surface without the crystal lattice distortions associated with a surface of ultralow roughness. In technical literature the kind of treatment resulting in such a surface often is called "strain-free, mirrorlike finishing."

Achievements in physical-chemical mechanics and mechanochemistry and rheology of solids dispersion make it possible to regard polishing of sapphire from other positions and to interpret the role of the medium, the polishing surface material, and mechanical factors in a new way. According to the second approach to sapphire polishing physics [5], the rate of polishing does not depend on the size and hardness of the abrasive grains utilized. The typically metallic polishing surface is much softer than the crystalline workpiece, and its mechanical action on the piece under treatment therefore is insignificant. Such circumstances cannot give rise to cutting or even fatigue wear of the workpiece surface. The role of mechanical factors during polishing is reduced mainly to chemical–mechanical activation of fragments projecting over the polished surface. The process of polishing is so fine that chemical reactions on the treated surface would not proceed without this activation and removal of material would not occur.

On the foundation of this new understanding of the physical processes of polishing, fundamentally new procedures and methods have been developed, especially for the stage of ultrafinishing. They provide achievement of the sapphire surface quality requirements necessary in modern devices and facilities. Like traditional treatment methods, these new procedures of material removal from the crystal surface are based on chemical and physical actions. Unlike traditional methods, these actions are used simultaneously, not individually and sequentially, and this has turned out to be highly efficient.

In this chapter the procedures of compound polishing, which combine the mechanical action of abrasive material on the treated surface with mechanically activated chemical reactions, are considered. The efficiency of these reactions is dependent on the activation of the treated surface by the mechanical energy derived from the action of abrasive materials (high pressure and temperature in the zone of contact, accompanying plastic deformation, etc.) [6, 7]. This activation changes the physical and chemical properties of the treated surface, which in turn enables and accelerates the chemical reactions in which the reagents and components participate. Under such conditions solid-phase, solid-liquid, and even solid-gas chemical reactions can proceed.

To summarize, during the polishing process material is removed by both mechanical and chemical reactions. Mechanical removal is realized due to friction, whereas chemical reactions dissolve material. By combining the action of these mechanisms, one can obtain the required surface quality, i.e., a smooth surface with undisturbed crystallinity and the complete absence of residual stresses. This task can be fulfilled only by removing several atomic layers of material, which requires use of other types of polishers and abrasive materials, as well as other kinematic processes. As mentioned above, the depth of the layer removed by polishing is insignificant. Abrasives affect only the surface-adjacent layer by smoothing out the surface relief left from lapping; they leave faint scratches and minutely cleaved surfaces (Fig. 6.3).

There are two kinds of the processes that occur during polishing:

- Mechanical removal of the damaged layer resultant from lapping
- Plastic deformation created under this damaged layer and formation of a new layer with elevated concentrations of dislocations, point defects, and residual stresses

The unique combination of polishing procedures and materials used lead to different quantities and densities of defects in the layers adjacent to the polished surface, but do not change this general tendency. The main types of treatment used in modern methods of finish polishing are shown in Fig. 6.4.

To minimize the thickness of the surface-adjacent layer containing elevated concentrations of point defects and residual stresses, different methods of polishing have been developed, specifically those using chemical reagents that *etch* the treated surface. Consider some of these polishing methods [8, 9].

Mechanical polishing. This name implies the action of only mechanical factors. The polishing surface materials typically used are soft metals such as cast iron, tin, lead, or copper, and occasionally even various resins and plastics. These materials favor penetration of abrasive materials into the polishing surface. Because of this, only a portion of the abrasive grains work, thereby reducing the load on the treated surface and removing the material while producing thin shallow scratches (Fig. 6.5).

Wet chemical–mechanical polishing. This is an evolution of the above-considered method of mechanical polishing. It is characterized by the use of special substances added to the polishing suspension, which activate the polishing process through chemical interactions with the workpiece surface. As a general rule, alkaline SiO_2 solution (colloidal silica) is used for this purpose. Authors describing this process [10–15], as well as analogous polishing methods, proceed from the assumption that the chemical interaction of aluminum oxide with silica is followed by the formation of an aluminosilicate:

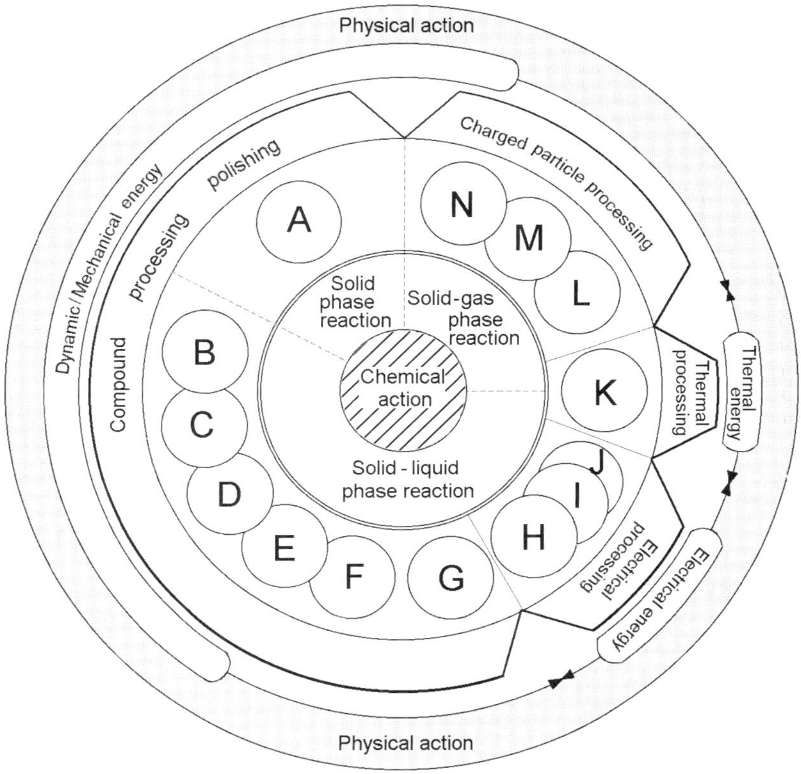

Fig. 6.4 The main types of treatment used in finish polishing, physical–mechanical, and chemical compound processing methods

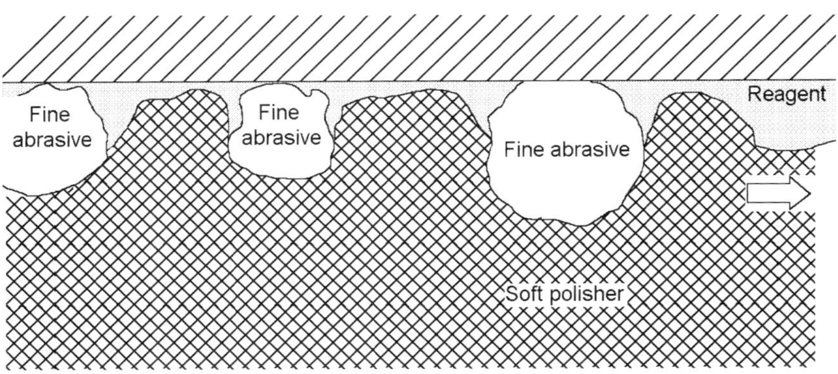

Fig. 6.5 Processing mechanism of mechanical polishing

$$Al_2O_3 + 2SiO_2 + 2H_2O \rightarrow Al_2Si_2O_7 \cdot 2H_2O$$

This reaction product is drawn by friction of the tool against the workpiece, thus providing removal of the material. The above chemical reaction proceeds at relatively low temperatures. Finish polishing using an abrasive paste with a grain size of 1 μm results in microasperities with average height of about 0.03 μm, whereas with chemical–mechanical polishing using an Aerosil suspension of SiO_2 microasperities do not exceed 0.01 μm.

The earlier considered mechanisms of chemical–mechanical polishing are complemented by a mechanism of chemical etching of the surface, which diminishes the incumbent defective layer. Figure 6.6 demonstrates the polishing factors essential for wet chemical–mechanical polishing. For all referenced crystals the removal rate of material increases as specific pressure rises; note that sapphire polishing generally requires particularly high pressures (Fig. 6.7).

Dry chemical–mechanical polishing. In this case, mechanical polishing is activated through the action of chemical solid-phase reactions between the sapphire workpiece and the abrasive. Surface porosity develops, which enables removal of the outer layer of the crystal by soft polishers, and consequently diminution of the surface-adjacent defective layer. A typical example of this case is polishing of sapphire by fine SiO_2 abrasive [16]. When the SiO_2 particles interact with the sapphire surface, the zone of contact undergoes localized reactions of high pressure and high temperature. As a result, solid-phase reactions proceed between the SiO_2 and the Al_2O_3; the sapphire surface becomes porous and the reaction products are carried away by a soft polisher.

As is known [11, 16], the efficiency of sapphire polishing by colloidal silica in the presence of water is low. Dry polishing using SiO_2 gel is more effective, but not

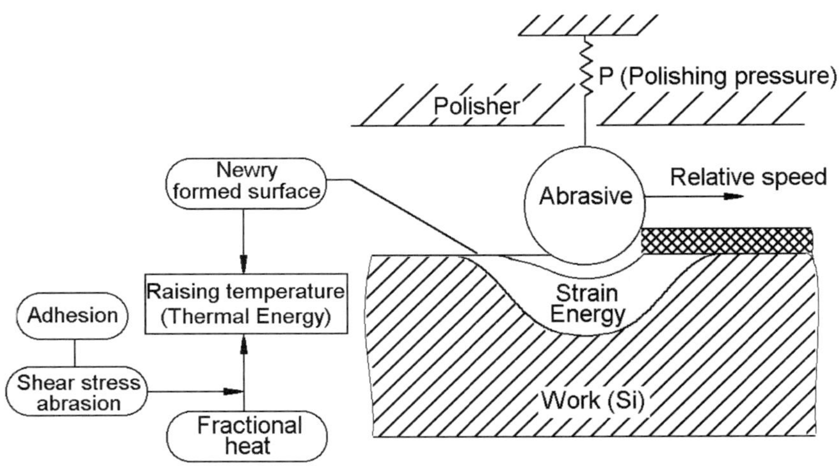

Fig. 6.6 Wet-type chemical–mechanical polishing

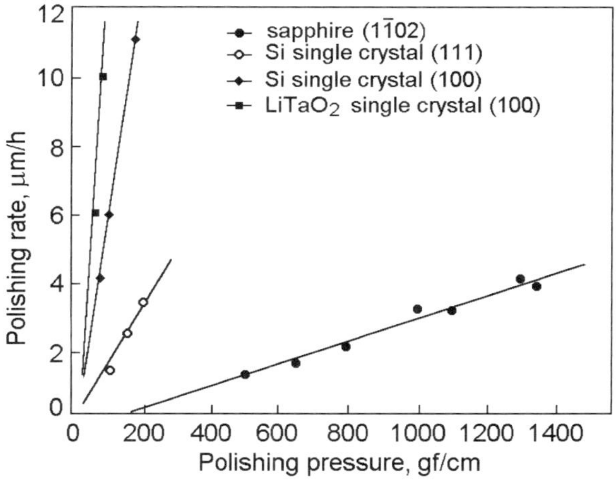

Fig. 6.7 Removal of material vs. applied specific pressure for different materials. Polishing surface: artifical leather

easily realized as it requires high temperatures and protection of the working zone from sputtering of the gel. A procedure of polishing based on fixed colloid particles of Aerosil-175 in the composite Siopol-1 has been proposed [17, 18]. The polishing tool used has a specially designed working surface and is constantly kept in silocozole suspension. This arrangement provides continuous supply of the reagents to the polishing zone. The efficiency of such a polishing progresses into the range of 0.02 mg/min, ~40 times higher than the procedure of polishing on a quartz faceplate with a colloidal silica water solution (0.0005 mg/min). The use of these methods allows a sufficiently high surface quality to be obtained, but the problems of stress and crystal lattice distortions remain unsolved.

Colloidal silica polishing. This method is based on the typical colloidal phenomenon which arises when atomically small particles of silica (10–100 Å) are used in a slurry of preset alkalinity (e.g., with pH 4–12). Very fine colloidal silica is supplied to a soft polisher, and as soon as pressure is applied to the workpiece, a gelling phenomenon peculiar to colloidal solutions starts working ("dilatancy phenomenon") (Fig. 6.8) [19].

The previously considered methods of wet and dry chemical–mechanical polishing to a certain extent depend on the chemical properties of the workpiece surfaces, and therefore can be effectively applied to a limited set of materials. Conversely, colloidal silica polishing is applicable to practically all materials and provides achievement of mirrorlike surfaces free from stresses.

Contactless polishing. This method also is called elastic emission machining (EEM) [20] and results in reduction of material removal by a factor of 10–100 in comparison with mechanical polishing. Particles with a diameter of about 100 Å interact with the workpiece surface, removing only several tens of atoms.

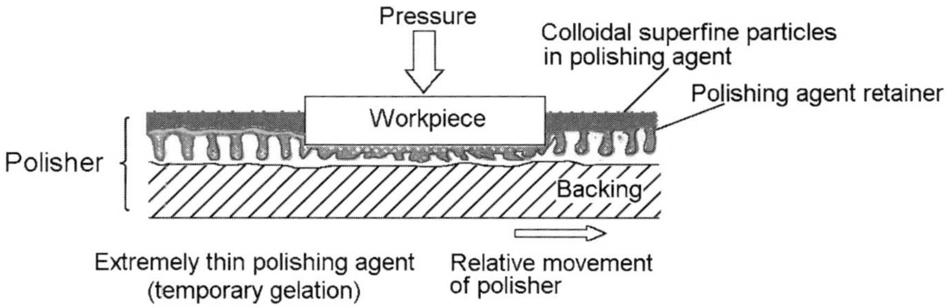

Fig. 6.8 Constitutional diagram of colloidal silica polishing

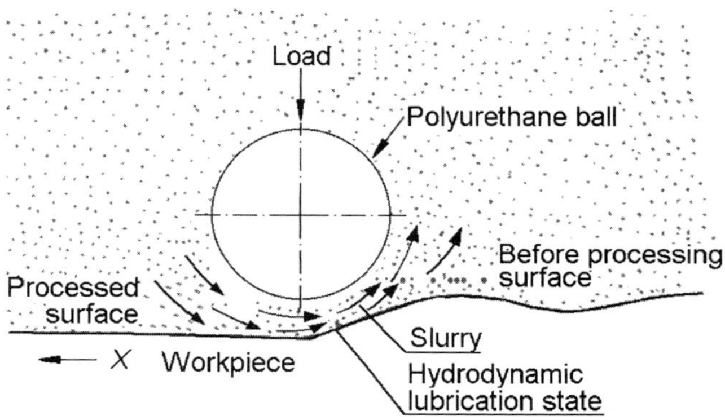

Fig. 6.9 Principle of elastic emission machining

Figure 6.9 demonstrates the action of a "polishing float." When the distance between the polishing facility and the workpiece surface approaches values on the atomic or molecular order, the slurry particles and the surface atoms join together. By this mechanism, separation of the adjoined particles from the workpiece surface does not cause plastic flow. So, the surface is not damaged as these particles are removed.

The considered examples of polishing show that the treatment efficiency (the rate of material removal), the surface quality, the thickness or absence of the defective layer, and the precision and accuracy of surface shape are interrelated. As a rule, chemical–mechanical polishing is characterized by a relatively high treatment efficiency and insignificant resultant defective layer. However, the increasing role of chemical reactions results in more difficulty controlling process precision. The utilization of soft polishing surface materials raises the surface quality, but deteriorates surface shape control.

The above methods of finishing sapphire articles to a low surface roughness and absence of the defective surface-adjacent layer are less economical and typically are reserved for production of expensive products. For these polishing methods, the relative contribution of labor expenses to the cost of the articles is insignificant. Recent increases in surface quality requirements sharply raise the cost of sapphire articles Fig. 6.10.

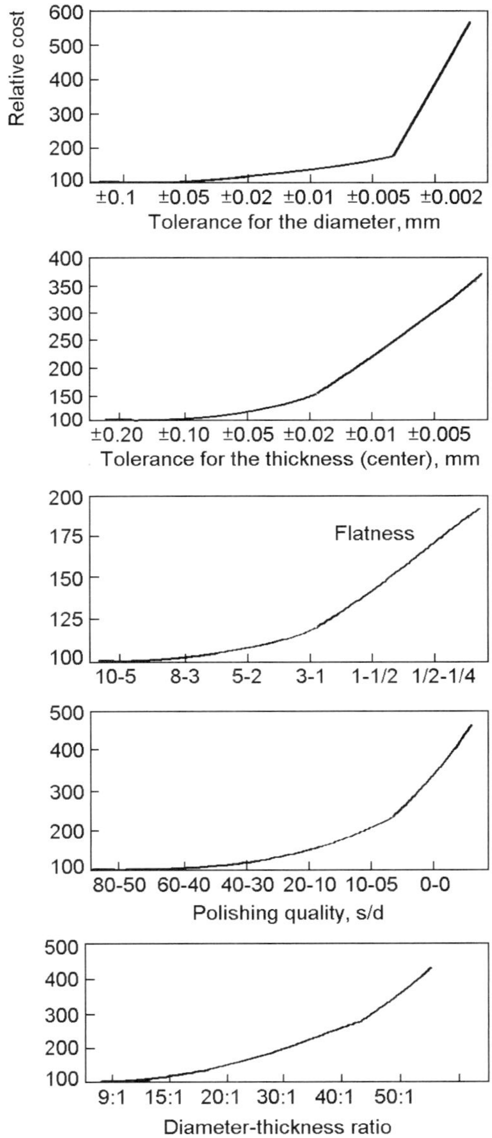

Fig. 6.10 Relative cost of articles with different parameters following treatment quality and allowances. 100 – cost of one ingot

The use of chemical–mechanical polishing is expedient and necessary in cases where the usage and longevity of sapphire articles are defined by the roughness and optical accuracy of their surfaces, as well as by the absence of factors that may distort a wave front.

6.3 Structure of Mechanically Treated Sapphire Surfaces

Several schemes exist that describe the structure of the surface-adjacent layer damaged by mechanical treatment of sapphire. According to one of these schemes, such a layer has four zones. The first, lying immediately on the surface, has serious exhibited damage including cracks and cleaveds. The second zone contains both cracks and traces of plastic deformation; the third zone contains only plastic deformation; and the fourth zone contains a field of residual stresses. In consequence of abrasive treatment, the depth of the plastically deformed layer does not exceed 2 μm. However, it also has been shown that in mechanically treated sapphire a layer of elevated dislocation density forms that can be up to 70 μm in depth [21]. Currently, there is not an elaborate understanding of the structure of a single crystal surface or the processes that occur on the surface and inside its surface-adjacent layer during mechanical treatment. For instance, the publications that report the distribution of defects inside the said layer contain no data on the state of the surface itself, and vice versa. In addition, the surface-adjacent layer itself primarily has not been considered in relation to the previous history of the crystal.

An attempt to describe the surface structure has been made on the basis of a fractal model in which the fractal dimension of the transient layer structure diminishes from $D_f = 3$ in the crystal bulk to $D_f = 2$ on the crystal surface [22]. The surface-adjacent layer has an elevated density of dislocations that arise due to the absorption of accumulated elastic energy. This zone is characterized by compressive stresses and its fractal dimension is $D_f = 3$ as well. However, the long-range order within this layer is disturbed by the high dislocation density. Located closer to the surface is a porous, amorphous zone with a fractal dimension D_f of 2.5–3.0 that also contains tensile stresses. These stresses increase the lattice period and energy accumulation as compared to the crystal bulk. The next zone ($D_f = 2.0$–2.5) contains compressive stresses that are compensated by vacancies on the surface and dislocations within the surface-adjacent layer. In this zone, the higher the density of dislocations, the higher their energy and the lesser the fractal dimension value. The crystal surface itself is considered to be a two-dimensional film subjected to the action of compressive stresses. This layer acts as a dissipative structure characterized by a certain structural state, its region of localization, and its fractal dimension. The surface film provides elastic reaction to weak mechanical disturbances and protects the transient layer, and consequently the crystal bulk from the influence of environmental chemical factors. The depth of the transient layer and its fractal dimension have a fundamental significance, since altering these factors makes it possible to control the properties of the surface-adjacent layer and the overall mechanical characteristics of the material. Although not all statements of the fractal model can be agreed with, such a model enables consideration of the surface structure and of the surface-adjacent layer of mechanically treated sapphire from a consistent position.

Elementary simulation of brittle material dispersion under the action of abrasive particles has shown that the repeated formation and coalescence of cracks leads to removal of small particles. The surface acquires a relief consisting of protrusions and hollows under which cracks propagate into the bulk of the crystal (Fig. 6.11). Therefore, the ground surface of a brittle material consists of a visible relief layer with depth h_p and an adjoining layer containing cracks with depth h_d. These layers are called the damaged layer of a surface treated by grinding (H_p).

The crack-containing layer is formed in brittle materials during all types of abrasive treatment and with any size abrasive grains. The depth of this layer increases as the rate of dispersion rises. This is correlated to the magnitude of forces on the material, the crystallographic orientation of the crystal, and the direction of the cutting forces during grinding. The crack-containing layer ultimately limits the strength of sapphire, especially during tension and bending. The crack-containing layer borders the defective surface-adjacent layer and its depth is the distance l_0 from the crystal surface where the dislocation density and the residual stresses become equal to the corresponding values of the crystal bulk.

On the basis of results reported in literature [24, 25] a conclusion can be made concerning the structure of the surface-adjacent layer that arises in sapphire during grinding and mechanical polishing processes. Crystals with $\rho = (3\text{–}5) \cdot 10^6\ \text{cm}^{-2}$ and block boundary lengths of 0.5–1.5 mm^{-1} were investigated. Cleaved surfaces were studied on the $(11\bar{2}0)$ surface treated by means of free or fixed abrasive. In the latter case, diamond faceplates were used with diminishing grain size; at the final stage of treatment the grain size was 7/5 μm (R_a = 0.32 μm). To investigate the structure of the surface-adjacent layer after free-abrasive treatment (rough and fine grinding), two series of experiments were carried out. In the first, samples extracted from the same crystal ingot were treated by boron carbide with a grain size of 40, 20, and

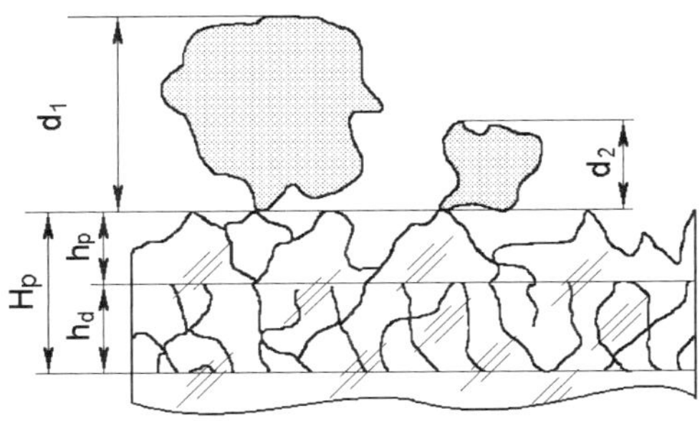

Fig. 6.11 Schematic of the main elements of the abrasive dispersion process for brittle materials: d_1 and d_2 are abrasive grains [23]

14 μm, and then by diamond 7/5, 5/3, 3/2, and 1/0 μm. In the second series, samples were sequentially subjected to all the stages of mechanical treatment and then chemically polished in a borax melt at 1,200 K to completely remove the defective surface-adjacent layer. Afterward each sample was treated by one kind of abrasive. Such a succession of operations made it possible to establish the influence of each type of treatment on the structure of the surface-adjacent layer.

The state of the crystal surface was investigated by electron diffraction analysis. The distribution of dislocations in the surface-adjacent layer was studied by selective etching and layer-by-layer chemical polishing. The defective surface-adjacent layer was removed by a polishing melt of $Na_2B_4O_7 \cdot 10H_2O$ at 1,200–1,300 K or in an etching $KHSO_4$ melt at 1,000 K. The length of block boundaries and the block disorientation angles were determined by X-ray diffraction analysis and the method of random secants. Throughout the study of the mechanically treated surfaces cleaved facets with a structure similar to that of the crystal bulk were used as primary standards.

The interpretation of results from diffraction investigations do not permit unambiguous characterization of the structure of the defective layers studied. To obtain more precise models of the surface structure, the method of interference functions was used [26]. This method is based on the idea that the sample is an aggregate of small, ~10 Å crystallites in which the atomic coordination bond is similar to that of single crystals. Such an approach is used for systems with rigid bonding and small disorientation of valence angles, in particular for carbon structures, and enables the creation of three model levels of the surface-adjacent layer structure [27]. As a first approximation, the surface-adjacent layer is taken as a collection of medium crystallites of equivalent size. As a second approximation, boundaries between the crystallites are taken into account by modeling simple structures consisting of several atoms. As a third approximation, variances in crystallite size about the perfect medium crystallite are taken into consideration. The first and second approximations are realized by simulating the structure of the defective surface-adjacent layer arising during mechanical treatment. Electron diffraction patterns make it possible to confirm the structural models of medium crystallites and medium crystallites with boundaries as the basic structural elements of the surface-adjacent layer formed during treatment.

During fixed-abrasive treatment of the crystal surfaces a thin, polycrystalline surface-adjacent layer is created, manifesting itself as rings in the electron diffraction patterns. In many cases, it is impossible to determine the cracking resistance of such a surface due to the large quantity of scratches, which distort the shape of the typical diamond indenter imprint and change the length of cracks around this imprint. Electron diffraction patterns obtained from the crystal surface exposed by removing a 1.5-μm-deep layer by chemical polishing reveal certain changes in the bulk crystal structure as the surface-adjacent layer is removed. In addition to a textured polycrystalline structure, a clearly discernable block-containing structure with disorientation angles of several degrees is found. The presence of the former structure can be judged from texture arcs, whereas a regular two-dimensional network of reflexes indicates the block structure. The electron diffraction pattern contains reflexes up to the tenth order, which are obtained from reflecting planes with low indices. Therefore, in accordance with Bragg's law, the angles of block disorientation must reach several degrees. Removal of a 2-μm-deep layer by chemical polishing reveals electron diffraction patterns corresponding to those of the reference samples. The density of dislocations on such a surface is 2–2.5 times higher than in the crystal bulk. Further removal of the surface-adjacent layer up to 60 μm does not influence the character of the electron diffraction patterns. The density of dislocations on the surface can reach $1.5 \cdot 10^6$ cm^{-2}, whereas at a depth of 30–60 μm it is equal to the value ρ_0 typical of the reference samples. The contribution of mechanical treatment to the magnitude of internal stresses turns out to be significant. These stresses reach 24 MPa in the crystal, with such an increase characteristic of a layer up to 1- to 2-mm thick.

Treatment of the (0001) plane by free abrasives with a grain size of 40, 20, or 14 μm for 10 h leads to the formation of a thin ($3 \cdot 10^{-5}$ to $3 \cdot 10^{-4}$ cm) polycrystalline layer. The size of a "medium" crystallite a_c is ~60 Å (Table 6.1). For this layer, the intensities of the diffraction lines differ from the tabulated data for polycrystalline structures. For instance, in comparison to a reference sample, the relative intensity of the lines *(hkl)* with small values of l is lower, whereas that with large values of l is higher. So, the maximum reflection intensity is characteristic of planes close to the (0001) plane. In other words, the predominating orientation of the [0001] axis of the crystallite lattice is observed normally to the treated surface.

Typical of depths exceeding 10^{-4} cm is a transition from a textured polycrystalline layer to a plastically deformed single crystalline layer with an elevated dislocation density. The depth of this layer depends on the kind of abrasive. When a rough abrasive with a grain size of 40 μm is used, the density of dislocations near the

Table 6.1 Influence of the type of surface treatment on the structural perfection and mechanical characteristics of the surface-adjacent layer

Kind of surface treatment	Surface structure	$(\rho-\rho_0)/\rho_0$	l_0 (μm)	$(\tau^*-\tau_0^*)/\tau_0^*$	k_c (MN m$^{-3/2}$)	a_c (Å)
Grinding $R_a = 0.32$, AM 14/10, 7/5	Polycrystalline layer	2.8	30	0.7	–	50
$R_a = 0.1$, AM 5/3	Fine dispersed polycrystalline layer	2.1	20–25	0.5	4.8–5.0	60
Polishing $R_z = 0.05$, AM 1/0	Quasiamorphous layer	1.0	5–15	0.4	4.8–5.4	4
Termal treatment, $T = 2,300$ K	Single crystal	0	0	0	3–3.5	4
Chemical polishing	Single crystal	0	0	0	2.8–3.5	4
Cleaved surface	Single crystal	0	0	0	2.8–3.6	4

surface turns out to be so high ($\rho > 3 \cdot 10^6$ cm^{-2}) that it is impossible to reliably determine its value by selective etching methods. In this case, the initial data on density of dislocations only can be obtained for a depth of 15–20 μm, whereas the thickness of the layer with elevated density of dislocations reaches ~50 μm. As the size of the abrasive decreases, the depth of the said layer decreases. For a grain size of 14 μm, this layer is close to 30 μm. The use of AM 7/5 abrasive also leads to the formation of a polycrystalline layer. In this case, the relative intensities of the diffraction lines are closer to the tabulated values, pointing to an absence of a pronounced texture. It should be noted that the reduction of the abrasive grain to 7/5 gives rise to an insignificant increase in the degree of the diffuseness of the diffraction lines, which corresponds to the reduction of the "medium" crystallite size down to $a_c = 50$ Å. The thickness of the plastically deformed layer ($l_0 \sim 20$ μm) also diminishes (Table 6.1).

When surface treatment is realized using AM 5/3 abrasive at greater than 2 h, the size of the formed crystallites sharply decrease. Further increases in the treatment duration result in the transition to a quasiamorphous layer. As seen from weakly diffused reflexes from the single crystal, the thickness of this layer is ~100 Å. The quasiamorphous layer and the size of its related coherent scattering region correspond to a model of an anisotropic "medium" crystallite consisting of 15 atoms of Al and O. The experimental and theoretical integral functions $I(S)$ are most accurate (where $S = 4\pi \sin\theta/\lambda$, θ is the angle of scattering and λ is the incident radiation wavelength) if it is assumed that some part of the volume of the quasiamorphous layer (about 20%) is occupied by the "boundary" between crystallites. Such a boundary may be modeled as an Al_2O_3 cluster consisting of five atoms with characteristic distances O–O (0.2–74 nm), Al–O (0.2–0.6 nm), and Al–Al (0.2–64 nm). Similar results are obtained when the surface is treated by the abrasives AM 3/2 and AM 1/0.

When using the abrasive AM 5/3, the depth of the plastically deformed layer does not exceed 15 μm. It should be noted that for crystals grown by different methods or by the same method but under different conditions, the thickness of the defective layer may vary within wide limits. In particular, for sapphire grown by the Verneuil method the layer thickness is 10–15 μm, whereas for sapphire grown by the HDS method it is 3–5 μm. Essential differences also may be observed in the structure of the surface-adjacent layer of different facets of sapphire after mechanical polishing. For instance, polishing the plane (0001) with AM 3/2 and 1/0 leads to formation of a quasiamorphous surface-adjacent layer. After the same treatment, the diffraction pattern of the plane $(11\bar{2}0)$is typical of a finely dispersed polycrystalline layer. Considering these different facets, the distinction in the ability to form amorphous surface-adjacent layers upon mechanical treatment is related to the fact that the plane (0001) is an active plane of dislocation slip. This hampers the motion of dislocations inward in the crystal and favors accumulation of a large quantity of defects in the surface-adjacent layers.

Structural evolution of the surface-adjacent layer upon changing the type of mechanical treatment is caused by the following factors. As is known, the dimension of the stress relaxation region during the interaction between an abrasive grain and crystal is proportional to the grain size *d*. Consequently, when *d* is reduced, the excessive energy transferred to the crystal must act within a thinner surface-adjacent layer. This has been observed in experiments, while using finer abrasives the rate of material abrasion diminishes. As a result, the excessive energy is accumulated in the thin layer not removed from the crystal surface during treatment. This seems to be accompanied by a certain disturbance of the interatomic bonds and subsequent formation of new boundaries between the crystallites. In other words, an ultrafine dispersed layer (i.e., quasiamorphous) is formed. Energy accumulation also is favored by the fact that the thermal conductivity of the quasiamorphous layer is 15 times lower than that of the crystal. The higher the excessive energy in the layer, the larger the share of the disturbed interatomic bonds, and consequently the level of its amorphous character.

The behavior of crystals during the mechanical treatment process essentially is defined by a characteristic point defect distribution, such as the grain structure, which influences plastic deformation. With an increase in the size of the grain structure elements, specifically with growth in the local point defect density, the material's plasticity diminishes. In the process of treatment using the abrasive AM 5/3, crystals with a small size of grain structure elements are plastic enough that the depth of their defective layer extends 10–15 μm. In crystals with larger-size grain structure elements treated under the same conditions, the defective layer depth may not be larger than 5 μm. Nonuniform distribution of grain structure in the crystal bulk leads to structural anomalies, such as the presence of zones containing no dislocations, a nonuniform dislocation distribution, characteristic impurity distributions (e.g., ultramicrostriation), differences in the relaxation of thermoelastic stresses, and differences in machineability of the crystals. In other words, these crystals exhibit different reaction under the action of equivalent forces.

Thus, the structure of a mechanically treated sapphire surface is the following: a damaged layer consisting of a relief and crack-containing layer followed by a transient surface-adjacent layer with a depth of $\sim 2 \cdot 10^3$ μm characterized by elevated residual stresses, dislocation density, and values and angles of block disorientation. The defective surface-adjacent layer further can be segregated into several zones. The first, with a depth of 10–100 Å (depending on the type of mechanical treatment), can be either polycrystalline or amorphous. The second, with a depth of 1.5–2 μm, is distinguished by the presence of both a textured polycrystalline state and a single crystalline state in which block disorientation angles run into several degrees. The third zone, with a depth of 30–50 μm, is a single crystal with an elevated density of vacancies and dislocations. The fourth zone, with a depth of 1–2 mm, is characterized by elevated residual stresses. Such a scheme is variable, as the the structure of each zone is defined by the characteristic action of the cutting tool on the crystal; this action is dependent on the treatment regimen, size and type of abrasive grain, kind of binder, and LCL, as well as the initial crystal structure. Therefore, each treatment stage affects the behavior and structure of the surface-adjacent layer at subsequent stages.

6.4 Control of the Defective Layer during Mechanical Treatment of Sapphire

In brittle materials such as sapphire, the ultimate strength of the material and the magnitude of forces acting on the material are the main factors that define the intensity of dispersion, the depth of the layer containing cracks, and the structure of the surface-adjacent layer formed. By influencing the rate, load, and character of the movement of abrasive grains with respect to the treated surface, one can change the structure of the defective layer and the removal rate during treatment. While choosing sapphire treatment regimens that minimize development of the defective layer, one must take into account the ratio of the tangential component of the cutting force, P_z, to the radial component, P_y. In the literature, the ratio P_z/P_y is called the "coefficient of abrasive cutting," and for sapphire it is equal to 0.13. This means that the value P_y has the greatest influence on the development of the defective layer as it is directed inside the treated material, exceeding by approximately an order of magnitude P_z in providing removal of the material from the surface. Therefore, in order to obtain a minimal defective layer at sapphire grinding, it is necessary to choose those treatment regimens that allow reduction of P_y and at the same time provide an economically justified level of treatment efficiency.

According to abrasive wear theory [28], the volume of detached material is given by the equation

$$V = NP^{5/4} l_a k_c^{-3/4} H_T^{-1/2}, \tag{6.1}$$

where N is the quantity of abrasive particles, l_a is the path length of an abrasive particle along the surface, and P is the load on the particle.

As follows from (6.1), the intensity of abrasive wear rises with increasing load on a particle and its path length along the surface; the physical properties of the material are taken into account by means of the values H_T and k_c. The expression (6.1) allows for estimation of the volume of material detached at each process stage.

To compare different types of mechanical treatment, the concept of surface grindability has been introduced [29]. The influence of the applied load and velocity of tool movement on the coefficient of material removal during optical polishing is described by the expression [30]

$$V_p = fSPv_t t, \tag{6.2}$$

where V_p is the volume of material removed from the area S during the time t, f is the friction coefficient, P is the applied normal pressure, and v_t is the velocity of the tool's movement (rotation). The friction coefficient can be substituted by the proportionality coefficient, μ, which depends on the properties of the material, abrasive, and so forth, as in the equation

$$V_p = \mu SPv_t t, \tag{6.3}$$

where μ is grindability. When this property is high, the material can be removed rapidly without application of considerable loads. Grindability changes at the transition from brittle failure to plastic flow. The linear dependence of the coefficient of material detachment on the rotational velocity of the tool and on the applied pressure is described by the function $P^{9/8}$ [31].

As a rule, sapphire is treated by *fixed or free abrasives*, and in some cases a combination is used. In addition to treatment conditions, detachment of sapphire depends on the characteristics of the crystallographic planes (Table 6.2). In the table, d is the interplanar distance, n is the quantity of free bonds, and E_f is the surface formation energy.

The minimum removal rate is observed on the basal plane with treatment by both fixed and free abrasive. The maximum detachment rate for different treatment regimens depends on the type of abrasive. In the case of free-abrasive treatment it is the plane $\{10\bar{1}0\}$, while for fixed-abrasive treatment it is the plane $\{10\bar{1}2\}$. The essential differences observed in the detachment rate seem to be caused by the influence of the grinding regimen on the cleaved formation mechanism, which, in turn, depends onthe crystallographic orientation. With free-abrasive treatment the detachment rate correlates to the reciprocal value of the reticular atomic density in the plane of interest and with the reciprocal of the quantity of free bonds per unit of surface.

Fixed-abrasive treatment provides the maximum removal efficiency, but also maximizes formation of the surface-adjacent defective layer. The applied loads may exceed the limit of elasticity, and then a branched network of microcracks arises on the crystal surface. Even if the surface has undergone high-quality mechanical polishing, the depth of microcracks reaches 2 μm, and a large number of cracks are formed with an elevated density of dislocations around them. In the interval of abrasive size from 4 to 80 μm, the dependence of μ on the surface roughness is cubic (Fig. 6.12).

The structural perfection of the surface-adjacent layer depends on such factors as the rotation frequency of the tool, the size of the abrasive grain, its concentration, the specific pressure experienced by the sapphire crystal at the place of contact with

Table 6.2 Sapphire attrition rate (mm/h) [32]

Sample characteristics				Abrasive characteristics					
				Free abrasive, boron carbide N4				Fixed abrasive ACM80/63	
				Load × 10^{-2} (kg/mm^2)					
Plane	d (Å)	n	E_f (J/m^2)	3	5	7	11	2.2	5
(0001)	2.165	6.6	5.44	1.97	2.52	3.48	3.84	2.12	7.12
$\{10\bar{1}2\}$	3.479	3.5	~1.3	2.49	3.38	4.22	3.76	6.23	17.8
$\{10\bar{2}0\}$	2.379	4.8	1.27	3.45	4.83	6.48	7.83	2.31	8.29
$\{11\bar{2}0\}$	1.374	–	1.09	3.26	4.97	7.48	8.11	3.59	11.8

E_f is surface formation energy

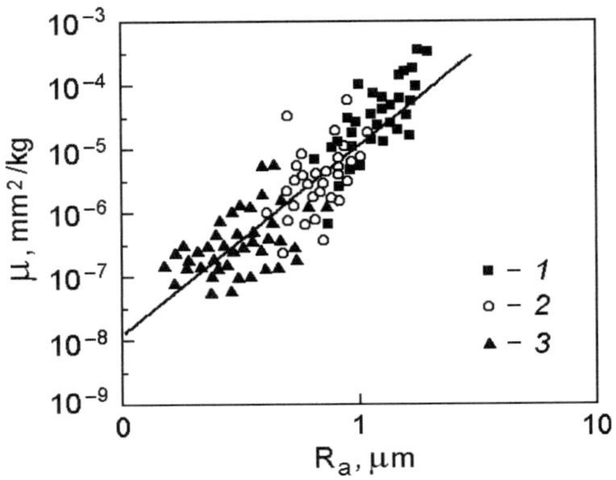

Fig. 6.12 Correlation between the grindability, μ, and surface roughness of sapphire with varying size fixed abrasive [29]. (*1*) Rough abrasive; (*2*) medium abrasive; (*3*) fine abrasive

the tool, the interval between dressings, and so forth. The most essential parameters were established by means of a mathematical model using the fundamental tools of dispersion and regression analysis [33]. The anomalous birefringence of sapphire (proportional to the value of stresses) depends mainly on the concentration of diamond powder in the diamond-containing layer, the rotation frequency of the tool, and the specific load on the tool. Increasing the rotation frequency from 4 to 40.4 m s^{-1} diminishes the initial value of anomalous birefringence by 18%; increasing the concentration of diamond by 25–150% reduces this value by 14%. When sapphire is treated by fixed abrasive on diamond faceplates, increasing the peripheral velocity of the abrasive diamond tool from 4 to 20 m s^{-1} results in a reduction in the difference in optical path length between the ordinary and extraordinary rays. However, further increase of the velocity does not lead to appreciable changes in this value (Fig. 6.13) [34].

Such behavior in the anomalous birefringence is obviously correlated to more intense local heating in the contact area resulting from increasing rotation frequency of the tool. The local temperature reaches values at which a partial relaxation of stresses may occur. The same situation is observed if the concentration of diamond in the diamond-containing layer or the specific load on the tool increase, but the effect is less pronounced.

The process of sapphire treatment is multistage, and the influence of successive treatment stages on the defectiveness of the surface-adjacent layer can be established [34, 35] (Fig. 6.14). Following from Fig. 6.14, each subsequent batch of samples is treated by an abrasive with a smaller grain size.

To explain the observed maximum (curve *1*, Fig. 6.14a), the dependences $\Delta\Gamma/\Gamma_0$ of the defective layer depth on the abrasive grain size are presented (curves *3* and *4*). Each batch of samples was treated by the same abrasive grain size. As is seen, for grain sizes of 160 μm or more the treatment makes a contribution to the value of Γ, which diminishes only with grain sizes smaller than 160 μm. Therefore, the

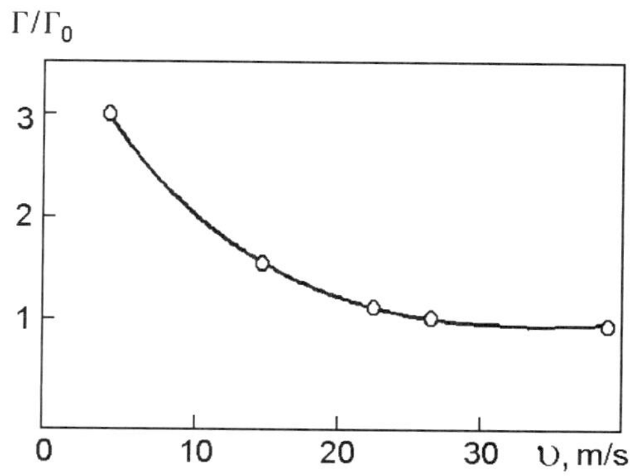

Fig. 6.13 Relative value of anomalous birefringence vs. the rotational velocity of the tool. The value Γ is the anomalous birefringence in the mechanically treated crystal, Γ_0 is the difference in the ray paths in the same crystal prior to mechanical treatment

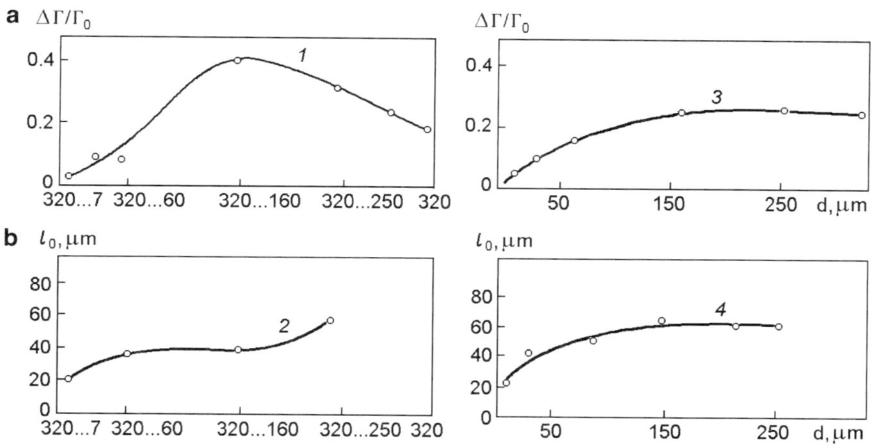

Fig. 6.14 Relative value of the anomalous birefringence (**a**) and defective layer depth (**b**) depending on successive mechanical treatment stages and abrasive grain size for the plane (0001)

maximum of curve 1 is caused by the fact that for treatment stages using a grain size of 320–160 μm the total value of internal stresses rises, since each subsequent stage makes its own, sufficiently large contribution to the value of stresses. The earlier formed defective layer is only partially removed. When using faceplates with a smaller grain size the mentioned layer also is removed, but the subsequent contribution of treatment with grain size smaller than 160 μm is insignificant. Through this connection, the samples that have undergone all stages of treatment exhibit a diminishing value of birefringence. The depth of the disturbed layer gradually increases as the grain size increases (Fig. 6.14, curve *4*) and such a rise is especially significant for 7–60-μm grains.

Thus, for grain sizes up to 60 μm the depth of the layer with an elevated density of dislocations turns out to be comparable to the grain size. The use of abrasives with a grain size exceeding 60 μm does not appreciably change the depth of this layer. Shown for comparison in Fig. 6.15 is the distribution of the density of individual dislocations in ground samples treated with 250 μm grains and in polished samples with ACM 7/5 [36]. In the ground samples, the density of dislocations at the surface is so high that its value can be reliably measured only at depths exceeding 40 μm.

Free abrasive allows a higher surface quality and lower density of defects in the surface-adjacent layer with application of lesser force (Table 6.3). The same level of perfection also can be achieved through treatment using a combination of free and fixed abrasives with subsequent mechanical polish finishing with free abrasive on a boxwood polisher.

Treatment by boron carbide allows sufficiently high perfection of the surface-adjacent layer, yet the surfaces become contaminated with boron carbide particles,

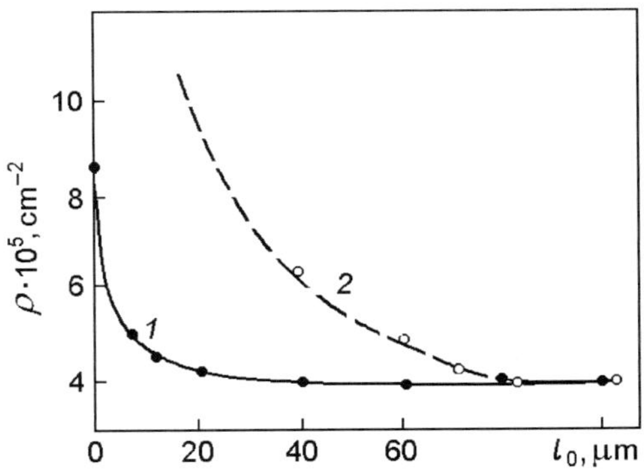

Fig. 6.15 Distribution of the density of dislocations vs. depth in a polished sample (*1*) and in a sample subjected to grinding (2) for the plane (0001)

Table 6.3 Types of treatment and defectiveness of the surface-adjacent layer

Kind of treatment	$\Delta\Gamma/\Gamma_0$	$\Delta\rho/\rho_0$	l_0 (μm)	Optical strength (a.u.) (0001)	Optical strength (a.u.) $(11\bar{2}0)$
Free abrasive	0.4	1.0	15	0.8	0.5
	–	0.7	20	–	0.3
	–	1.7	25	–	–
Combination of free	0.7	2.3	35	0.6	0.4
and fixed abrasive	0.7	1.7	20	0.5	0.4
	0.6	1.0	15	1.0	0.5
	0.65	1.3	30	0.6	0.4
Fixed abrasive	0.7	2.8	30	–	–

$\Delta\rho/\rho_0$ – is the relative variation of density of dislocations, where ρ is the density of single dislocations on the surface of a mechanically treated sample; ρ_0 is the density of dislocations in the volume; l_0 is thickness of damaged layer which are difficult to clean. The resulting optical strength of such surfaces is lower than that of surfaces finished with diamond micropowders.

A correlation between the grindability of the surface and its roughness was observed with free-abrasive treatment (Fig. 6.16). The study was carried out on ruby and optical glass.

The slope of the straight line in logarithmic coordinates is equal to 3, and the roughness decreases with decreasing grindability. Sapphire surfaces subjected to durable free-abrasive treatment are characterized by the presence of chaotically spaced, short dislocation boundaries propagating to a depth of 5–10 μm. In other words, there is observed mechanical polygonization (Fig. 6.17).

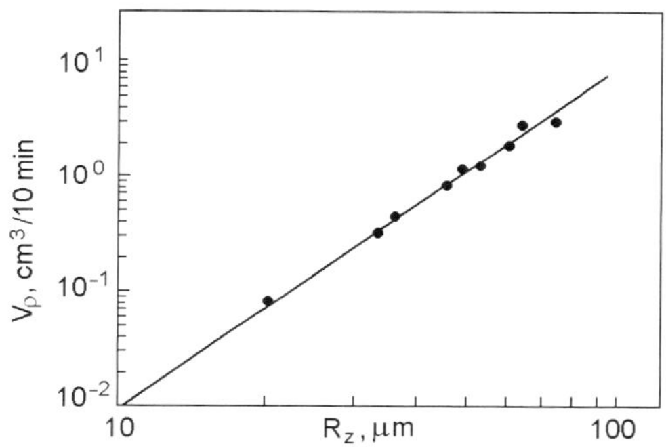

Fig. 6.16 Correlation between the coefficient of material removal and the surface roughness at free-abrasive grinding

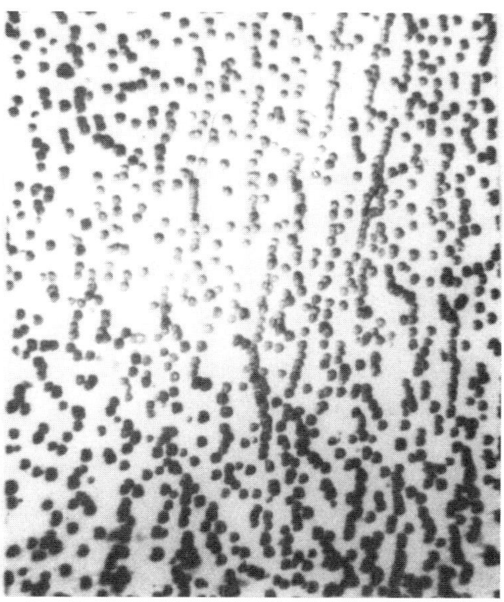

Fig. 6.17 Mechanical polygonization arising during polishing of sapphire

Investigation of the surface-adjacent layer by electron microscopy, X-ray topography, and so on [37], reveals that ground surfaces are characterized by an extremely defective surface-adjacent layer with a depth exceeding 10 μm. This layer contains cracks, distorted zones with high stresses, and has a high density of dislocations. After polishing, a network of dislocation half-loops, which propagate to a depth not less than 100 nm, remain within the surface-adjacent layer under cracks (Fig. 6.18).

The anisotropy of sapphire properties manifests itself during both free-and fixed-abrasive treatment (Fig. 6.19, Table 6.4). As follows from this figure, the planes $(11\bar{2}0)$ are more easily treated than the basal planes (0001), and this fact is especially apparent when using boron carbide. The rate of removal on the plane (0001) ± 10′ is less by half than for the (0001) plane with a deviation of ~2°.

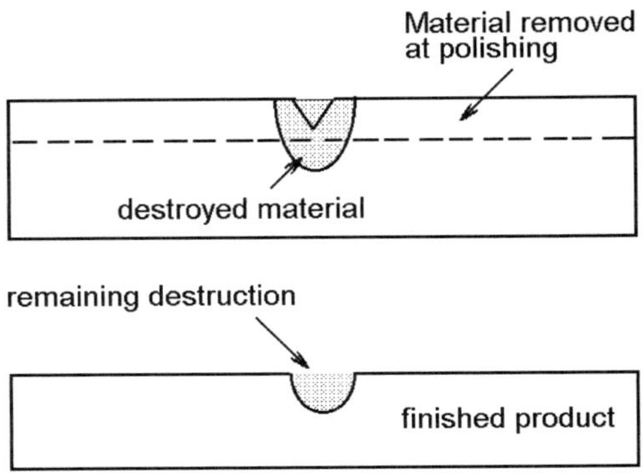

Fig. 6.18 Schematic of the defective zone remaining after polishing

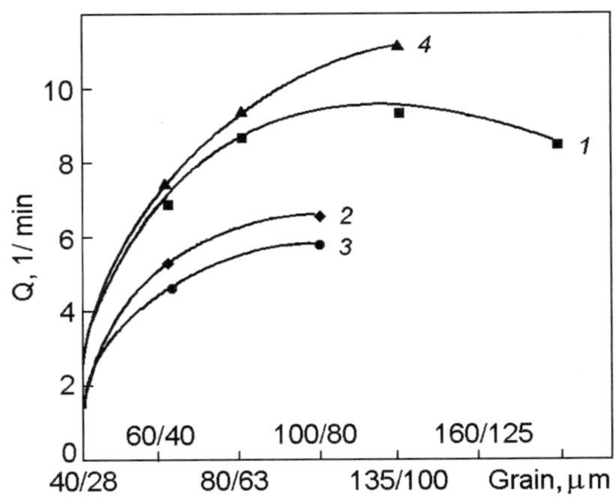

Fig. 6.19 Efficiency of grinding by diamond wheels (*Q*) depending on the fraction size of diamond grains and sample orientation (*1*) orientation $(10\bar{1}0)$, (*2*) orientation (0001) ±2°, (*3*) (0001) ± 10′, (*4*) orientation $(11\bar{2}0)$. Testing was carried out on the samples grown by the method of HDSM

Table 6.4 Influence of lattice anisotropy on surface roughness [69]

Type of treatment	R_a (μm)	
	(0001)	$(10\bar{1}2)$
Multiwire fixed-abrasive cutting	0.25	0.53
Grinding by free boron carbide, 40 μm	1.11	0.75
Grinding by free boron carbide, 20 μm	0.73	0.54

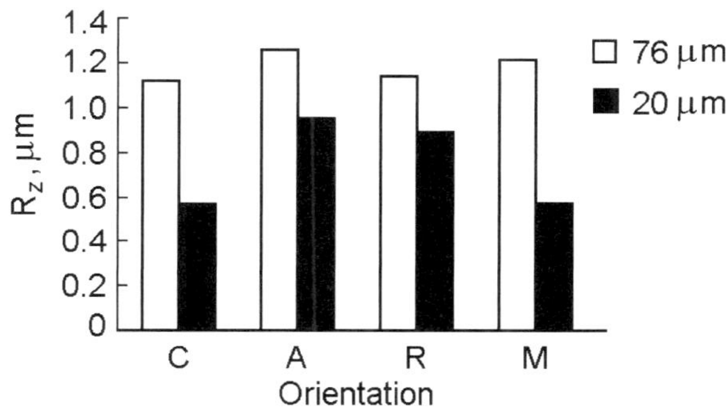

Fig. 6.20 Roughness of sapphire crystals of different orientations treated by diamond wheels with grain size 76 and 20 μm

Samples with orientations (0001), $(11\bar{2}0)$, $(10\bar{1}0)$, $(10\bar{1}2)$ cut from crystals grown by the HEM method have been investigated [38] (Fig.6.20). The samples underwent grinding with constant cutting conditions by different grit diamond wheels. Analogous results are obtained with crystals grown by other methods.

The distinction in grindability observed between the basal and prism planes seems to be a result of cleavage planes, which in sapphire are the prism and rhombohedral planes $(10\bar{1}0)$. During grinding, basal plane cracks are formed at a large angle to the treated surface, while on the prism planes cracks usually arise at a substantially lesser angle and often are almost parallel to the treated surface. In the former case there appear deep pits and in the latter case these pits are shallow and wide, which favors better treatment of the surface. Bending of the (0001), $(11\bar{2}0)$, and $(10\bar{1}0)$ sample surfaces treated under the same conditions also testifies to high residual stresses within the surface-adjacent layer.

The lubricating-cooling liquid. The structure of the damaged layer is greatly influenced by the liquid component of the abrasive suspension used, which provides heat withdrawal from the zone of cutting, favors improvement of the surface quality, and decreases the density of defects in the surface-adjacent layer. When choosing a lubricating-cooling liquid for mechanical polishing of crystals one must take into account the necessity of mechanical activation of the surface, i.e., the creation of a metastable active form of the surface. This can be achieved through consideration of the mechanical action on the crystal, the action duration, the temperature and composition of the surrounding medium, and the strength characteristics of the material. For sapphire treatment water, kerosene, watchmaker's oil, sunflower and olive oils, and polyatomic alcohols have been used as lubricating-cooling liquids. Water is conventionally used during fixed abrasive and boron carbide treatment, sunflower and olive oils have been applied for free-abrasive treatment, and kerosene is efficient while working with boron carbide.

The liquid phase must possess high-heat conductivity, low wedging action, high boiling temperature, minimal pressure of saturated vapors at room temperature, and sufficient viscosity for holding the abrasive grains on the flat horizontal surface of the rotating polisher. Moreover, it must sufficiently moisten the polisher surface, be easily removed from the crystal surface, and remain neutral to any materials used to fix the crystals in place during polishing. Finally, the liquid phase should be nontoxic, inflammable, nonoxidizing to the surface of metallic polisher, and not able to polymerize during the polishing process. The polishing efficiency with the liquid must be suitably high, while at the same time the liquid itself should not be cost-prohibitive. It often is difficult to choose a lubricating-cooling liquid that simultaneously satisfies all the above requirements, since very often these are mutually exclusive. For example, to raise the efficiency of sapphire treatment, it is necessary to increase its brittleness, so the lubricating-cooling liquid must possess a strong wedging action. But, on the other hand, this will lead to the appearance of a large number of microcracks as well as a reduction in the purity and perfection of the treated surface.

To a considerable extent these requirements can be satisfied by polyatomic alcohols such as glycerine, ethylene glycol, and their mixtures. The use of viscous glycerine slows down the process of polishing due to formation of a rather thick layer of liquid between the treated surface and the polisher, preventing a large part of diamond grains from participating in polishing. The use of ethylene glycol raises the efficiency of polishing by diamond, but causes the appearance of individual scratches on the crystal surface. This is explained by the low viscosity of ethylene glycol and the resulting thin layer of liquid between the polisher and the treated surface. As a result, individual large-size grains leave deep scratches on the surface. The choice of the optimum abrasive suspension makes the efficiency

Table 6.5 Optimal concentrations of polyatomic alcohols for different abrasive grain sizes

	C (%)		
Grain size	Glycerine	Ethylene glycol	C of suspension (%)
AM 10/7	75	25	1:20
AM 7/5	80	20	1:30
AM 5/3	85	15	1:40
AM 3/2	90	10	1:60
AM 1/0	95	5	1:125

of mechanical polishing sufficiently high, while maintaining surface quality. Finish polishing using polyatomic alcohols provides a value of $R_a < 0.03$ μm, a minimal depth of the damaged layer, and no visible scratching of the crystal surface. Although in some cases the density of dislocations increases on the surface, the defective layer depth is not larger than 5 μm. The absence of the defective surface-adjacent layer provides an increase in the value of optical surface strength by 1.5–2 times.

The role of the liquid phase of the abrasive suspension can be reduced to a chemical interaction with the treated material. Such a chemical–mechanical treatment of sapphire allows high-quality surfaces without a defective surface-adjacent layer. The use of this mixture for fixed-abrasive treatment is inexpedient, since it does not permit high-quality surfaces.

Certain possibilities for controlling the structural perfection of the surface-adjacent layer and for raising the efficiency of mechanical treatment are provided by the use of electrolytically dissociated water. Water is widely used as a lubricator-coolant during mechanical treatment of sapphire; however, the treatment efficiency is typically low. Extraordinary properties of electrolytically disassociated water recently have been revealed. For instance, if the water subjected to electrolytic dissociation (leading to an alkaline reaction with pH 8–10) is used for the preparation of an abrasive suspension, the efficiency of sapphire treatment by this suspension may be raised by 1.2–2 times [39]. The ability of dissociated water to influence the quality of the treated surface and the treatment efficiency seems to be associated with the fact that during the process of mechanical treatment of sapphire (which is a dielectric) an excessive electric charge arises. This leads to a decrease in the removal rate of the material. A positive effect is achieved when using only the water taken from the cathode zone of electrolysis apparatus, i.e., water enriched with H^+ ions.

Sapphire is known to be resistant to the action of aggressive media and one would think that its chemical–mechanical polishing would be hindered. However, owing to the above-mentioned peculiarities of the surface-adjacent layer, sapphire is amenable to chemical and especially chemical–mechanical action. Chemical polishing is carried out in a continuous laminar liquid flow that removes the damaged layer, particularly that of spherical and ellipsoidal surfaces. Chemical polishing in a melt of borax ($Na_2B_4O_7 \cdot 10H_2O$), in V_2O_5, or in metavanadates of alkaline metal at 1,120–1,620 K results in high-quality surfaces. Orthophosphoric acid H_3PO_4 (670–770 K), fluoric lead melt PbF_2 (1,120 K), and potassium bisulfate melt (720 K) also are used as polishing and etching solutions. The rate of polishing and the quality of the surface can be varied by temperature and by melt viscosity, as well as by the degree of melt saturation with aluminum oxide. Removal of the defective surface-adjacent layer raises the crystal strength. Chemical polishing in a borax melt may lead to embrittlement of mechanically treated surfaces, as it removes the layer with "fresh" mobile dislocations. Removal of this layer leads to relaxation of the stresses that arise in the sample as a result of the action of concentrated load by brittle failure.

Gas polishing of sapphire also is promising. For polishing the planes $(10\bar{1}0)$, $(11\bar{2}0)$, and (0001), chlorotrifluoromethane ($CClF_3$) may be used. The etching (polishing) temperature is 1,670–1,770 K, and the highest rate of material removal is 0.6 μm/min at 1,770 K. Commonly used for polishing is a hydrogen flow with temperature intervals of 1,370–1,470 and 1,570–1,670 K. The rate of sapphire etching in hydrogen flow logarithmically depends on the temperature. The drawbacks of this method include the high temperature of the process and the possibility of selective etching of the surface. To lower the polishing temperature to 470–770 K, ionization of the gaseous medium and ionic flame etching are used. However, chemical and gas polishing do not result in sapphire surfaces with the required flatness, although the defective surface adjacent layer is removed.

It should be noted that sapphire crystals grown by different methods have different grindability. Thus, in sapphire grown by the Verneuil method plastic deformation is observed during free-abrasive treatment and a layer with a thickness up to 15 μm and an elevated density of dislocations is formed. In crystals grown by the HDS method under the same treatment conditions, stresses relax mainly due to brittle failure, and often such a layer cannot be revealed. During treatment by means of large-size diamond grains a branched network of cracks is formed in the crystals, but at subsequent stages of treatment, cracks are removed and the layer with an elevated density of dislocations is either absent or has a depth not exceeding 5 μm. As mentioned above, the crystals grown by the Verneuil method are more plastic. At initial treatment stages they contain lesser quantities of microcracks and cleaveds, but after finish polishing the depth of the layer with an elevated density of dislocations is larger.

Thus, by varying polishing parameters, such as the rate of cutting, the size of abrasive grains, the order of polishing stages, and the composition of the liquid phase of the abrasive suspension, it is possible for chemical–mechanical polishing to control the structure of the surface-adjacent layer, i.e., to strengthen the surface (or make it less strong) and to increase or diminish the depth of the defective layer. The character of such control depends on the problem stated. The solution of this problem defines the service life and normal operation of sapphire articles, their optical homogeneity, strength, and other parameters.

6.5 Prediction of Sapphire Strength Characteristics by Microindentation Methods

As is known, the deformation characteristics of sapphire depend on crystallographic orientation. Based on an average creep activation energy of 7.8 eV, diffusion dislocation creep has been suggested to be the most probable deformation mechanism. In the opinion of some authors, the activation energy diminishes as the tension increases from 5.6 eV at 7 MPa to 4.35 eV at 45.5 MPa. Sapphire creep has been attributed to thermally activated displacement of dislocations over the Payerls barriers. The power dependence of creep velocity on the load ($\upsilon \sim P^{4.2}$) and consistency of the activation energy in the high-temperature region suggest that at $T > 2{,}100$ K deformation is provided by nonconservative displacement of dislocations. In this case, the activation energy value of 8.6 eV corresponds to that of self-diffusion for the least mobile O^{2-} ion and is tension independent.

The activation parameters of dislocation motion and the characteristic deformation temperature in the crystal can be determined by the prevalent method of local loading of the crystal by indenter throughout a broad temperature range with subsequent study of dislocation configurations ("rosettes") around the indenter imprints.

The configuration of the dislocation rosettes depends on the temperature, the load on the indenter, and the load duration. As a rule, on close examination of the rosettes a substantial quantity of growth dislocations fall within the field of view. Such dislocations differ from those formed under the indenter, exhibiting a larger size of etch pits. In the 1,370–1,770 K temperature range, a hexagonal rosette with vertices directed along the $(10\bar{1}0)$ arises at relatively low load and short loading duration. As the load increases, the rosette takes the contour of a six-pointed star with the vertex directions $\langle 11\bar{2}0\rangle$ formed by the superposition of two triangles with sides $\langle 10\bar{1}0\rangle$. With further increases of the load, wide dislocation aggregations appear at the star vertices approximately in the directions $\langle 10\bar{1}0\rangle$. They are especially apparent with low-growth dislocation densities. Upon long-duration loading, the contour of the star under the indenter becomes blurred, the pits regroup, evidently due to the development of creep, and hardness decreases. After indentation at $T \geq 1{,}870$ K, the rosettes look like a cloud of etch pits with slight rays. The configurations of the rosettes are similar to those observed in alkali halide and semiconductor crystals. The ray configuration of the rosette is caused by dislocation slip in the planes normal to the indented surface (in the planes of the $\{10\bar{2}0\}1$ prism for sapphire). The configuration of the star is defined by slip in the planes inclined to the surface (in the planes $\langle 10\bar{1}1\rangle$ for sapphire).

Dislocation loops broadening in the inclined planes intersect along the directions $\langle 11\bar{2}0\rangle$ and form sessile dislocations. The latter also can be formed in reactions while interacting with the dislocations slipping in the basal plane. The strongest deformation under the indenter occurs on the plane (0001) due to slipping over the basal planes parallel to the surface. However, upon etching the plane (0001) these dislocations are not revealed.

For alkali halide and metallic crystals the transformation of the rosette into a cloud of pits is connected with a sharp intensification of the tendency to transverse slip and probably to creep of the dislocations. If it is assumed that the mechanisms of slip in sapphire are analogous, then the change in the configuration of the rosette indicates a strong tendency to cross slip in sapphire at $T \geq 1{,}870$ K. At lower temperatures, cross slip (and probably creep) become essential when the loading duration increases.

On the premise that the dislocations move in flat rows, the length of the dislocation row l generated by one source on the surface is described by the expression [40]

$$l = cP^{m/2(m+1)}\, t^{1/(2m+1)} \exp[-E^*/kT(2m+1)], \tag{6.4}$$

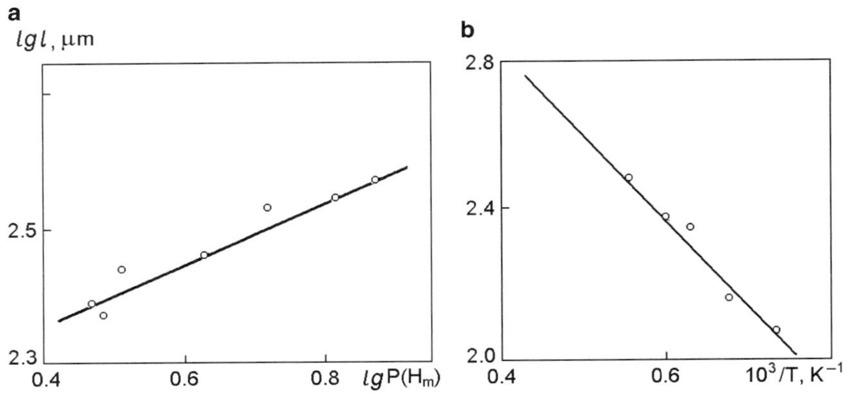

Fig. 6.21 The ray length of the rosette on the plane (0001) of sapphire vs. load on the indentor at $T = 1{,}670$ K (**a**) and temperature of indentation at $P = 3.8$ N (**b**) ($t = 15$ s)

where c is a material constant; P is the load on the indenter; t is the loading duration; E^* is the activation energy of the dislocation motion; and m is the exponent in the empirical dependence of the dislocation motion velocity v_r on the shear stress τ^* ($v_r \sim \tau^{*m}$). The values of m and E^* can be determined from the dependences of l at certain T and t: $l_{T,t} = f(P)$ and $l_P = f(T)$ when t is not very large. The optimum condition is $t = 15$ s. It is necessary to use moderate loads ($P \leq 7$ N at 1,670 K), which do not change the configuration of the rosette. From the slope of the curves lg$l$–lg$P$ (Fig. 6.21a) one can estimate the value $m \approx 2.1$. Such a small value of m, close to $m = 1$–1.5 for purely covalent crystals [41], gives grounds to conclude that the main contribution to resistance to dislocation motion in Al_2O_3 belongs to the Payerls stresses.

In the temperature range resulting in the same configuration of the rosette, the dependence lg$l = f(1/T)$ also turns out to be linear (Fig. 6.21b). This allows estimation of the activation energy for the known m as $E^* \approx 2.4$. Such a value is somewhat higher than $E^* = 1.9$ eV, which others have obtained [42] using the results of testing sapphire for creep [43]. This distinction is possibly connected to the dominating role of slip in the basal plane during testing for creep, whereas the dislocation rosette is formed mainly due to prismatic or pyramidal slip. For the latter case, the activation energy of dislocation motion in Al_2O_3 is $E^* \approx 6$–7 eV. The conformity of these values should be considered satisfactory if the following consideration is taken into account. On the basis of direct and inverse jumps being used to describe dislocation motion velocity [41], it turns out that the activation energy value E^* obtained from experimental data of the temperature dependence of mechanical properties is about three times higher than the value of E^* determined from dislocation mobility.

The value of E^* also can be used to calculate the parameter $w = E^*/kT_m$, which characterizes the crystal lattice rigidity with respect to dislocation motion [41]. At

$T_m = 2{,}323$ K, the value w equals 12. For purely covalent crystals $w \approx 15$. Thus, the degree of crystal lattice rigidity for sapphire is close to that of purely covalent crystals.

From the value of the parameter w, it is possible to determine the characteristic temperature of deformation T^* below which the Payerls stress becomes essential and the temperature dependence of the yield limit sharply rises [42].

$$T^* \approx \sqrt{w/21}T_m \tag{6.5}$$

The estimation of T^* in (6.6) gives the value $T^* \approx 1{,}770$ K. This agrees much better with the data on the temperature dependence of the critical shear stress of sapphire compared with $T^* = 1{,}520$ K [41], since a sharp rise of the critical shear stress corresponds to $T < 1{,}770$–$1{,}870$ K [44]. The value T^* has been estimated from the study of yield stress after a 100-h loading. The value $T^* \approx 1{,}770$ K also is confirmed by the above-described change of the configuration of the dislocation rosette around the indenter imprint at $T > 1{,}770$ K. In the range of homological temperatures, the characteristic deformation temperature in sapphire T/T_m is 0.76 K. Such a value is closer to that for covalent crystals (0.82–0.83 K), for partially covalent crystals T/T_m does not exceed 0.67.

Thus, the study of the dislocation spread on indentation of sapphire in the plane (0001) at 1,370–1,970 K testifies that the mechanism of dislocation motion in sapphire at such temperatures is close to the mechanism in covalent crystals. Specifically, it is controlled by high Payerls barriers ($E^* \approx 2.4$ eV) at a high value of the characteristic temperature of deformation.

A dimensionless parameter, defined as the fraction of plastic deformation to total deformation under the indenter, has been suggested as a characteristic of plasticity

$$B^* = \varepsilon_p / \varepsilon_d = \varepsilon_e / \varepsilon_d = 1, \tag{6.6}$$

where ε_p, ε_e, and ε_d are the values of plastic, elastic, and total deformation, respectively, averaged over the area of the indenter-sample contact. The ratio of plastic deformation to total deformation characterizes the ability of a material to reduce

Crystal	H_h (GPa)	σ	ε_e (%)	B^*
Diamond	80	0.1	6.34	0.161
Sapphire	20	0.23	4.41	0.417

stress through geometric or structural change under load. In contrast, elastic deformation raises the stresses, including normal stresses, responsible for the nucleation and development of cracks. An increase in ε_e upon loading the sample indicates a weak stress relaxation by plastic deformation, i.e., a low plasticity of the material. Therefore, the plasticity can be expressed as $B^* = \varepsilon_p/\varepsilon_e$.

The plasticity parameter of sapphire at room temperature is expressed by the equation

$$B^* = 1 - 4.13(1 - \sigma - 2\sigma^2)H_T / E \tag{6.7}$$

It is interesting to compare these values with the corresponding parameters of diamond:
Diamond has the largest value of elastic deformation and the lowest B^*. Indeed, on the surface of diamond the plastic imprint is very weak; it is seen mainly along the indenter diagonals where the local concentration of stresses is high. The plasticity of sapphire is relatively much higher, but is still low on an absolute scale, since the covalent component of the atomic bonding in this crystal is large. The measurement of the parameter B^*, defined mainly by the ratio of the hardness to Young's modulus, seems to be the only way for direct comparison of the plasticity of materials, which are conventionally considered brittle but exhibit elastic-plastic deformation proceeding without macroscopic damage upon loading by the indenter.

As the load on the indenter increases, small radial cracks appear on the surface. Then the so-called medial cracks are formed, symmetric with respect to the load axis. With further increases of the load, both types of cracks join together. At a certain depth, lateral cracks appear under the surface and then propagate almost parallel to it. Radial cracks are formed near the edge of the indenter at locations of stress concentration. This scheme takes into account not only elastic but also residual stresses arising as a consequence of local microplastic deformation.

Production of crystalline products requires express methods for the determination of strength characteristics of the treated surfaces. The above-considered dependencies permit use of the microindentation method for the measurement of viscosity of damage, which characterizes the resistance of the material to crack propagation. The value of viscosity of damage is defined by the effective energy of damage or the coefficient of cracking resistance, k_c. The measurement of k_c is based on loading of the crystal by an indenter with subsequent correlation between the size of the indenter imprint and the length of radial cracks near the print [45].

The study of the cracking resistance of sapphire crystals grown by different methods shows that k_c does not belong in tabulated lists of material properties. It is a structure-sensitive characteristic dependent on the density of low-mobility growth dislocations and of mobile dislocations introduced by mechanical treatment [44]. The value of k_c also is defined by the type and concentration of impurities (Table 6.6).

Mechanical treatment of the surface, which raises the dislocation density in the surface-adjacent layer, leads to an increase in k_c. The history of the crystal (in particular, the method of its growth) influences the process of plastic deformation during the interaction of chip surfaces and the mechanically treated surface, but does not strongly affect the overall picture of the rise of k_c.

The data in Table 6.7 demonstrate the role of the density of mobile dislocations in the surface-adjacent layer irrespective of the crystal growth method. If the cleaved surface is polished by chemical methods, the value of cracking resistance is

Table 6.6 Effect of impurity on the coefficient of fracture toughness

Corundum crystal	Impurity	C (% by mass)	k_c (MN m$^{-3/2}$)
Sapphire		–	4.0
	Yttrium	$1 \cdot 10^{-2}$	3.7
	Potassium	$1 \cdot 10^{-2}$	3.3
	Molybdenum	$1 \cdot 10^{-3}$	2.8
Ruby			
Pink		$1 \cdot 10^{-3}$	3.8–3.9
Deep-red	Chromium	1.0	2.9–3.0
Sapphire			
Blue	Titanium, iron	$1 \cdot 10^{-2}$	2.6
Yellow	Nickel	$1 \cdot 10^{-2}$	3.0
Alexandrite	Vanadium	$1.4 \cdot 10^{-1}$	3.2
Amethyst	Chromium, iron, titanium	$1 \cdot 10^{-2}$	2.4

unchanged. At the same time, chemical polishing of mechanically treated surfaces, which completely removes the layer of elevated dislocation density, diminishes k_c to the value typical of the cleaved surface.

Comparison of the influence of growth dislocation densities and mobile dislocation densities formed during mechanical treatment shows that the value of dislocation density itself cannot unambiguously characterize the cracking resistance of a crystal. With an increase in the density of growth dislocations, the value of k_c diminishes. With an increase in the density of mobile dislocations, the value of k_c increases. The effect of an impurity is ambiguous as well. An impurity poorly dissolved in the crystal lattice and segregating on dislocations lowers the cracking resistance. But if the impurity is readily dissolved and forms a continuous series of solid solutions with the matrix, its influence on the value of k_c is much less significant. For instance, if the concentration of readily dissolved chromium increases by three orders of magnitude, k_c is lowered approximately to 30%. At the same time, molybdenum introduced into crystals in even small quantities (~10^{-3} mol%) fundamentally reduces the value of k_c (see Table 6.6).

The cracking resistance of the crystal surface changes when passing from a cleaved surface to polycrystalline or quasiamorphous surfaces. If the cracking resistance of the cleaved surface is not high, then after fine grinding and the associated formation of a fine-grained polycrystalline layer and a layer with an elevated density of mobile dislocations, the value of k_c will sharply rise (see Table 6.7). The highest cracking resistance is characteristic of crystals surfaces containing a quasiamorphous layer, such as appears due to durable treatment by a micron-grit diamond abrasive.

The presented data testify that the informative ability of the parameter k_c is high enough to allow conclusions concerning the mechanical characteristics of crystalline products, in addition to determining the factors that define the formation of the crystals. Such data are indicative of the chain of interrelated, multifactor phenomena that in the end define the quality of crystalline articles. In particular, the

Table 6.7 Effect of chemical–mechanical treatment on the coefficient of fracture toughness

	k_c (MN m$^{-3/2}$)			
		Kind of treatment		
Growth method	Cleaved surface	First chemical polishing	Mechanical treatment	Second chemical polishing
Verneuil	2.8	2.6	4.6	3.0
Stepanov	3.0	2.9	4.4	2.9
Czochralski	3.5	3.3	4.6	3.4
Kyropoulos	4.0	4.0	5.4	3.8

processes that precede the crystallization front in the melt and are responsible for the distribution of point defects in the crystal depend on the state of the liquid phase and influence the crystal's machineability, as well as the serviceability and longevity of crystalline articles. Such an influence manifests itself in the structure of the grain elements and in the density of point defects at their intersections, which define the density of growth dislocations and correspondingly the value of the coefficient of cracking resistance. This dependence and the correlation between the value of internal stresses and k_c have been used to create a method for prediction of the suitability of sapphire crystals to mechanical treatment from the value of k_c [46]. The basis of this method is the criterion $k_c = 2$ MN m$^{-3/2}$. At $k_c \geq 2$ MN m$^{-3/2}$, crystals can be mechanically treated without additional annealing, and articles of any size and configuration can be manufactured on their base. At $k_c < 2$ MN m$^{-3/2}$ the probability of damage in the process of mechanical treatment sharply increases. Such crystals require high-temperature annealing in order to reduce residual stresses.

The value of cracking resistance measured by surface microindentation methods may differ from the values of k_c determined by other techniques. However, this method is widely used since it does not require any special equipment, is applicable for fast measurements, and most significantly is informative enough for comparative analysis of the influence of growth conditions, treatment history, and density and type of defects on the strength character.

References

1. Poduraev V.N. Cutting of Hard-to-Machine Materials. Moscow: Higher School Publication, 1974, 587 [in Russian].
2. Rebinder P.A. Physico-Chemical Mechanics. 1958, 75 [in Russian].
3. Caseate, Horio K., Doy K.T. Sensors Mater. MYU 2, 1989, 301.
4. Caseate Lapping and Polishing, Ultraprecision Machining Technology, Kumagaya
5. Yakovleva T.P., Khodakov G.S. Opt. J. 6, 1994, 32 [in Russian].
6. Fink M. et al. ASLE-ASME Lubricantin Conference. 6, 1956, 65.
7. KubeT. Sci. Mach. 23(11), 1961, 1500 [in Japanese].
8. Sakurai T. J. Jpn. Soc. Lubrication Energy. 11(1), 1966, 26 [in Japanese].
9. Marinescu I. D., Tonshoff H. K., Inasaki I. (eds.). Handbook of Ceramic Grinding and Polishing. Noyes, Park Ridge, NJ; William Andrew, LLC Norwich, New York1998.

10. Guiche H., Moody J. J. Electro Chem. Soc Solid State Sci. Technol. 125, 1978, 136.
11. Ydsumiya N., Imanaka O. Technocrat. 8, 1975, 15–21.
12. Prikhodko V.L., Parfenova A.V., Lomakin A.S. Polishing Composition for Semiconductor Materials Inventor's Certificate No 513413, 1973.
13. Zunzficcht P.J. Polishing Composition for Semiconductor Materials, Pebruafy. 125 1978 299.
14. Alabyev I.V., Pankov V.S., Surovtsev M.V., Chumak V.D. Electron. Ind. 94, 1980, 28–35 [in Russian].
15. Rogov V.V., Ph.D. thesis Finish Precision Treatment Corundum Single Crystals. Moscow. 1992, 71 [in Russian].
16. Karaki T., Vatanabe Dzhundzy.J. Jpn. Soc. Precis. Eng. 48, 1981, 1458–1463.
17. Rogov V.V. et al.. Superhard Mater. 5, 2002, 35 [in Russian].
18. Rogov V.V., Rublev N.D., Troyak O.V., Krotenko T.L., A method of finish precision treatment of components from corundum single crystals and polishing tool. Patent of Ukraine 48581 A. 2002.
19. Karaki-Doy T. Sensors Mater. 3, 1988, 153–166.
20. Meri Y., et al. Proceedings of the Third International Conference on Production Energy. 1961, 336.
21. Babiychuk I.P., Dobrovinskaya E.R., Litvinov L.A. et al. Phys. Chem. Mater. Treat. [Collected Volume]. 2, 1973, 89–91 [in Russian].
22. Kulikov D.V., Mekalova N.V., Zakirpichnaya M.M. Destruction of Composition Materials. Moscow. 1999, 230 [in Russian].
23. Tatarinkov A.I., Episherlova K.L., Rusak T.F., Gridnev V.I. Methods of Control of Damaged Layers during Mechanical Treatment of Single Crystals. Moscow: Energiya, 1978, 63 [in Russian].
24. Gurevich D.M., Deryugin Yu V., Dobrovinskaya E.R. et al. Corundum Single Crystals in Jewelry Industry. Leningrad: Mashinostroenie, 1984, 150 [in Russian].
25. Torchuk N.M., Gridneva I.V., Dobrovinskaya E.R., Tsaiger A.M. Phys. Chem. Mater. Treat [Collected Volume]. 1984, 54–55 [in Russian].
26. Chaikovskiy E.F., Kostenko A.B., Rozenberg Kh G., et al. Single Cryst. Mater. [Collected Volume]. 11, 1983, 123 [in Russian].
27. Kostenko A.B., Puzikov V.M., Rozenberg Kh G., Semenov A.V. Abstracts of All-Union Conference "New Potentialities of Diffraction, X-Ray and Electron", Moscow. 1987, 49 [in Russian].
28. Milman Yu, V., Lotsko D.V., X11 International plansee seminar, Austria. 1, 1989, 281–300.
29. Funkenbusch P.D., Takahashi T. SPIE. 3134, 1997, 293.
30. Preston J. Soc. Glass Technol. 11, 1927, 214–256.
31. Evans A., Marshall D. Fundamentals of Friction and Wear of Materials Ed. by D.A. Rignay. Washington: ASM. 1981, 439–452.
32. Voloshin A.V., Livinov L.A. Funct. Mater. 9(3), 2002, 555 (7, 333).
33. Babiychuk I.P., Dobrovinskaya E.R., Chukaev V.I. Chem. Eng. Treat. Methods Manuf. Prod. 1, 1971, 177–194 [in Russian].
34. Babiychuk I.P., Dobrovinskaya E.R., Litvinov L.A. Phys. Chem. Mater. Treat. Called. 2. 1973. 89–91 [in Russian].
35. Babiychuk I.P., Dobrovinskaya E.R., Litvinov L.A., Chukaev V.I. Single Cryst. Eng. [Collected Volume]. 5, 1971, 124–130 [in Russian].
36. Babiychuk I.P., Dobrovinskaya E.R., Litvinov L.A., Chukaev V.I. Single Cryst. Eng. [Collected Volume]. 4, 1971, 191–197 [in Russian].
37. Back D., Polvani R., Braun L. et al. SPIE. 3060, 1997, 102.
38. Anikin A.V., Litvinov Yu M, Raskin A.A. , Moscow. 2005, 447-450 [in Russian].
39. Smith M. Schmid Khattak P. SPIE. 3705, 1999, 85.
40. Babiychuk I.P., Dobrovinskaya E.R., Komilich O.N., Litvinov L.A., A Treatment Method for Single Crystals, Inventor's certificate 1042968 (USSR). 1983 [in Russian]
41. Gridneva I.V., Milman Yu V., Trefilov V.I. et al. Phys. Stat. Sol. A 54, 1979, 195–206.
42. Trefilov V.I., Milman Yu V., Firstov S.A. Physical Foundations of Hardness of High-Melting Metals. Kiev: Naukova dumka, 1975, 315pp [in Russian].

43. Wachtman J.D., Maxwell L.H. Am. Ceram. Soc. 40, 1957, 377–385.
44. Belaya A.N., Dobrovinskaya E.R., Litvinov L.A., Tsaiger A.M. Phys. Chem. Mater. Treat. 1, 1984, 128–131 [in Russian].
45. Evans A.G., Charles E.A. J. Am. Ceram. Soc. 59, 1976, 371.
46. Dobrovinskaya E.R., Pishchik V.V. Corundum Single Crystals: Problems of Obtaining and Quality. Moscow: NIITEKhim. 1988, 74pp.; Part 11. 62pp [in Russian].

Chapter 7
The Effect of Thermal Treatment of Crystals on Their Structure Quality and Mechanical Characteristics

The modern technological processes of manufacturing single-crystal articles usually include thermal treatment of the already-made products or their blanks as one of the necessary stages. Thermal treatment presumes exposure of crystals at certain temperature for the specified time, those being sufficient for the course of relaxation processes at a necessary velocity. At the same time the changes in the structure quality of crystals is connected with a decrease of their free energy due to structural reconstruction: motion of single atoms and dislocations, assemblies of dislocations, block boundaries, and so forth. There is an opinion that the higher the annealing temperature, the closer it is to the melting temperature; and the longerthe duration of the isothermal exposure, the higher the guarantee of service time and efficiency of single crystal articles. However, this not always corresponds to the real situation.

The first experiments on annealing corundum crystals were carried out at the Institute of Crystallography of the Academy of Sciences of the USSR (1940–1942). The annealing was a part of necessary technological operations because in crystals grown by the Verneuil method (on that level of this method development) the values of the internal stresses were 100–150 MPa and during preparation of articles of them these stresses relax by means of brittle fracture. In the works of these years and in the later ones [1–3] it was shown that for the removal of the internal stresses thermal treatment at high temperatures, close to the premelting ones, is necessary. At present in the crystals of colored corundum grown by the Verneuil method with the technology kept to the stresses do not usually exceed 50 MPa; a special thermal treatment is not obligatory for mechanical treatment of such crystals and the residual stresses are removed by a simpler and more efficient way – by splitting a crystal into two half-boules along the cleavage plane. Only in case of crystals with the concentration of chromium about 1%, used for manufacturing watch stones, a high-temperature annealing is still conducted prior to mechanical treatment. For corundum crystals grown by the Czochralski, Kyropoulos, and HDS methods annealing is not an obligatory condition the fulfillment of which would allow these crystals to be subjected to mechanical treatment since stresses in them do not exceed 30 MPa. For crystals grown by the Stepanov method a necessity in the annealing for the relaxation of the residual stresses arises when growing large crystals or those of complicated shape.

E.R. Dobrovinskaya et al., *Sapphire: Material, Manufacturing, Applications*,
DOI: 10.1007/978-0-387-85695-7_7,

Despite the fact that crystals can be treated without preliminary high-temperature annealing they are annealed for the improvement of structure quality of the volume or subsurface layer. As it has already been mentioned, a defective subsurface layer is formed in the process of mechanical treatment; this being so, the stresses in it are such that the available microcracks lead when developing the appearance of cleaveds, especially at sharp edges. A durable, high-temperature annealing of crystals does not give any positive results because the stresses that lead to the formation of cleaveds are formed later – during mechanical treatment. A problem arises of how to remove the defective subsurface layer or to lower the stresses in it.

The defective layer can be removed by different techniques: chemical polishing, gas or ionic etching, and so forth. But in practice a decrease of density of defects in this layer and removal of stresses in it are basically provided by thermal treatment.

7.1 Dislocation and Block Structure

Volume stresses in crystals relax only at rather high temperatures ($T > 2{,}000$ K). Increase of the annealing temperature promotes acceleration of the velocity of stress removal and lowers its limiting value at durable exposures. The higher the annealing temperature, the quicker one realizes the relaxation provided the initial stresses are equal. At different initial stresses and the same annealing temperature (T_a) the stress relaxation is higher for those samples in which the initial stresses are higher. It should be taken into account that the stresses relax during the first hours of annealing; adding more time has no influence on the value stress.

At high-temperature annealing with the decrease of the internal stresses simultaneously changes density of single dislocations, extension of block boundaries, and angles of their disorientation [4]. An extremely important circumstance attracts attention: the fact that in crystals of different compounds that are related to different crystal systems and that differ in the character of bonding forces, melting temperature (T_m), and other physical–chemical characteristics, the time and temperature dependences of the defects' density are similar; the only changes are temperature interval and time required for the run of the processes of annihilation, polygonization, and "scattering" of dislocation boundaries. The velocity of these processes' runs depends largely on temperature. The higher the T_a, the closer it is to T_m, and the higher the velocity of their realization.

Temperature dependence of dislocations density in sapphire grown by the Verneuil method (curve *1*) and HDSM (curve *2*) is shown in Fig. 7.1. For crystals with higher density of dislocations, their more intensive decrease compared to those with lower density of defects and low residual stresses is typical. Decrease of dislocation density is observed at $T \geq 2{,}000$ K.

Of the most interest are the time dependences of density of dislocations and extension of block boundaries (Fig. 7.2). At the initial stage of annealing ρ decreases and $\Sigma\rho$ increases owing to the actively running processes of annihilation of the opposite sign dislocations and polygonization.

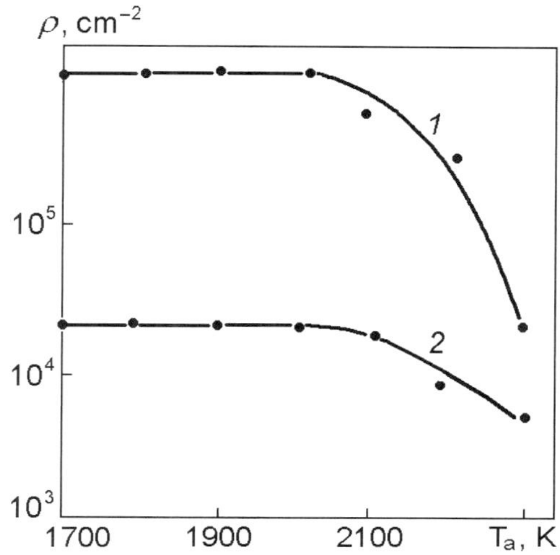

Fig. 7.1 Temperature dependence of density of single dislocations in sapphire; time of annealing, 6 h

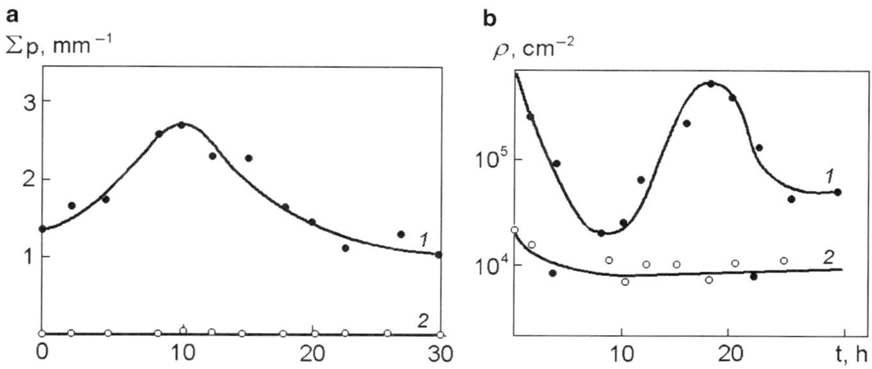

Fig. 7.2 Time dependence of block boundary extension (**a**) and density of single dislocations (**b**) sapphire at $T_m = 100$ K: (*1*) Verneuil method; (*2*) HDS method

As the time of annealing lengthens, the extension of boundaries lessens, which is explained by their "scattering" [5]. It is significant that while in the alkali halide crystals slowly cooling from the premelting to room temperatures "scattering" begins at the initial stage of annealing in the stressed crystals grown by the Verneuil method; polygonization takes place first and only with time the opposite process – "scattering" of dislocation boundaries – comes into effect. This process is energetically less advantageous than polygonization, but at high-temperature annealing unstable configurations of dislocation boundaries appear in crystals and the low internal stresses turn out to be sufficient for the creation of the "scattering" centers, for instance, in the places of the meeting-point of dislocation boundaries.

The character of the dislocation structure strongly changing depends on the original structure quality of crystals. Irrespective of the growth method if the crystals are free of blocks and are characterized by low internal stresses at $\rho < \rho_{cr}$ at the initial stage of annealing (during the first hour), density of dislocations slightly decreases and further lengthening of the annealing time does not provide any noticeable variation of the value ρ [4]. Figure 7.3 shows time dependences of the relative density of dislocations and block boundary extensions for three groups of crystals that differ in the original density of defects. As a reference point taken on the abscissa axis were ρ and Σp in the samples heated to the annealing temperature and cooled down to room temperature without isothermal exposure. In the samples of the first group block boundaries may not be formed; they may form only in separate areas if ρ approaches ρ_{cr}. Low-angle dislocation boundaries appear. In the samples of the second group the dependences $\rho/\rho_0(t)$ and $\Sigma p(t)$ are similar to those shown in Fig. 7.2. It is noteworthy that the transition from $T = T_m - 100$ K to $T = T_m - 10$ K calls forth a dramatic shortening of the run time of the processes of annihilation, polygonization, and "scattering" of dislocation boundaries. Annihilation of dislocations and "scattering" of the boundaries proceed practically two times

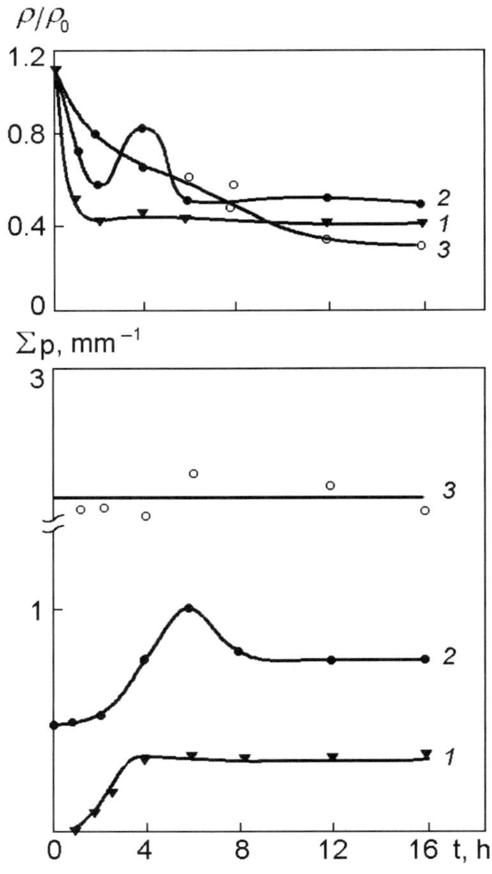

Fig. 7.3 Time dependence of ρ/ρ_0 and Σp in crystals at $T = T_m - (10 \pm 5)$K: (*1*) $\Sigma p = 0$, $\rho_0 = (2–5) \cdot 10^5$ cm^{-2}; (*2*) $\Sigma p = 0.5–2$ mm^{-1}, $\rho_0 = (5–7) \cdot 10^5$ cm^{-2}; (*3*) $\Sigma p = 2.5–3$ mm^{-1}, $\rho_0 = (7–10) \cdot 10^5$ cm^{-2}

quicker. Annealing of the third group samples has no significant influence on the values ρ and Σp. In such block-containing crystals the annihilation and "scattering" of the boundaries take place in the first hours of annealing. Still, at high original density of defects polygonization and "scattering" of boundaries compensate each other, owing to which the value Σp remains practically unchanged.

Hence, essential change of the structure at high-temperature annealing is characteristic of only crystals irrespective of their growth method with density of dislocations about the critical one and extension of block boundaries to 2.5 mm^{-1}. At annealing more perfect crystals and those with $\Sigma p \geq 2.5$ mm^{-1} under conditions of a premelting temperature one can decrease the density of the considered structure defects only during the first hour of isothermal exposure. Lengthening the exposure time does not result in the change of structure quality of crystals.

Along with the original structure quality content of impurity, particularly chromium, has an effect on the behavior of defects during thermal treatment. Thus, all other conditions being the same, in ruby ($C_{Cr} \approx 3 \cdot 10^{-2}\%$) the dislocation structure changes at earlier stages of annealing than in sapphire. This result is somewhat unexpected since it is known that chromium impurity strengthens corundum crystals [6]. The latter circumstance supported the supposition that the presence of chromium decreases mobility of dislocations and the dislocation structure changes at high-temperature annealing in ruby later than in sapphire. However, it turned out that annihilation and polygonization in ruby come about earlier than in sapphire. This is perhaps connected with high internal stresses in the impurity-containing crystal.

The measurements show that the internal stresses in ruby are higher than in sapphire; since mobility of dislocations is to a high extent defined by stresses, it is no wonder that dislocations in ruby at the initial stage of annealing are more mobile than in sapphire despite of the presence of additional stoppers. With lengthening of the annealing time when the internal stresses sharply decrease, the mobility of dislocations in ruby becomes lower than in sapphire. After a durable annealing density of dislocations in ruby both on the base plane and on the plane of the prism still remains higher than in sapphire. Very often increase of the single dislocation density during annealing is not accompanied by the diminishing of the extension of boundaries. Increase of ρ is connected with the fact that the boundaries emit single dislocations at high temperature isothermal exposure, at the same time the disorientation angles of the neighboring blocks become lower.

7.2 Evolution of Impurity Striation

There is no doubt that not only dislocation structure but also distribution of impurity in the crystal volume change at high-temperature annealing. From Fig. 7.4 one can see that fuzzying of the impurity peaks takes place and their amplitude changes. Provided that the impurity concentration changes in the crystal according to the sinusoidal law along the x-axis (crystal growth direction) the impurity concentration at any point can be presented as

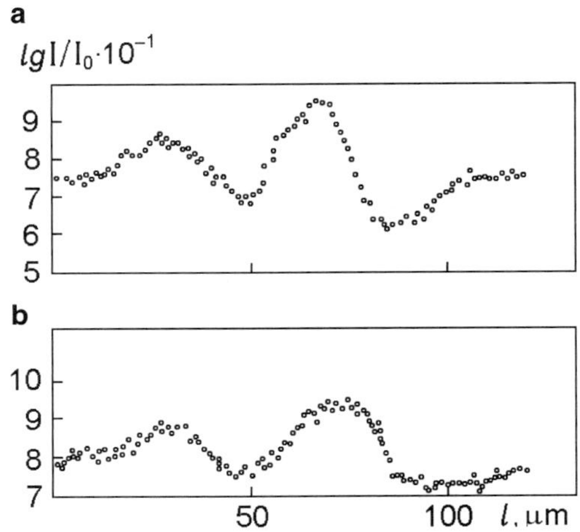

Fig. 7.4 Distribution of optical density in ruby with concentration of chromium about 1%: (**a**) before and (**b**) after the gradient annealing

$$C_{(x)} = C_0 + C_1 \sin 2\pi / \lambda_{\mu s} \cdot x, \tag{7.1}$$

where C_0 is the mean impurity concentration, and C_1 is the amplitude of the impurity microstria.

The impurity concentration for ultramicrostriation can be presented analogously:

$$C'_{(x)} = C'_0 + C'_1 \sin 2\pi / \lambda'_{\mu s} \cdot x. \tag{7.2}$$

Here C' and $\lambda'_{\mu s}$ are the amplitude and period of ultramicrostriation, respectively.

At annealing the values C and C' decrease to the extent of the diffusion coefficient. Since $\lambda_{\mu s} \gg \lambda'_{\mu s}$ and $C_1 > C'_1$ it can be considered that the amplitude variation of two types of striation occurs independently. Using the solution of the second Fick equation for the case of the diffusion from a plate λ_n in thickness the equation for the determination of the impurity concentration is written in the form:

$$C_{(x,t)} = C_0 + C_1 \exp\left[-\left(\frac{2\pi}{\lambda_{\mu s}} \right)^2 D(T)t \right] \sin \frac{2\pi}{\lambda_{\mu s}} \cdot x. \tag{7.3}$$

Correspondingly, for ultramicrostriation:

$$C'_{(x,t)} = C'_0 + C'_1 \exp\left[-\left(\frac{2\pi}{\lambda'_{\mu s}} \right)^2 D(T)t \right] \sin \frac{2\pi}{\lambda'_{\mu s}} \cdot x. \tag{7.4}$$

The given dependences that allow for the character of impurity distribution and evolution of the impurity striae to the extent of the diffusion coefficient at high-temperature annealing supported the proposal of a new method of determination of the diffusion constants in crystal bodies, the method being distinguished by simplicity and lower labor-intensiveness as compared to the known methods. A conception on the broadening of the impurity stria depending on temperature and time of annealing forms the basis of this method.

The method consists of the following [7]. In crystals with impurity microstriation distribution of concentration of the diffusing impurity is measured; the sample is placed in the gradient thermal zone of the furnace with the known temperature distribution so that one of the ends of the sample was in the zone of the maximum temperature. The temperature of the furnace is raised until fusion of the end; the sample is exposed to these conditions for a specified period and then is cooled to room temperature. Distribution of the diffusing impurity concentration is measured for the second time. Fusion of one of the sample ends in the gradient thermal zone of the furnace allows measurement of a precise temperature binding to the change in the annealing profile of the impurity nonuniformity, i.e., gives a possibility if obtaining a series of curves of impurity distribution at different temperatures after only one experiment. A double measurement of distribution of the diffusing impurity concentration (before and after annealing) allows to register the evolution of the impurity profile in the given temperature range (Fig. 7.5).

Having a pattern of impurity distribution at the same points inthe samples before and after annealing and knowing the time of the diffusion annealing and the gradient of temperature at which the sample was annealed (annealing temperature of each part of the crystal), it is possible to determine the diffusion coefficient at the given

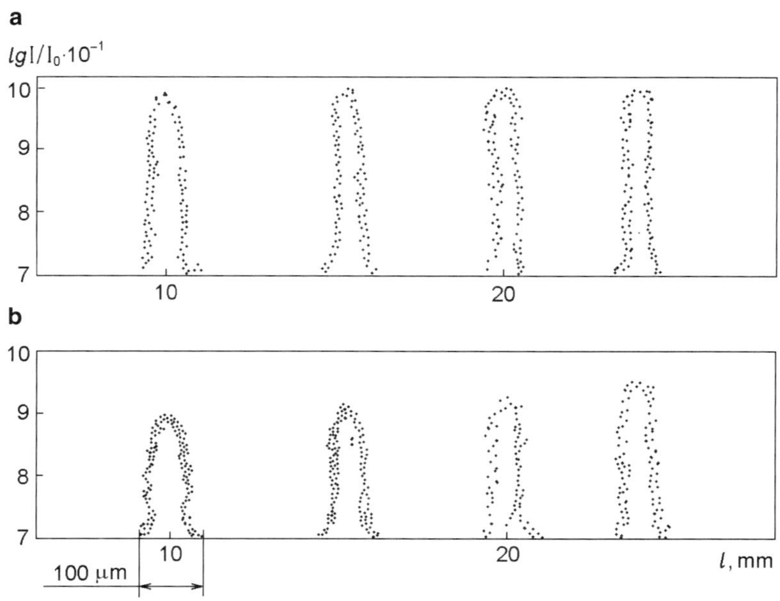

Fig. 7.5 Evolution of impurity striation during annealing process: (**a**) before and (**b**) after the gradient annealing

temperature. For this purpose a half-width of the impurity microstria is measured after (Δh_2) and before (Δh_1) annealing. Knowing the time of annealing one can calculate the diffusion coefficient value:

$$D = (\Delta h_2 - \Delta h_1)^2 / 4t. \tag{7.5}$$

Having determined the value D at different points of the sample, i.e., at different temperatures, one can plot the dependence $D(T)$ and calculate the activation energy of the diffusion process E^* and the pre-exponential factor D_0. At chromium diffusion to the crystal the temperature dependence of the diffusion coefficient has the form:

$$D = 4.10^{-4} \exp(-3.5 / kT). \tag{7.6}$$

7.3 Difference in the Behavior of the Dislocation Ensemble in the Volume and in the Subsurface Layer

During thermal treatment of crystals not only density of bulk defects decreases but essentially changes the structure of the defective subsurface layer. At annealing at 1,500–1,600 K in the inert gas and in vacuum a recrystallization of the hardened layer can occur which appears at mechanical treatment [8]. The recrystallization favors partial removal of stresses in the subsurface layer. Increase of T_a to 1,800–1,900 K allows improvement of structure quality of the subsurface layer, but it does not eliminate all the defects formed at cutting, grinding, and polishing of the surface which can be judged by the presence of extraspots on the electron diffraction patterns. Further increase of temperature to 2,000–2,200 K can lead to a partial evaporation of the recrystallized surface.

There are assumptions that the subsurface layer after annealing of crystals has another chemical composition as compared to the crystal volume [9]. Thus, annealing at 1,550 K results in a depletion of the surface in oxygen owing to which the aluminum cation is in the state of lower valence (Al^{2+} or Al^{+}), i.e., the subsurface layer contains lower oxides AlO and Al_2O.

At a consideration of the behavior of structure defects in the subsurface layer at thermal treatment two problems arise: behavior of dislocations and block boundaries in the layer preliminarily mechanically treated and in the layer not mechanically treated. Let us consider first of all the first problem and compare the behavior of defects in the subsurface layer and in the crystal volume.

Often the data obtained by different authors are contradictory. Thus, it is shown in ref. [10] that aggregations of dislocations in the subsurface layer turn out to be unstable, a recrystallization takes place, there appear block boundaries, and the thickness of the defective layer enlarges. The latter raises serious doubts since the analysis of the depth of the layer with the increased density of dislocations showed that at reaching 2,000 K the defective layer fully disappears (Fig. 7.6). This disappearance

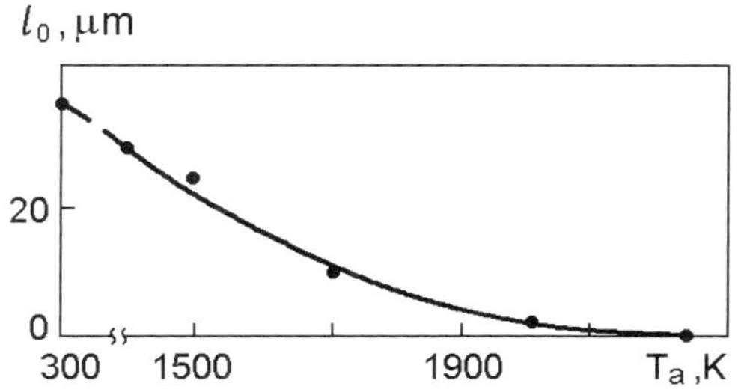

Fig. 7.6 The depth of the defective layer versus the annealing temperature

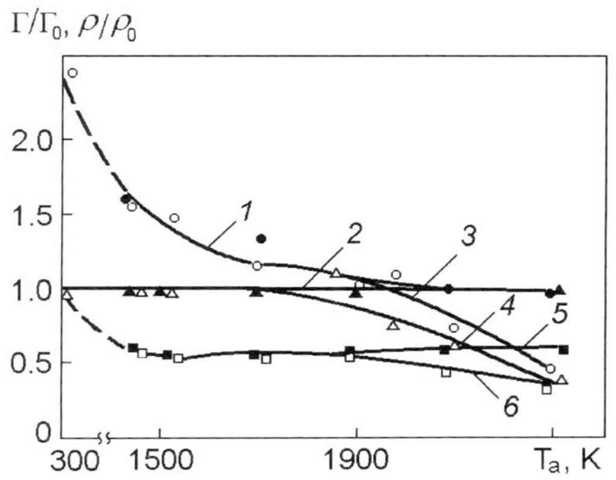

Fig. 7.7 Temperature dependences of the relative density of dislocations on the surface (*1, 3*) and in the volume (*2, 4*) of the crystal and anomalous birefringence (*5, 6*) for the samples annealed without isothermal exposure (*1, 2, 5*) and with 7-h exposure (*3, 4, 6*)

is observed even at annealing without isothermal exposure; this being so, the surface quality does not change since the process of thermal etching does not have time.

For a comparison of the behavior of the bulk and surface defects during annealing, Fig. 7.7 shows the temperature dependence of the relative density of single dislocations on the surface of sapphire where ρ_0 is the density of dislocations in the volume of nonannealed crystals. The data are presented for two cases: annealing without isothermal exposure (curve *1*) and with 8-h exposure (curve *3*). Also, the temperature dependence of the relative density of bulk dislocations (curves *2* and *4*) is shown. The dependences presented show that at 1,400–2,000 K density of dislocations in the volume remains the same while on the surface it sharply drops; note that in this temperature range the densities of dislocations for the samples annealed with the isothermal exposure and without it coincide. Increase of the annealing temperature to 2,300 K without exposure does not lead to significant changes in the bulk and surface densities of dislocations. The isothermal exposure at $T > 2{,}050$ K conditions decrease density of single dislocations both in the volume and on the surface of crystals. For a comparison, Fig. 7.7 shows the temperature dependences of the relative value of the anomalous birefringence Γ/Γ_0 equal to the relative value of the residual stresses τ^*/τ^*_0. The behavior of the dependences $\rho/\rho_0 = f(t)$ and $\Gamma/\Gamma_0 = f(t)$ is equal for the samples annealed during 8 h and without isothermal exposure (Γ is the anomalous birefringence before annealing, Γ_0 – after annealing).

The measured values of the differences in the path in the samples annealed with and without isothermal exposure keep within one curve up to $T = 2{,}000$ K. Only at $T >$ 2,050 K with the lengthening of isothermal exposure time the internal stresses decrease. It should be assumed that annealing at 1,400–2,000 K allows the internal stresses to decrease basically owing to the decrease of their values in the subsurface layer. In the reference samples that were not subjected to mechanical treatment, thermal treatment in this temperature range at different isothermal exposures does not change the differences of the path. The lengthening of the isothermal exposure time at $T \geq 2{,}000$ K offers the possibility of lowering the anomalous birefringence due to the decrease of bulk stresses. While the mechanically treated surface has an insignificant roughness after annealing, the total decrease of density of single dislocations is observed. If the original surface is characterized by such large numbers of scratches that it is not practically etched (such surface often is met at treatment of gems), in the process of high-temperature annealing the surface roughness also becomes much lower; however, strongly pronounced gliding bands appear. It is noteworthy that short, low-angle boundaries of polygonization origin often appear on the annealed surfaces (Fig. 7.8).

The data received as a result of annealing of crystals that underwent different kinds of mechanical treatment at 1,800 and 2,300 K without isothermal exposure are presented in Table 7.1.

A relative variation of the anomalous birefringence Γ/Γ_0, density of single dislocations $\Delta\rho/\Delta\rho_0$ and the depth of the defective layer l_0 is shown here. Annealing at 1,800 K leads to the fact that the values of the anomalous birefringence and the

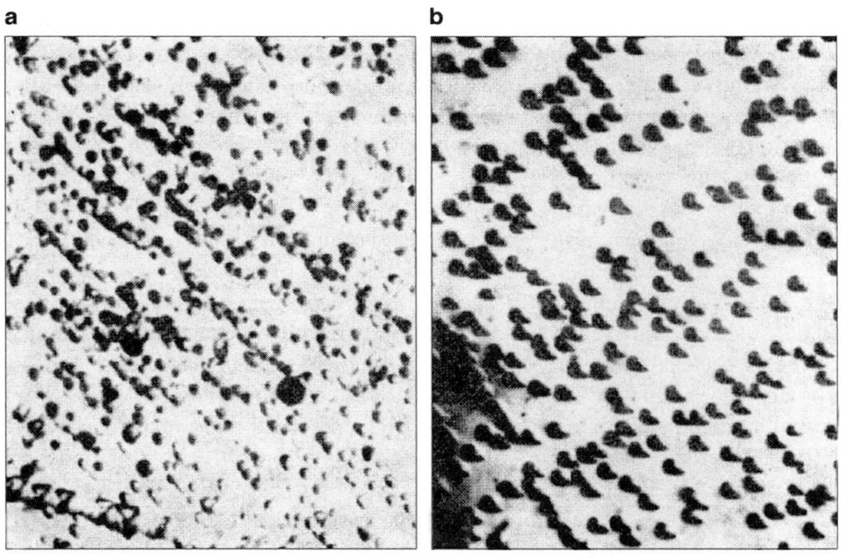

Fig. 7.8 A microphotograph of the etched surface: (**a**) before and (**b**) after high-temperature annealing (×200)

depth of the defective layers become practically equal for all the samples and equal to the corresponding values for the best nonannealed samples. After thermal treatment at 2,300 K the defective subsurface layer disappears in all kinds of samples, and the values of the anomalous birefringence become lower. Such changes in the structure of the subsurface layer are first explained by high internal stresses at relaxation of the processes of annihilation of the opposite sign dislocations and dislocationbuildup in boundaries. It also is possible that the dislocations exiting from the crystal under the action of the "mirror reflection" force may contribute to the changes. The role of the last factor is confirmed by the presence of a maximum on the curve $\rho = f(l_0)$ at a distance 5–10 μm from the surface in crystals annealed at 1,500–1,700 K (Fig. 7.9). At the final decrease of density of dislocations in the subsurface layer, the value ρ right near the surface becomes lower. No doubt the action of the "mirror reflection" force is noticeable only in a very narrow layer directly adjacent to the surface owing to which a dislocation depleted zone can appear right at the surface.

Table 7.1 The effect of the temperature of annealing mechanically treated crystals on the structure quality of their subsurface layer

Type of the abrasive material	1,800 K			2,300 K		
	$\Delta\Gamma/\Delta\Gamma_0$	$\Delta\rho/\Delta\rho_0$	l_0 (μm)	$\Delta\Gamma/\Delta\Gamma_0$	$\Delta\rho/\Delta\rho_0$	l_0 (μm)
Free abrasive	0.05	–	15	0.15	0.23	0
Combination of the	0.13	0.35	20	0.26	0.49	0
free and fixed	0.17	0.02	15	0.25	0.34	0
abrasive	0.12	0.07	15	0.19	0.15	0
	–	–	20	0.41	0.65	0
Fixed abrasive	0.09	–	15	0.32	–	0

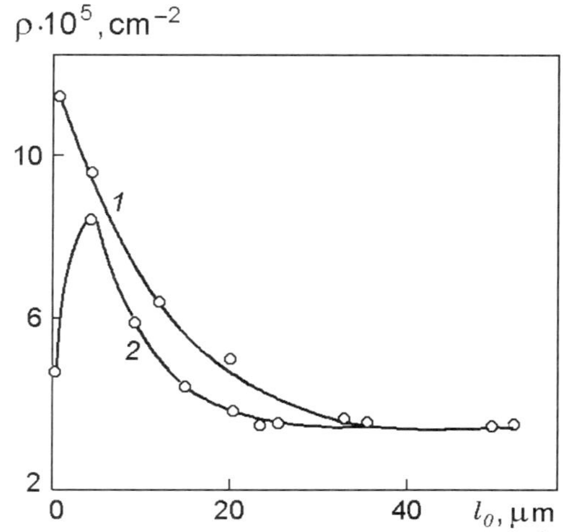

Fig. 7.9 Distribution of density of dislocations over the depth in the samples: (*1*) nonannealed; (*2*) annealed at T = 1,700 K

Exiting to the surface and annihilation in the annealing process of the most mobile dislocations result in the strengthening of the subsurface layer. It can be achieved even at annealing at 1,500 K for 1 h.

Hence, if high-temperature annealing allows the improvement of the bulk structure, perfection of the "medium quality" crystals $\rho > \rho_{cr}$, and extension of the block boundaries to 2.5 mm^{-1}, such annealing in any case changes the structure of the defective subsurface layers. The internal stresses decrease 1.3–1.5 times, and density of dislocations in the subsurface layers drops irrespective of the kind of mechanical treatment owing to the actively running process of annihilation of the opposite sign dislocations. While a polycrystalline layer is formed on the surface due to mechanical treatment, it is recrystallized at high-temperature annealing. Annealing at 2,100 K does not lead to a full removal of defects in the subsurface layer, which is confirmed by fuzziness of *K*-lines on the electron diffraction patterns and their splitting connected with the formation of low-angle boundaries. The depth of the layer with the increased density of dislocations is not more than 10 μm, and gliding bands can be revealed on the surface by the selective etching method. Annealing at 2,300 K completely removes the layer with the increased density of dislocations; this means that high-temperature premelting annealing leads to the evolution of the defective subsurface layer or to its full disappearance. If the scheme of the defective subsurface layer suggested in Chapter 6 is kept, the structure of the first two zones is changed: the first transforms into a crystal with high angles of block disorientation, and in the second the disorientation angles become lower. The evolution of the third and fourth zones proceeds on the way of decrease of dislocation density, and a sharp drop of the residual stresses often with the formation of clearly pronounced gliding bands.

Naturally, the disappearance of the quasiamorphous layer, decrease of density of mobile dislocations in the subsurface layer reveals the value of the fracture toughness coefficient. While annealing the cleaved surface does not change its fracture toughness high-temperature annealing of the mechanically treated surface lowers the value k_c 1.4–1.8 times (Table 7.2). Multiple repetition of the mechanical treatment and annealing at 2,250 K procedures does not change the general picture: mechanical treatment increases the fracture toughness of the surface compared with the reference sample (cleaved surface); the annealing at which density of mobile dislocations introduced by mechanical treatment becomes lower brings back the value of k_c to those characteristic of the reference samples.

Table 7.2 Fracture toughness coefficient of crystals after thermal and mechanical treatment

Kind of treatment	ρ/ρ_0	k_c (MN $m^{-3/2}$)
Cleavage	1	2.8
I Mechanical treatment	2.5	4.9
I Annealing	1.2	3.1
II Mechanical treatment	2.1	4.0
II Annealing	1.2	3.0
III Mechanical treatment	2.5	4.3

The vacuum annealing of Ti-sapphire results in diffusion of the anionic vacancies from the matrix and dissolution of aluminum atoms (being contained in the submicrometer-sized inclusion) in the crystal matrix. At a certain annealing stage, those inclusions become transformed into submicrometer pores containing saturated aluminum oxide vapor. The pores in a crystal are known to be diffusively dissolved or to be cured according to the dislocation mechanism. The diffusive dissolution of a pore is due to an increased vacancy concentration at its surface. In its turn, a constant uniaxial load applied to the crystal favors the diffusion processes and the diffusion-dislocation creeping, thus increasing the pore dissolution rate. The destruction dynamics of the submicrometer-sized inclusions in the crystal was monitored using the decrease of optical scattering intensity that is in proportion to the scattering center concentration.

The dislocations in corundum single crystals show a maximum mobility under cleaving stresses directed along the (0001) plane belonging to the easiest sliding system. Under uniaxial stresses along the [$11\bar{2}0$] and [0001] directions, those cleaving stresses are minimal, the dislocation motion in the (0001) plane is hindered, and the pore is cured mainly due to the vacancy-by-vacancy dissolution. This is confirmed by the fact that the submicrometer-sized inclusions are dissolved at the same rate during the sample annealing without any loading and under uniaxial compressive stresses acting along [$11\bar{2}0$] and [0001] (Fig. 7.15). The submicrometer pores are destroyed most intensely at the high-temperature vacuum annealing of the crystal under uniaxial compressive stresses applied along the [$1\bar{1}20$] crystallographic direction, since in this case, both mechanisms of pore dissolution are active.

7.4 Formation of a Dislocation-Free Zone

The appearance of the dislocation-depleted zone in the subsurface layer at annealing was found in the alkali halide crystals. Density of dislocations in it is two orders lower than in the volume. This zone is characterized by a small number of block boundaries. The size of this zone and distribution of dislocations in it depend on temperature and time of the isothermal exposure. At high temperatures, with lengthening of the annealing time the subsurface dislocation layer becomes wider, but with the next lengthening of the annealing time this zone narrows somewhat. The main reason for the formation of this low dislocation layer is the action of the "mirror reflection" force $F_c \sim x^{-1}$ (x is the distance from the surface); it is assumed at the same time that dislocations emerge on the surface due to their creep. The low dislocation layer begins to form in these crystals at $T \geq 1{,}300$ K. Very often right near the surface a zone appears which is 200 μm in width with no dislocations. Such a zone forms in sapphire in a much shorter time than in alkali halide crystals [11].

The study of the kinetics of the origination and "obliteration" of low dislocation and dislocation-free layers demonstrated that at $T \sim 0.5T_m$ a dislocation-free layer forms, which grows as the exposure time becomes longer (Fig. 7.10). When the annealing temperature is raised to 1,600 K at an insignificant exposure time, a

Fig. 7.10 Time dependence of the width of the low dislocation zone formed at annealing: (*1*) $T = 1{,}400$ K; (*2*) 2,200 K; (*3*) 1,600 K

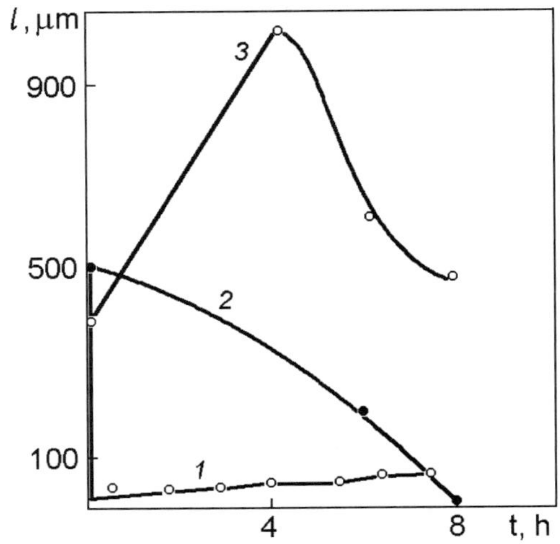

gradual increase of the width of the low dislocation layer is observed; with further lengthening of the annealing time the layer becomes narrower. The distance from the surface to $x = l$ at which density of dislocations is equal to the bulk density is taken for the width of the low dislocation layer. With the increase of the annealing temperature the size of the dislocation-free zone dramatically diminishes. At 2,200 K the low dislocation layer origination (dislocation-free area is absent at all) can be observed only after annealing without isothermal exposure; with the lengthening of time the size of the zone (layer) gradually becomes smaller. It is insufficient to describe the behavior of the assembly of dislocations in the subsurface layer, in particular of the effect of "obliteration" of the dislocation-free zone, only to a consideration of "mirror reflection" forces.

It can be supposed that the processes in the subsurface layer develop in the following succession. At the initial moment the drift of dislocations from the surface is caused only by the action of the "mirror reflection" force, leading to the formation of the dislocation-free layer. After some time some gradient of ρ appears in the subsurface layer and the motion of dislocations toward the surface depends on this gradient value. Perhaps, during annealing at 1,300–1,500 K, mobility of dislocations in the subsurface layer is not high and if at the initial stage of annealing a dislocation-free layer is formed the lengthening of the annealing time will not lead to a significant redistribution of dislocations in this layer and decrease of their gradient. At $T \leq 1{,}800$ K such redistribution occurs: the dislocation-free zone size diminishes and a region with gradually varying density of dislocations appears behind it. At temperatures close to the premelting, owing to high mobility of dislocations the origination of the dislocation-free zone and its "obliteration" proceed so quickly that it is almost impossible to observe it.

A mathematical description of the variation of the dislocation density in the subsurface layer of the crystal during its annealing along with the flow of dislocations caused by their directed motion under the influence of forces of "mirror reflection," which lead to the formation of the dislocation-free layer, it is necessary also to bear in mind the flow caused by the arising dislocation density gradient which was not taken into account before. In this case the equation defining the concentration of dislocations can be written in the usual form:

$$\frac{\partial C(x,t)}{\partial t} = \frac{\partial}{\partial x} \upsilon(x) C(x,t) + \frac{\partial}{\partial x}\left[D(C) \frac{C(x,t)}{\partial x} \right], \tag{7.7}$$

where υ_ρ is the dislocation motion velocity, and D is the effective coefficient of dislocation diffusion.

Unfortunately it is impossible to determine simultaneously the values of υ_ρ and D. However, for the estimation of these values one can consider a case when only one summand plays the main role in the equation: convective or diffusive. For example, when annealed a crystal with an a priori formed dislocation-free layer, a distinct boundary of which is at such distance from the crystal boundary that the action of the "mirror reflection" force is negligibly weak, the main role in the variation of concentration of dislocations will probably play their diffusion toward the external surface of the crystal. In this case having specified the initial and boundary conditions $C(x, 0) = 0$; $0 \le x < l^*$; $\partial C(\infty, t)/\partial x = 0$; $C(0, t) = 0$; $C(x, 0) = C_0$; $l^* < x < \infty$, where l^* is the dislocation-free zone size and in the assumption of the diffusion coefficient constancy, one can arrive at the solution of the diffusion equation by the known methods [12]

$$\frac{\partial C(x,t)}{C_0} = \frac{1}{2}\left(\operatorname{erf} \frac{l+x}{2\sqrt{Dt}} - \operatorname{erf} \frac{l-x}{2\sqrt{Dt}} \right). \tag{7.8}$$

Of interest was a comparison of the results of the calculation of dislocation distribution by the formula with the experimental curve (Fig. 7.11). The experimental values $l^* = 50$ μm and $t = 6$ h were chosen for the calculation. The diffusion coefficient can be determined so that the experimental and theoretical curves were the closest. The calculated curve corresponding to the diffusion coefficient $D = 8 \cdot 10^{-9}$ cm^2 s^{-1} is shown in strokes in Fig. 7.11. At this value of D the experimental points keep within the calculated curve.

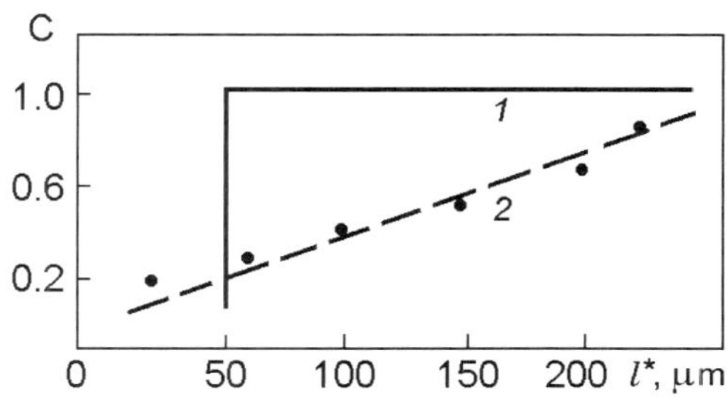

Fig. 7.11 Distribution of density of dislocations in the subsurface layer: (*1*) dislocation-free zone at 1,400 K; (*2*) annealing at 2,200 K (the calculated curve); $t = 6$ h

A comparison of the obtained value with the literature, diffusion coefficient of vacancies in corundum crystals at $T \approx 2{,}200$ K ($D_{\upsilon} \approx 10^{-6}$ cm^2 s^{-1}) demonstrates that the diffusion coefficient of dislocations is approximately two orders lower than the diffusion coefficient of vacancies. From the point of view of physics this is probably explained by the fact that diffusion of dislocations is connected with the motion of not one but a whole complex of vacancies arranged along the dislocation line. The "obliteration" of the dislocation-free layer near the crystal boundary at high-temperature annealing can be explained by the flow of dislocations toward the crystal boundary conditioned by the gradient of density of dislocations.

The reason for the wide dislocation-free zone formation remains unclear since, undoubtedly, it cannot be explained only by the "mirror reflection" force acting in the vicinity of the surface. The physical sense of the effective diffusion coefficient of dislocations also is obscure. Despite this fact the regularities obtained can be used for development of articles of sapphire, which has no defective layer or has a dislocation-free subsurface layer.

7.5 Anomalies of the Crystal Behavior at High-Temperature Annealing

The attempts to use the above-described experimental data received on the basis of an empiric approach to the problem of thermal treatment of crystals for controlling structure quality of crystal articles, i.e., imparting to them the necessary exploitation characteristics to encounter essential difficulties. The practice shows that thermal treatment sometimes does not improve structure quality, but on the contrary favors formation of new macro- and microdefects on the surface and in the volume of crystals. Thus, crystal blanks grown by different methods under practically equal thermodynamic conditions and simultaneously annealed can behave quite differently during mechanical treatment. In some, cleaveds, micro-, and macrocracks appear, correspondingly increasing the share of spoilage, while in others these phenomena are absent. Moreover, often crystals grown by one method behave inadequately. These paradoxes of "good" and "bad" batches of crystals are known to all the specialists involved in the solution of the problem of treatment of crystal.

The effect of "unpredictability" is manifested not only at the stage of annealing the crystals but at the finishing stage of thermal treatment of ready-made products as well. Thus, at 1,400–2,200 K local microfractures can appear on the polished ($R_z = 0.05$ μm) crystals. Usually they consist of a central cavity surrounded by an ulcerated surface. In a number of cases a colored halo can be seen around thermal fractures using optical microscopy methods. The size of the central cavity varies in the range of 10–20 μm, the total size of the damaged area can reach 300 μm, and the depth is about 6 μm. It is not always that thermal fracture is accompanied by the formation of cavities and ulcerated regions. At thermal treatment lacelike defects can appear in the subsurface layers; these defects sometimes come out with one end to the surface. At 2,200 K fractures with fused edges appear.

The appearance of the said microfractures is promoted by the defects of the subsurface layer that have been formed at mechanical treatment. The mechanism microfracture origination consists of the following. It usually is supposed that each next stage of mechanical treatment fully removes the defective layers formed at the previous stages. The practice shows that if the technological process does not anticipate strict fulfillment of this requirement, the microcracks filled with the leavings of abrasive left under the treated surface from the previous operations become the sources of thermal damage up to the formation of cavities. Such polished out microcracks can be the reason not only of thermal damage but also of that which appears on the surface of articles under the effect of laser radiation.

Density of defects can increase in crystals when in the short run they happen to occur in the zone of elevated temperature, i.e., are subjected to thermal shock. Thermal shock at premelting temperatures (about 2,300 K) can result in the crystal destruction; this being so, a part of the crystals from one batch is damaged while the other remains intact. Thermal shock at much lower temperatures, e.g., 1,100 K, does not evoke destruction and formation of new dislocations; at the same time, however, a part of a dislocation breaks off and their impurity clouds and shifts. The selective etching of crystals after thermal shock at $T < 1{,}300$ K allows the location of the initial position of dislocations fixed by a flat-bottomed etching pit and a new position fixed by a faceted pit with the top in the center. The shift of the base dislocations at such thermal shock varies in a wide range (5–20 μm).

The mentioned data testify to a different distribution of stresses in the crystal volume, to the difference in the possibility of stress relaxation by plastic flow, and to the difference in the plastic flow itself at thermal reaction. To explain the reasons for the observed differences in the behavior of crystals, let us consider the driving forces of these processes. The thermoactivation character of plastic deformation in crystals is connected with dislocations that overcome the short-range-acting obstacles under the influence of stress [13],

$$\tau^{*} = \tau_{T}^{*} + \tau_{\mu}^{*} \tag{7.9}$$

where τ^{*}_{T} is the component that depends on temperature; τ^{*}_{μ} is the athermic component of stress. Under the influence of this stress plastic flow occurs in the crystal, its velocity is

$$\varepsilon_{\mathrm{p}} = \varepsilon_{\mathrm{p}0} \exp[-E^{*}(\tau^{*})/kT], \tag{7.10}$$

where $\varepsilon_{\mathrm{p}0}$ is the pre-exponential factor depending on τ^{*} and on the defectiveness of crystal structure.

It is shown in [14] that the velocity of the shear deformation in corundum crystals in the range of 1,200–2,300 K for crystals of 60° orientation allowing for easy glide can be written in the form

$$u = A/T\left(\tau^{*}_{\mathrm{ef}}\right)^{n} \exp\left[-E^{*}\left(\tau^{*}_{\mathrm{ef}}\right)/kT\right] \quad (7.11)$$

or as

$$u = B\exp\left[-E^{*}\left(\tau^{*}_{\mathrm{ef}}\right)/kT\right], \quad (7.12)$$

for crystals of 90° orientation:

$$u = B\left(\tau^{*}_{\mathrm{ef}}\right)\exp\left[-E^{*}\left(\tau^{*}_{\mathrm{ef}}\right)/kT\right], \quad (7.13)$$

where A, B, n are the constants, and τ^{*}_{ef} is the effective stress.

The given dependences explain the experimentally observed phenomena. The value $\tau^{*}_{\mathrm{ef}} = \tau^{*} - \sum J\varepsilon_{g}$, where ε_{g} is the shear deformation, J depends on the crystal substructure. From the general physical conceptions it is clear that the value J is higher the more stoppers are in the way of dislocation motion. It is known that behavior of crystals at each next stage of manufacture of an article is defined by the distribution and density of structure defects that have been formed at a previous technological stage, i.e., is characterized by its own value of the deformation-strengthening coefficient. So, if in crystals during growth distribution of point defects of the grain structure type appears, depending on the size of this structure elements, character of distribution of point defects the internal stresses relax in the crystal volume, and plastic deformation takes place. Apparently this alone explains the difference in the behavior of crystals at mechanical and thermal treatment as well as at thermal shock. Damaged first are less plastic crystals in which the size of the grain structure elements is large and consequently the density of point defects at the meeting-points of these elements is high. If crystals are plastic enough (the size of the grain structure elements is small) at such thermal shock the processes of plastic deformation actively run in them, the gliding bands appear. The glide usually proceeds along the base plane but the glide in the prism plane is not excluded.

Summing up all the cumulative experimental material and analyzing it one can speak with confidence about the unambiguous influence of thermal treatment on structure quality and mechanical characteristics of crystals. High-temperature annealing by decreasing density of the growth dislocations in the volume and density of mobile dislocations introduced by mechanical treatment to the subsurface layer lowers the internal stresses. This being the case, the fracture toughness of the crystal volume becomes somewhat higher and that of the surface essentially decreases. Annealing at lower temperatures (about 1,800 K) by lowering the value of the internal stresses preserves the zone with the elevated density of dislocations to 20 μm in depth, and fracture toughness of such surface remains rather high.

It should be noted that the processes running at thermal and mechanical treatment depend not only on the original structure of crystals and treatment regimens but also are largely defined by the medium in which they are realized.

7.6 Influence of the Annealing Medium on the Crystals' Structure and Their Machinability

The atmosphere of annealing with different redox potential is meant as medium at thermal treatment; at mechanical treatment it is a lubricating-cooling liquid not only participating in heat removal from the zone of contact of the cutting instrument with the part but also defining physical–chemical interaction between the abrasive grain and the processed crystal. Simultaneously, while in the process of mechanical treatment the appropriate choice of the medium facilitates control of structure quality of only the subsurface layer, at thermal treatment one can control the structure of both subsurface layer and crystal volume. The effect of the character of the lubricating-cooling liquid on the variation of temperature and pressure in the contact zone, store of the excess energy, degree of the material dispersion, and other accompanying phenomena has been sufficiently well studied. There are much fewer data about the effect of the annealing medium and the available ones are contradictory. Not analyzing the totality of the published papers and not considering their contradictions allowed us to focus on the most important results that have an application significance [15].

A redox potential of the annealing medium. The main driving factor of the charge exchange processes at annealing corundum is violation of stoichiometry due to the formation of Schottky defects at thermal dissociation of aluminum oxide and evaporation of aluminum and oxygen atoms from the crystal surface. As a result, there are six components in the vapor phase above the crystal: Al, O, O_2, AlO, Al_2O, Al_2O_2, the content of which can be calculated from the system of equations [16]. The calculated data are given in Table 7.3.

The stoichiometric evaporation of corundum crystals can be described by the following quasichemical reaction:

$$Al_2^{3+}O_3^{2+} \xleftrightarrow{T} Al_{2-x}^{\;3+}O_{\left(3-\frac{3}{2}x\right)}^{\;2-}V_{Al}^{\;3-}V_{O\left(\frac{3}{2}x\right)}^{\;2+} + xAl\uparrow + \frac{3}{2}xO\uparrow$$

or in a simplified form:

$$O \xleftrightarrow{T} 2Al\uparrow + 3O\uparrow + 2V_{Al}^{\;3-} + 3V_{o}^{\;2+} \tag{7.14}$$

with the equilibrium constant:

$$K^{*} = \left(P_{Al}^{0}\right)^{2}\left(P_{o}^{0}\right)^{3}\left[V_{Al}^{\;3-}\right]_{P}^{2}\left[V_{o}^{\;2+}\right]_{P}^{3}, \tag{7.15}$$

Table 7.3 Partial pressure of the components at dissociation in the annealing process

T (K)	$-\lg P$(Al)	$-\lg P$(O)	$-\lg P(O_2)$	$-\lg P$(AlO)	$-\lg P(Al_2O)$	$-\lg P(Al_2O_2)$
1,873	9.25	9.08	10.9	10.78	11.36	13.42
2,000	8.18	8.00	9.66	9.52	9.97	11.91
2,303	6.11	5.92	7.24	7.07	7.30	8.97

where 0 is an index of ideal defect-free crystal; P_{Al}^0, P_O^0 is the equilibrium partial pressure of oxygen and aluminum vapors in the products of thermal dissociation of Al_2O_3 of stoichiometric composition; V_{Al}^{3-}, V_O^{2+} is the equilibrium concentration of anionic and cationic vacancies on the crystal surface.

It follows from these equations that variation of partial pressure of aluminum and oxygen in the vapor phase changes the concentration of the corresponding point defects in the crystal. High temperature medium is neutral with respect to the annealed crystal if the partial pressure of aluminum and oxygen vapors P_{Al}^*, P_O^* is quantitatively equal to P_{Al}^0 and P_O^0. Deviation from this condition leads to an intensification of evaporation of the corresponding ions from the crystal surface and to the crystal stoichiometry violation, i.e., to the redox process. Oxidation is characterized by saturation of the crystal with cationic vacancies (cationic stoichiometry violation) and its reduction by saturation with anionic vacancies (atomic stoichiometry violation).

Since the neutral high-temperature medium is characterized by dynamic equilibrium of oxygen and aluminum atoms in the solid and gaseous phases the redox potential of the medium ε^* can be expressed through chemical potentials of aluminum and oxygen in the products of thermal dissociation of corundum μ_{Al}^0, μ_O^0 and the annealing medium μ_{Al}^*, μ_O^*: [15]:

$$\varepsilon^* = 2/3\left(\mu_{Al}^0 - \mu_{Al}^*\right) - \left(\mu_O^0 - \mu_O^*\right) = RT\ln\left[\left(P_{Al}^0/P_O^*\right)^{2/3} P_O^*/P_O^0\right]. \quad (7.16)$$

The symbol ε^* points to the direction of the reduction-oxidizing process: at $\varepsilon^* > 0$ the crystal oxidizes, at $\varepsilon^* < 0$ it reduces.

Using the results [16] calculated were the values ε^* for the conditions of vacuum annealing and additional coloration of sapphire (Table 7.4). It can be seen that in

Table 7.4 Redox potential (ε^*, kV/mol) of the annealing medium and content of aluminum and oxygen in the products of thermal dissociation [16]

Characteristics of the annealing medium		T (K)			
		2,000	2,100	2,200	2,300
Al_2O_3	P_O	$9.9 \cdot 10^{-4}$	$4.8 \cdot 10^{-3}$	$2.3 \cdot 10^{-2}$	$1.2 \cdot 10^{-1}$
	P_{Al}	$6.7 \cdot 10^{-4}$	$3.3 \cdot 10^{-3}$	$9.5 \cdot 10^{-3}$	$7.9 \cdot 10^{-2}$
W–Al_2O_3	P_O	$8.3 \cdot 10^{-3}$	$9.7 \cdot 10^{-3}$	$1.9 \cdot 10^{-2}$	$8.0 \cdot 10^{-2}$
	P_{Al}	$8.0 \cdot 10^{-3}$	$9.8 \cdot 10^{-3}$	$2.4 \cdot 10^{-2}$	$9.3 \cdot 10^{-2}$
	ε^*	−26	−16	−29	−12
Mo–Al_2O_3	P_O	$9.8 \cdot 10^{-4}$	$5.2 \cdot 10^{-3}$	$9.6 \cdot 10^{-3}$	$5.8 \cdot 10^{-2}$
	P_{Al}	$9.8 \cdot 10^{-4}$	$5.0 \cdot 10^{-2}$	$2.0 \cdot 10^{-2}$	$9.7 \cdot 10^{-2}$
	ε^*	−10	−9.5	−36	−19.7
Al–Al_2O_3	P_O	$7.1 \cdot 10^{-8}$	$9.5 \cdot 10^{-7}$	$9.6 \cdot 10^{-2}$	$4.6 \cdot 10^{-5}$
	P_{Al}	$1.2 \cdot 10^{3}$	$4.2 \cdot 10^{3}$	$7.2 \cdot 10^{3}$	$1.2 \cdot 10^{4}$
	ε^*	−317	−310	−306	−302

the interval 2,000–2,300 K the vacuum annealing is weakly reducing and the additional coloration corresponds to the condition of intense reduction of corundum crystals. At the same time the potential ε^* does not depend on the annealing temperature.

It is possible to estimate the value of effective chemical potential of growth medium

$$\varepsilon = -[2E_{\upsilon} + 2RT\ln(A^{-1}V_{\mathrm{O}}P_{\mathrm{O}})],$$

where R is the universal gas constant; T is the temperature of the technological process; P_{O} is the equilibrium partial pressure of oxygen vapors in the products of thermal dissociation of sapphire with stoichiometric composition; and V_{O} is the equilibrium concentration of anionic vacancies in the crystal. The coefficients $E_{\upsilon} = 1.12 \cdot 10^3$ J/mol, $A = 5.5 \cdot 10^{33}$ at·cm^{-3} were determinated by the anionic vacancy concentration in the sapphire annealed in the saturated vapor of aluminum. For carbon-contained medium, $\varepsilon = -235$ kJ/mol [17].

In this way the influence of the high-temperature medium and the character of crystal stoichiometry violation depend on the redox potential of the annealing medium which is equal to the difference of chemical potentials of aluminum and oxygen in the products of thermal dissociation of the crystal and medium. By the values of the oxidizing potential of the annealing medium the following can be arranged in the succession:

Saturated vapors of aluminum ($\varepsilon^* \approx 310$ kJ/mol)
Carbon-containing medium ($\varepsilon^* \approx -200\ldots -50$ kJ/mol)
Vacuum (residual pressure about 10^{-2} Pa, $\varepsilon^* \approx 15$ kJ/mol)
Saturated vapors of aluminum oxide ($\varepsilon^* = 0$)
Air ($\varepsilon^* \approx 270$ kJ/mol)

Annealing of crystals in a neutral medium. Annealing of corundum crystals in a neutral medium ($\varepsilon^* = 0$) allows the attainment of crystals with minimal stoichio-metry violation. This kind of thermal treatment is possible when saturated vapors of aluminum oxide are used as medium. Since the potential is independent of temperature thermal treatment, for thermal activation of diffusion and relaxation processes it should be preferentially performed at premelting temperatures. As the pressure of saturated aluminum oxide vapors at 2,273 K is 0.1 Pa [18] crystals should be annealed at the residual pressure in the working zone 0.1–0.2 Pa.

Oxidizing annealing of crystals. The driving force of the charge exchange processes at the oxidizing annealing is cationic violation of stoichiometry, which appears at thermal treatment of crystals in the medium with a higher content of oxygen than in the vapors of thermal dissociation of α-Al_2O_3 of the stoichiome-tric composition:

$$1/2\mathrm{O}_2 \leftrightarrow 2/3V_{\mathrm{Al}}^{\ 3-} + \mathrm{O}_0^{\ 2-} + 2h^+ \tag{7.17}$$

$$\mathrm{Me}^{n+} + 2h^+ \leftrightarrow \mathrm{Me}^{(n+1)+} \tag{7.18}$$

The intensity of oxidation of the activator ions (Me^{n+}) is defined by the concentration of charge carriers–holes (h^+) formed at the origination of cationic vacancies V^{3+}_{Al}. Using the equilibrium constant of pseudochemical reactions

$$K^* = \left[V_{Al}^{3-}\right]^{2/3} \left[h^+\right]^2 / \left(P_{O_2}\right)^{1/2}, \tag{7.19}$$

one can determine the concentration of "holes" participating in the charge exchange processes at annealing of corundum crystals in the oxygen-containing medium:

$$\left[h^+\right] = 1.32\left(K^*\right)^{0.375}\left(P_{O_2}\right)^{0.188}. \tag{7.20}$$

Consequently, the control for the oxidizing potential of the annealing medium by partial pressure of oxygen P_{O_2} allows one to purposefully affect the intensity of the charge exchange processes in corundum.

At the oxidizing annealing the displacement of the solid-phase reaction front is described by the equation

$$x^2 = D_f t = 2N_0 Dt / C, \tag{7.21}$$

where D_f is the diffusion coefficient of the solid-phase reaction front; N_0 is the concentration of point defects on the sample surface; and C is the impurity concentration. The kinetics of heterogeneous reactions running during the oxidizing annealing was studied using the doped crystals at $T > 1{,}770$ K [16]. Prior to the oxidizing annealing, the laser ruby samples of various thicknesses were annealed in a carbon-containing atmosphere to saturate them with color centers responsive for the absorption at 315-nm wavelength. Then the samples were annealed in an oxidizing atmosphere at 2,020 K for 1, 2, 4, 6, 10, and 20 h. The Cr^{3+} content and the absorption at 315 nm selected as a criterion of the oxidation process were measured after each annealment stage. The oxidation of 1-, 2.5-, 5-, and 7.5-mm thick samples is completed after 2, 4, 6, and 20 h, respectively (Table 7.5).

The reaction front displacement parameter is defined as

$$A = 2N_0 D = x^2 C / t. \tag{7.22}$$

The value of parameter A is determined from the system of inequalities of $x^2_1 C_1/t_1 < A < x^2_2 C_2 </t_2$ kind derived using the data from Table 7.5. The kinetics of solid-phase interaction reaction between the point defects and impurity ions depends only weakly on the impurity nature. The value A increases by one decimal order as the annealing temperature is elevated from 1,750 to 2,050 K (Table 7.6). That is, the D_f quantity also increases with the temperature, the impurity content in the crystal being the same, and thus the oxidation process rate is increased considerably.

The annealing in a reducing atmosphere causes a reduced charge state of some impurity ions and results in anionic stoichiometry violation in sapphire. The efficient reducing atmospheres for sapphire include aluminum vapor, carbon-containing atmosphere being formed in a furnace with graphite equipment, and hydrogen.

Table 7.5 Time dependence of displacement of the solid-phase reaction front (x) at the oxidizing annealing

Anneal duration (h)	Parameter	Sample thickness (mm)			
		1	2.5	5	7.5
1	$C_{Cr^{3+}}$ (vol%)	0.029	0.036	0.033	0.023
	Absorption at 315 nm	–	+	+	+
	x	$0.5 < x < 1.25$			
2	$C_{Cr^{3+}}$ (vol%)	0.040	0.035	0.032	0.023
	Absorption at 315 nm	–	+	+	+
	x	$0.5 < x < 1.25$			
4	$C_{Cr^{3+}}$ (vol%)	0.040	0.037	0.042	0.027
	Absorption at 315 nm	–	–	+	+
	x	$1.25 < x < 2.5$			
6	$C_{Cr^{3+}}$ (vol%)	0.032	0.032	0.026	0.027
	Absorption at 315 nm	–	–	–	+
	x	$2.5 < x < 3.75$			
10	$C_{Cr^{3+}}$ (vol%)	0.036	0.029	0.026	0.027
	Absorption at 315 nm	–	–	–	–
	x	$2.5 < x < 3.75$			
20	$C_{Cr^{3+}}$ (vol%)	0.037	0.034	0.032	0.029
	Absorption at 315 nm	–	–	–	–
	x	$x < 3.75$			

Table 7.6 The parameter A, depending on the impurity and annealing conditions

Crystal	A (cm²/s) 1,750 K	A (cm²/s) 2,050 K	P^*_O (Pa)
Al_2O_3:Ti	$2.2 \cdot 10^{-11}$	$3.8 \cdot 10^{-10}$	28
Al_2O_3:Ti	$3.5 \cdot 10^{-11}$	–	140
Al_2O_3:Cr	–	$7.56 \cdot 10^{-10}$	20

The anionic stoichiometry violation in the high-temperature carbon-containing atmosphere is defined by:

- Interaction of oxygen atoms with carbon monoxide in the vapor phase
- Reduction of oxygen partial pressure in the vapor phase and formation of anionic vacancies at the crystal surface
- Increase of the aluminum equilibrium partial pressure in the vapor phase
- Partial annihilation of anionic vacancies at the crystal surface due to aluminum atom pulling to graphite
- Diffusion of anionic vacancies into the crystal bulk

The impurity ions are reduced during the sapphire annealing in hydrogen due to the lattice saturation with hydrogen atoms. The reduction efficiency is defined by hydrogen solubility in sapphire. During the annealing in hydrogen, anionic vacancies are formed in the graphite presence only according to the mechanism described above.

The surface phase changes under reducing conditions take place when the crystal structure losses its stability at a considerable sapphire stoichiometry violation, so the phases with other structures become more energetically favorable. After annea-ling at 1,970–2,220 K in Ar atmosphere with reducing additives of CO and H_2 (5–10%), the corundum phase transition to the spinel $AlAl_2O_4$ (polycrystalline phase with $a \sim 7.933$–7.948 Å) or mixed phase was observed in the near-surface layer [19]. The corundum structure is recovered at annealing in air at about 1,570 K. In ref. [20], the spinel phase is explained by carbochemical reduction of Al_2O_3. The formula $x\mathrm{Al_2OC}\cdot\mathrm{Al_2O_3}$ has been proposed to characterize the compound composition.

Diffusion and interaction of point defect in sapphire under annealing. The arising vacancies form characteristic optical absorption centers in the UV spectral region. A high-temperature annealing in controlled gas atmospheres is an effective way to influence those defects [18, 21]. The kinetics of anionic stoichiometry violation in sapphire caused by high-temperature carbon-containing annealing atmosphere and the interaction character of anionic vacancies with cationic vacancies have been studied [22].

Under thermodynamic equilibrium between the crystal and high-temperature gas medium, $P^*_{\mathrm{Al}} = P_{\mathrm{Al}}; P^*_{\mathrm{O}} = P_{\mathrm{O}}$. This makes it possible to transform the expression (7.17) to the form

$$P^*_{\mathrm{O}} = P^0_{\mathrm{O}} \exp(\varepsilon/2RT), \tag{7.23}$$

$$P^*_{\mathrm{Al}} = P^0_{\mathrm{Al}} \exp(-3\varepsilon/4RT). \tag{7.24}$$

Under annealing in the carbon-containing atmosphere or in aluminum vapor, the reduction of sapphire is accompanied by the crystal lattice saturation with anionic vacancies.

The transformation of equations from [22] allows the characterization of the functional relationship between the reducing potential ε and the equilibrium concentration of anionic vacancies:

$$\left[V^x_{\mathrm{O}}\right] = \frac{K_3}{P^0_{\mathrm{O}}} \exp(-\varepsilon/2RT) = K_6 \left(P^0_{\mathrm{Al}}\right)^{2/3} \exp(-\varepsilon/2RT). \tag{7.25}$$

Since the equilibrium constants of the reactions describing the anionic vacancy formation can be presented in the form [23]

$$K = A \exp(-E_{\mathrm{V}} / 2RT), \tag{7.26}$$

(7.25) can be written as follows:

$$\begin{aligned} \left[V^x_{\mathrm{O}}\right] &= A_1 \left(P^0_{\mathrm{O}}\right)^{-1} \exp[-(2E_{\mathrm{V}} + \varepsilon) / 2RT] \\ &= A_2 \left(P^0_{\mathrm{Al}}\right)^{\frac{-2}{3}} \exp[-(2E_{\mathrm{V}} + 3\varepsilon) / 6RT], \end{aligned} \tag{7.27}$$

where E_V = 1,048 kJ/mol is the anionic vacancy formation energy in sapphire; $A_1 = 2.3 \cdot 10^{3/2}$ atm/cm^3; and $A_2 = 2.6 \cdot 10^{27}$ atm$^{3/2}$/cm^3. The coefficients A_1, A_2, and E_V have been determined in experiment [22] from the anionic vacancy concentration in a crystal with zero effective charge (*F*-centers) $[V_O^x]$, obtained after the sapphire anneal in saturated aluminum vapor at 2,043 and 2,153 K.

The gas medium chemical potential was estimated from its effect on the crystal, since the equilibrium concentration of anionic vacancies is established at the crystal surface under the reducing anneal. The effective reducing chemical potential value for a carbon-containing atmosphere is

$$\varepsilon = -\left[2E_V + 2RT\ln\left(A_1^{-1}\left[V_O^x\right]P_O^0\right)\right]. \tag{7.28}$$

The point defect concentration at the surface being constant, the defect distribution in the diffusion layer (Fig. 7.12) is described [24] as

$$[V_O]_{(x,\tau)} = [V_O]_{max}\left\{1-\mathrm{erf}\left[x_i/\left(2D_{ef}\tau\right)^{-1/2}\right]\right\}, \tag{7.29}$$

where $[V_O]_{max}$ is the equilibrium concentration of anionic vacancies at the annealed crystal surface; and D_{ef} is the effective diffusion coefficient of anionic vacancies.

Based on the determined point defect concentration in various sites of the diffusion layer, the D_{ef} and the anionic vacancy concentration at the crystal surface can be estimated, while the temperature dependence of D_{ef} makes it possible to estimate the pre-exponential factor D_0 and the activation energy E.

The crystals grown by the Kyropoulos technique had the initial cationic stoichiometry violation that was evidenced by the broad optical absorption band about 220 nm (Fig. 7.12). The samples were annealed for 3, 6, 12, and 16 h in a furnace with

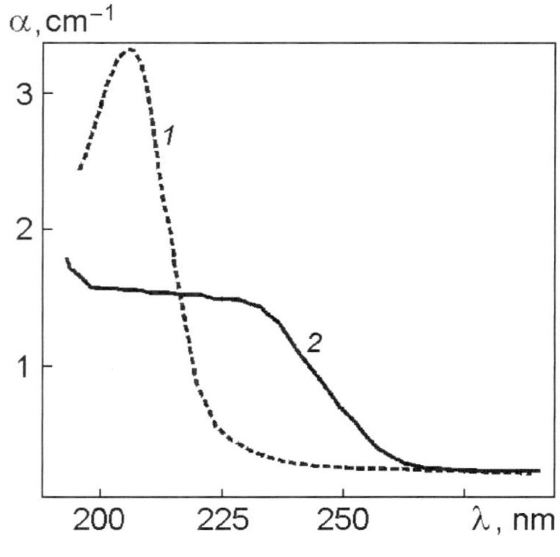

Fig. 7.12 UV absorption spectrum typical of sapphire with violated stoichiometry: *dashed line*, anionic nonstoichiometry; *straight line*, cationic nonstoichiometry

graphite heater at 2,170 and 2,270 K under residual pressure of 5×10^{-1} mmHg. The 2.5-mm-thick plates were cut out of the central part of the annealed samples and the planes thereof were polished to the roughness of $R_z = 0.5–1.0$ μm. Using an automated spectral setup provided with a scanning device with the $1.5 \cdot 0.03$ mm² probing beam, the distribution of *F* centers over the plate volume along the $[10\bar{1}0]$ direction was examined.

The anionic vacancy concentration as calculated using the Smacula formula [23, 24] from the optical absorption band intensity at 206-nm wavelength was found to be

$$[V_0] = (1.86\alpha_{206}) \times 10^{16}\,[\text{cm}^{-3}]. \qquad (7.30)$$

Under reducing annealing in a carbon-containing atmosphere, the anionic vacancies are formed at the surface and diffuse into the crystal volume (Fig. 7.13). The vacancy distribution kinetics is monitored using the distribution of the optical absorption intensity at 206-nm wavelength over the diffusion layer.

As the crystal volume is saturated with anionic vacancies, the intensity and character of the optical absorption in the UV region varies over the cross section of the sample being annealed (Fig. 7.14). This demonstrates the annihilation of the initial color centers based on cationic vacancies. As early as 3 h after annealing, a region (point C) with the minimum absorption on 206-nm wavelength is observed within the crystal volume (Fig. 7.14b), which corresponds to the $[V_{Al}]:[V_O] = 2:3$ condition and to the annihilation of vacancy color centers. The anneal duration increase up to 6 h is accompanied by broadening of that zone and its propagation toward the crystal center; the further annealing results in anionic stoichiometry violation in the entire crystal volume (Fig. 7.15). The sample saturation with anionic vacancies is complete when the concentration thereof in the crystal volume becomes equal to that at the surface. Thus, at a certain annealing duration, a stoichiometric crystal matrix is formed with a minimum optical absorption in the UV spectral region.

According to experimental data obtained within the 1,970–2,270 K temperature range, $D_0 = 486 \cdot 10^{-4}$ cm²/s; $E = 470$ kJ/mol. The effective diffusion coefficient of anionic vacancies and the concentration thereof at the surface of the crystals being annealed are as follows: $D_{ef}^{2,170} = 2 \times 10^{-7}\,\text{cm}^2/\text{s}$; $[V_O]_{max} = 1.5 \cdot 10^{17}$ cm⁻³ at 2,170 K and $D_{ef}^{2,270} = 6.5 \times 10^{-7}\,\text{cm}^2/\text{s}$; $[V_O]_{max} = 1.2 \cdot 10^{17}$ cm⁻³ at 2,270 K; the effective reducing

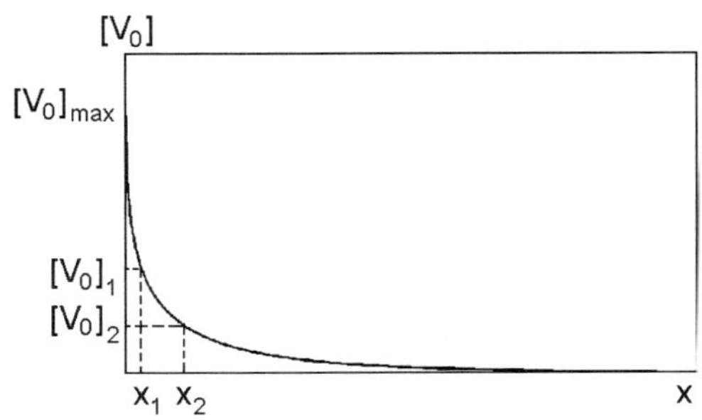

Fig. 7.13 Distribution character of anionic vacancies in the diffuse layer

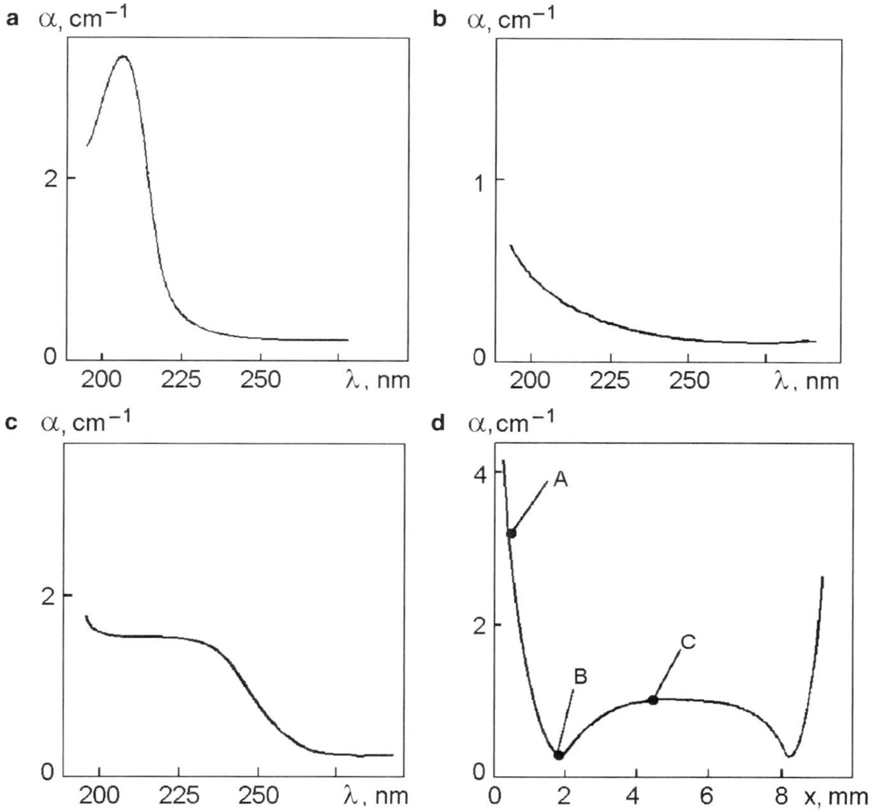

Fig. 7.14 Distribution of the optical absorption at 206-nm wavelength in sapphire over the cross section of a 10-mm thick sample annealed under a carbon-containing atmosphere at 2,170 K for 3 h: (**a, b, c**) the optical absorption spectra in the corresponding points (A, B, C) of the crystal cross section

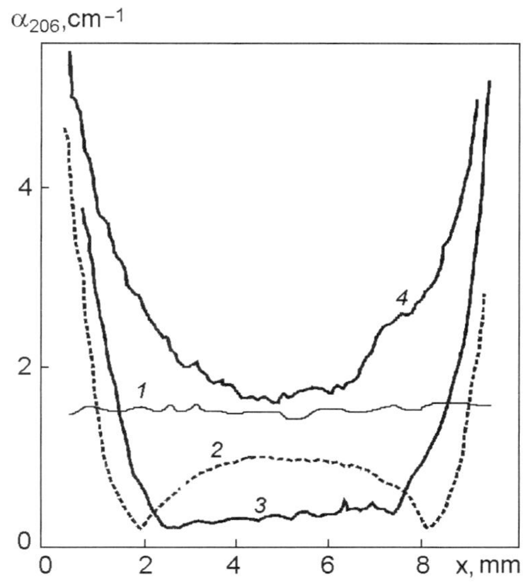

Fig. 7.15 Distribution of the optical absorption at 206-nm wavelength in sapphire over the cross section of 10-mm thick samples annealed under a carbon-containing atmosphere at 2,170 K: (*1*) the initial sample; (*2*) the sample annealed for 3 h; (*3*) that annealed for 6 h; (*4*) that annealed for 12 h

chemical potential of the carbon-containing annealing atmosphere is $\varepsilon_{2,170} = -264$ kJ/mol and $\varepsilon_{2,270} = -230$ kJ/mol, respectively.

Making use of the vacancy color center annihilation in the point C after 3-h annealing (see Fig. 7.14 and 7.23 and 7.24 to determine the anionic vacancy concentration within the local stoichiometric zone), the cationic vacancy concentration in the initial crystal can be calculated as

$$[V_{\mathrm{Al}}] = 0.67[V_{\mathrm{O}}]_{pC} = 0.67[V_{\mathrm{O}}]_{pC} \times \left\{1 - \operatorname{erf}\left[x_{pC}/2\left(D_{\mathrm{ef}}\tau\right)^{1/2}\right]\right\} = 9.5 \cdot 10^{15}\,\mathrm{cm}^{-1}, \quad (7.31)$$

where the index "pC" denotes the values in the point C (Fig. 7.15). That concentration of defects results in the initial sample absorption value at 220-nm wavelength ($\alpha_{220} = 1.6$ cm^{-1}), thus making it possible to determine the optical absorption cross section of cationic vacancy-based color centers: $\alpha_{220} = \alpha_{220}[V_{\mathrm{Al}}]^{-1} = 1.7 \cdot 10^{16}$ cm^2.

Based on these results, the manufacturing regimen of a sapphire window having a minimum absorption in the UV spectral region can be calculated. (For example, a 5 mm thick blank of 22 mm in diameter was annealed at 2,220 K in a carbon-containing atmosphere under 30 Pa residual pressure. Then the 1.5 mm thick material layer was removed from each blank side and the working planes were polished.) The articles so obtained show an optical spectrum typical of the stoichiometric matrix with a low absorption level in the UV region (Fig. 7.16).

Thus, a stoichiometric matrix with a reduced absorption level in the UV region can be formed as a result of interaction between anionic and cationic vacancies in sapphire and in finished workpieces under reducing annealing. The vacancy defects are formed not only under annealing, but also under irradiation. Some characteristics of the vacancy centers, the formation and destruction conditions thereof are presented in the Appendix Table 7.15.

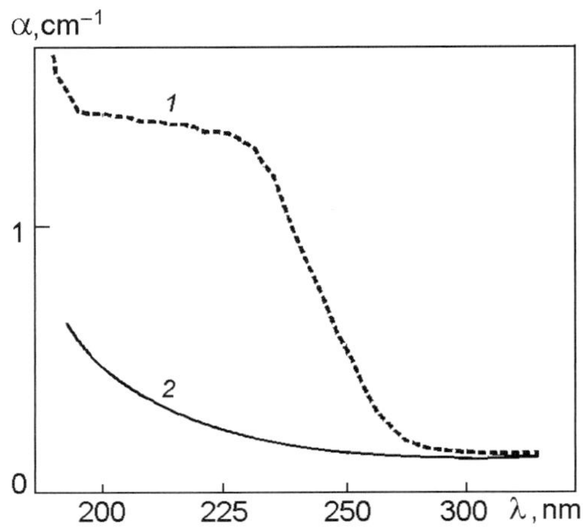

Fig. 7.16 Optical absorption spectra of sapphire windows of 20 mm in diameter and 2 mm in thickness: *dashed line*, without heat treatment; *straight line*, after a special annealing

7.7 Effect of the Annealing Atmosphere on Mechanical Properties

The crystal mechanical properties depend to a considerable extent on the type and density of vacancies and vacancy complexes. So, sapphire wear resistance is the least after annealing in hydrogen (1,800 K, 10^4 Pa, 10 h); the abrasion rate increases after annealing in a carbon-containing atmosphere (1,800 K, 66 Pa, 2.5 h), and increases even more after air annealing (1,800 K, 2.5 h). The maximum wear resistance is shown by the vacuum-annealed crystals (2,100 K, 10^4 Pa, 2.5 h). The data in Table 7.7 are related to crystals grown by Stepanov technique treated by free abrasive M40 under 25 kPa loading.

The wear resistance is influenced considerably by the structure defects responsible for the optical absorption within the 205- to 210-nm range, that is, the anionic vacancies. The vacancy complexes hinder the crack propagation. It is just the impurity-vacancy complexes of [$2Al^{3+}$–Ti^{4+}–V_{Al}–O^{2-}–($2O^0 + 2e^-$)]; [Al^{3+}–$2Ti^{4+}$–V_{Al}–$2O^{2-}$–O^-], [$3Ti^{4+}$–V_{Al}–O^{2-}–$3O^{2-}$], and [Ti^{4+}–V_{Al}] types that are especially efficient from this point of view.

The vacancy complexes are arranged in the following sequence according to the crack-stopping effect [16]:

- The cationic vacancy-based impurity–vacancy complexes
- The anionic vacancy-based impurity–vacancy complexes

7.8 Effect of the Annealing Atmosphere on Optical Properties

As to leucosapphire, the optical properties depend on the annealing atmosphere mainly in the region of absorption and luminescence of the *F*-centers. In doped sapphire, the atmosphere affects much more than the optical properties and the functional properties related thereto. Let Ti:Al_2O_3 and Cr:Al_2O_3 be taken as examples.

Ti:Al_2O_3. The optical absorption spectra contain the absorption bands of Ti^{3+} (495 nm) and [Ti^{3+}–*F*] complex (268 nm) along with those of *F* and F^+-centers (206, 225 nm). The vacuum annealing normalizes the lattice but the absorption in the UV region is increased (Fig. 7.17) [25].

The intensity of absorption induced by UV radiation is defined by the content of impurity–vacancy complexes [$2Al^{3+}$–Ti^{4+}–V_{Al}–O^{2-}– ($2O^0 + 2e^-$)]. The annihilation of

Table 7.7 Abrasion rate and optical absorption intensity (D_λ) of sapphire annealed in various atmospheres [16]

Parameters	Hydrogen		Carbon-containing atmosphere		Air		Vacuum	
υ (mm h^{-1})	0.4		0.5		0.6		0.8	
λ (nm)	205–210	240–245	205–210	240–245	205–210	240–245	205–210	240–245
D_λ (cm^{-1})	0.09	0.47	0.50	0.80	0.85	1.85	1.7	1.35

Fig. 7.17 Optical absorption spectra of sapphire (*1*) and Ti:Al_2O_3 (*2, 3*) annealed under carbon-containing atmosphere (*1, 2*) and in vacuum (*3*)

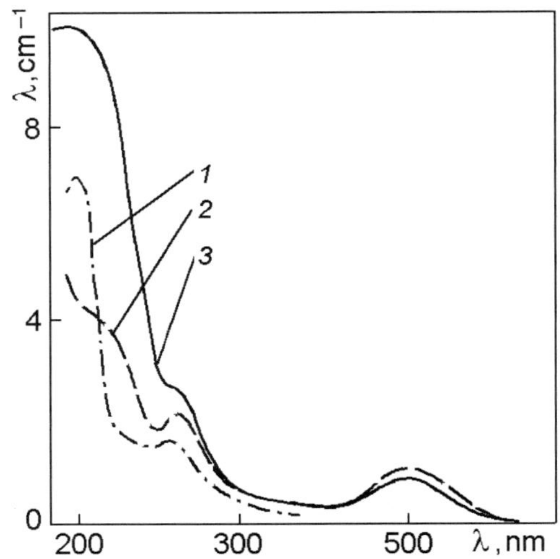

Table 7.8 Effect of the annealing atmosphere on the radiation resistance of Ti:Al_2O_3 [25]

	$\Delta D_{410} = D^*_{410} - D$		
C_{Ti} (wt%)	Air (T = 2,000 K)	Vacuum (T = 2,200 K)	Carbon-containing atmosphere (T = 2,200 K)
0.015	1.39	0.10	0.02
0.018	1.38	0.15	0.01
0.070	–	0.10	0.00
0.100	0.36	0.00	0.00
0.130	0.32	0.00	0.00

cationic vacancies and improvement of the crystal radiation resistance is due to the annealing in a reducing atmosphere (Table 7.8).

Four characteristic bands are present in the radioluminescence spectrum of Ti:Al_2O_3; their intensity depends on the annealing atmosphere and titanium content (Fig. 7.18). The crystal optical resistance against the laser-induced deterioration depends on the content and type of absorption centers (Table 7.9).

Cr:Al_2O_3. As in the preceding case, the selection of an appropriate annealing atmosphere favors an increased content of active laser centers Me^{3+} at the expense of reduced Me^{4+} fraction. The point defects influence the inhomogeneous spreading of the ruby R_1 line. The laser ruby samples were annealed in a carbon-containing atmosphere at 2,220 K and in oxidizing atmosphere at 2,020 K for 20 h. The oxidation extent was monitored using the optical absorption intensity at 315-nm wavelength.

The observed doublet structure 0.38 cm^{-1} (Fig. 7.19) is due to splitting of four times degenerated ground state of chromium into two Kramers' doublets. The heat treatment in oxidizing atmosphere is seen to result in an appreciable (by about 25%) width reduction of the main components and appearance of a fine structure that has been explained before by the presence of Cr^{3+} isotopes [26]. The inhomogeneous

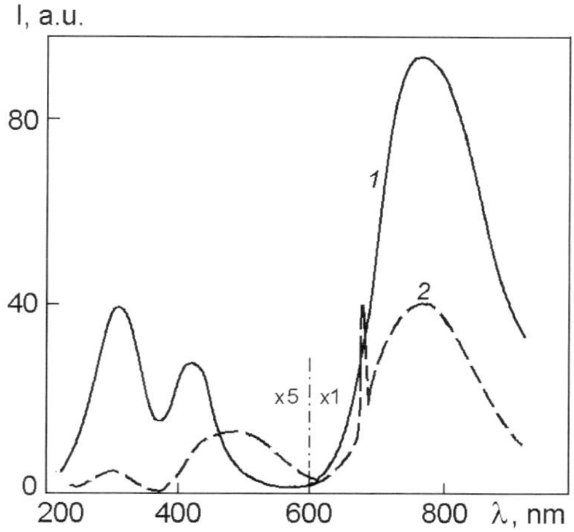

Fig. 7.18 X-ray luminescence of Ti:Al_2O_3 annealed under atmospheres with redox potential (kJ/mol): (*1*) 230 and (*2*) 30

Table 7.9 Ti:Al_2O_3 resistance against the laser breakdown [25]

		Breakdown threshold (relative units)		
Ti^{3+} (wt%)	Crystallographic orientation	Carbon-containing atmosphere (growth)	Vacuum (annealing at 2,200 K)	Hydrogen (annealing at 2,000 K)
0.02	(0001)	1.0	1.2	1.52
	$(11\bar{2}1)$	0.9	1.2	1.52
	$(10\bar{1}0)$	0.8	1.2	1.52
0.03	(0001)	0.86	1.15	1.47
	$(11\bar{2}1)$	0.82	1.15	1.12
	$(10\bar{1}0)$	0.83	1.14	1.08

spreading of the R_1 line decreases along with the optical absorption reduction at 315 nm and attains a minimum as that band disappears (Fig. 7.20) [27].

That effect cannot be explained by stress relaxation, since the oxidizing annealing was preceded by annealing at a considerably higher temperature. The annealing in a carbon-containing atmosphere causes an increased density of point defects, thus intensifying the local field distortion near Cr^{3+} and resulting in a spread of R_1 line components. Those distortions are weakened under oxidizing annealing, since, as a result of the reaction, the structural oxygen atoms and cationic vacancies arising at the crystal surface cause a reduced content of oxygen vacancies in the crystal volume.

$$1/2O_2 \leftrightarrow 2/3V_{Al}^{*} + O^{*}$$

This fact explains the variations in the half-width of R_1 line components synchronous with the reduction of optical absorption intensity at 315-nm wavelength.

Thus, the inhomogeneous R_1 line spreading in the ruby luminescence spectrum is sensitive enough to the point defect content and when estimated analytically can

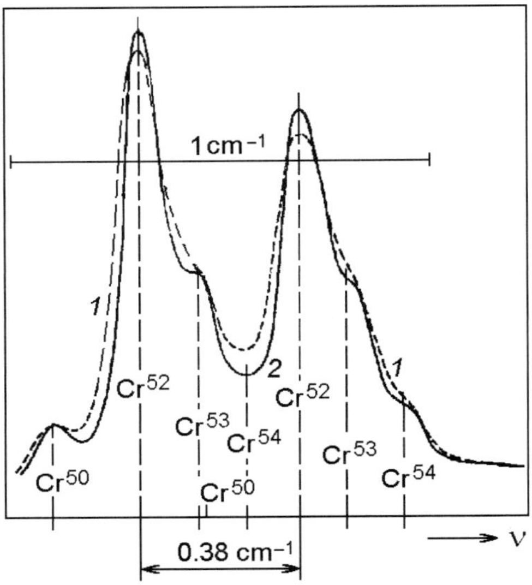

Fig. 7.19 R_1 line of ruby after annealing under reducing (*1*) (*dashed line*) and oxidizing (2) atmospheres. The interferometer free dispersion 1.25 cm^{-1}, the apparatus width 0.04 cm^{-1}

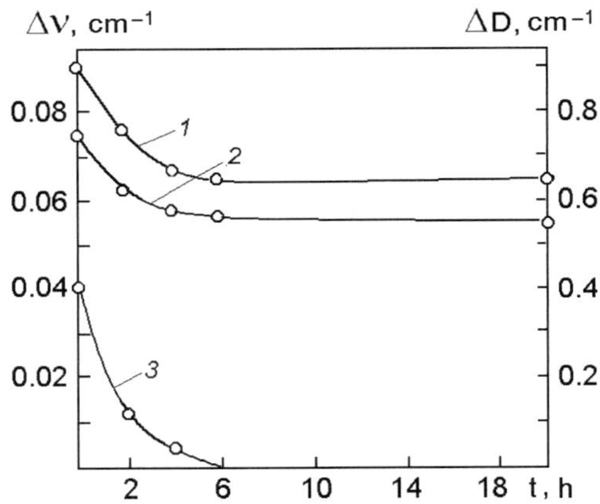

Fig. 7.20 Dependence of the half-width ($\Delta\nu$) of short-wavelength (*1*) and long-wavelength (2) components of the ruby R_1 line and of optical absorption (ΔD) at 315 nm wavelength (*3*) on the oxidizing anneal duration (*t*)

be used to determine the concentration thereof and to optimize the redox potential of the annealing atmosphere.

7.9 Effect of Annealing on Laser Characteristics of Ruby and Sapphire Articles

Various annealing regimens make it possible to enhance some characteristics of laser elements, including:

- To improve of the optical homogeneity, and thus to lower the generation threshold, increase the energy output and efficiency

Table 7.10 The wave channel parameters

	Ground channel		Polished channel	
Parameter	Prior to annealing	After annealing	Prior to annealing	After annealing
R_z (μm)	12.2	6.4	0.43	1.48
R_a (μm)	1.75	1.05	0.087	0.21
Average spacing of roughness (μm)	76.57	82.29	387	145
R_{max} (μm)	1.27	12.95	3.59	3.9
Depth of the layer with increased dislocation density (μm)	70	2–15	15	0
Surface finish class	6	7	11	9

R_{max} is the maximum roughness height measured within the base length

- To reduce the light scattering and to enhance the radiation directivity
- To transform the activating ions and impurity ions into the required valence state
- To enhance the optical resistance of the element butt surfaces and volume
- To improve the radiation resistance

In waveguide-type CO_2 lasers, sapphire waveguides with a channel of 2 to 4 mm diameter are used. The channel parameters (surface roughness, shape irregularity, etc.) influence considerably the competition of transversal vibrations in resonators, thus providing the main transversal mode separation and reducing the radiation divergence. Increased roughness results in increased losses. The waveguide annealing improves the channel surface quality (Table 7.10) [28].

It follows from Table 7.10 that it is only the postgrinding anneal of waveguides that is reasonable. The annealing of polished waveguides causes deterioration of the functional parameters due to the surface thermal etching.

7.10 Stress Relaxation under Annealing

The stress relaxation in sapphire already is observed at 1,970 K (that is, at 0.83 T_m). The annealing at 2,170–2,223 K provides the stress reduction to the level allowing the machining of any crystal, independently of the initial residual stress level. The anomalous birefringence in sapphire grown by the Verneuil method is reduced from 700 nm/cm (after annealing at 1,973 K) to 70 nm/cm (as a result of annealing at 2,223 K). The stresses are relaxed due to thermal activation processes and the residual stress level depends on the temperature closeness to T_m as well as on athermal relaxation of structure defects. The relaxation may be hindered by impurities or dopants.

The developed premelting temperature sensor [29] connected with the temperature monitoring and control system made it possible to elevate the annealing temperature T_a up to 0.995 T_m. (In this chapter, the T_a is indicated intentionally in relation to T_m since the absolute melting temperature of doped crystals may vary within 10°. Therefore, the sensor signal is related to the premelting temperature of the individual crystal.)

In Table 7.11, the residual stress values σ (determined from the anomalous birefringence) depending on the annealing temperature approaching the premelting temperature and the results of statistical processing of the data obtained are

Table 7.11 Dependence of residual stresses on the annealing temperature

	Prior to annealing		After annealing					
			0.95 T_m		0.995 T_m			
σ (MPa)	$\bar{x}$	S^2	$\bar{x}'$	$(S')^2$	$\bar{x}''$	$(S'')^2$	$\frac{\bar{x}''}{\bar{x}'}$	$\frac{(S'')^2}{(S'')^2}$
σ	51.37	32.83	3.138	0.852	1.520	0.161	0.5	0.072
σ_{max}	87.95	65.14	4.440	2.359	2.158	0.361	0.49	0.153
σ_{min}	34.62	24.66	3.048	2.555	1.403	0.244	0.45	0.095

$\bar{x}$ arithmetic mean; S^2 dispersion

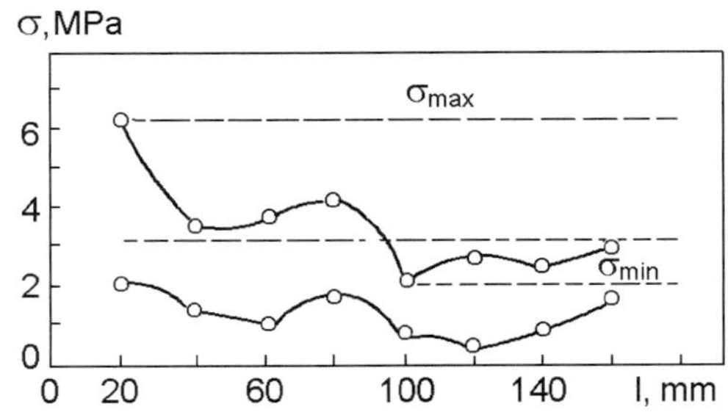

Fig. 7.21 Distribution of residual stresses over the length for 160 mm long ruby laser elements of 7 mm in diameter resulting from annealing at (*1*) 0.95 T_m and (2) 0.995 T_m

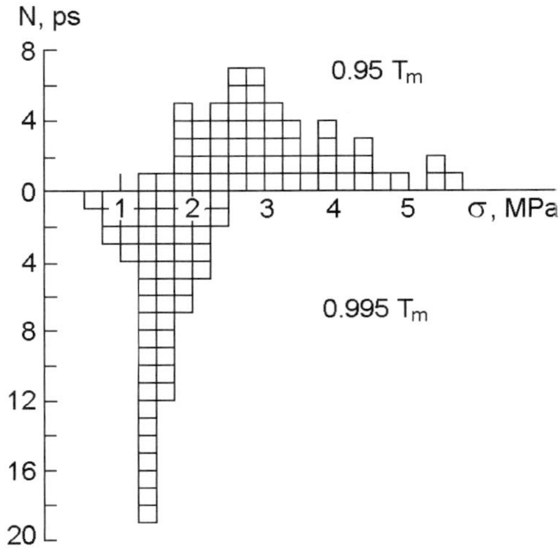

Fig. 7.22 Distribution of the samples in the residual stress levels after annealing at 0.95 T_m and 0.995 T_m

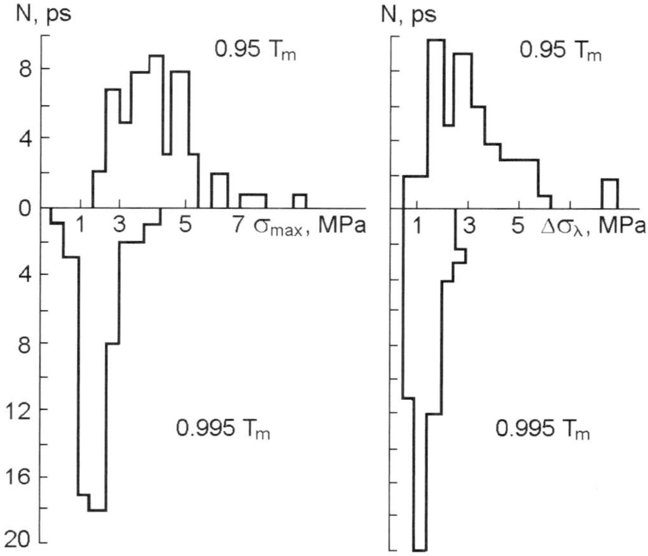

Fig. 7.23 Distribution of the samples in σ_{max} values and the residual stress inhomogeneity $\Delta\sigma$ after annealing at 0.95 T_m and 0.995 T_m

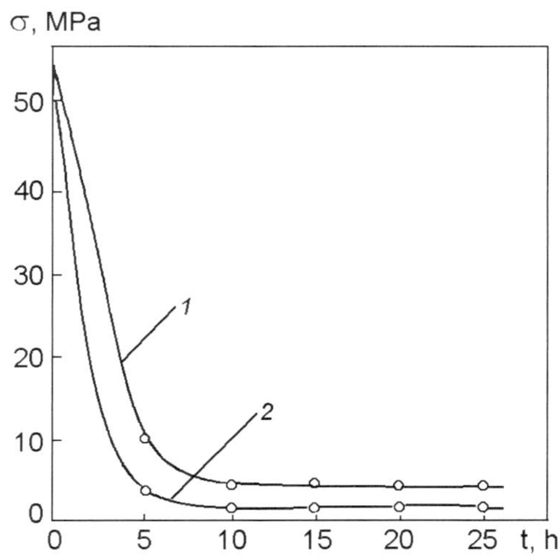

Fig. 7.24 Dependence of the residual stresses in ruby laser rods on the annealing duration at (*1*) 0.95 T_m and (*2*) 0.995 T_m

presented. Fifty laser ruby samples of 8 mm in diameter and 120 mm in length grown using the Verneuil technique were annealed [30].

Figures 7.21–7.24 show the effect of T_a approach to the premelting approach on the stress distribution over the crystal length as well as the statistical distribution of the residual stresses in laser ruby elements [30]. The residual stress levels and the homogeneity of characteristics are within rather narrow limits ($1 \le \sigma \le 2$ MPa; $1.5 \le \sigma_{max} \le 3$ MPa; $0.5 \le \Delta\sigma \le 2$ MPa) for most samples annealed at 0.995 T_m.

7.11 Effect of Annealing on the Crystal Strength

The available data on the annealing influence on the variation of strength properties are ambiguous. It follows from Fig. 7.25 [35] that the vacuum annealing causes an increased H_υ value but does not result in an appreciable change in the crystal strength properties. The maximum attained values of flexural strength limit are 90–95 kg/cm². The tensile strength limit variation under heating is illustrated in Table 7.12 for fibers of sapphire and its eutectics grown using the μ-PD technique [37]. A series of experiments have been carried out to study the strength characteristics of sapphire grown mainly by the HEM technique [38–44]. In some cases, the crystals grown by Czochralski and EFG methods were studied. The studies can be subdivided into two groups. The first is aimed at the preliminary annealing effect on the strength characteristics of sapphire, and the results obtained therein are in correlation with the results presented in the preceding paragraphs of this chapter. In the second group, the variation of strength characteristics at elevated temperatures was studied. The influence of temperature, space orientation, and the material purity was considered at various straining types (tension, compression, and four-point bending). These two groups of experiments are interrelated, since not only the sapphire strength at elevated temperature was studied in the second group of experiments, but the strength characteristics of premachined and annealed sapphire were compared.

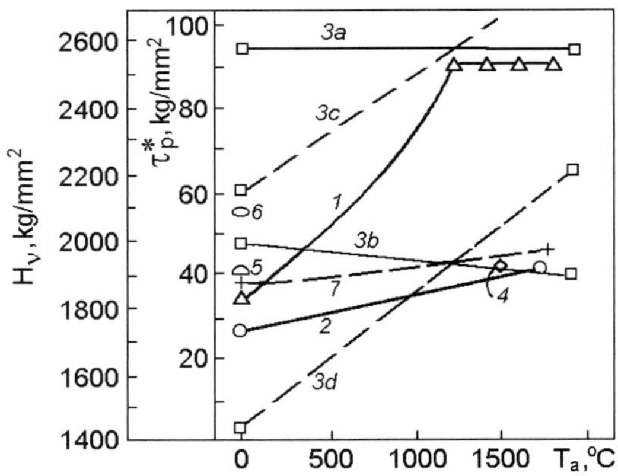

Fig. 7.25 Temperature dependence of sapphire microhardness (*dashed lines*) and strength limit (*solid line*). The curve numbers correspond to references cited: (*1*) [31]; (2) [32]; (*3a*) [33]; (*3b*) [33]; (*3c*) [33]; (*3d*) [33]; (*4*) [34]; (*5, 6*) [35] (*7*) [36]

Table 7.12 Strength characteristics (MPa) of sapphire and eutectics [37]

Material	20°C	1,500°C
Sapphire	2,900	200
$Al_2O_3^{81.3\%}YAG^{18.7\%}$	927	576
Al_2O_3/ZrO_2	2,000	–
$Al_2O_3/ZrO_2/TiO_2$	4,000	–

In the first experimental series, annealing has been shown to result in reduced residual stress level [42] and improved strength of polished sapphire plates [40]. The flexural strength of 2 mm thick polished plates of 38 mm in diameter was examined. The average strength has been determined to amount to 528 MPa. The strength value of scatter attained was 17.7%. Annealing provides the average strength enhancement up to 692 MPa, that is, by 31%, and the strength scatter reduction down to 14%. In Fig. 7.26, the failure probability of unannealed and annealed polished sapphire plates as a function of the loading applied is shown.

The plates annealed preliminarily at 870, 1,170, and 1,720 K were studied. In the authors' opinion, a 4-h annealing often is sufficient to reduce or remove the stresses arising due to machining. The reduction of residual stresses also provides an improved planeness of thin sapphire plates, especially when the plate surfaces are treated in different manners, e.g., when one plate side is polished and another ground.

The strength improvement of the annealed plates makes it possible to diminish the thickness thereof, thus reducing the cost and weight and improving the transmittance.

The second group of experiments has shown [39, 44] (Fig. 7.27) that when the compressive force is applied along the *C* direction, the strength drops considerably more steeply as the temperature increases than in the case of loading along the prism direction *A*-prism. This is due to the rhombohedral twinning.

The twinning along one rhombohedron plane does not cause any substantial change in the crystal strength; however, the twinning in two crossing rhombohedron planes results in a crack origination in the intersection point. At four-point bending for case 3 (when the sample length is parallel to the *C*-axis), the strength drops most steeply in the temperature interval 290–1,270 K. In case 4 (when there are no compressive stress components along the *C*-axis), the strength rises within the same temperature interval. For one and two orientations, the strength starts to decrease at $T \geq 770$ K.

The data presented are related to pure and rather perfect crystals where the twinning occurs easily. Defects (in first turn, the Mg^{2+} and Ti^{4+} impurities) cause a substantial increase of the compressive strength along the *C*-axis. Introduction of Cr^{3+} results in an increased flexural strength within the temperature range of 290–1,270 K. The same results have been obtained for introduction of boron, silicon, iron, and so forth. The impurities reduce the twinning probability.

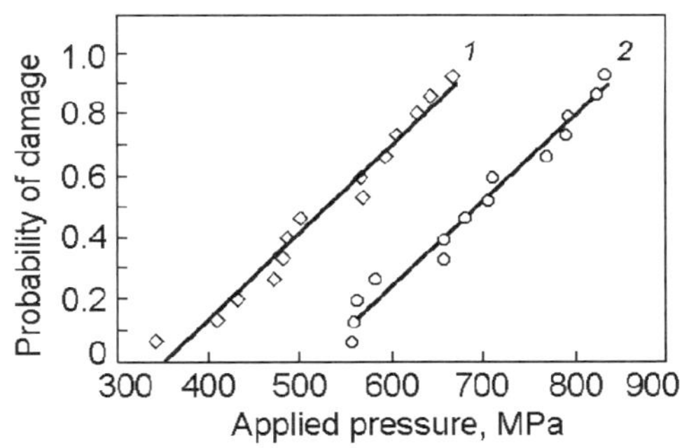

Fig. 7.26 Variation of sapphire breakdown probability at increasing stress applied to (*1*) unannealed and (*2*) annealed polished plates

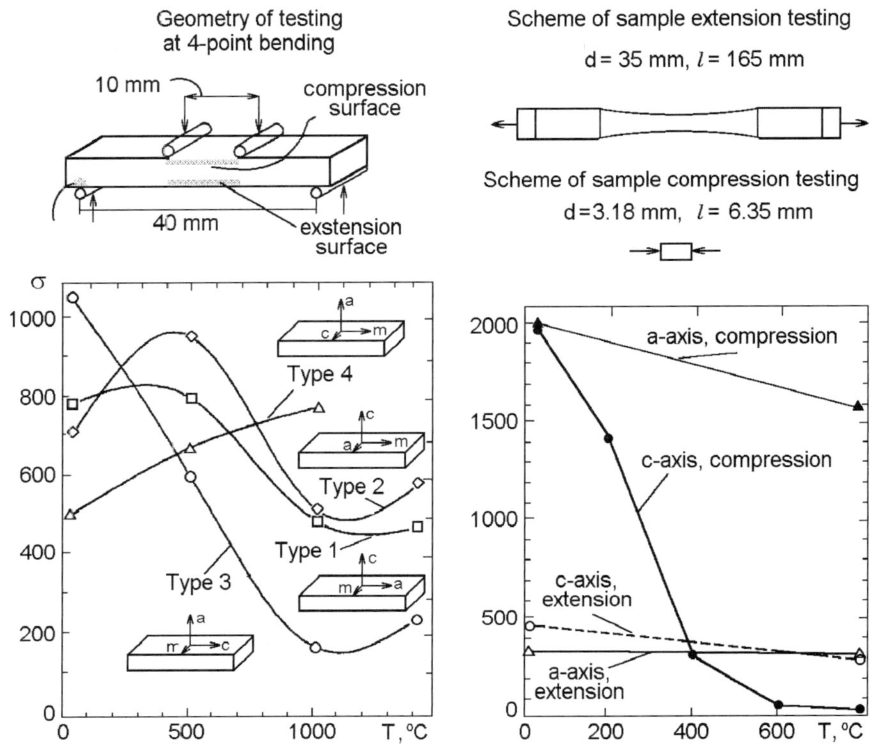

Fig. 7.27 Mechanical strength (σ_s) of sapphire measured under four-point bending (*left*), tension and compression (*right*)

The surface state influences the strength, too [41]. For example, the ground samples show the compressive strength at 870 K 2.2 times higher than those polished on all surfaces. The preliminary annealing increases the strength measured at elevated temperatures. The samples were annealed under hydrogen, argon, and oxygen as well as in air at 1,070, 1,720, and 2,020 K, the annealing duration being varied from 4 to 48 h. Annealing in oxygen flow at 1,720 K for 48 h results in the maximum strength improvement. Such annealing provides the sapphire compressive strength increased by 60%.

7.12 Effect of Annealing on the Optical Inhomogeneity

The optical inhomogeneity in doped crystals is due to the dopant distribution gradient and to structure defects. The dopant distribution is connected with the crystallization front shape [45]. To establish that connection, three sets of laser ruby rods have been grown by the Verneuil technique at various isotherm shapes at the crystallization front:

I – Concave–convex isotherm
II – Convex isotherm
III – Almost flat isotherm

The crystals have been annealed in vacuum at temperatures 100 K and 10 K lower than T_m. As the T_a approaches the T_m, the structure rebuilding processes are activated nonlinearly. In the crystals annealed at $T_2 = T_m$–10 K, the monotonous inhomogeneity of refractive index (the difference of indices at T_2 and T_1) is insignificantly changed compared with crystals annealed at $T_1 = T_m$–100 K for the group I crystals (Δn = 1.04); it becomes somewhat less for the group II ones (Δn = 0.88) and decreases considerably for the group III samples (Δn = 0.63). The samples of group I annealed at T_1 show asymmetrical interference rings (Fig. 7.28). A considerable discrete inhomogeneity of refractive index is observed at the sample periphery, which is especially visible in the extraordinary ray (Fig. 7.28b). After the annealing at T_2, the rings become axially symmetric and the discrete inhomogeneity of refractive index decreases, particularly at the periphery (Fig. 7.28).

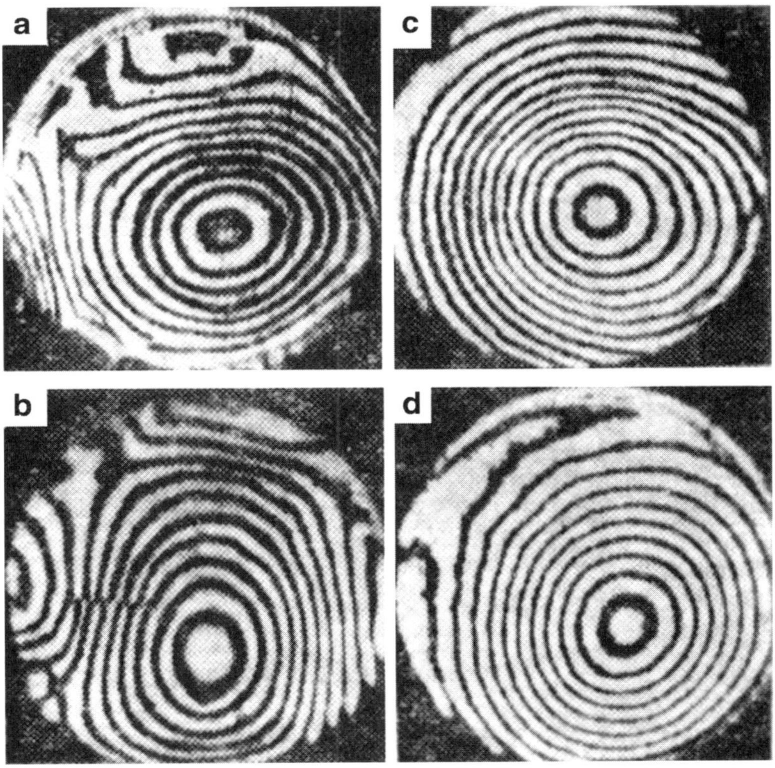

Fig. 7.28 Optical inhomogeneity of the first group ruby after annealing at $T = T_m - 100$ K (**a, b**) and $T = T_m - 10$ K (**c, d**) in the (**b**) ordinary and (**d**) extraordinary rays

For the group II crystals annealed at T_1, the discrete inhomogeneity of refractive index is clearly pronounced (Fig. 7.29a, b). After annealing at T_2, the number of rings is reduced, their asymmetry drops sharply, and the discrete inhomogeneity of refractive index is decreased considerably. In the group III crystals that are characterized by an insignificant "chromium lens," the optical inhomogeneity after annealing at T_1 (Fig. 7.30) mainly is due to residual stresses that are responsible for the ring asymmetry and the discrete inhomogeneity of refractive index. After annealing at T_2, the number of rings, their asymmetry, and the discrete inhomogeneity of refractive index are reduced considerably (Fig. 7.31).

The optical inhomogeneity changes correlate well with the Cr^{3+} distribution. In the group I, the maximum concentration is observed in the center and at the periphery of the sample cross section; in group II, it is at the periphery; while in group II, chromium is distributed rather homogeneously over the entire sample cross section.

It is the residual stresses and the Cr^{3+} ions that contribute to the optical inhomogeneity of ruby; furthermore, the chromium ions strengthen the crystal lattice. Therefore, as the T_a approaches the premelting value, the thermally activated relaxation

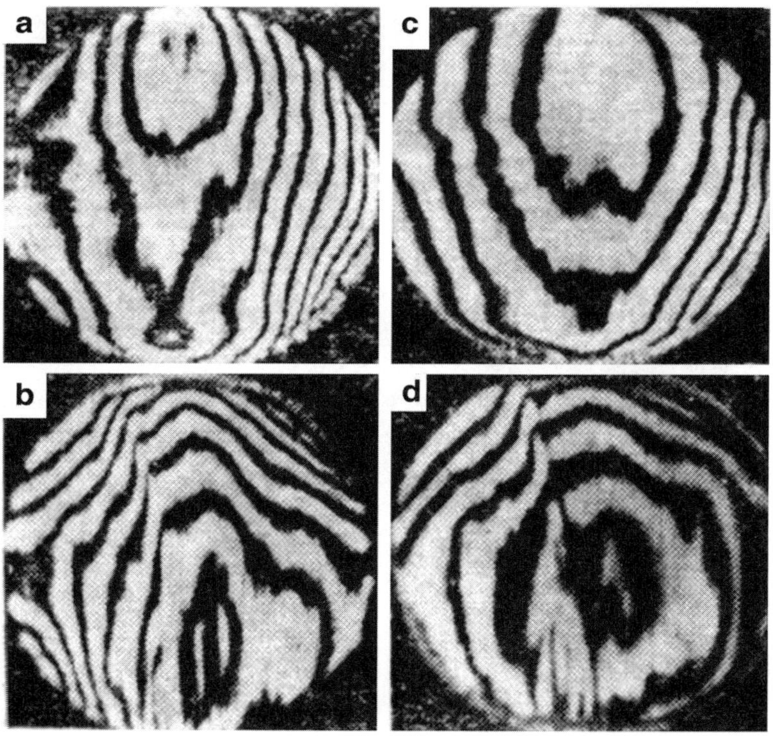

Fig. 7.29 Optical inhomogeneity of the second group ruby after annealing at $T = T_m - 100$ K (**a, b**) and $T = T_m - 10$ K (**c, d**) in the (**b**) ordinary and (**d**) extraordinary rays

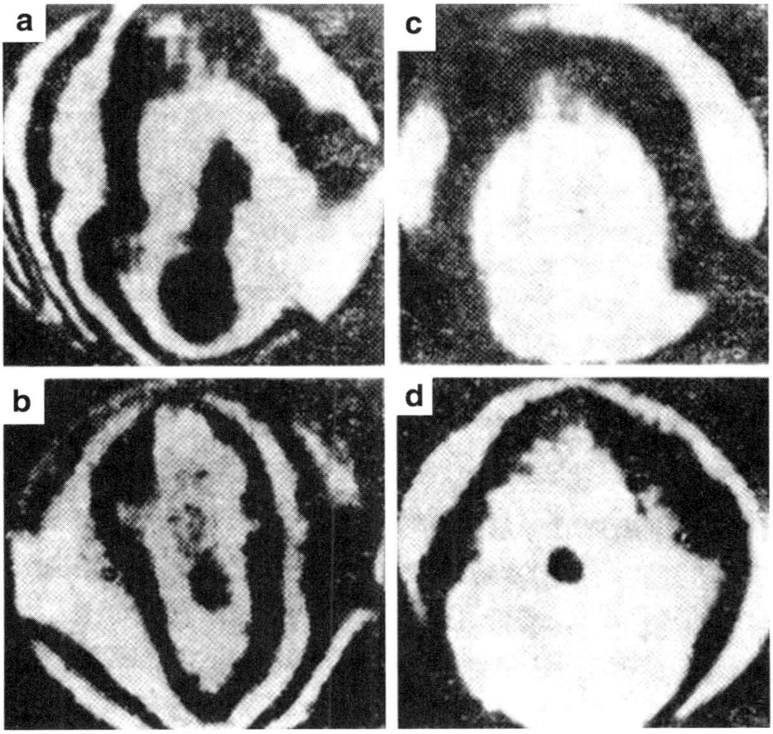

Fig. 7.30 Optical inhomogeneity of the third group ruby after annealing at (**a, b**) $T = T_m - 100$ K and (**c, d**) $T = T_m - 10$ K in the (**b**) ordinary and (**d**) extraordinary rays

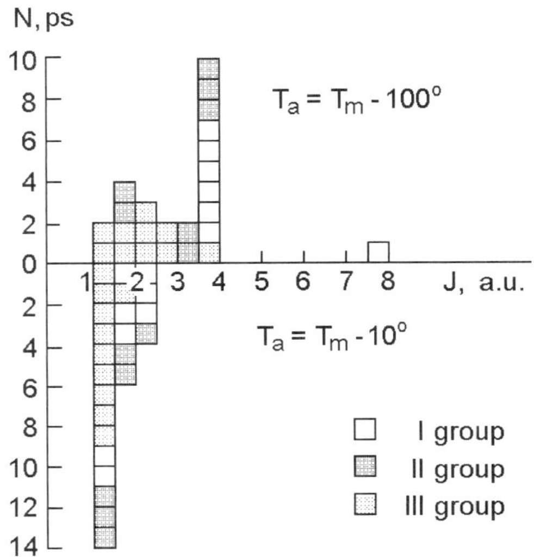

Fig. 7.31 Discrete inhomogeneity of refractive index for ruby annealed at various temperatures

processes are manifested most effectively in the crystal parts with minimum Cr^{3+} content. The residual stresses are more than halved after annealing at T_2 compared with that at T_1 (Fig. 7.22).

In the crystals with homogeneous chromium distribution (group III), the substantial improvement in optical homogeneity after annealing at T_2 is due to the residual stress reduction. As for the samples of other groups, the monotonous inhomogeneity of refractive index due to inhomogeneous chromium distribution is changed by the same quantity as for the group III ones; however, the relative change Δn is much less, since there is no substantial equalizing of chromium content over the cross section caused by diffusion.

7.13 Effect of Annealing on the Small-Angle Light Scattering in Crystals

The small-angle scattering (SAS) defines to a great extent the laser characteristics crystals. The SAS is connected with the crystal optical inhomogeneity and can be reduced by vacuum annealing at a premelting temperature. The reduction extent also depends on the initial crystal optical quality (Table 7.13).

In the crystals with low optical inhomogeneity ($\Delta n < 5 \cdot 10^{-6}$, $j_0 < 2$), the σ and j values remain almost unchanged, while the monotonous inhomogeneity of refractive index decreases considerably. The constancy of σ and j is explained by the relatively low initial density of defects.

In the ordinary optical quality crystals ($5 \cdot 10^{-6} \leq \Delta n < 14 \cdot 10^{-6}$, $2 \leq j < 3.5$ the sample cross-section area fraction (S) where the monotonous inhomogeneity of refractive index is $(1–2) \cdot 10^{-6}$ is about 20–30%), the σ, Δn, and j values are reduced appreciably after annealing. The contribution of inhomogeneous Cr^{3+} distribution to Δn is still not too large; therefore, the residual stress relaxation causes an appreciable decrease of the monotonous inhomogeneity of refractive index.

The maximum reduction of σ and j is typical of optically imperfect crystals ($\Delta n \geq 14 \cdot 10^{-6}$, $j > 3.5$, $S > 40\%$), although Δn does not change appreciably in that

Table 7.13 Effect of annealing at premelting temperature $T_2 = T_m - 10°$ (±3°) on the residual stresses (σ), optical inhomogeneity (Δn), and discrete inhomogeneity of refractive index (j) [46]

Prior to annealing			After annealing					
σ_0 (cm^{-1})	$\Delta n_0 \cdot 10^6$	j_0	σ_2 (cm^{-1})	$\Delta n_2 \cdot 10^6$	j_2	σ_2/σ_0	$\Delta n_2/\Delta n_0$	j_2/j_0
0.008	2.6	1.8	0.009	1.76	1.8	1	0.63	1
0.026	4.4	1.9	0.030	2.64	1.78			
0.039	8.8	2.83	0.031	5.28	2.24	0.75	0.63	0.67
0.049	10.6	3.0	0.035	7.04	2.0			
0.053	14.1	3.8	0.038	14.1	1.87	0.65	1	0.5
0.064	17.6	4.0	0.037	15.84	2.0			

case. The Δn constancy is due to the fact that the main contribution to the monotonous inhomogeneity of refractive index is caused by the inhomogeneous Cr^{3+} distribution that remains unchanged under annealing even at premelting temperatures.

The most probable structural defects in crystal responsible for the small-angle light scattering are the block boundaries that are manifested in the interference patterns in the extraordinary ray as local distortions of the interference bands that correspond to the refractive index changes by about $2 \cdot 10^{-6}$. After annealing at a premelting temperature, the local band distortion becomes weakened, thus demonstrating the reduced discrete inhomogeneity of refractive index connected with the residual stress relaxation along the block boundaries and reduced length of the boundaries themselves.

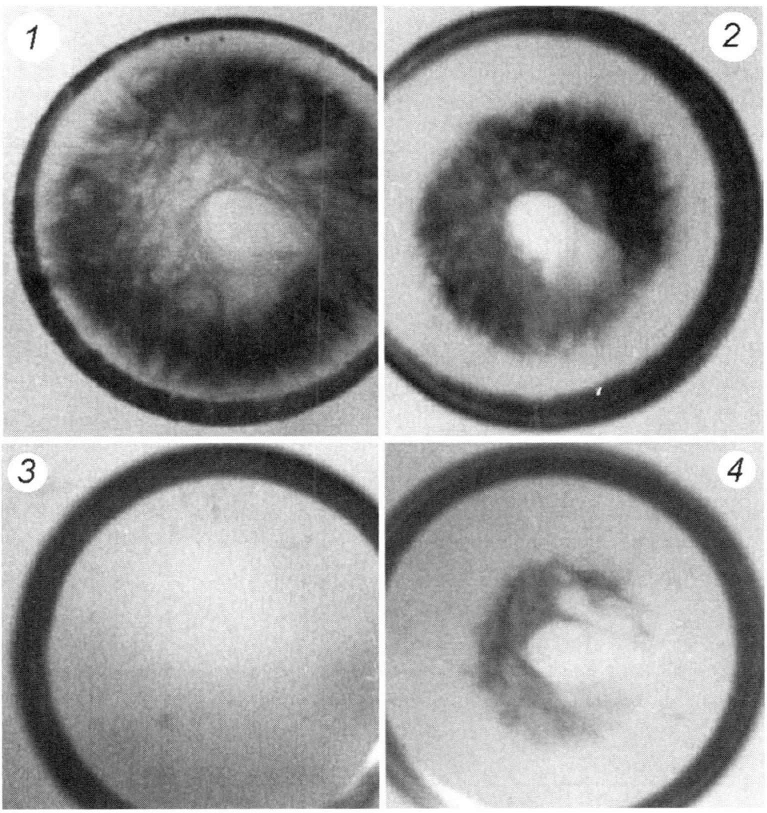

Fig. 7.32 Distribution of foreign-phase inclusions in sapphire rod of 14 mm diameter: (*1*) prior to annealing; (*2*) after 7 h of annealing; (*3*) after 12 h of annealing; and (*4*) after 15 h of annealing

The sapphire rods grown using the Stepanov technique under carbon-containing atmosphere may contain a high density of scattering centers (Fig. 7.32), especially when the growing has been carried out along the *C*-axis [17].

The presence and intensity of the color center (CC) depended on the crystallization rate and crystallographic direction. Therefore, the absence of Tindall light scattering testified to the absence of foreign-phase inclusions. In these crystals CC had optical absorption in 330–450 nm wavelength region with an absorption maximum at 387 nm. Annealing of the crystals in saturated vapors of the thermal dissociation products of Al_2O_3 ($\varepsilon = 0$ kJ/mol) at 2,070–2,300 K allow the destruction of the defects and achievement of transparency of the crystals in the visible region of the spectrum. The kinetics of the destruction of the inclusion and CC at such annealing has a diffusive character. As the crystal becomes transparent, starting from its surface, then the transparency extends into its bulk in the process of annealing. This allows the conclusion that the inclusions and CC with an absorption band at 330–450 nm are based on the complexes formed by impurity ions and anion vacancies. The foreign-phase inclusions appear in the crystals with anionic-type stoichiometry violation when the impurity concentration is higher than a critical value. Annealing of the crystals in the medium with low chemical potential ($-50 \leq \varepsilon \leq 50$ kJ/mol) reduces the degree of the stoichiometry violation and favors the destruction of the foreign-phase inclusions and CC based on the impurity-vacancy complexes [17].

7.14 Effect of High-Temperature Annealing on the Light Transmittance of Machined Surfaces

The defect surface layer becomes as if it were "recrystallized" under high-temperature annealing, so that the transmittance increases. In Table 7.14, the effect of annealing in saturated vapors of aluminum oxide and its thermal dissociation products at 2,250 K under residual press of $7 \cdot 10^{-4}$ mmHg on the optical characteristics of sapphire samples with different surface roughness is presented. The annealing atmosphere can prevent the sapphire evaporation.

In all the samples, one surface was polished up to $R_z = 0.05$ μm and the second surface roughness was varied. At the measurements, the incident light beam ($\lambda = 0.633$ μm)

Table 7.14 Dependence of transparence (T_t) and integral light transmission (T_i) of sapphire samples ($d = 20$ mm, thickness 3 mm) with various roughness on the annealing duration [47]

	Imperfect layer	$t = 0$		$t = 1.5$ h		$t = 8$ h	
R_z (μm)	depth (μm)	T_t	T_i	T_t	T_i	T_t	T_i
6.3–10	70	0.233	1.00	0.413	2.836	0.476	3.540
1.1–2.5	50	0.440	2.973	0.641	4.305	0.676	4.975
1.1–0.2	35	0.958	4.448	0.954	3.801	0.944	3.323

was directed onto the polished surface. The annealing resulted in a considerable improvement of the transparence and integral transmittance (the scattering indicator area of the light beam passed through the sample) of the samples with 1.1- to 10-μm roughness, the maximum changes occur in the samples with the maximum roughness; the samples with 1.1- to 0.2-μm roughness show some deterioration of characteristics.

The surface layer recrystallization results in a reduced number of multiple reflections that contribute mainly to the transparence reduction of the ground surface. The largest changes are observed during the initial annealing stages (up to 1.5 h) when the evaporation of abrasive inclusions proceeds in parallel to the recrystallization. In the low-roughness samples, the surface thermal etching competes with those processes. After 8 h of annealing at T_i, the maximum roughness is attained (R_z= 1.1–2.5 μm). This fact becomes understandable when taking into account the intensity variation of the incident light beam reflected from the ground surface due to differences in roughness and orientation of elementary reflecting areas. At $R_z \leq$ 0.75 μm and the area element tilt exceeding 45°, the intensity of light reflected from the sapphire surface is several fractions percent, while the Fresnel reflection for polished surfaces of that sample is 7.5%. The profile patterns show that it is just the samples with R_z= 1.1–2.5 μm where up to 50% of area elements are oriented at 50° to the surface. No dependence of the optical parameter variations on the sample crystallographic orientation has been revealed.

7.15 Annealing under Loading

The vacuum annealing of Ti-sapphire results in diffusion of anionic vacancies from the matrix and dissolution of aluminum atoms (being contained in the submicrometer-sized inclusions) into the crystal matrix. At a certain annealing stage, those inclusions become transformed into submicrometer pores containing saturated aluminum-oxide vapor. The pores in a crystal are known to be diffusively dissolved

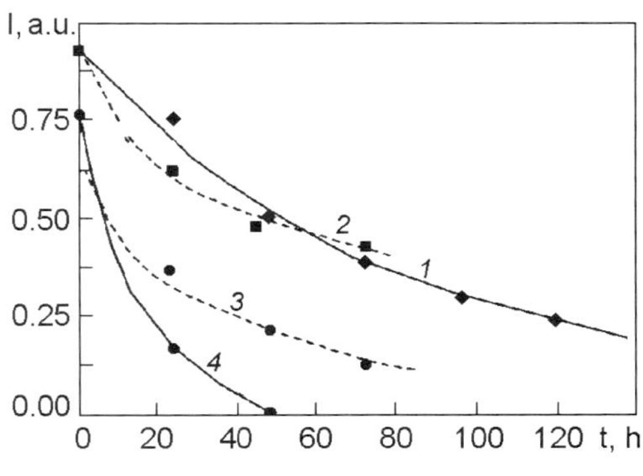

Fig. 7.33 Destruction dynamics of submicrometer inclusions at Ti-sapphire vacuum annealing without loading (*1*) and under uniaxial compressive stress = 7.8105 N/m² applied along different crystallographic directions: (*2*) [11$\bar{2}$0], (*3*) [0001] (*4*) [1$\bar{1}$02]

or to be cured according to the dislocation mechanism. The diffusive dissolution of a pore is due to an increased vacancy concentration at its surface. In its turn, a constant uniaxial load applied to the crystal favors the diffusion processes and the diffusion-dislocation creeping, thus increasing the pore dissolution rate. The destruction dynamics of the submicrometer-sized inclusions in the crystal was monitored using the decrease of optical scattering intensity that is in proportion to the scattering center concentration [36].

The dislocations in corundum single crystals show a maximum mobility under cleaving stresses directed along the (0001) plane belonging to the easiest sliding system. Under uniaxial stresses along the $[11\bar{2}0]$ and [0001] directions, those cleaving stresses are minimal, the dislocation motion in the (0001) plane is hindered, and the pore is cured mainly due to the vacancy-by-vacancy dissolution. This is confirmed by the fact that the submicrometer-sized inclusions are dissolved at the same rate during the sample annealing without any loading and under uniaxial compressive stresses acting along $[11\bar{2}0]$ and [0001] (Fig. 7.33). The submicrometer pores are destroyed most intensely at the high-temperature vacuum annealing of the crystal under uniaxial compressive stresses applied along the $[110\bar{2}]$ crystallographic direction, since in this case, both mechanisms of pore dissolution are active.

References

1. Dobrovinskaya E.R. Control of Structural Perfection of Single Crystals of a Corundum. Moscow: Niitehim, 1979, 68p [in Russian].
2. Ruby and Sapphire. Moscow: Nauka, 1974, 235p [in Russian].
3. Klassen-Nekludova M.V. ZhETF. 12(9), 1942, 519 [in Russian].
4. Dobrovinskaya E.R. Rost i defekty metallicheskikh kristallov [Growth and Defects of Metal Crystals]. Kiev, 1972, 284.
5. Geguzin Ya.E., Matsokin V.P. Fiz. Tverd. Tela. 8(9), 1966, 2558.
6. Klassen-Nekludova M.V., Govorkov V.G., Urosovskaya A.A. Single Crystals. 39(2), 1970, 679 [in Russian].
7. Dobrovinskaya E.R., Pishchik V.V., Litvinov L.A. Method of Definition of Diffusion Constants in Crystalline Bodies with Impurity Inhomogeneity. USSR Patent No. 1548709, MKI G 01 N 13/00. 1990. No. 9.
8. Semiletov S.A., Bagdasarov X.S., Popkov V.S., Magomedov A. Fiz. Tverd. Tela. 10(1), 1968, 71 [in Russian].
9. French T.M., Somorjal C.A. J. Phys. Chem. 74(12), 1970, 2489.
10. Pavilainen S.V., Timofeev V.N. Influence of Annealing on Depth of the Broken Stratum of Single Crystals of Silicon after an Abrasion Fiz. i Khim. Obrab. Mater. 4, 1973, 80 [in Russian].
11. Timan B.L., Dobrovinskaya E.R., Babiichuk I.P. Fiz. Tverd. Tela. 16(6), 1974, 1792 [in Russian].
12. Lykov A.V. The Theory of a Thermal Conduction. Moscow: Gosanergoizdat, 1952, 344p [in Russian].
13. Evans A.G., Raivlings R.D. Phys. Stat. Sol. 34(1), 1969, 9.
14. Kulikov I.S. Thermodynamics of Oxides. Moscow: Metallurgiya, 1986, 342p [in Russian].
15. Krivonosov E.V. Funct. Mater. 1(2), 1994, 6670.

16. Krivonosov E.V. Influence of Conditions of Annealing on the Optical and Mechanical Performances of Single Crystals of a Corundum. Abstract of Ph.D. (Engineering) Thesis. Kharkiv, 1989, 20p [in Russian].
17. Andreev Y.P., Kryvonosov Y.V., Lytvynov L.A., Vyshnevskiy S.D. Funct. Mater. 12(1), 2005, 142–145.
18. Konevsky V.S., Krivonosov E.V., Litvinov L.A. Izv. AN SSSR, Ser. Neorg. Mater. 11, 1983, 1939–1940 [in Russian].
19. Dan`ko A. Ya., Rom. M.A., Sidelnikova N.S. et al. Abstract of 12th National Conference on Crystal Growth, Moscow. 2006, p. 162.
20. Grass V.E., Sitnikov P.A., Istomin P.V. et al. Abstract of 11th National Conference on Crystal Growth, Moscow. 2004, c. 503.
21. Konevsky V.S., Krivonosov E.V., Litvinov L.A. Izv. AN SSSR, Ser. Neorg. Mater. 25(9), 1989, 1486–1490 [in Russian].
22. Vishnevsky S.D., Krivonosov E.V., Litvinov L.A. Funct. Mater. 10, 2003, 238–241 [in Russian].
23. Kofstad P. Stoichiometry Deviations, Diffusion and Electric Conductivity in Simple Metal Oxides. Moscow: Mir, 1975, 396pp.
24. Shumon P. Diffusion in Solids. Moscow: Metallurgia, 1966.
25. Krivonosov E.V., Litvinov L.A. Funct. Mater. 3, 1996, 77–80.
26. Evans B.D., Stapelbrock M. Phys. Rev. B. 18, 1978, 7089–7098.
27. Gorokhovsky A.A., Konevsky V.S., Krivonosov E.V. et al. Zh. Prikl. Spektr. 42, 1985, 670–672 [in Russian].
28. Dobrovinskaya E.R., Litvinov L.A. Optiko-Mekhan. Prom. (7), 1991, 32-34 [in Russian].
29. Konevsky V.s., Litvinov L.A. Zavod. Lab. (9), 1982, C.70-72 [in Russian].
30. Konevsky V.S., Krivonosov E.V., Litvinov L.A. Funct. mater. 5, 1998, 521–524.
31. Askulyonok E.M. et al. Neorg. Mater. 6, 1970, 158 [in Russian].
32. Berezina I.E. et al. Izv. AN SSSR, Ser. Fiz. 49, 1985, 2398 [in Russian].
33. Starostin M.Y.U. et al. Izv. RAN. Ser. Fiz. 63, 1999, 1747 [in Russian].
34. Ivanov V.I. et al. Izv. RAN. Ser. Fiz. 58, 1994, 63 [in Russian].
35. Nosov Yu. G., Antonov P.I. Abstract of 10th National Conference on Crystal Growth, Moscow. 2002, 251.
36. Vyshnevskiy S.D., Kryvonosov Y.E.V., Lytvynov L.A., Vyshnevskiy S.D. Funct. Mater. 13(2), 2006, 238–244.
37. Yoshikawa A., Hasegawa K., Fukuda T. Book of Lecture NOtes. First International School on Crystal Growth Technology, Switzerland. 1998, 769.
38. Harris D.C., Schmid F., Black D.R. et al. SPIE. 3060, 1997, 226.
39. Harris D.C. SPIE. 3705, 1999, 2.
40. Borden M.R., Askinazi J. SPIE. 3060, 1997, 246.
41. Schmid F., Schmid K., Khattak C.P. et al. SPIE. 3705, 1999, 36.
42. Smith M.B., Schmid K.A., Schmid F. et al. SPIE. 3134, 1997, 284.
43. McClure D.R., Cayse R., Black D., Goodrich S. SPIE. 4375, 2001, 20.
44. Harris D.C., Johnson L.F. SPIE. 3705, 1999, 44.
45. Konevsky V.S., Krivonosov E.V., Litvinov L.A. Optiko-Mekhan. Prom. (9), 1982, 35–37 [in Russian].
46. Konevsky V.S., Krivonosov E.V., Litvinov L.A. Optiko-Mekhan. Prom. (11), 1982, 30–31 [in Russian].
47. Konevsky V.S., Krivonosov E.V., Litvinov L.A. Optiko-Mekhan. Prom. (11), 1983, 25–27 [in Russian].

Appendix

Table 7.15 Vacancy centers in sapphire

Defect type	Optical absorption, luminescence, polarization	Formation conditions	Observation conditions	Decomposition temperature (K)
$V_O^{2+} + e^- \equiv F^+$	Absorption 4.8 eV; luminescence 3.75 eV; A_a = 1.25; polarization A_p = 1.2; 5.4 eV; A_p <0.6	Neutron irradiation; additive coloring	77 K; 300 K; optical excitation at 4.8 eV; 5.4 eV; 6.1 eV; X-ray excitation	950
$V_O^{2+} + 2e^- \equiv F$	Absorption 6.1 eV; A_a = 0.96	Neutron irradiation; additive coloring	77–1,600 K	1,600
$V_{Al}^{3-} + h^+ \equiv V^{2-}$	Absorption 3.05 eV; A_a > 0.96	Oxidizing annealing and irradiation	After heating above 350 K	550
$V_{Al}^{3-} + 2h^+ \equiv V^-$	Absorption 3.05 eV; A_a > 0.96	Oxidizing annealing and irradiation	350 K	380

Note: A_a is the polarization absorption ratio, $A_a = k_{\gamma\|}/k_{\gamma\perp}$; A_l is the polarization luminescence ratio, $A_l = I_\|/I_\perp$, where $k_{\gamma\|}$ and $k_{\gamma\perp}$ are absorption coefficients; $I_\|$ and $I_\perp$ are luminescence intensities when the vector **E** is parallel to the crystal *C* axis and perpendicular thereto, respectively.

Chapter 8
Methods for Obtaining Complex Monolithic Sapphire Units and Large-Size Crystals

The present-day state of crystal growth technologies makes it possible to obtain sapphire products of rather large size and complex configuration. However, demand has arisen for super-large sapphire crystals and complex single-piece units, which cannot be grown in practice but must be assembled from separate components. In addition, sapphire-metallic and sapphire-ceramic joints are demanded, the main requirements for which include vacuum, electrical, and mechanical strength, high transparency, wear resistance, chemical stability, and so forth.

Naturally, the opportunity to increase the size of grown crystals is limited. In the next few years, there clearly will be a rise of some 10–15%. However, the growth of these crystals may turn out to be economically inexpedient. So, a search has begun for new technologies yielding large-size crystals and units based on the fabrication of single-piece joints.

8.1 Creation of Single-Piece Crystalline Joints

Single-piece joints can be obtained in the following ways: gluing, soldering, and welding. The third method is the creation of single-piece joints by means of local fusion/melting or simultaneous deformation resulting in the formation of strong bonds between the atoms of the joined components. There are two types of joints created by welding. One of these implies melting in the contact zone (zone melting by external heaters with different heating mechanisms, such as electron beam [1, 2] or laser [3]). The other is that obtained in solid-state under the action of various forces on the components to be joined, such as thermal diffusion or thermal compression, explosion, magnetic pulse, and friction. These methods are based on plastic deformation of the material in the junction zone.

E.R. Dobrovinskaya et al., *Sapphire: Material, Manufacturing, Applications*,
DOI: 10.1007/978-0-387-85695-7_8, © Springer Science+Business Media, LLC 2009

8.2 Gluing of Sapphire

Gluing of sapphire has performed well in making constructions meant for protection from bullets and shell splinters, as well as for laser epilation devices. Strong seams are achieved by special treatment of the component surfaces to be joined (mechanical cleaning, degreasing, etc.), by the development of components with large cementing surfaces, and by ensuring favorable conditions for distribution of shear and break-off forces on the seams. As a rule, adhesion of glue to the cementing surface exceeds cohesion inside the cementing film, so the joints/seams must possess minimal thickness.

According to the nature of the main component, adhesives are classified as inorganic and organic. Inorganic adhesives include frits (aqueous suspensions of compositions containing oxides of alkaline and alkaline-earth metals), while organic adhesives are organosilicon, organoboron, or organometallic compositions. These possess rather high-heat resistance and thermal stability, strong junction at short-term heating up to $T \sim 1{,}000°C$, and ability to withstand durable heating at 400–600°C.

The main adhesive components are synthetic polymers [4] which have a number of advantages over natural products. These include high strength, short drying time, high elasticity and transparency of adhesive film, and the absence of odor or other fumes (excluding some synthetic adhesives).

Constructional and special adhesives exist [5]. The former are the compositions that provide transfer of dynamic and static loads. These must possess high strength and the absence of creep under durable load. Special adhesives are those compositions that have additional properties, such as optical adhesives or medical adhesives used with complex sapphire implants. Optical-quality joints are obtained by cementing together sapphire components using an adhesive with a refractive index equal to that of sapphire. Some optical adhesives solidify under the action of ultraviolet radiation; as a rule, they are viscous. There is no universal adhesive for sapphire. So, in each case the composition of adhesive and the technology of cementing are specially chosen by application.

The tendency to cement sapphire components rather than soldering or welding has become ever more pronounced [6], as cementing does not require heating and does not change the structure or break continuity of the material. Moreover, it allows soft and vibration-resistant joints to be obtained.

8.3 Soldering of Sapphire

Soldering involves a liquid solder interlayer between the surfaces to be joined. The solder wets these surfaces and spreads within a thin gap between the components. After crystallization a joint appears where chemical bonds are formed between the surface atoms (ions) of the components to be joined and the solder [7–10]. The joint strength depends greatly on the gap thickness between the components (0.03–0.2 mm), the surface roughness, and the uniformity of heating of the components. To remove oxide film and provide protection from the atmosphere, fluxes are used. The process of soldering is complicated by the presence of surface substances, which impede wetting of solder over the gap between the components. Liquid solder metal also oxidizes and may not spread over the surfaces. Thus, nonsoldered areas may arise. Only gold and sometimes silver can be used without fluxes to remove oxide layers. Often, prior to soldering sapphire articles, their surfaces are coated with thin gold layers. Although such a method is expensive, it is still used, particularly in the production of semiconductor devices.

During soldering [11] gaps to be filled with liquid solder must be provided. Capillary soldering is based on the ability of liquids to spontaneously penetrate into capillaries due to surface tension forces in the process of wetting the surface (Fig. 8.1). Solder does not penetrate into either large or small gaps. In the case of noncapillary gaps, solder is confined within gaps by using compositional solders containing filling agents for the formation of capillaries in the gaps. Such solders are obtained by sintering powders or fibers of a high-melting material with subsequent impregnation of a liquid-fusible phase (Fig. 8.2).

Joining of sapphire with titanium, steel, copper, and other metals is realized by means of contact-reactive soldering. This technique is based on the ability of some metals to form contact alloys (eutectics or solid solutions), which have melting temperatures lower than those of the materials to be joined. If the latter do not form alloys of the said type, they can be joined by means of an interlayer or special surface coating. This method also is used for joining sapphire components. At the initial step of contact soldering a physical contact is formed between the surfaces through active centers. Throughout the course of soldering, the quantity of such contacts increases. Squeezing removes the surplus liquid phase together with oxide particles, which may reduce the seam strength. The liquid interlayer is more active than solder: it wets and dissolves the main material more readily. Soldered joints have low levels of residual stress, as the entire construction undergoes heating with

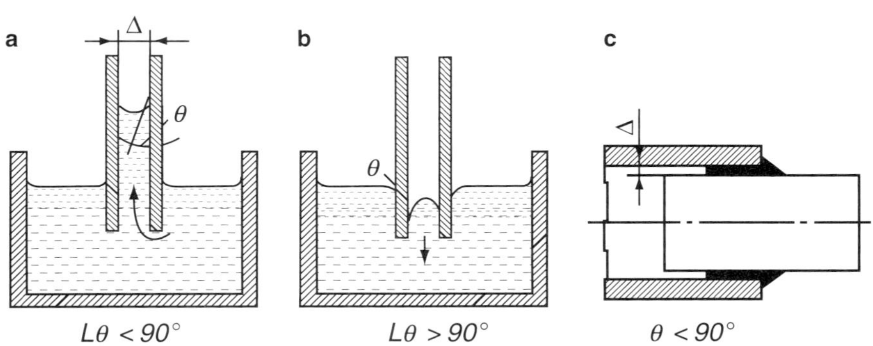

Fig. 8.1 Interaction of liquid solder with solid gap or capillary surfaces (**a**) with wetting and (**b**) without wetting. (**c**) Scheme of soldered joint. θ is the angle of wetting

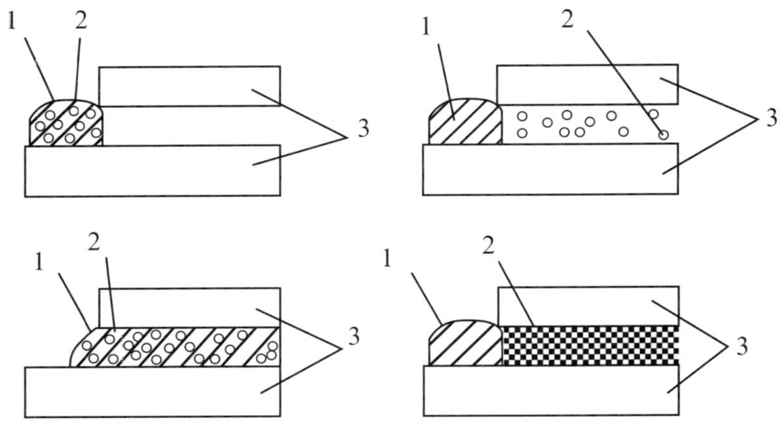

Fig. 8.2 Location of the filling agent (*1*) and the fusible component (*2*) of compositional solder between the plates (*3*)

subsequent cooling. The methods of soldering are simple and do not change the structure of the joined components, but the resulting products have low mechanical and electric strength. Low-temperature soldering implies the use of solders with $T_m < 820$ K. For high-quality soldering the temperatures of the soldered surfaces must be equal. As a rule, the temperature of soldering does not exceed 1,800 K. Diffusion of solder components into the crystal, particularly during thermal cycling of the products, worsens the working characteristics of the finished products [12].

The distinction between the processes of soldering and welding consists of the fact that the former implies melting of the edges of the components to be joined, whereas the latter process requires only melting of the solder, which T_m is lower than the sapphire components. Moreover, with soldering, overlapping seams are predominantly made, and this means elevated consumption of material.

8.4 Welding of Sapphire

During the past few years, special attention has been paid to the methods of welding, which result in a seam strength equal to that of the main material. We shall not consider all the existing trends and technical approaches (the number of recently granted patents devoted to this problem already has exceeded 200). Rather, we would like to focus on the technologies that seem to be the most efficient: diffusion welding and the method of welding via displacement of the local melted zone.

Mechanical treatment of ingots leads to the appearance of macroscopic and microscopic inhomogeneities on their surfaces. Such surfaces can make contact only at separate points (Fig. 8.3).

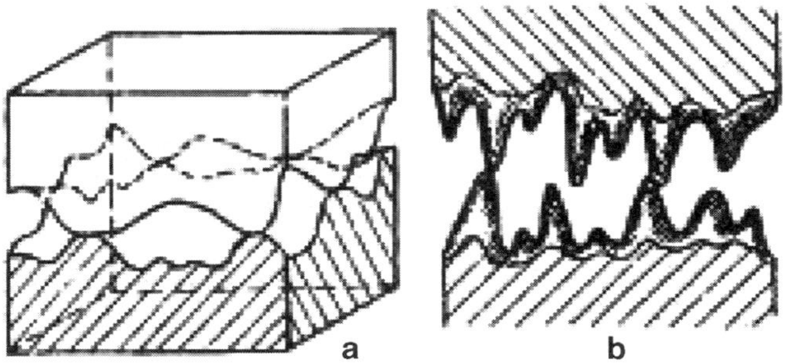

Fig. 8.3 Model of contact between solids at (**a**) macroscopic waviness and (**b**) microscopic roughness of their surfaces

The process of welding is complicated by the presence of oxides, adsorbed gases, and moisture and organic (fatty) contaminants on the contact surfaces. It should be noted that on cleaned surfaces kept in air monomolecular gas layer is formed within $2.4 \cdot 10^{-9}$ s.

Even after polishing, the size of surface microasperities exceeds the crystal lattice parameters. Therefore, welded joints can be formed only through plastic deformation of the welded surfaces under the action of compression pressures or through their melting with subsequent formation of a common welding bath. Diffusion welding techniques exist in which the joined components are heated up to $T \geq T_m$, and solid-phase welding techniques (without melting the joined components), such as pressure or explosion welding, point welding, and so forth, exist.

8.4.1 Diffusion Welding

In 1973, the International Welding Institute (Dusseldorf) gave the following definition of diffusion welding: "Diffusion bonding of materials in the solid state is a process for making a monolithic joint through the formation of bonds at the atomic level, as a result of closure of the mating surfaces due to the local plastic deformation at elevated temperature which aids interdiffusion of the surface layers of the materials being joined." The above-mentioned method makes it possible to obtain a bond with strength equal to that of the main material.

During diffusion welding a good deal of effort is undertaken to prepare the surfaces to be joined [13]. However, even the most thorough preparation cannot provide ideal contact. This is explained by the fact that mechanical treatment results in the formation of surface roughness with a microasperity height of $(0.3–1.0) \cdot 10^{-4}$ cm, whereas the interatomic attraction forces act when the distance between the surfaces is equal to $(1–5) \cdot 10^{-8}$ cm. This is why the contact of surfaces without plastic deformation leads to their fusion only at separate and distinct points, which is insufficient.

The formation of a strong joint may be considered a topochemical reaction of substitution or addition [14], which proceeds on the surface of a solid and implies the breakage of bonds in the initial substances and the formation of new bonds leading to the appearance of a new substance. The kinetics of this process depend on temperature, time, and pressure, as well as the presence of microdefects. Such reactions are stepwise [15].

Joining materials in the solid state takes place at separate active points. At these points, called active centers, the formation of physical contact is accompanied by the appearance of interaction seats where chemical bonds are formed between the atoms of the joined materials. In terms of the theory of phase transformations [16], such interaction seats may be considered to be analogous to the nuclei or centers of a new phase.

As the concentration of interaction seats is limited [17], the expansion rate of the interaction area, $S(t)$, in the process of formation, growth, and confluence of seats at first increases and then diminishes (Fig. 8.4a). Accordingly, the kinetics of the increase in strength will have the form of a curve with saturation (Fig. 8.4b), where $\sigma(t)/\sigma_m$ is the ratio of interaction area strength to total surface strength.

Joining homogeneous materials is always thermodynamically advantageous, since the energy of the boundary is essentially lower than the free surface energy. The process of diffusion welding is described in many papers [17–19]. Irrespective of the solid-state welding method, the mechanism of joint formation is to be considered to be a three-stage, topochemical reaction.

The first stage is the formation of physical contact, i.e., the approach of the atoms of the bonding surfaces by plastic deformation to a distance that will provide either physical interaction caused by van der Waals forces or weak chemical interaction. The latter is realized when the materials possess starkly differing properties and their approach is achieved by the plastic deformation of at least one of them.

The second stage is the formation of active centers. During welding of different materials, active centers arise on the surface of the material with the higher hardness.

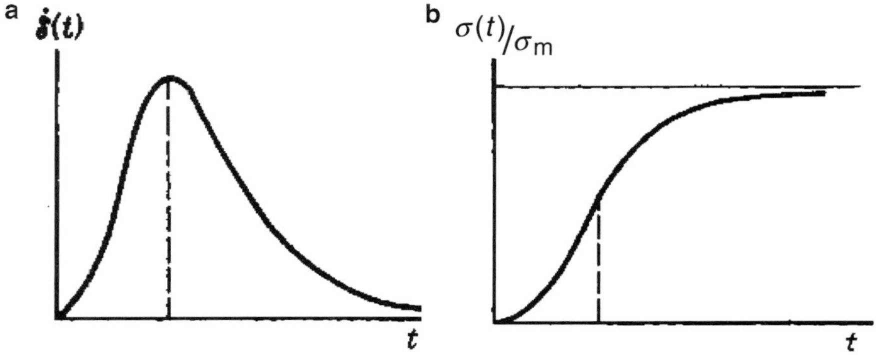

Fig. 8.4 Increase in the interaction area (**a**) and the relative joint strength (**b**) with time

The presence of such a stage and its duration depend on peculiarities of plastic deformation in the harder material. During welding of two sapphire components, the first and second stages are practically indistinguishable, since activation of the two contacting surfaces begins with flattening of microasperities through plastic deformation.

The third stage is bulk interaction, which immediately follows the formation of active centers and occurs both in the contact plane (where strong chemical bonds are formed) and in the bulk of the contact zone. This process is localized on the active centers, such as dislocations within a stress field. In the contact plane it ends when discrete interaction seats merge; in the bulk of the material its termination is characterized by relaxation of stresses. However, the formation of strong joints requires further development of relaxation processes such as heterodiffusion. With welding of homogeneous materials, recrystallization leading to the formation of common/mutual grains in the contact zone may be a criterion for the end of the third stage. With welding of inhomogeneous materials, the necessity to develop or restrain heterodiffusion depends on the properties of the diffusion zone and of the phases formed in it.

The methods of pressure welding [20–22] imply the formation of contact by deformation of the surfaces. However, complete contact cannot be achieved, since cavities between microasperities on the welded surfaces do not disappear. When the contact is being formed, the rate of microplastic deformation diminishes and may reach a magnitude corresponding to the formation of complete physical contact. During solid-phase welding not only the contact-adjacent regions undergo plastic deformation, but also the entire welded product. Naturally, increase of the surface roughness means diminution of the angles at the microasperity vertexes. If pressure is applied to such surfaces, the deformation rate is higher than with a surface containing microasperities with large angles at the vertexes. Therefore, after rough surface treatment deformation strengthening is more essential and is accompanied by a sharp decrease in the rate of microasperity deformation and contact formation intensity. As the surface roughness diminishes and the angles at the vertexes of microasperities increase, not only microasperities, but also contact-adjacent regions of the material become involved in plastic flow. When the degree of deformation localization decreases, the formation of physical contact is more intense compared to the case of rough treatment of the contacting surfaces.

Since strengthening is a structure-sensitive process, the choice of the initial structure will make it possible to decrease the degree of deformation strengthening in the contact-adjacent region, if not suppress it. Plastic deformation always leads to the formation of a characteristic surface relief: the traces of slipping along the grain boundaries or in their bulk. The geometric parameters of the relief are defined by the mechanisms of deformation.

As mentioned above, physical contact formed as a result of microplastic deformation always has microcavities that may be filled through diffusion processes. Usually, the notion of physical contact applies to surfaces separated by a distance of interatomic interaction. During the initial stages of the joint formation, the contact area increases. Such a deformation is limited since the contact-adjacent region of the material strengthens, whereas mechanical strength at the contact diminishes. When the stress achieves a minimum and becomes insufficient for the action of threshold deformation mechanisms, the joint zone will be a structure with defects lying in the plane of the original contact. During welding of homogeneous materials such defects are located along a boundary that has properties analogous to those of a large-angle grain boundary. If the migration-driving force of such a boundary is sufficient to start its displacement out of the contact plane, then complete contact may be achieved through dissolution of the defects located inside the grains. In the case when the migration-driving force is insufficient to initiate the movement of the grain boundary, complete physical contact is accompanied by dissolution of the mentioned defects at the grain boundary.

As the temperature increases, microplastic deformation and the process of contact formation become more intense, the resistance to deformation diminishes, and conditions favorable to the development of deformation relief are created. Other possible effects include an increase in the number of dislocations reaching the zone of physical contact; an increase in the rate of dislocation movement and consequently an increase in the quantity of energy introduced into the zone of physical contact by each dislocation; diminution of the shear modulus of the material; a decrease in the height of the potential energy barrier, which restricts the formation of chemical bonds; and a change in the number of atoms participating in the chemical interaction at one active center, i.e., a change of the active center area.

Pressure also has a significant role in the development of physical contact. Upon heating without pressure application, the contact is formed by plastic flow of the material under the action of surface tension forces or diffusion mass exchange. Increasing pressure favors the microplastic deformation of microasperities and deformation relief development.

Thus, the formation of physical contact at diffusion welding is connected with mutually related and competing processes. If physical contact has been formed between two components to be joined, then activation of the contacting surfaces is a necessary and sufficient condition for the formation of a welded joint in the solid phase.

8.4.2 Methods of Diffusion Welding Intensification

The physical–chemical processes that occur in the zone of contact during each stage of diffusion welding can be intensified by means of certain actions. To date, the specific influence of welding parameters and the properties of welded materials on the joint strength have not been studied. Nevertheless, some tendencies are obvious:

- Small temperature changes fundamentally change the kinetics of joint formation.
- Increasing the holding time of welded products at a preset temperature and pressure raises the joint strength.

- Increasing pressure improves the joint quality at any value of temperature and time, but excessive pressure may cause plastic deformation of the components or even their destruction.
- High-quality joining requires optimal roughness of the welded surfaces, which provides deformation of microasperities and realization of mass exchange.

The formation of physical contact during the initial stage can be intensified by means of cyclic pressure changes, ultrasonic vibrations introduced into the zone of contact, torsional vibrations, and so forth. Any physical or chemical factor that provides breakdown and subsequent renewal of the surface atomic bonds intensifies the processes during the second stage. Application of electrostatic and magnetic fields, irradiation of the materials by high-energy ionizing particles, and other actions make it possible to intensify the processes of mass exchange and stress relaxation during the third stage.

The simplest method of physical contact intensification is cyclic change of the welding pressure according to the following scheme: a period of compression under force, removal of the pressure for a set duration, return to compression, and so on. Such a scheme raises the degree of plastic deformation in the zone of contact, as compared to simple static loading. Each new loading directed perpendicularly to the plane of contact is accompanied by a period of active microasperity deformation and creep connected with the development of recovery processes during load removal. This regimen permits an increase in the joint strength with a reduction in the temperature of the process.

Another method for intensifying diffusion processes is based on the application of a stretching force perpendicular to the action of pressure on the welded components, which initiates elastic deformation. Crystal lattice distortions that arise under these conditions not only speed up the diffusion process, but also facilitate the formation of vacancies, which in turn intensify the interaction of the materials to be welded. The use of ultrasonic vibrations for stimulating the development of physical contact, due to an increased rate of creep in the material, was proposed by N.F. Kazakov, the author of the invention in the field of diffusion welding [23]. Ultrasonic vibrations are introduced into the interaction zone perpendicular to the plane of welding. Alternating stresses give rise to the generation of new dislocation sources and increase the dislocation mobility, thus intensifying the process of mutual diffusion.

To prevent the formation of brittle intermetallide layers in the contact zone when joining different materials, the process duration is reduced by impact pressure. Single impulses of force are applied to locally heated regions in the zone of contact and the duration of the welding process is reduced to 1–10 ms. The welded components are heated to different temperatures to provide uniform plastic deformation in the zone of contact. Vacuum-tight joints with good mechanical properties are formed in vacuum under impact pressure, and the joint strength is comparable to that of the weakest material.

These methods of diffusion welding intensification do not solve the set problem completely, since the decrease in temperature or the duration of loading necessitates an increase of plastic deformation in the welded components. This is not always permissible. At lower temperatures it is impossible to obtain welded joints with strength equal to that of the initial material.

8.4.3 Welded Seam Structure upon Diffusion Welding

The dependencies of structural perfection and welded seam strength on the conditions of welding are very complex. As stated in papers devoted to the quality of welded joints [24–26], optimal temperature, load, and duration of loading provide the absence of local destruction in welded components. However, to choose the regimen of welding, only the temperature-force dependence of creep in the components was taken into account in these investigations. No dislocation aggregates were observed in the regions adjacent to the joint boundary, which testified to plastic deformation in the surface-adjacent layers. Meanwhile, on the basis of data obtained in recent years it should be concluded that the mentioned results are a particular case. In fact, the joint quality is defined mainly by the initial structure of the welded components [27–29] and the thermodynamic parameters of welding.

It has been shown [30, 31] that upon diffusion welding the structure of the surface-adjacent sapphire layers changes. In particular, the polycrystalline layer arising in the process of mechanical treatment recrystallizes, the level of residual stress and dislocation density change, and deformation relief may appear. The character of the phenomena observed with the action of thermal and mechanical factors is determined by the fine structure of the crystals. The mentioned papers are the only reports where plastic flow is shown to be connected to the prehistory of the crystals and their initial structural perfection. The role of the initial crystal structure was investigated for crystals grown by the Verneuil, Stepanov, and HDS methods. Sapphire was welded at 1,900°C under an initial load of 0.5 kg/mm^2 along the (0001) planes, with surface roughness $R_Z = 0.05$ μm. The seam quality was studied on prism planes; therefore, the samples were cut in directions perpendicular to the seam. The mechanical characteristics of the zone of welding were estimated from the k_c value.

According to the character of plastic flow in the surface-adjacent crystal layers during diffusion welding, sapphire crystals fall into two groups. The first group contains the crystals grown by the Verneuil and Stepanov methods. In these crystals the action of thermal and mechanical factors brings about deformation relief: the welding seam kinks as microquantities of the crystal move relative to one another within 20- to 150-μm distance (Fig. 8.5). This movement arises where the boundaries of blocks intersect the welding seam. Deformation relief may give rise to corrugation of the welded surfaces (Fig. 8.5b). This arises even in cases when the orientation of the welded crystals and the direction of loading are chosen such as to minimize plastic deformation. The latter arises first in blocks with a disorientation of 1–3°, which have reached the zone of contact. Another cause is slip over block boundaries, or in some cases the formation of new microblocks. Sometimes

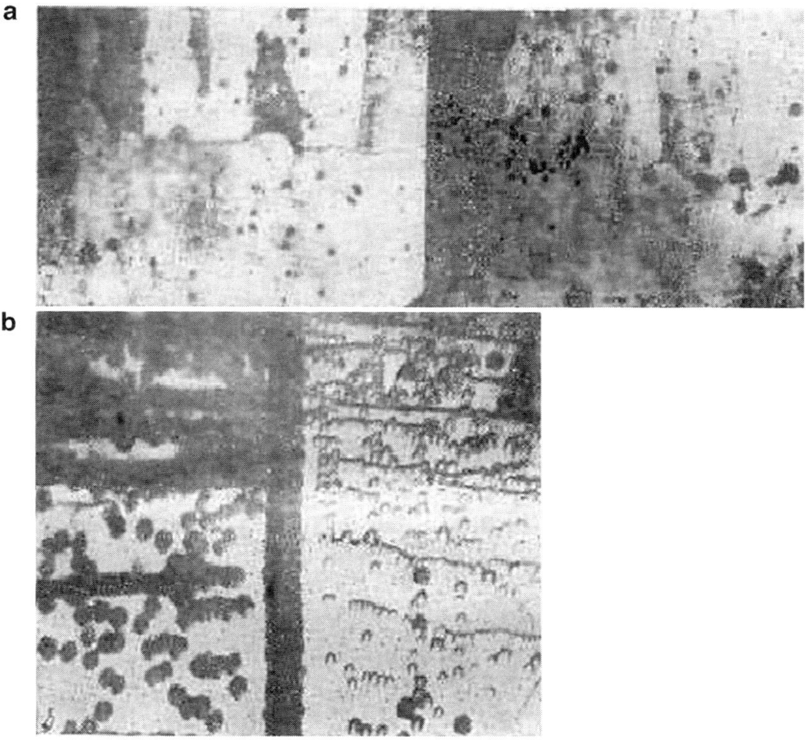

Fig. 8.5 Welding seam formed by diffusion welding of an etched surface of sapphire samples made from the crystals grown by (**a**) Stepanov method (×70), (**b**) HDS method (×120)

deformation relief is formed only due to the action of thermal factors, in which case it is equivalent to the rise of surface roughness with heating. The strength characteristics of the welding seam for the crystals of this first group depend on the stage when deformation relief manifests itself. If it is formed upon heating of the samples before loading, the strength of the seam turns out to be lower than that of the main material by 40–70%. If the relief is caused by the action of thermal and force factors, the welding seam is revealed only while etching the crystals and the strength of the joint is close to that of the bulk material.

The second group contains crystals grown by the Czochralski, Kyropoulos, and HDS methods. In such crystals deformation relief does not occur and the welding seam remains straight, resembling a block boundary (Fig. 8.5b). For such crystals, obtaining joints with uniform strength requires higher temperatures and a longer process duration in comparison with the crystals from the first group, with the values of specific loading being the same or higher by 20–50%.

For crystals from the first group, cracking resistance starts to fall at a distance of 50–60 μm from the seam and reaches its minimum at a distance between ~30 μm from the seam to the seam itself. However, in some cases the value of cracking resistance on the seam does not differ from the bulk value of k_c. For crystals from the second group, the change in k_c value in the seam zone does not exceed the measurement error. The behavior of the crystals grown by different methods in the process of diffusion welding testifies to the different character of plastic flow and dependence on the initial substructure. The degree of plasticity of the material is inversely proportional to the size of grain structure elements and consequently to the density of point defects on their boundaries. In crystals from the first group, the size of the elements is small (50–200 Å), plastic deformation is not hindered, the dislocation density is high, and in the process of welding deformation relief is readily formed. In the crystals from the second group, the size of the elements may reach 500–600 Å, the density of point defects at their boundaries is high, and plastic deformation is hindered. Such crystals are characterized by low dislocation density and deformation relief does not manifest itself (Fig. 8.6) with diffusion welding.

8.4.4 Diffusion Welding Using Interlayers [32–40]

Interlayers are used for the following purposes:

- To decrease chemical inhomogeneity in the joint region
- To relieve residual stresses and to remove the differences in the linear expansion of welded materials
- To prevent plastic deformation
- To diminish welding temperature and pressure as well as duration

There exist melting and nonmelting interlayers. The former are solders and oxides of alkali metals.

Welding is carried out at a temperature close to the melting point of the interlayer to provide for diffusion between welded materials and the interlayer compo-

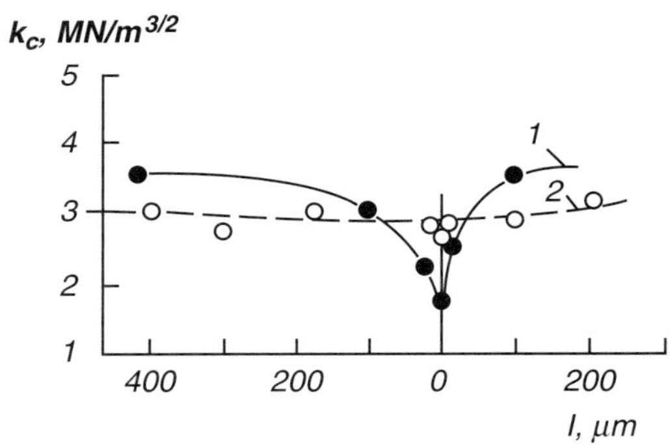

Fig. 8.6 Coefficient of cracking resistance in the welding zone: (*1*) the first group of crystals; (*2*) the second group of crystals

nents. Liquid phase promotes the processes of separation, dispersion, and dissolution of oxide films. In many cases the interlayers provide cleaning of the surface, and they are pressed out of the contact zone during the process of joint formation. The main disadvantage of such layers is the low strength of the welded joints, comparable to that of soldered joints. Taking into account the fact that this technology is similar to soldering [39], we will not discuss it here.

Joints can be created by means of a liquid interlayer formed by melting together with the contact-adjacent parts of the welded components. Similar to solders, liquid interlayers wet the joined surfaces for a short period of time and form permanent joints upon cooling. The advantage of this method over soldering resides in the fact that the system "basic material – eutectic interlayer – basic material" has no clear interface, and the resultant sharp change in properties is typical of the system "basic material – solder – basic material." The eutectic alloy of the basic material with the interlayer possesses the properties of both of them.

For nonmelting interlayers, plastic metals (gold, silver, nickel, aluminum, etc.) are used in the form of foils, wires, powders, or films. The diffusion processes are intensified through the medium of interlayers applied on the surfaces by vacuum deposition. The thickness of the deposited layer ranges from a fraction of a micrometer to tens of micrometers, and the layers have a fine-grained structure. They dissolve in the welded materials and do not influence the joint strength.

The conditions and parameters of diffusion welding for metals fundamentally differ from those of nonmetallic materials. Although this process is a topochemical reaction proceeding on the active centers, the type of bonds that arise in the compound is defined by the nature of the materials. Joining of sapphire with metals by diffusion welding is realized through microscopic deformation of the interlayer or of the metallic component.

Strong joints can be obtained through the interaction between lower oxides and the oxide system of a nonmetal. Therefore, the practice of diffusion joining of a nonmetal to a metal is reduced to the choice of a metallic interlayer with an appropriate surface oxide layer. For joining nonmetals, the most often used interlayer materials are aluminum, copper, Kovar 29 NK, niobium, and titanium foils with a thickness not exceeding 0.2 mm.

During recent years, active investigations aimed at obtaining large-area sapphire products by the methods described above have been carried out, and impressive results have been achieved [41]. Contact between the surfaces to be joined is achieved by means of a thin interlayer of foreign material which melts at the temperature of welding, followed by the formation of a liquid transition phase [42–44]. At subsequent stages this layer crystallizes under isothermal conditions, and on the surface of the components there arises a thin polycrystalline layer that is homogenized by additional thermal treatment of the joint. Contact of the surfaces (i.e., their approach to the interaction distance of interatomic forces) is practically instantaneous, as they are wetted by liquid alloy. However, such a seam has poor mechanical characteristics, since the foreign-phase interlayer is bonded with the basic material only by a weak chemical interaction.

The authors of the here-cited paper made an attempt to combine the advantages of earlier proposed approaches in order to create a method of diffusion welding using a polycrystalline interlayer [35]. The achievement of permanent joints between sapphire components is based on the use of those mechanisms that act at the boundaries of polycrystals. In the general case, with solid-state welding all the procedures performed on the welded components (heating, plastic flow) are aimed at concentration of the interaction energy in the zone of contact to provide rearrangement of the surface-adjacent layers. Such a rearrangement comprises diffusion, a recrystallization that proceeds spontaneously at lower energies compared to deformation. As a rule, such a high concentration of energy on the surface is easily achieved in welding polycrystals. In this connection, there arose the idea to create a thin (1.5–2 μm) polycrystalline layer on the contacting surfaces of single crystals. This layer can be obtained by mechanical treatment of the surfaces, as well as by deposition or implantation of polycrystalline aluminum oxide. The surfaces are brought into contact, then medium-intensity compressive force is applied, and the temperature of the components is raised to a value higher than the recrystallization temperature but lower than their melting point. The components to be joined are kept under isothermal conditions and afterward moved through a gradient zone ($T' \sim$ 200–400 K/cm) at a velocity of 10–150 mm/h. In the process of compression and heating the contact between the single crystals and the polycrystalline layer leads to the formation of common grains. This is caused by the fact that the pores at the joint boundary diminish to such an extent that the energy of the joint boundary turns out to be close to the energy of the grain boundary of the basic material. As the zone of contact moves in the temperature gradient, the grain boundaries migrate until the interface between the surfaces disappears. In other words, such a displacement leads to secondary recrystallization in the zone of the joint: the single crystals "swallow up" the fine-grained polycrystalline layer. This process is initiated by a decrease in the energy of the grain boundaries caused by diminution of the boundary length and non-equilibrated surface tension at the grain boundaries. Thermal stresses arising in the joint during its motion through the gradient zone are an additional driving force, which prevents attenuation of the process. The welding temperature, loading, and time of exposition are mutually related parameters to be chosen depending on the size and initial structure of the welded components.

The structure of the seam obtained by means of a polycrystalline layer is distinguished by the presence of two processes at the joint boundary. The first of them is primary recrystallization accompanied by the creation of grains common for the two components, i.e., a common polycrystalline layer. The other process is secondary recrystallization characterized by migration of the grain boundaries leading to the disappearance of the polycrystalline layer. The technological parameters that provide these processes are mutually related: the grain boundary mobility increases with the temperature. Therefore, the motion velocity can be increased and the value of T' diminished, since the grain boundary mobility manifests itself at lower thermoelastic stresses. On the contrary, decrease of the

Table 8.1 Ratios of the ultimate tensile strength in the welding seam to bulk sapphire depending on the motion velocity v and the temperature gradient T'

	T' (K/cm)					
v (mm/h)	50	200	300	350	400	450
8	0.5	0.6	0.8	0.7	0.8	0.6
10	0.6	0.9	1.0	1.0	1.0	0.8
100	0.7	0.9	1.0	1.0	1.0	0.7
150	0.6	1.0	1.0	1.0	1.0	0.7
160	0.7	0.8	0.8	0.7	0.6	0.5

temperature requires diminution of the motion velocity, i.e., an increase in the time during which the joint stays in the gradient zone, and rise of the axial temperature gradient. These dependencies are saddlelike and the interval of the optimum parameter values is rather narrow, in particular, with a motion velocity $v <$ 10 mm/h and an axial temperature gradient $T' <$ 200 K/cm, a coarse-grained inhomogeneous structure of the polycrystalline layer is formed in which the ratio of the surface area for different grains in the layer may reach 1:10. This inhomogeneity of the polycystalline matrix sharply limits the possibility of the transformation of the polycrystalline layer into the single crystalline in the process of secondary recrystallization, and consequently decreases the welding seam strength. For a motion velocity $v >$ 150 mm h^{-1} and $T' >$ 400 K/cm, thermoelastic stresses arise the value of which is higher than the ultimate strength of the welded components and the latter crack during welding (Table 8.1).

The value of cracking resistance of the welding seam obtained by the discussed method at the optimum technological process parameters is practically equal to k_c in the bulk of the material. When these parameters deviate from the optimum, the value of k_c sharply decreases.

8.4.5 Technique of Diffusion Welding

The components to be welded are assembled in a special facility, treated under vacuum, heated, and kept under compressive pressure. In some cases, the products are further maintained at temperature after releasing the pressure to allow completion of recrystallization processes, and subsequently the formation of high-quality joints.

Diffusion welding of sapphire does not require the use of solders, electrodes, fluxes, or protective gases. Subsequent mechanical treatment also is unnecessary since scale, slag, and flash are absent. The mass of the product does not increase, as occurs with other types of welding, soldering, and gluing; the components do not warp. The use of vacuum makes it possible to obtain joints with a minimal content of impurities or complex-configuration products from same and different materials.

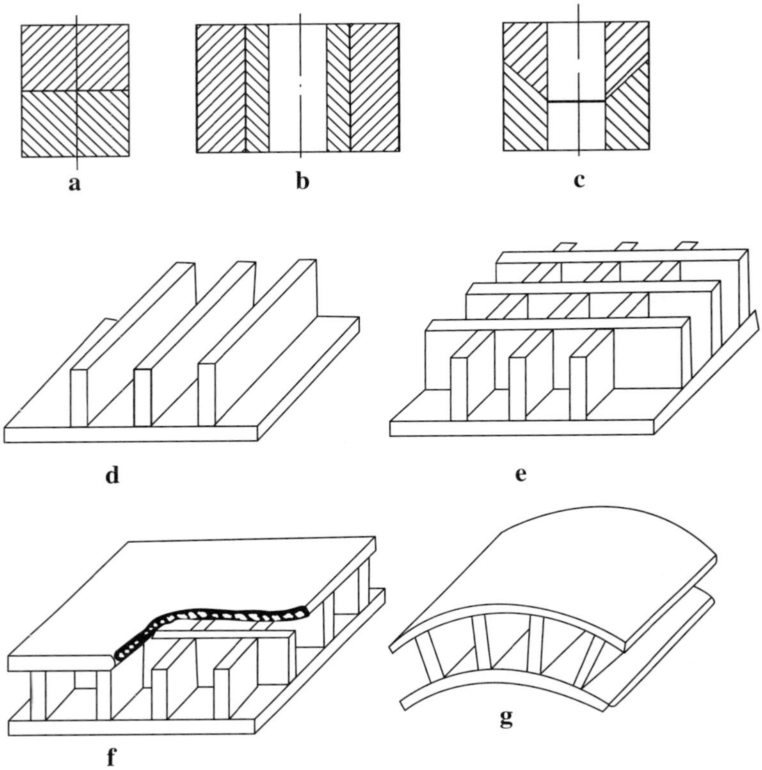

Fig. 8.7 Some types of structures obtained by diffusion welding

Diffusion welding also is an efficient method of group treatment in published articles [45]. It is used to obtain joints of sapphire with Kovar, copper, titanium, and heat-resistant and high-melting metals and alloys in vacuum. The components to be joined may have compact (Fig. 8.7a–c) or intricate (Fig. 8.7d–g) contacting surfaces. The dimensions of these components range from several microns (for the production of semiconductor devices) to several meters (for the making of stratified structures).

Diffusion welding is distinguished by relatively high temperatures of heating (0.5–0.7 of T_m) and low specific compression pressures (<0.5 MPa); the length of time that the products are under isothermal conditions varies from several minutes to several hours.

The solid-phase method of joining under the action of intense force factors is promising for obtaining joints of sapphire with plastic materials (metals, plastics), since in such cases the structure of the sapphire crystal remains practically unchanged. The joint is achieved through plastic flow in the "soft" component.

At present, investigations of microwave heating are being carried out. This method is economically promising, since the time required to heat aluminum oxide samples to 1,600°C can be reduced from 480 min to less than 30 min when heating by microwave versus traditional radiant methods [41–47].

8.5 Welding by Contact Zone Melting

In this case the components to be welded are placed edge to edge or with a narrow clearance, and a heat source is used to melt small amounts of the edges of the components. More often an additionally introduced material is melted together with these edges, and the whole of the melted substance is placed into a bath. The process of melting leads to the destruction of contaminants present on the surface and formation of atomic and molecular bonds. As the heater moves along the product, the melted bath solidifies and forms crystalline compound.

Welding of sapphire by means of contact zone melting is the fastest method of joining, but is rather complicated [48]. The described procedure, which implies the use of external heaters for the motion of the local melt zone, is not widely applied due to the fact that it is technically difficult to keep the liquid zone stable over the cross section of the joined components. The liquid zone remains stable when the thickness of the joined components does not exceed 1–2 mm [49].

At first sight, the physics of such welding is simple enough. Upon melting of sapphire the bonds between ions are partly preserved. They also are preserved in the interphase solid–melt boundaries. Therefore, a sufficient condition for the formation of interatomic bonds is the formation of the common bath, and as a consequence the disappearance of the boundary between the joined surfaces (Fig. 8.8).

Since this type of welding is essentially directed crystallization of the melt, it is characterized by the corresponding dependencies of seam quality on the thermodynamic parameters of crystallization and regularities of defect formation in the crystallization zone (i.e., the welding seam). But, unlike the conventional directed crystallization, this process is realized from two opposite sides.

The main effects observed in sapphire upon its melting and subsequent crystallization are the following:

- Changes in the chemical composition of the seam due to segregation of impurities
- Formation of residual stresses caused by nonuniform heating, which may give rise to deformation of the components and even to destruction of the welded joint
- An increase in point and linear defects in the zone of welding

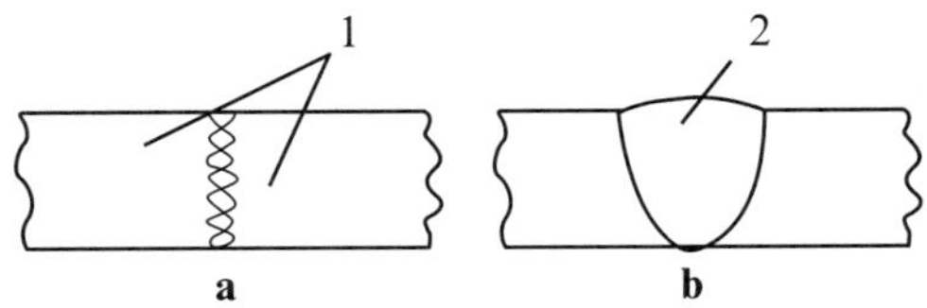

Fig. 8.8 Schematic of the formation of a joint at welding by means of contact zone melting: (**a**) initial state; (**b**) state after welding; (*1*) welded components; (*2*) welding seam

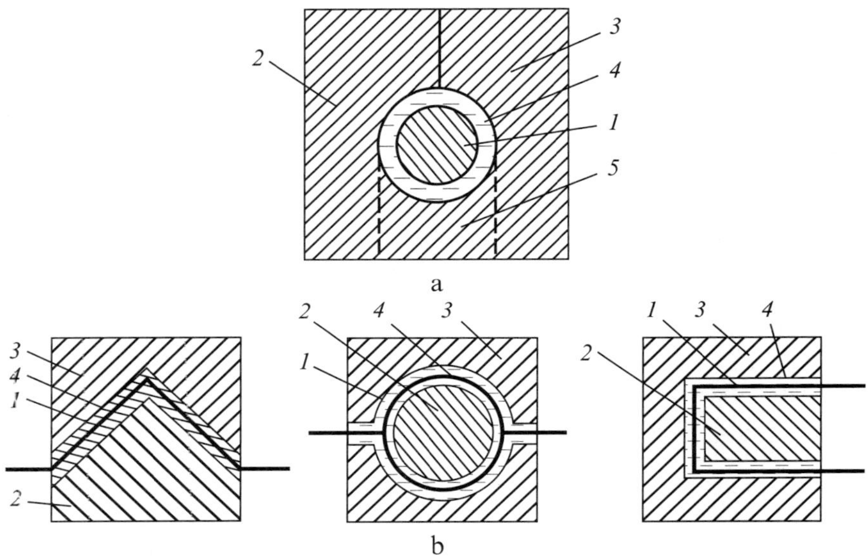

Fig. 8.9 (**a**) Schematic of crystal welding apparatus and (**b**) examples of welding of complex-configuration crystalline components: (*1*) heating element, (*2*–*3*) welded samples, (*4*) melted zone, (*5*) welding seam

The method of welding by means of a local melted zone created and moved along the joint by means of an additional heating element (Fig. 8.9) seems to be promising.

This method can be used to weld sapphire as well as other single crystals and polycrystals. Welding of single crystals is more difficult to realize, since it requires precise mutual crystallographic orientation of welded components. It is practically impossible to achieve ideal orientation of the single crystals to be welded, and in most cases low-angle block boundaries arise along the welding seam. For polycrystals such orientation is not necessary, and after welding performed under optimum conditions the structure and strength of the welding seam does not differ from the initial material. This method permits joining of components of large thickness and complex configurations. The heater used to create the melted zone also may have intricate shape. The melted zone is confined by surface tension of the melt. As a rule, this moves upward and carries gas bubbles formed in the melt out of the melted zone, preventing their capture by the crystallization front (Fig. 8.10).

At low temperature gradients and high crystallization rates the crystallization front is drop-shaped (Fig. 8.10). As a result, a zone with high density of defects is formed in the center of the seam. The width of this zone is 0.4–1 mm if the velocity of the melted zone, v, is 6–24 mm/h. The content of defects in the welding seam (block boundaries, dislocations, and vacancies) increases at low temperature gradients with increasing crystallization rate; at high gradients and $v < 6$ mm/h it is comparable with that of the seam-adjacent regions of the single crystals. The temperature gradients at the crystallization front are not varied due to technical difficulties. However, as crystal growth experience shows, the temperature gradients at the crystallization front may be on the order of 10–50 K/cm.

Transmission spectra measured in the 200–600 nm wavelength region show that the transparency of single crystals with welding seams introduced perpendicularly to the beam does not decrease. Without preliminary preparation of the contacting surfaces, the joined components are placed in a fixing holder, which then is located in the main heater and heated to T ~ T_m. Another heater is placed along the joint boundary and moves along it at 2,350–2,600 K until it reaches the opposite side of the components. The process of welding can be realized in vacuum or in protective atmosphere.

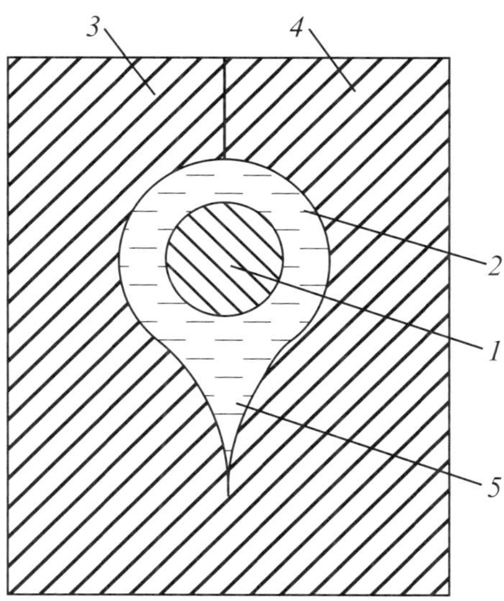

Fig. 8.10 Shape of crystallization front: (*1*) heating element; (*2*) melted zone; (*3–4*) welded components; (*5*) drop-shaped crystallization front

The joint quality is defined by the initial substructure of the components [50]. If the size of the components' grain structure elements does not differ significantly, then under conditions of local zone melt overheating a permanent joint with strength equal to that of the basic material can be obtained. When the said difference is significant, no joint can be obtained under these conditions and the welding zone is destroyed upon cooling. However, after overheating the joint zone "forgets" the initial structure of the contacting single crystals, and permanent joints can be obtained for components whose grain structure elements essentially differ in size. The level of residual stresses in such a joint is higher by at least 30–50%.

If the conditions of welding have been chosen properly, the joint is not revealed even through selective etching. It can be found only by a detailed analysis of the impurity distribution and of the coefficient of cracking resistance. As a rule, in the melted zone the value of k_c increases. For instance, if in the bulk of the material $k_c \sim 3$ MN m$^{-3/2}$, in the joint zone $k_c \sim 3.2 \ldots 3.6$ MN m$^{-3/2}$.

References

1. Pat. USA 3390019, MKJ B23K 15/00
2. Pat. USA 3518400, MKJ B23K 15/00
3. Pat. USA 4263495, MKJ B23K 27/00
4. Kapelyushnik I.I. et al. Technology of Glueing Components in Aircraft Making. Moscow: Mashinostroenie, 1998 (in Russian).
5. Mikheev I.I. et al. Technology of Glueing Components. Moscow: Mashinostroenie, 1965 (in Russian).
6. Zeving R. (ed.). New Technological Processes in Precise Instrument Making, 1993 (in Russian).
7. Lopatko A.P., Nikiforova Z.V. New Methods of Welding and Soldering (Collected Volume), Moscow, 1979, p. 88 (in Russian).
8. Lyubimov M.L. Joints of Metal with Glass. Moscow: Energiya, 1988 (in Russian).
9. Patent of France 2173673
10. Patent of France 2234661
11. Petrushenko. I.E. (ed.). Soldering Reference Book. Moscow: Mashinostroenie, 2003 (in Russian).
12. Dobrovinskaya E.R., Kozhushko G.M., Litvinov L.A. et al. Svetotekhnika. 4, 1979, 8–9 (in Russian)
13. Patent of Great Britain 1365403
14. Delmon V. Kinetics of Heterogeneous Reactions. Moscow: Mir, 1972 (in Russian).
15. Gleston S., Laidler K., Airing G. Theory of Absolute Reaction Rates. Moscow: Inostrannaya literatura, 1948, 240pp (in Russian).
16. Ya V. Kinetic Theory of Phase Transformations. Moscow: Metallurgiya, 1973 (in Russian).
17. Diffusion Welding of Titanium: Reference Book. Moscow: Metallurgiya, 1977, pp. 18–28 (in Russian)
18. Shorshorov M.Kh., Krasulin Yu.L. Welding Engineering. 12, 1967, p. 1 (in Russian).
19. Krasulin Yu.L. Solid-Phase Interaction of Metal with Semiconductor. Moscow: Nauka, 1971 (in Russian).

20. Kazakov I.F., Krivoshey A.V., Studenkov E.G. USSR Inventor's Certificate 178658. A Method of Diffusion Welding of Materials, 1976.
21. Karakozov E.S., Ternovskiy A.P., Zamidchenko S.S., Tarlavskiy V.E. USSR Inventor's Certificate 617209 A Method of Pressure Welding of Different Materials, 1978.
22. Sergeev A.V., Kazakov I.F. USSR Inventor's Certificate 880669 A Method of Diffusion Welding of Different Materials, 1981.
23. Kazakov N.F. Diffusion Welding of Materials. Moscow: Mashinostroenie, 1976, p. 312 (in Russian).
24. Regel V.R. Stepantsov E.A. Influence of External Factors on Real Structure of Ferro- and Antiferroelectrics (Collected Volume). Russia: Chernologolovka, 1981. p. 50 (in Russian).
25. Stepantsov E.A., Regel V. R. Ferroelectric Crystals in Different Fields (Collected Volume), Leningrad, 1031, pp. 17–23 (in Russian).
26. Regel V. R., Stepantsov E.A. Growth of Semiconductor Crystals and Films (Collected Volume), Novosibirsk, 1984, Part 2, pp. 44–48 (in Russian).
27. Dobrovinskaya E.R., Pishchik V.V. Growth of Crystals from the Melt. Yerevan: Armenian Academy of Sciences, 1985, pp. 96–97 (in Russian).
28. Dobrovinskaya E.R., Pishchik V.V. Izvestiya AN SSSR. Ser. fiz. 49(12), 1985, pp. 2386–2389 (in Russian).
29. Bagdasarov Kh.S., Dobrovinskaya E.R., Litvinov L.A., Pishchik V.V. Izvestiya AN SSSR. Ser. Fiz. 37, 1973, pp. 2362–2366 (in Russian).
30. Dobrovinskaya E.R., Zvyagintseva I.F., Litvinov L.A., Pishchik V.V. Izvestiya AN SSSR. Ser. Fiz. 49(12), 1985, pp. 2390–2392 (in Russian).
31. Dobrovinskaya E. R., Zvyagintseva I.F., Litvinov L.A., Pishchik V.V. Capillary and Adhesive Properties of Melts (Collected Volume), Kiev. 1987, pp. 140–143 (in Russian)
32. Axelson, et al. USA Patent No 6,012,303.
33. Patent of Great Britain 2132050
34. Patent of Japan 58–47947
35. Dobrovinskaya E.R., Litvinov L.A., Pishchik V.V. Patent USSR No. 1315199. Method of Diffusion Welding of Single Crystals of a Corundum, 1987.
36. USSR Inventor's Certificate 544 36 58. 22.08.95
37. USSR Inventor's Certificate 173 417. 21.07.65
38. Inventor's Certificate of Russian Federation 2131798. 20.06.99
39. Achievements and Prospects of Diffusion Welding (Collected Volume), Moscow. 1987. p.170 (in Russian).
40. Axelson, et al. United States Patent. 6,012,303, 11 January 2000.
41. McGuire P., Pazof B., Gentilmarf R., Askinazi J., Lochef J. Large Area Edge-Bonded Flat and Curved Sapphire Windows. Proceedings of SPIE AeroSense Symposium, Orlando, FL, 16 April 2001.
42. Gentilman R. et al. High Strength Edge-Bonded Sapphire Windows, vol. 3705, SPIE, Orlando, FL, 5–6 April 1999, p. 282.
43. Patent of Great Britain 2132050 B 23 K 19/00.
44. USA Patent 6,012,303, 11 January 2000.
45. Lyushinskiy A.V. Diffusion Welding of Different Materials. Moscow: Academiya, 2006, 208pp.
46. Lewinsohn C.A., Colombo P., Reimanis I., Ünal O. J.Am. Ceram. Soc.. 84(10), 2001, p. 2240.
47. Ahmed A., Siores E. J. Mater. Process. Technol. 118, 2001, 88–95.
48. Stores E., DoRego D. J.Mater. Process. Technol. 48, 1995, 619–625.
49. Beale G.O., Li M. Robust Temperature Control for Microwave Heating of Ceramics. IEEE Trans. Industr. Electron. 44(1), February 1997, pp. 124–131.
50. Dobrovinskaya E.R., Pishchik V.V. Corundum Single Crystals. Problems of Growth and Quality. Moscow. 1988, Part 1, 74p; Part 2, 62 p. (in Russian).

51. Aravindan S., Krishnamurthy R. Mater. Lett. 38, 1999,245–249.
52. Beale G.O. Controller Robustness Analysis for Microwave Heating of Ceramics. Proceedings of Korema '96, Opatija, Croatia, 18–20 September 1996, pp. 13–16.
53. Li M., Beale G.O., Tian Y.L. Automatic Control During Microwave Heating of Ceramics. Proceedings of Microwaves: Theory and Application in Materials Processing III, American Ceramic Society, Cincinnati, May 1995.

Conclusion

Tendencies of the development of high-temperature crystallization from the melt clearly show that knowledge of the crystallization kinetics, especially of the one at the growth front, is insufficient. It also is necessary to consider the physicochemical processes that accompany melting of the initial substance and the mechanism of its crystallization (taking into account external conditions). Moreover, one must have information on those phenomena that occur in the grown crystal and in the vicinity of the crystallization front particularly. These high-temperature processes strongly influence the formation of the real crystal structure. Nowadays such a comprehensive approach can be realized, since the arsenal of methods for investigating high-temperature crystallization has greatly widened. Further studies will comprise consideration of heat and mass transfer at the crystallization front and establishment of the relation of these processes to the real structure of sapphire.

High-temperature crystallization from the melt is distinguished from the low-temperature one by a variety of physicochemical processes of the interaction of the melt with the ambient medium essentially influencing the real crystal structure, as well as by kinetic phenomena that arise in the grown crystals at cooling. Therefore, complete description of the crystal growth process requires consideration of the physical and chemical kinetics as a single whole. This necessitates further development of the theory of crystal growth based on new experimental data. A comprehensive approach to the consideration of high-temperature crystallization from the melt, taking into account the state of the initial substance and the mechanisms of its melting and crystallization, will help to work out more complete substantiation of the growth methods and to establish the main trends of their further development.

Nowadays it is obvious that the level of the technology for the making of sapphire products as well as their output to a great extent define the development of quantum electronics, radio electronics, microradioelectronics, radio engineering, radiolocation, high-temperature and superhigh-resolution optics, aviation, nuclear power engineering, chemical instrument-making, space and microwave engineering, and so forth. Sapphire is the only material that provides the scope for visual observation of different processes occurring in chambers under the conditions of superhigh pressures, temperatures, and aggressive media. Moreover, it becomes more and more widely used in the systems of bar code scanning for conveyors and commercial networks as the main material for the windows of stationary and hand-operated scanners. The production of mobile phones requires some 6 billion sapphire glasses annually.

Today, sapphire is primarily utilized in the following applications:

- Fabrication of substrates for light-emitting diodes, new generations of TV receivers, projectors, and microwave devices
- Fabrication of windows for civilian and military equipment
- Production of bearings and windows for watches and devices
- Making of precious jewels for the jewelery industry

The present-day market of synthetic stones is estimated to exceed $6 billion. In the world ~300 tons of rubies and colored sapphires and 100 tons of sapphire are produced. In the opinion of specialists, by 2008 the world market of synthetic crystals will reach $11.3 billion; the share of sapphire beingabout a quarter of this value. The growth of the rate of sapphire production is extremely high. One should expect that during the next 20 years the world production of sapphire will increase approximately tenfold. Nowadays the leading manufacturers of sapphire are the United States and Russia. In the United States sapphire crystals are grown mainly by the Kyropoulos, HEM, and EFG methods. In Russia the Kyropoulos and Stepanov methods dominate. The rate of sapphire production is being stepped up in Ukraine and Japan. China is endeavoring to hit the world market, too. At present the annual increment of the rate of the production of sapphire and especially of sapphire articles in China is ~20%.

Wide use of sapphire is restrained by its rather high cost, as well as large capital outlays necessary for the improvement of the existing production capacities and especially for the creation of the new ones. In the near future the crystal growth methods will be modernized with the purpose of increasing the size and weight of the crystals. The HEM, Kyropoulos, and Czochralski methods will facilitate obtaining crystals with a weight of 60–70 kg, 500 × 500-mm^2 plates will be produced by the method of HDS. However, the crystal size cannot be increased unlimitedly without using basically new technical approaches. For instance, the method of HDS seems to provide a scope for crystallization by transporting an echelon of containers through the crystallization zone.

Another way is making larger products in the form of permanent joints of sapphire elements. Combination of the technologies of the growth and welding of sapphire will revolutionize both the technologies of sapphire production and materials science as a whole, as it will permit attainment of compound sapphire units and crystals of practically unlimited size.

Thus, it is to be concluded that in the near future sapphire, as with silicon and germanium, will become one of the strategic materials for materials science.

Index

N

O

T

Printed in the United States of America